Studies in Systems, Decision and Control

Volume 582

The series “Studies in Systems, Decision and Control” (SSDC) covers both new developments and advances, as well as the state of the art, in the various areas of broadly perceived systems, decision making and control–quickly, up to date and with a high quality. The intent is to cover the theory, applications, and perspectives on the state of the art and future developments relevant to systems, decision making, control, complex processes and related areas, as embedded in the fields of engineering, computer science, physics, economics, social and life sciences, as well as the paradigms and methodologies behind them. The series contains monographs, textbooks, lecture notes and edited volumes in systems, decision making and control spanning the areas of Cyber-Physical Systems, Autonomous Systems, Sensor Networks, Control Systems, Energy Systems, Automotive Systems, Biological Systems, Vehicular Networking and Connected Vehicles, Aerospace Systems, Automation, Manufacturing, Smart Grids, Nonlinear Systems, Power Systems, Robotics, Social Systems, Economic Systems and other. Of particular value to both the contributors and the readership are the short publication timeframe and the world-wide distribution and exposure which enable both a wide and rapid dissemination of research output.

Indexed by SCOPUS, DBLP, WTI Frankfurt eG, zbMATH, SCImago.

All books published in the series are submitted for consideration in Web of Science.

Oscar Castillo · Gökhan Çuvalcıoğlu ·
Feride Tuğrul
Editors

Advanced Mathematics for the Modeling and Solution of Challenging Problems in Engineering

Springer

Editors
Oscar Castillo
Division of Graduate Studies and Research
Tijuana Institute of Technology, TecNM
Tijuana, Baja California, Mexico

Gökhan Çuvalcıoğlu
Department of Mathematics
Mersin University
Mersin, Türkiye

Feride Tuğrul
Department of Computer Engineering
Munzur University
Tunceli, Türkiye

ISSN 2198-4182 ISSN 2198-4190 (electronic)
Studies in Systems, Decision and Control
ISBN 978-3-031-82341-1 ISBN 978-3-031-82342-8 (eBook)
https://doi.org/10.1007/978-3-031-82342-8

This Springer imprint is published by the registered company Springer Nature Switzerland AG
The registered company address is: Gewerbestrasse 11, 6330 Cham, Switzerland

Preface

In this book, some current problems in the field of mathematics and engineering, solutions to solve these problems and existing engineering methods for these problems are included.

Contributions have been made to engineering problems such as thermal oxidation problems of some alloys, tribological behavior of some systems, system stability problems, optimization modeling, traffic signaling modeling with mathematical studies such as data analysis, fuzzy modeling, relations with algebraic structures, system creation with Laplace transforms, geometric modeling.

Basically, machinery, chemistry, electronics, computers, construction, environment, etc., studies in both the application and theoretical fields of mathematics are needed to solve optimization, modeling, encryption methods and thermal problems that are currently encountered in engineering fields. In this book, some current problems are presented and related theoretical and applied mathematical studies are included.

In addition, this book also includes current studies on topological spaces, vector spaces, geometric spaces, and algebraic structures.

In this book, not only the solution of theoretical problems, but also studies such as signaling, the effects of Li-ion batteries on nature, mathematical modeling of toxicological factors affecting the marine fauna can be found as engineering problems to be used in daily life.

As a result, the edited book includes studies that include mathematical solution methods and mathematical modeling to support these methods for solving problems in the current application area of engineering, as well as studies that include solutions for problems that are also related to current life areas.

We hope that the book will serve as a resource for researchers and graduate students working in mathematics and engineering.

Tijuana, Mexico — Oscar Castillo
Mersin, Türkiye — Gökhan Çuvalcıoğlu
Tunceli, Türkiye — Feride Tuğrul
May 2024

Contents

Gap Intervals of Eccentricity Matrices of Chain Graphs

Abdullah Alazemi and Milica Anđelić

Abstract We show that the eccentricity matrix of a chain graph has a gap interval, i.e., an interval with no eccentricity eigenvalues in it. Our considerations are grounded on the row equivalence of two matrices: the characteristic matrix of the quotient matrix of the eccentricity matrix of a chain graph and a particular polynomial matrix whose underlying graph is a cycle.

Keywords Eccentricity matrix · Chain graph · Eigenvalue interval

1 Introduction

We consider a simple unoriented connected graph G with vertex set $V(G) = \{w_1, w_2, \cdots, w_n\}$. The distance $d(w_i, w_j)$ is the length of the shortest path between the vertices w_i and $\mathrm{w_j}$.

The eccentricity of a vertex w, denoted by ecc(w), is defined as $\mathrm{ecc}(w) = \max\{d(w, v) : v \in V(G)\}$. A vertex v is called an eccentric vertex of the vertex w if $d(v, w) = \mathrm{ecc}(w)$. The eccentricity matrix $\varepsilon(G)$ of a connected graph G, is the $n \times n$ matrix whose entries are as follows:

$$(\varepsilon(G))_{ij} = \begin{cases} d(w_i, w_j), & \text{if } d(w_i, w_j) = \min\{\mathrm{ecc}(w_i), \mathrm{ecc}(w_j)\}, \\ 0, & \text{otherwise.} \end{cases}.$$

If in the distance matrix of a connected graph G we set the largest distances in each row and 0 for the remaining entries, then we obtain $\varepsilon(G)$. The resulting matrix $\varepsilon(G)$ is symmetric, with real eigenvalues that are indexed in non-increasing order as $\varepsilon_1(G) \geq \varepsilon_2(G) \geq \cdots \geq \varepsilon_n(G)$. By $\sigma(A)$ we will denote the multiset of eigenvalues

A. Alazemi (✉) · M. Anđelić
Department of Mathematics, Kuwait University, Al-Shadadiyah, Kuwait
e-mail: abdullah.alazemi@ku.edu.kw

M. Anđelić
e-mail: milica.andelic@ku.edu.kw

O. Castillo et al. (eds.), *Advanced Mathematics for the Modeling and Solution of Challenging Problems in Engineering*, Studies in Systems, Decision and Control 582,
https://doi.org/10.1007/978-3-031-82342-8_1

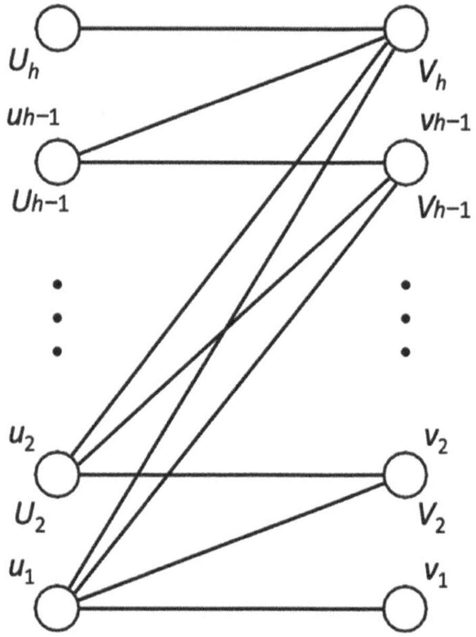

Fig. 1 The chain graph $G = \mathrm{CG}(u_1, \ldots, u_h; v_1, \ldots, v_h)$

of an arbitrary symmetric matrix A and by φ_A its characteristic polynomial. For more detailed studies on eccentricity matrices the reader is referred to [1, 2].

Here we provide some new spectral properties of ε-eigenvalues of chain graphs. These graphs are also known as double nested graphs [3]. The second name is related to the "nesting" property that the neighbourhoods of the vertices within the same colour class possess (see Fig. 1).

A chain graph does not contain neither $2K_2$ nor C_3, C_5 as induced subgraphs. Consequently, any chain graph is bipartite. Moreover, chain graphs can be seen as a bipartite counterpart of threshold graphs due to their property that in the class of connected bi-partite graphs of fixed order and size they maximize the largest eigenvalue of adjacency and (signless) Laplacian matrices.

In [3, 10], a procedure for generating chain graph from a given binary code of the form

$\beta = (\beta_1\beta_2 \cdots \beta_n)$ was introduced. It reads the following:

i. If i = 1, then $G_1 = G(\beta_1) = K_1$, i.e. G_1 is a single (black) vertex.
ii. If i $\geq$ 2, and $G_{i-1} = G(\beta_1\beta_2 \cdots \beta_{i-1})$, then $G_i = G(\beta_1\beta_2 \cdots \beta_{i-1}\beta_i)$ is constructed from G_{i-1} by adding an isolated black vertex if $\beta_i = 0$ or a dominating white vertex that dominates all black vertices in G_{i-1} if $\beta_i = 1$.

The structure of a connected chain graph is depicted in Fig. 1.

If $u_i = |U_i|$ and $v_i = |V_i|$, for $i = 1, 2, \cdots, h$, then G is denoted by $\mathrm{CG}(u_1, \ldots, u_h; v_1, \ldots, v_h)$.

The abbreviation CG comes from chain graph.

In the sequel we clarify several important facts on binary generating codes. The resulting chain graph does not depend on choice of β_1, due to (i). In order to standardize the procedure, we assume that β_1 is always zero. The corresponding graph is connected if and only if $\beta_n = 1$. Eccentricity matrix is considered only for connected graphs. Therefore, we always assume that $\beta_n = 1$.

If we assume that u_i's and v_i's are maximum lengths runs of consecutive zeros and ones, then a generating binary code that we will consider throughout is of the form

$$\beta = (0^{u_1}1^{v_1})(0^{u_2}1^{v_2})\cdots(0^{u_h}1^{v_h}) \quad \text{where} \quad u_i, v_i > 0 \quad \text{for } 1 \leq i \leq h. \tag{1}$$

The content of the remaining part is as follows. Section 2 contains several prerequisites. The characteristic polynomial of the quotient matrix $E(G)$ of the eccentricity matrix of a chain graph G is a subject of Sect. 3. We obtain that there exist finite intervals where this polynomial is either strictly positive or strictly negative. These intervals are eigenvalue-free intervals, i.e., they do not contain any eigenvalue of $\varepsilon(G)$. We obtain this result as a consequence of an intensive application of elementary row operations on $E(G) - tI_{2h}$. We show that for some $t \in \mathbb{R}$ the matrix $E(G) - tI_{2h}$ is row equivalent to the positive definite matrix. This result extends existing results on eigenvalue-free intervals (or gap intervals) for chain graphs regarding other types of graph matrices, such as adjacency matrix [3, 4], distance matrix [5] and Seidel matrix [6].

2 Preliminaries

Let $S = \{1, 2, \cdots, n\}$, and let $\pi : S_1 \sqcup S_2 \sqcup \cdots \sqcup S_k$ be a partition of S. We consider an $n \times n$ real symmetric matrix A with rows and columns ordered according to $S_1 \sqcup S_2 \sqcup \cdots \sqcup S_k$. This induces a block partition $[A_{i,j}]$ of A. Provided that each block $A_{i,j}$ has a constant row sum $b_{i,j}$, the partition π is called equitable, and the $k \times k$ matrix $\hat{B} = [b_{i,j}]$ is called the quotient matrix of A.

The quotient matrix retains some of the eigenvalues of the initial matrix.

Theorem 1 *Cvetković et al.* [7], [Theorem 3.9.5] *Let* $\hat{B}$ *be a quotient matrix of a real symmetric matrix* A. *Then the characteristic polynomial of* $\hat{B}$ *is a factor of the characteristic polynomial of* A.

We include an important argument on a real symmetric tridiagonal matrix with strictly positive main diagonal. It provides sufficient conditions for the positive definiteness of such matrices.

Theorem 2 [8] Let

$$A = \begin{bmatrix} a_1 & b_1 & & & \\ b_1 & a_2 & b_2 & & \\ & b_2 & \ddots & \ddots & \\ & & \ddots & & b_{n-1} \\ & & & b_{n-1} & a_n \end{bmatrix}$$

be a real symmetric tridiagonal matrix with positive diagonal. If $b_i^2 < \frac{1}{4}a_i a_{i+1}\frac{1}{\cos^2(\frac{\pi}{n+1})}$, for $1 \le i \le n-1$, *then A is positive definite.*

3 The Eccentricity Spectrum of Chain Graphs

Let $G = G(\beta)$ be a chain graph generated by (1), h > 1. Then

$$\pi : U_1 \cup V_1 \cup \cdots U_h \cup V_h$$

induces an equitable partition of ε(G). The resulting quotient matrix is denoted by $\hat{E}(G)$. It reads

$$\hat{E}(G) = \begin{bmatrix} 2(u_1-1) & 0 & 2u_2 & 0 & \cdots & 2u_h & 0 \\ 0 & 0 & 3u_2 & 0 & \cdots & 3u_h & 2v_h \\ 2u_1 & 3v_1 & 0 & 0 & \cdots & 0 & 0 \\ 0 & 0 & 0 & 0 & \cdots & 3u_h & 2v_h \\ \vdots & \vdots & \vdots & \vdots & \ddots & \vdots & \vdots \\ 2u_1 & 3v_1 & 0 & 3v_2 & \cdots & 0 & 0 \\ 0 & 2v_1 & 0 & 2v_2 & \cdots & 0 & 2(v_h-1) \end{bmatrix}. \quad (2)$$

More precisely,

$$\hat{E}(G)_{1,j} = \begin{cases} 2(u_1-1) & \text{if } j = 1 \\ 2u_k & \text{if } j = 2k-1, \\ 0 & \text{if } j \text{ is even,} \end{cases}$$

$$\hat{E}(G)_{2i,j} = \left\{ \begin{array}{ll} 0 & \text{if } j < 2i < 2h \text{ or } j \text{ is even} \\ 3u_k & \text{if } j = 2k-1,\ i \le k \le h \\ 2v_h & \text{if } j = 2h \end{array} \right\}, \text{ for } i < h,$$

$$\hat{E}(G)_{2i-1,j} = \left\{ \begin{array}{ll} 2u_1 & \text{if } j = 1 \\ 3v_k & \text{if } j = 2k,\ 1 \leq k < i, \\ 0 & \text{if } j \geq 2i-1 \text{ or } j \text{ is odd} \end{array} \right\}, \quad \text{for } i > 1, \text{ and}$$

$$\hat{E}(G)_{2h,j} = \begin{cases} 0 & \text{if } j \text{ is odd} \\ 2v_k & \text{if } j = 2k,\ k \neq h \\ 2(v_h - 1) & \text{if } j = 2h. \end{cases}$$

For $h = 1$ we consider the corresponding spectrum separately.

Lemma 1 *If* h $= 1$, *then* G *is complete bipartite graph generated by* $(0^{u_1} 1^{v_1})$, *and*

$$\sigma(\varepsilon(G)) = \left\{-2^{u_1+v_1-2}, 2(u_1-1), 2(v_1-1)\right\}$$

In case $h > 1$, we first deduce what $\sum_{i=1}^{h}(u_i + v_i) - 2h$ eigenvalues of $\varepsilon(G)$ are. The proof reads similar lines as analogue results for Laplacian matrix (see [9]) of chain graphs and therefore it is omitted here.

Theorem 3 *Let* $G = \mathrm{CG}(u_1, \cdots, u_h; v_1, \cdots, v_h)$, h > 1 *be a chain graph, and* ε(G) *its quotient matrix defined in* (2). *Then the following hold:*

i. *If* $u_1 \geq 2$ or $v_h \geq 2$, *then* $-2 \in \sigma(\varepsilon(G))$ *with multiplicity* $u_1 + v_h - 2$ *and the corresponding eigenvectors are of the form*

$$x_k = (\underbrace{1, 0, \cdots, -1}_{k}, \underbrace{0, \cdots, 0}_{u_1-k}, 0, 0, \cdots, 0)^\top, \quad \text{for } 2 \leq k \leq u_1$$

$$y_k = (0, \cdots, 0, \underbrace{1, 0, \cdots, -1}_{k}, \underbrace{0, \cdots, 0}_{v_h-k})^\top, \quad \text{for } 2 \leq k \leq v_h$$

ii. *If* $u_i \geq 2$ or $v_j \geq 2,\ i \neq 1,\ j \neq h$ *then* $0 \in \sigma(\varepsilon(G))$ with multiplicity $\sum_{i=1}^{h}(u_i + v_i) - (u_1 + v_h - 2) - 2h$ *and the corresponding eigenvectors have the form*

$$z_k = (0, \ldots, 0, \underbrace{\underbrace{1, 0, \ldots, 0, -1}_{k}, \underbrace{0, \ldots, 0}}_{u_i\ (v_j)}, 0, \ldots, 0)^\top, \quad \text{for } 2 \leq k \leq u_j \ (2 \leq k \leq v_j)$$

iii. $\hat{E}(G)$ *has no repeated eigenvalues.*

For the sake of simplification of the computation of $\det(\hat{E}(G) - tI_{2h})$, *we pursue the following elementary row operations*:

- $r_{2i-1} \leftarrow r_{2i-1} - r_{2i+1}$, for $1 \leq i \leq h-1$,
- $r_1 \leftarrow r_1 - \frac{2}{3} r_2$,
- $r_{2i} \leftarrow r_{2i} - r_{2i+2}$, for $1 \leq i \leq h-2$,
- $r_{2h-1} \leftarrow \frac{3}{2} r_{2h}$.

The resulting matrix is equal to

$$\begin{bmatrix} -2-t & -3v_1 + \frac{2}{3}t & -t & & & & & -\frac{4}{3}v_h \\ & -t & 3u_2 & t & & & & \\ & & -t & -3v_2 & t & & & \\ & & & \ddots & \ddots & \ddots & & \\ & & & & -t & -3v_{h-1} & t & \\ & & & & & -t & 3u_h & 2v_h \\ 2u_1 & & & & & & -t & -3v_h + 3 + \frac{3}{2}t \\ -\frac{4}{3} - \frac{2}{3}t + \frac{4}{3}u_1 & \frac{4}{9}t & & & & & & -\frac{8}{9}v_h \end{bmatrix}.$$

Next, we preform.

- $r_{2i-1} \leftarrow -r_{2i-1}$, for $1 \leq i \leq h$.

and we obtain

$$\begin{bmatrix} 2+t & 3v_1 - \frac{2}{3}t & -t & & & & & \frac{4}{3}v_h \\ & -t & 3u_2 & t & & & & \\ & & t & 3v_2 & -t & & & \\ & & & \ddots & \ddots & \ddots & & \\ & & & & t & 3v_{h-1} & -t & \\ & & & & & -t & 3u_h & 2v_h \\ -2u_1 & & & & & & t & 3v_h - 3 - \frac{3}{2}t \\ -\frac{4}{3} - \frac{2}{3}t + \frac{4}{3}u_1 & \frac{4}{9}t & & & & & & -\frac{8}{9}v_h \end{bmatrix}.$$

In the sequel, we apply.

- $r_1 \leftarrow \frac{3}{2} r_{2h} + r_1$;

and we obtain

$$\begin{bmatrix}
2u_1 & 3v_1 & -t & & & & & \\
 & -t & 3u_2 & t & & & & \\
 & & t & 3v_2 & -t & & & \\
 & & & \ddots & \ddots & \ddots & & \\
 & & & & t & 3v_{h-1} & -t & \\
 & & & & & -t & 3u_h & 2v_h \\
-2u_1 & & & & & & t & 3v_h - 3 - \frac{3}{2}t \\
-\frac{4}{3}-\frac{2}{3}t+\frac{4}{3}u_1 & \frac{4}{9}t & & & & & & -\frac{8}{9}v_h
\end{bmatrix}.$$

Finally, by consequent switching of the rows, we can move the last row to the top, i.e. by

- $r_{2h} \leftrightarrow r_{2h-1}$,
- $r_{2h-1} \leftrightarrow r_{2h-2}$,
- $\dots$
- $r_2 \leftrightarrow r_1$

we end up with

$$E_t' = \begin{bmatrix}
-\frac{4}{3}-\frac{2}{3}t+\frac{4}{3}u_1 & \frac{4}{9}t & & & & & & -\frac{8}{9}v_h \\
2u_1 & 3v_1 & -t & & & & & \\
 & -t & 3u_2 & t & & & & \\
 & & t & 3v_2 & -t & & & \\
 & & & \ddots & \ddots & \ddots & & \\
 & & & & t & 3v_{h-1} & -t & \\
 & & & & & -t & 3u_h & 2v_h \\
-2u_1 & & & & & & t & 3v_h - 3 - \frac{3}{2}t
\end{bmatrix}.$$

Accordingly, we have

$$\det(\hat{E}(G) - tI_{2h}) = (-1)^h(-1)^{h-1}\det(E_t') = -\frac{4}{9}\det(E_t), \tag{3}$$

where

$$E_t = \begin{bmatrix} 3u_1-3-\frac{3}{2}t & t & & & & & & & -2v_h \\ 2u_1 & 3v_1 & -t & & & & & & \\ & -t & 3u_2 & t & & & & & \\ & & t & 3v_2 & -t & & & & \\ & & & \ddots & \ddots & \ddots & & & \\ & & & & & t & 3v_{h-1} & -t & \\ & & & & & & -t & 3u_h & 2v_h \\ -2u_1 & & & & & & & t & 3v_h-3-\frac{3}{2}t \end{bmatrix}.$$

In what follows, we consider when 0 is an eigenvalue of $\hat{E}(G)$.

Lemma 2 *For the chain graph G generated by (1), $h > 1$ let $\hat{E}(G)$ be its quotient matrix given in (2). Then$0 \in \sigma(\hat{E}(G))$ if and only if $5u_1v_h - 9u_1 - 9v_h + 9 = 0$. In addition, if $0 \in \sigma(\hat{E}(G))$ then its multiplicity is 1.*

Proof From

$$\det(\hat{E}(G) - 0I_{2h}) = -\frac{4}{9}\det(E_0) = \det\begin{bmatrix} 3u_1-3 & 0 & & & & & & -2v_h \\ 2u_1 & 3v_1 & & & & & & 0 \\ & 0 & 3u_2 & 0 & & & & \\ & & \ddots & \ddots & \ddots & & & \\ & & & 0 & 3v_{h-1} & 0 & & \\ & & & & 0 & 3u_h & 2v_h & \\ & & & & & 0 & 3v_{h-3} & \end{bmatrix}$$

it follows, by expansion along the first row that

$$\begin{aligned} \det(\hat{E}(G) - 0I_{2h}) &= (3u_1-3)(3v_h-3)3^{2h-2}v_1u_h\prod_{i=2}^{h-1} v_iu_i - 4\cdot 3^{2h-2}v_hu_1v_1u_h\prod_{i=2}^{h-1} v_iu_i \\ &= 3^{2h-2}v_1u_h\prod_{i=2}^{h-1} v_iu_i((3u_1-3)(3v_h-3) - 4u_1v_h) \\ &= 3^{2h-2}v_1u_h\prod_{i=2}^{h-1} v_iu_i(5u_1v_h - 9u_1 - 9v_h + 9). \end{aligned}$$

Therefore $0 \in \sigma\left(\hat{E}(G)\right)$ if and only if $5u_1v_h - 9u_1 - 9v_h + 9 = 0$. By Theorem 3. all eigenvalues of $\hat{E}(G)$ are simple. This completes the proof.

Remark 1 We point out that

$$5u_1v_h - 9u_1 - 9v_h + 9 = \det\begin{bmatrix} 3u_1-3 & -2v_h \\ -2u_1 & 3v_h-3 \end{bmatrix}.$$

Then $5u_1v_h - 9u_1 - 9v_h + 9 \leq 0$ if and only if $u_1 = 1$ *or* $v_h = 1$ *or* $u_1 = 2, v_h \in \{2, \ldots, 9\}$ *or* $v_h = 2, u_1 \in \{2, \ldots, 9\}$. In particular, $5u_1v_h - 9u_1 - 9v_h + 9 = 0$ if and only if $(u_1, v_h) \in \{(2, 9), (3, 3), (9, 2)\}$. For all other values $5u_1v_h - 9u_1 - 9v_h + 9 > 0$.

Example 1 Consider the chain graph with the generating code $(0^2 1^5)(0^3 1^9)$. Then

$$\sigma\left(\hat{E}(G)\right) = \{25.2, 8.19, 0, -15.4\}$$

i e.

$$\hat{E}(G) = \begin{bmatrix} 2 & 0 & 6 & 0 \\ 0 & 0 & 9 & 18 \\ 4 & 15 & 0 & 0 \\ 0 & 10 & 0 & 16 \end{bmatrix}$$

is singular.

We proceed with several observations on $\det(E_t)$, i.e., on $\det(E(G) - tI_{2h})$.

The expansion of $\det(E_t)$ along the first row, followed by the expansion across the first columns in the corresponding sub-matrices yields:

$$\det(E_t) = \left(\left(\frac{3}{2}t + 3 - 3u_1\right)\left(\frac{3}{2}t + 3 - 3v_h\right) - 4u_1v_h\right)\det S_1$$
$$+ 2v_h t\left(\frac{3}{2}t + 3 - 3u_1\right)\det S_2 + 2u_1 t\left(\frac{3}{2}t + 3 - 3v_h\right)\det S_3 + 2v_h t^2 \det S_4,$$

where S_i, $1 \leq i \leq 4$ are sub-matrices of E_t obtained by crossing some rows and columns in E_t:

- For S_1 the first and the last rows and columns;
- For S_2 the first and the last two rows and columns;
- For S_3 the first two and the last one rows and columns;
- For S_4 the first two and the last two rows and columns.

By Theorem 2, if

$$t^2 < \frac{9}{4} min_{2\leq i\leq h-1}\{v_1u_2, v_iu_i, v_iu_{i+1}\}\frac{1}{\cos^2(\frac{\pi}{2h+1})}$$

then each one of S_i, $1 \leq i \leq 4$ is positive definite.

Set

$$a_h = \begin{cases} \min_{2\leq i\leq h-1}\{v_1u_2, v_iu_i, v_iu_{i+1}\}\frac{1}{\cos^2(\frac{\pi}{2h+1})} & \text{if } h > 2 \\ v_1u_2\frac{1}{\cos^2(\frac{\pi}{2h+1})} & if \;\; h = 2. \end{cases}.$$

Then, for $t \in \left(-\frac{3}{2}\sqrt{a_h}, \frac{3}{2}\sqrt{a_h}\right)$, all S_i, $1 \leq i \leq 4$ are positive definite.

In the sequel we find all $t \in R$ such that

$$\left(\frac{3}{2}t + 3 - 3u_1\right)\left(\frac{3}{2}t + 3 - 3v_h\right) - 4u_1v_h, 2v_ht\left(\frac{3}{2}t + 3 - 3u_1\right),$$
$$2u_1t\left(\frac{3}{2}t + 3 - 3v_h\right), 2v_ht^2$$

are all positive.

Bearing in mind that.

- $\left(\frac{3}{2}t + 3 - 3u_1\right)\left(\frac{3}{2}t + 3 - 3v_h\right) - 4u_1v_h > 0$ if and only if $t \in (-\infty, t_2) \cup (t_1, \infty)$, where $t_{1/2} = u_1 + v_h - 2 \pm \sqrt{(u_1 + v_h - 2)^2 - \frac{4}{9}(5u_1v_h - 9u_1 - 9v_h + 9)}$, provided that $(u_1 + v_h - 2)^2 - 4(5u_1v_h - 9u_1 - 9v_h + 9) \geq 0$, otherwise $\left(\frac{3}{2}t + 3 - 3u_1\right)\left(\frac{3}{2}t + 3 - 3v_h\right) - 4u_1v_h > 0$ for any $t \in R$.
- $2v_ht\left(\frac{3}{2}t + 3 - 3u_1\right) > 0$ if and only if $t \in (-\infty, 0) \cup (2(u_1 - 1), \infty)$.
- $2u_1t\left(\frac{3}{2}t + 3 - 3v_h\right) > 0$ if and only if $t \in (-\infty, 0) \cup (2(v_h - 1), \infty)$.
- $2v_ht^2 > 0$ for any $t \neq 0$.

we conclude that the real numbers we are looking for are the elements of

$$\begin{cases} (-\infty, t_2) \cup (t_1, \infty) & \text{if } 5u_1v_h - 9u_1 - 9v_h + 9 < 0,\ D \geq 0; \\ (-\infty, 0) \cup (\max\{t_1, 2(u_1 - 1), 2(v_h - 1)\}, \infty) & \text{if } 5u_1v_h - 9u_1 - 9v_h + 9 \geq 0,\ D \geq 0; \\ (-\infty, 0) \cup (\max\{2(u_1 - 1), 2(v_h - 1)\}, \infty) & \text{if } D < 0 \end{cases}$$

where $D = (u_1 + v_h - 2)^2 - \frac{4}{9}(5u_1v_h - 9u_1 - 9v_h + 9)$.

All previous observations lead to our main result.

Theorem 4 *Let G be a chain graph generated by a binary sequence* (1), $h > 1$ *and $\hat{E}(G)$ its quotient matrix given in* (2), along with

$$a_h = \begin{cases} \min_{2 \leq i \leq h-1}\{v_1u_2, v_iu_i, v_iu_{i+1}\}\frac{1}{\cos^2(\frac{\pi}{2h+1})} & \text{if } h > 2 \\ v_1u_2\frac{1}{\cos^2(\frac{\pi}{2h+1})} & \text{if } h = 2. \end{cases}.$$

Then

- *If* $(u_1 + v_h - 2)^2 - \frac{4}{9}(5u_1v_h - 9u_1 - 9v_h + 9) \geq 0$, *then*

$$\left(-\frac{3}{2}\sqrt{a_h}, \frac{3}{2}\sqrt{a_h}\right) \cap ((-\infty, min\{0, t_2\}) \cup (B, \infty)),$$

is a gap interval of $\hat{E}(G) - tI_{2h}$

where $B = \max\{t_1, 2(u_1 - 1), 2(v_h - 1)\}$ *and*

$$t_{1/2} = u_1 + v_h - 2 \pm \sqrt{(u_1 + v_h - 2)^2 - \frac{4}{9}(5u_1 v_h - 9u_1 - 9v_h + 9)}.$$

- *Otherwise,* $\hat{E}(G) - tI_{2h}$ *as no* ε*-eigenvalues in*

$$\left(-\frac{3}{2}\sqrt{a_h}, 0\right) \cup \left(\left(0, \frac{3}{2}\sqrt{a_h}\right) \cap (C, \infty)\right),$$

where $C = \max\{2(u_1 - 1), 2(v_h - 1)\}$.

If all $u_i, v_i \geq 2$, then $a_h \geq 9$. This implies the following.

Lemma 3 *Let G be the connected chain graph generated by* (1), $h > 1, u_i, v_i > 1$. *Then* -2 *is not an eigenvalue of* $\hat{E}(G)$ *given in* (2).

We conclude with the following proposition.

Proposition 1 *Let G be the chain graph generated by* (1),

$h > 1, u_i, v_i > 1, u = \sum_{i=1}^{h} u_i, v = \sum_{i=1}^{h} v_i$ *and* $\hat{E}(G)$ *the quotient matrix of* $\varepsilon(G)$ *given in* (2). *Then*

$$\phi_{\varepsilon(G)}(x) = x^{u_1+v_h-2}(x+2)^{v+u-(u_1+v_h-2)-2h}\phi_{\hat{E}(G)}(x).$$

We wrap this section by illustrating the results in two comprehensive examples.

Example 2 Fort the chain graph G generated by $(0^4 1^3)(0^2 1^5)$. Then

$$\sigma\left(\hat{E}(G)\right) = \{14.4, 9.11, 0.51, -10\}$$

Since $(u_1 + v_h - 2)^2 - 4(5u_1 v_h - 9u_1 - 9v_h + 9) < 0$, we obtain that the gap interval is

$$\left(-\frac{3}{2}\sqrt{a_h}, 0\right) \cup \left(\left(0, \frac{3}{2}\sqrt{a_h}\right) \cap (C, \infty)\right),$$

for $a_h = 9.17$. Consequently, it can be confirmed that no eigenvalues of $\hat{E}(G)$ are contained in $(-4.54, 0)$.

We now provide another example for the first case of Theorem 4, that is, the case where

$(u_1 + v_h - 2)^2 - \frac{4}{9}(5u_1 v_h - 9u_1 - 9v_h + 9) \geq 0$.

Example 3 If G is the chain graph with the binary code $(0^2 1^2)(0^3 1^3)(0^4 1^5)$, then

$$\sigma\left(\hat{E}(G)\right) = \{-17.57, 20.07, -5.94, -0.12, 8.89, 4.68\}$$

where

$$\hat{E}(G) = \begin{bmatrix} 2 & 0 & 6 & 0 & 8 & 0 \\ 0 & 0 & 9 & 0 & 12 & 10 \\ 4 & 6 & 0 & 0 & 0 & 0 \\ 0 & 0 & 0 & 0 & 12 & 10 \\ 4 & 6 & 0 & 9 & 0 & 0 \\ 0 & 4 & 0 & 6 & 0 & 8 \end{bmatrix}$$

Since $(u_1 + v_h - 2)^2 - 4(5u_1v_h - 9u_1 - 9v_h + 9) = 41 \geq 0$, where $h = 3$, we obtain that the gap interval is $(-4.08, -0.17)$, for $a_h = 7.39$.

References

1. Wang, J., Lu, M., Belardo, F., Randić, M.: The anti-adjacency matrix of a graph: Eccentricity matrix, Discrete Appl. Math. **251**, 299–309 (2018)
2. Wang, J., Lu, M., Lu, L., Belardo, F.: Spectral properties of the eccentricity matrix of graphs. Discrete Appl. Math. **279**, 168–177 (2020)
3. Alazemi, A., Anđelić, M., Simić, S.K.: Eigenvalue location for chain graphs. Linear Algebra Appl. **505**, 194–210 (2016)
4. Anđelić, M., Du, Z., da Fonseca, C.M., Simić, S.K.: Tridiagonal matrices and spectral properties of some graph classes. Czechoslovak Math. J. **70**(4), 1125–1138 (2020)
5. Alazemi, A., Anđelić, M., Koledin, T., Stanić, Z.: Eigenvalue-free intervals of distance matrices of threshold and chain graphs. Linear Multilinear Algebra **69**, 2959–2975 (2021)
6. Mandal, S., Mehatari, R., Ch. Das, K.: On the spectrum and energy of Seidel matrix for chain graphs. https://doi.org/10.48550/arXiv.2205.00310
7. Cvetković, D., Rowlinson, P., Simić, S.: An introduction to the theory of graph spectra. Cambridge University Press, Cambridge (2010)
8. Anđelić, M., da Fonseca, C.M.: Sufficient conditions for positive definiteness of tridiagonal matrices revisited. Positivity **15**, 155–159 (2011)
9. Alazemi, A., Anđelić, M., Ch. Das, K., da Fonseca, C.M.: Chain graph sequences and Laplacian spectra of chain graphs. Linear Multilinear Algebra **71**, 569–585 (2023)
10. Anđelić, M., Koledin, T., Stanić, Z.: Nested graphs with bounded second largest (signless Laplacian) eigenvalue. Electron. J. Linear Algebra **24**, 181–201 (2012)

On Hermite-Hadamard Type Inequalities for Co-ordinated Convex Functions via Proportional Fractional Integrals

Ali Muhammedoğlu, Abdullah Akkurt, and Hüseyin Yıldırım

Abstract In this paper, we define proportional fractional integrals for continuous functions of two variables. Then, we prove the Hermite-Hadamard type inequalities for coordinated convex functions by using proportional fractional integrals. Many results obtained in this paper provide significant extensions of other related results given in the literature. Finally, we give some examples of our results.

Keywords Hermite-Hadamard inequality · Fractional integrals · Proportional fractional integral

1 Introduction

Let $f : I \subset \mathbb{R} \to \mathbb{R}$ be a convex function, then,

$$f\left(\frac{a+b}{2}\right) \le \frac{1}{b-a}\int_a^b f(x) \le \frac{f(a)+f(b)}{2}, \tag{1}$$

is known in the literature as Hermite-Hadamard inequality for convex mappings. Note that some of the classical inequalities for means can be derived from (1) for appropriate particular selections of the mapping f.

It is well known that the Hermite–Hadamard's inequality plays an essential role in mathematics. In recent years, this classical inequality has been extended and

A. Muhammedoğlu · A. Akkurt (✉) · H. Yıldırım
Department of Mathematics, Faculty of Science, Kahramanmaraş Sütçü İmam University, Kahramanmaraş, Turkey
e-mail: abdullahmat@gmail.com

A. Muhammedoğlu
e-mail: alidna1992@gmail.com

H. Yıldırım
e-mail: hyildir@ksu.edu.tr

O. Castillo et al. (eds.), *Advanced Mathematics for the Modeling and Solution of Challenging Problems in Engineering*, Studies in Systems, Decision and Control 582,
https://doi.org/10.1007/978-3-031-82342-8_2

generalized in several ways; there have been many studies on Her-mite–Hadamard's inequality [see 1–4, 7, 8, 9, 11–16].

Let us now consider a bidemensional interval $\Delta =: [a, b] \times [c, d]$ in $\mathbb{R}^2$ with $a < b$ and $c < d$. A mapping $f : \Delta \to \mathbb{R}$ is said to be convex on Δ if the following inequality:

$$f(tx + (1-t)z, ty + (1-t)w) \leq tf(x, y) + (1-t)f(z, w) \tag{2}$$

holds for all $(x, y), \ (z, w) \in \Delta$ and $t \in [0, 1]$. A function $f : \Delta \to \mathbb{R}$ is said to be convex on the co-ordinates on Δ if the partial mappings $f_y : [a, b] \to \mathbb{R}, \ f_y(u) = f(u, y)$ and $f_x : [c, d] \to \mathbb{R}, \ f_x(v) = f(x, v)$ are convex where defined for all $x \in [a, b]$ and $y \in [c, d]$ [3].

Definition 1 *A function $f : \Delta \to \mathbb{R}$ will be called co-ordinated convex on Δ, for all $t, s \in [0, 1]$ and $(x, y), (u, w) \in \Delta$, then the following inequalty holds;*

$$\begin{aligned} f(tx + (1-t)y, su + (1-s)w) &\leq tsf(x, u) + s(1-t)f(y, u) \\ &+t(1-s)f(x, w) + (1-t)(1-s)f(y, w). \end{aligned} \tag{3}$$

Clearly, every convex function is co-ordinated convex. Furthermore, there exist co-ordinated convex function which is not convex [3].

Earlier, Dragomir [3] establish the following inequality of Hermite-Hadamard type for co-ordinated convex mapping on a rectangle from the plane $\mathbb{R}^2$. Later, Sarikaya and Yaldiz gave the another proof of a special version of the below teorem by using the definition of the co-ordinated convex function.

Theorem 1 *Suppose that $f : \Delta \to \mathbb{R}$ is co-ordinated convex on Δ. Then one has the inequalities;*

$$\begin{aligned} f\left(\frac{a+b}{2}, \frac{c+d}{2}\right) &\leq \frac{1}{2}\left[\frac{1}{b-a}\int_a^b f\left(x, \frac{c+d}{2}\right)dx + \frac{1}{d-c}\int_c^d f\left(\frac{a+b}{2}, y\right)dy\right] \\ &\leq \frac{1}{(b-a)(d-c)}\int_a^b\int_c^d f(x, y)dydx \\ &\leq \frac{1}{4}\left[\frac{1}{b-a}\int_a^b f(x, c)dx + \frac{1}{b-a}\int_a^b f(x, d)dx \right. \\ &\left. + \frac{1}{d-c}\int_c^d f(a, y)dy + \frac{1}{d-c}\int_c^d f(b, y)dy\right] \end{aligned}$$

$$\leq \frac{f(a,c)+f(a,d)+f(b,c)+f(b,d)}{4}. \tag{4}$$

In the following, we will give some necessary definitions and mathematical preliminaries of fractional calculus theory which is used in this paper. For more details, one can consult Gorenflo and Mainardi [5], Kilbas et al. [6], Samko et al. [10].

Definition 2 *Let $f \in L_1[a,b]$. The Riemann Liouville fractional integrals $I^{\alpha}_{a+}f$* and *$I^{\alpha}_{b-}f$ of order $\alpha > 0$, $a \geq 0$ are defined by* [10]

$$I^{\alpha}_{a+}f(x) = \frac{1}{\Gamma(\alpha)}\int_a^x (x-t)^{\alpha-1}f(t)dt, \ x > a, \tag{5}$$

and

$$I^{\alpha}_{b-}f(x) = \frac{1}{\Gamma(\alpha)}\int_x^b (t-x)^{\alpha-1}f(t)dt, \ x < b, \tag{6}$$

respectively. Here, $\Gamma(\alpha)$ is the Gamma function and
$I^{0}_{a+}f(x) = I^{0}_{b-}f(x) = f(x)$.

Definition 3 *Let $f \in L_1([a,b]\times[c,d])$. The Riemann–Liouville integrals $I^{\alpha,\beta}_{a^+,c^+}$, $I^{\alpha,\beta}_{a^+,d^-}$, $I^{\alpha,\beta}_{b^-,c^+}$ and $I^{\alpha,\beta}_{b^-,d^-}$ of order $\alpha, \beta > 0$ with $a, c \geq 0$ are defined by* [11]

$$I^{\alpha,\beta}_{a^+,c^+}f(x,y) = \frac{1}{\Gamma(\alpha)\Gamma(\beta)}\int_a^x\int_c^y (x-t)^{\alpha-1}(y-s)^{\beta-1}f(t,s)dsdt, \tag{7}$$

$$I^{\alpha,\beta}_{a^+,d^-}f(x,y) = \frac{1}{\Gamma(\alpha)\Gamma(\beta)}\int_a^x\int_y^d (x-t)^{\alpha-1}(s-y)^{\beta-1}f(t,s)dsdt, \tag{8}$$

$$I^{\alpha,\beta}_{b^-,c^+}f(x,y) = \frac{1}{\Gamma(\alpha)\Gamma(\beta)}\int_x^b\int_c^y (t-x)^{\alpha-1}(y-s)^{\beta-1}f(t,s)dsdt, \tag{9}$$

$$I^{\alpha,\beta}_{b^-,d^-}f(x,y) = \frac{1}{\Gamma(\alpha)\Gamma(\beta)}\int_x^b\int_y^d (t-x)^{\alpha-1}(s-y)^{\beta-1}f(t,s)dsdt, \tag{10}$$

respectively. Here, $\Gamma(\alpha)$ is the Gamma function and

$$I^{0,0}_{a^+,c^+}f(x,y) = I^{0,0}_{a^+,d^-}f(x,y) = I^{0,0}_{b^-,c^+}f(x,y) = I^{0,0}_{b^-,d^-}f(x,y) = f(x,y).$$

Also,

$$I^{1,1}_{a^+,c^+}f(x,y) = \frac{1}{\Gamma(\alpha)\Gamma(\beta)} \int_a^x \int_c^y f(t,s)dsdt. \tag{11}$$

Similar to Definitions 2 and 3, we give the following fractional integrals:

$$I^{\alpha}_{a^+}f\left(x, \frac{c+d}{2}\right) = \frac{1}{\Gamma(\alpha)} \int_a^x (x-t)^{\alpha-1} f\left(t, \frac{c+d}{2}\right)dt, \tag{12}$$

$$I^{\alpha}_{b^-}f\left(x, \frac{c+d}{2}\right) = \frac{1}{\Gamma(\alpha)} \int_x^b (t-x)^{\alpha-1} f\left(t, \frac{c+d}{2}\right)dt, \tag{13}$$

$$I^{\beta}_{a^+}f\left(\frac{a+b}{2}, y\right) = \frac{1}{\Gamma(\beta)} \int_c^y (y-s)^{\beta-1} f\left(\frac{a+b}{2}, s\right)ds, \tag{14}$$

$$I^{\beta}_{a^+}f\left(\frac{a+b}{2}, y\right) = \frac{1}{\Gamma(\beta)} \int_y^d (s-y)^{\beta-1} f\left(\frac{a+b}{2}, s\right)ds. \tag{15}$$

It is remarkable that Sarikaya et al. [14] first give the following interesting integral inequalities of Hermite–Hadamard-type involving Riemann–Liouville fractional integrals.

Theorem 2 *Let $f : [a,b] \to \mathbb{R}$ be positive function with $0 \le a < b$ and $f \in L_1[a,b]$. If f is a convex functionon $[a,b]$ then the following inequalities hold:*

$$f\left(\frac{a+b}{2}\right) \le \frac{\Gamma(\alpha+1)}{2(b-a)^{\alpha}}\left[I^{\alpha}_{a^+}f(b) + I^{\alpha}_{b^-}f(a)\right] \le \frac{f(a)+f(b)}{2}, \tag{16}$$

with $\alpha > 0$.

Definition 4 *For $\rho > 0$ and $\alpha \in C$, $Re(\alpha) > 0$, the left and right proportional fractional integrals of f are respectively define as* [17]

$$({}_aI^{\alpha,\rho}f)(t) = \frac{1}{\rho^{\alpha}\Gamma(\alpha)} \int_a^t e^{\frac{\rho-1}{\rho}(t-\tau)}(t-\tau)^{\alpha-1} f(\tau)d\tau \tag{17}$$

and

$$\left(I_b^{\alpha,\rho} f\right)(t) = \frac{1}{\rho^\alpha \Gamma(\alpha)} \int_t^b e^{\frac{\rho-1}{\rho}(\tau-t)} (\tau - t)^{\alpha-1} f(\tau) d\tau. \tag{18}$$

The aim of this article is to first give a general definition of fractional integral and to establish Hermite-Hadamard type inequalities with the help of this definition. Then, we will give some theorem and results with the help of these fractional type integral inequalities. We will investigate some integral inequalities connected with the Hermite-Hadamard-type inequalities involving Riemann–Liouville fractional integrals.

2 Main Results

Now, let's introduce the definition of proportional fractional integrals for functions of two variables.

Definition 5 *Let $f \in L_1([a,b] \times [c,d])$ and $\rho > 0$. The Proportional Riemann–Liouville fractional integrals $I_{a^+,c^+}^{\alpha,\beta,\rho}$, $I_{a^+,d^-}^{\alpha,\beta,\rho}$, $I_{b^-,c^+}^{\alpha,\beta,\rho}$ and $I_{b^-,d^-}^{\alpha,\beta,\rho}$ of order $\alpha, \beta > 0$ with $a, c \geq 0$ are defined by;*

$${}^{\rho}I_{a^+,c^+}^{\alpha,\beta} f(x,y) = \Lambda \int_a^x \int_c^y (x-t)^{\alpha-1} e^{\frac{\rho-1}{\rho}(x-t)} (y-s)^{\beta-1} e^{\frac{\rho-1}{\rho}(y-s)} f(t,s) ds dt, \tag{19}$$

$${}^{\rho}I_{a^+,d^-}^{\alpha,\beta} f(x,y) = \Lambda \int_a^x \int_y^d (x-t)^{\alpha-1} e^{\frac{\rho-1}{\rho}(x-t)} (s-y)^{\beta-1} e^{\frac{\rho-1}{\rho}(s-y)} f(t,s) ds dt, \tag{20}$$

$${}^{\rho}I_{b^-,c^+}^{\alpha,\beta} f(x,y) = \Lambda \int_x^b \int_c^y (t-x)^{\alpha-1} e^{\frac{\rho-1}{\rho}(t-x)} (y-s)^{\beta-1} e^{\frac{\rho-1}{\rho}(y-s)} f(t,s) ds dt, \tag{21}$$

$${}^{\rho}I_{b^-,d^-}^{\alpha,\beta} f(x,y) = \Lambda \int_x^b \int_y^d (t-x)^{\alpha-1} e^{\frac{\rho-1}{\rho}(t-x)} (s-y)^{\beta-1} e^{\frac{\rho-1}{\rho}(s-y)} f(t,s) ds dt, \tag{22}$$

respectively. Where

$$\Lambda = \frac{\rho^{-(\alpha+\beta)}}{\Gamma(\alpha)\Gamma(\beta)}.$$

Here, $\Gamma(\alpha)$ is the Gamma function and

$$I^{0,0}_{a^+,c^+}f(x,y) = I^{0,0}_{a^+,d^-}f(x,y) = I^{0,0}_{b^-,c^+}f(x,y) = I^{0,0}_{b^-,d^-}f(x,y) = f(x,y).$$

also,

$$I^{1,1}_{a^+,c^+}f(x,y) = \frac{1}{\Gamma(\alpha)\Gamma(\beta)}\int_a^x\int_c^y f(t,s)dsdt. \tag{23}$$

In this part we will give the Hermite-Hadamard-type inequalities for Proportional Riemann–Liouville fractional integrals by using convex functions of two variables on the co-ordinates.

Theorem 3 *Let $f : \Delta \subset \mathbb{R}^2 \to \mathbb{R}$ be co-ordinated convex on $\Delta := [a,b] \times [c,d]$ in $\mathbb{R}^2$ with $\rho > 0$, $0 \le a < b, 0 \le c < d$ and $f \in L_1(\Delta)$. Then one has the inequalities:*

$$\begin{aligned} f\left(\frac{a+b}{2},\frac{c+d}{2}\right) &\le \frac{\Gamma(\alpha)\Gamma(\beta)\rho^{\alpha+\beta}}{\Omega(n,m)}\left[I^{\alpha,\beta,\rho}_{a^+,c^+} + I^{\alpha,\beta,\rho}_{a^+,d^+} + I^{\alpha,\beta,\rho}_{b^-,c^+} + I^{\alpha,\beta,\rho}_{b^-,d^-}\right] \\ &\le \frac{f(a,c)+f(a,d)+f(b,c)+f(b,d)}{4}. \end{aligned} \tag{24}$$

Here

$$\Omega(n,m) = 4(b-a)^\alpha(d-c)^\beta \sum_{n=0}^{\infty}\sum_{m=0}^{\infty}\left(\frac{\rho-1}{\rho}\right)^n\frac{1}{(n+\alpha)n!}\left(\frac{\rho-1}{\rho}\right)^m\frac{1}{(m+\beta)m!}f\left(\frac{a+b}{2},\frac{c+d}{2}\right).$$

Proof According to (2) with $x = t_1a + (1-t_1)b,\ y = (1-t_1)a + t_1b,\ u = s_1c + (1-s_1)d,\ w = (1-s_1)c + s_1d$ and $t = s = \frac{1}{2}$, we find that,

$$\begin{aligned} f\left(\frac{a+b}{2},\frac{c+d}{2}\right) \le \frac{1}{4}\big[&f(t_1a+(1-t_1)b, s_1c+(1-s_1)d) \\ &+ f((1-t_1)a+t_1b, s_1c+(1-s_1)d) \\ &+ f((1-t_1)a+t_1b, (1-s_1)c+s_1d) \\ &+ f(t_1a+(1-t_1)b, (1-s_1)c+s_1d)\big]. \end{aligned} \tag{25}$$

Thus, multiplying both side of (25) by $t_1^{\alpha-1}s_1^{\beta-1}e^{\frac{\rho-1}{\rho}t_1}e^{\frac{\rho-1}{\rho}s_1}$, then integrating with respect to (t_1, s_1) on $[0,1]\times[0,1]$, we obtain

$$\int_0^1\int_0^1 t_1^{\alpha-1}s_1^{\beta-1}e^{\frac{\rho-1}{\rho}t_1}e^{\frac{\rho-1}{\rho}s_1}f\left(\frac{a+b}{2},\frac{c+d}{2}\right)dt_1ds_1$$

$$\leq \frac{1}{4}\int_0^1\int_0^1 t_1^{\alpha-1}s_1^{\beta-1}e^{\frac{\rho-1}{\rho}t_1}e^{\frac{\rho-1}{\rho}s_1}$$

$$\times\big[f(t_1a+(1-t_1)b,s_1c+(1-s_1)d)+f((1-t_1)a+t_1b,s_1c+(1-s_1)d)$$

$$+f(t_1a+(1-t_1)b,(1-s_1)c+s_1d)$$
$$+f((1-t_1)a+t_1b,(1-s_1)c+s_1d)\big]dt_1ds_1$$

$$\sum_{n=0}^{\infty}\sum_{m=0}^{\infty}\left(\frac{\rho-1}{\rho}\right)^n\frac{1}{(n+\alpha)n!}\left(\frac{\rho-1}{\rho}\right)^m\frac{1}{(m+\beta)m!}f\left(\frac{a+b}{2},\frac{c+d}{2}\right)$$
$$\leq \frac{1}{4}\int_0^1\int_0^1 t_1^{\alpha-1}s_1^{\beta-1}e^{\frac{\rho-1}{\rho}t_1}e^{\frac{\rho-1}{\rho}s_1}$$
$$\times\big[f(t_1a+(1-t_1)b,s_1c+(1-s_1)d)$$
$$+f((1-t_1)a+t_1b,s_1c+(1-s_1)d)$$
$$f(t_1a+(1-t_1)b,(1-s_1)c+s_1d)$$
$$+f((1-t_1)a+t_1b,(1-s_1)c+s_1d)\big]dt_1ds_1$$

using the change of the variable, we get

$$\sum_{n=0}^{\infty}\sum_{m=0}^{\infty}\left(\frac{\rho-1}{\rho}\right)^n\frac{1}{(n+\alpha)n!}\left(\frac{\rho-1}{\rho}\right)^m\frac{1}{(m+\beta)m!}f\left(\frac{a+b}{2},\frac{c+d}{2}\right)$$
$$\leq \frac{1}{4(b-a)^{\alpha}(d-c)\beta}\left[\int_a^b\int_c^d (b-x)^{\alpha-1}(d-y)^{\beta-1}e^{\frac{\rho-1}{\rho}(b-x)}e^{\frac{\rho-1}{\rho}(d-y)}f(x,y)dxdy\right.$$
$$+\int_a^b\int_c^d (b-x)^{\alpha-1}(y-c)^{\beta-1}e^{\frac{\rho-1}{\rho}(b-x)}e^{\frac{\rho-1}{\rho}(y-c)}f(x,y)dxdy$$
$$+\int_a^b\int_c^d (x-a)^{\alpha-1}(d-y)^{\beta-1}e^{\frac{\rho-1}{\rho}(x-a)}e^{\frac{\rho-1}{\rho}(d-y)}f(x,y)dxdy$$
$$\left.+\int_a^b\int_c^d (x-a)^{\alpha-1}(y-c)^{\beta-1}e^{\frac{\rho-1}{\rho}(x-a)}e^{\frac{\rho-1}{\rho}(y-c)}f(x,y)dxdy\right].$$

Then,

$$\sum_{n=0}^{\infty}\sum_{m=0}^{\infty}\left(\frac{\rho-1}{\rho}\right)^{n}\frac{1}{(n+\alpha)n!}\left(\frac{\rho-1}{\rho}\right)^{m}\frac{1}{(m+\beta)m!}f\left(\frac{a+b}{2},\frac{c+d}{2}\right)$$
$$\leq\frac{\Gamma(\alpha)\Gamma(\beta)\rho^{\alpha+\beta}}{4(b-a)^{\alpha}(d-c)\beta}$$
$$\left[\frac{1}{\Gamma(\alpha)\Gamma(\beta)\rho^{\alpha+\beta}}\int_{a}^{b}\int_{c}^{d}(b-x)^{\alpha-1}(d-y)^{\beta-1}e^{\frac{\rho-1}{\rho}(b-x)}e^{\frac{\rho-1}{\rho}(d-y)}f(x,y)dxdy\right.$$
$$+\frac{1}{\Gamma(\alpha)\Gamma(\beta)\rho^{\alpha+\beta}}\int_{a}^{b}\int_{c}^{d}(b-x)^{\alpha-1}(y-c)^{\beta-1}e^{\frac{\rho-1}{\rho}(b-x)}e^{\frac{\rho-1}{\rho}(y-c)}f(x,y)dxdy$$
$$+\frac{1}{\Gamma(\alpha)\Gamma(\beta)\rho^{\alpha+\beta}}\int_{a}^{b}\int_{c}^{d}(x-a)^{\alpha-1}(d-y)^{\beta-1}e^{\frac{\rho-1}{\rho}(x-a)}e^{\frac{\rho-1}{\rho}(d-y)}f(x,y)dxdy$$
$$\left.+\frac{1}{\Gamma(\alpha)\Gamma(\beta)\rho^{\alpha+\beta}}\int_{a}^{b}\int_{c}^{d}(x-a)^{\alpha-1}(y-c)^{\beta-1}e^{\frac{\rho-1}{\rho}(x-a)}e^{\frac{\rho-1}{\rho}(y-c)}f(x,y)dxdy\right]$$

we obtain

$$\sum_{n=0}^{\infty}\sum_{m=0}^{\infty}\left(\frac{\rho-1}{\rho}\right)^{n}\frac{1}{(n+\alpha)n!}\left(\frac{\rho-1}{\rho}\right)^{m}\frac{1}{(m+\beta)m!}f\left(\frac{a+b}{2},\frac{c+d}{2}\right)$$
$$\leq\frac{\Gamma(\alpha)\Gamma(\beta)\rho^{\alpha+\beta}}{4(b-a)^{\alpha}(d-c)^{\beta}}\left[I_{a^{+},c^{+}}^{\alpha,\beta,\rho}+I_{a^{+},d^{+}}^{\alpha,\beta,\rho}+I_{b^{-},c^{+}}^{\alpha,\beta,\rho}+I_{b^{-},d^{-}}^{\alpha,\beta,\rho}\right]$$

by which the first inequality is prooved. For the proof of the second inequality in (24), we first note that if f is a co-ordinated convex on Δ, then by using (16) with $x=a,\ y=b,\ u=c,\ w=d$, it yields

$$f(ta+(1-t)b,sc+(1-s)d)\leq tsf(a,c)+t(1-s)f(a,d)$$
$$+s(1-t)f(b,c)+(1-t)(1-s)f(b,d)$$

$$f(ta+(1-t)b,(1-s)c+sd)\leq t(1-s)f(a,c)+tsf(a,d)$$
$$+(1-t)(1-s)f(b,c)+(1-t)sf(b,d)$$

$$f((1-t)a+tb,sc+(1-s)d)\leq(1-t)sf(a,c)+(1-t)(1-s)f(a,d)$$
$$+tsf(b,c)+t(1-s)f(b,d)$$

and

$$f((1-t)a+tb,(1-s)c+sd) \leq (1-t)(1-s)f(a,c)+(1-t)sf(a,d)$$
$$+t(1-s)f(b,c)+tsf(b,d)$$

by adding these inequalities we have

$$\begin{aligned} & f(ta+(1-t)b, sc+(1-s)d)+f((1-t)a+tb, sc+(1-s)d) \\ & +f(ta+(1-t)b,(1-s)c+sd)+f((1-t)a+tb,(1-s)c+sd) \\ & \leq \left[f(a,c)+f(a,d)+f(b,c)+f(b,d)\right]. \end{aligned} \tag{26}$$

Thus, multiplying both side of (26) by $t_1^{\alpha-1}s_1^{\beta-1}e^{\frac{\rho-1}{\rho}t_1}e^{\frac{\rho-1}{\rho}s_1}$, then integrating with respect to (t_1, s_1) on $[0,1]\times[0,1]$, we get

$$\begin{aligned} & \int_0^1\int_0^1 t^{\alpha-1}s^{\beta-1}e^{\frac{\rho-1}{\rho}t}e^{\frac{\rho-1}{\rho}s} \\ & \left[f(ta+(1-t)b, sc+(1-s)d)\right. \\ & \quad +f((1-t)a+tb, sc+(1-s)d) \\ & \quad +f(ta+(1-t)b,(1-s)c+sd) \\ & \quad +f((1-t)a+tb,(1-s)c+sd)dtds \\ & \quad \leq \left[f(a,c)+f(a,d)+f(b,c)+f(b,d)\right] \\ & \int_0^1\int_0^1 t^{\alpha-1}s^{\beta-1}e^{\frac{\rho-1}{\rho}t}e^{\frac{\rho-1}{\rho}s}dtds. \end{aligned}$$

Here, using the change of the variable we have

$$\frac{\Gamma(\alpha)\Gamma(\beta)\rho^{\alpha+\beta}}{(b-a)^{\alpha}(d-c)^{\beta}}\left[I_{a^+,c^+}^{\alpha,\beta,\rho}+I_{a^+,d^+}^{\alpha,\beta,\rho}+I_{b^-,c^+}^{\alpha,\beta,\rho}+I_{b^-,d^-}^{\alpha,\beta,\rho}\right]$$

$$\leq \sum_{n=0}^{\infty}\sum_{m=0}^{\infty}\left(\frac{\rho-1}{\rho}\right)^{n}\frac{1}{(n+\alpha)n!}\left(\frac{\rho-1}{\rho}\right)^{m}\frac{1}{(m+\beta)m!}f\left(\frac{a+b}{2},\frac{c+d}{2}\right)$$

$\times\left[f(a,c)+f(a,d)+f(b,c)+f(b,d)\right]$,
the proof is completed.

Remark 1 *If we choose $\rho = 1$ in (24), we can obtain Theorem 3 in [11].*

Remark 2 *If we choose $\alpha = \beta = \rho = 1$ in (24), we can obtain Theorem 1 in [3].*

Theorem 4 *Let $f : [a, b] \to \mathbb{R}$ be positive function with $0 \leq a < b$ and $f \in L_1[a, b]$. If f is a convex functionon $[a, b]$ then the following inequalities hold:*

$$f\left(\frac{a+b}{2}\right) \leq \frac{\Gamma(\alpha)\rho^{\alpha}\left[I_{a^+}^{\alpha,\rho} f(b) + I_{b^-}^{\alpha,\rho} f(a)\right]}{2(b-a)^{\alpha}\sum_{n=0}^{\infty}\left(\frac{\rho-1}{\rho}(b-a)\right)^{n}\frac{1}{(n+\alpha)n!}} \leq \frac{f(a)+f(b)}{2}$$

with $\alpha > 0$.

Proof Let f be a convex function on interval $[a, b]$, we have

$$f(\lambda x + (1-t)y) \leq \lambda f(x) + (1-\lambda)f(y). \tag{27}$$

If $\lambda = \frac{1}{2}$ selected the (27) expression will be

$$f\left(\frac{x+y}{2}\right) \leq \frac{f(x)+f(y)}{2}. \tag{28}$$

Here, using the change of the varible we have

$$x = at + (1-t)b, \;\; y = (1-t)a + tb$$

$$2f\left(\frac{a+b}{2}\right) \leq f(at + (1-t)b) + f((1-t)a + tb). \tag{29}$$

Thus, multiplying both side of (29) by $(t(b-a))^{\alpha-1}e^{\frac{\rho-1}{\rho}(b-a)t}$, then integrating t_1, on $[0, 1]$, we get

$$2f\left(\frac{a+b}{2}\right)\int_0^1 t^{\alpha-1}(b-a)^{\alpha-1}e^{\frac{\rho-1}{\rho}(b-a)t}dt$$

$$\leq \int_0^1 t^{\alpha-1}(b-a)^{\alpha-1}e^{\frac{\rho-1}{\rho}(b-a)t}f(at+(1-t)b)dt$$

$$+\int_0^1 t^{\alpha-1}(b-a)^{\alpha-1}e^{\frac{\rho-1}{\rho}(b-a)t}f((1-t)a+bt)dt$$

$$2f\left(\frac{a+b}{2}\right)(b-a)^{\alpha-1}\sum_{n=0}^{\infty}\left(\frac{\rho-1}{\rho}(b-a)\right)^n\frac{1}{(n+\alpha)n!}$$

$$\leq \int_0^1 t^{\alpha-1}(b-a)^{\alpha-1}e^{\frac{\rho-1}{\rho}(b-a)t}f(at+(1-t)b)dt$$

$$+\int_0^1 t^{\alpha-1}(b-a)^{\alpha-1}e^{\frac{\rho-1}{\rho}(b-a)t}f((1-t)a+bt)dt$$

sHere, using the change of the variable we have $u = at + (1-t)b$,

$$\int_0^1 t^{\alpha-1}(b-a)^{\alpha-1}.e^{\frac{\rho-1}{\rho}(b-a)t}f(at+(1-t)b)dt$$

$$=\frac{1}{(b-a)}\int_a^b (b-u)^{\alpha-1}e^{\frac{\rho-1}{\rho}(b-u)}f(u)du$$

$$=\frac{\Gamma(\alpha)\rho^\alpha}{(b-a)\Gamma(\alpha)\rho^\alpha}\int_a^b (b-u)^{\alpha-1}e^{\frac{\rho-1}{\rho}(b-u)}f(u)du$$

$$=\frac{\Gamma(\alpha)\rho^\alpha}{(b-a)}I_{a^+}^{\alpha,\rho}f(b)$$

thus, similarly, by integrating by parts we get $u = (1-t)a + bt$,

$$\int_0^1 t^{\alpha-1}(b-a)^{\alpha-1}e^{\frac{\rho-1}{\rho}(b-a)t}f((1-t)a+bt)dt = \frac{\Gamma(\alpha)\rho^\alpha}{(b-a)}I_{b^-}^{\alpha,\rho}f(a)$$

by which the first inequality is prooved. Considering the convexity of the function we have

$$f(at+(1-t)b) \le tf(a)+(1-t)f(b)$$

$$f((1-t)a+tb) \le (1-t)f(a)+tf(b)$$

by adding these inequalities we have

$$f(at+(1-t)b)+f((1-t)a+tb) \le f(a)+f(b) \tag{30}$$

Thus, multiplying both side of (30) by $(t(b-a))^{\alpha-1}e^{\frac{\rho-1}{\rho}(b-a)t}$, then integrating t_1, on [0, 1], we get

$$\int_0^1 t^{\alpha-1}(b-a)^{\alpha-1}e^{\frac{\rho-1}{\rho}(b-a)t}f(at+(1-t)b)dt$$
$$+\int_0^1 t^{\alpha-1}(b-a)^{\alpha-1}e^{\frac{\rho-1}{\rho}(b-a)t}f((1-t)a+bt)dt$$
$$\le (f(a)+f(b))\int_0^1 (t(b-a))^{\alpha-1}e^{\frac{\rho-1}{\rho}(b-a)t}dt$$

and so,

$$\frac{\Gamma(\alpha)\rho^{\alpha}}{(b-a)}\left[I_{a^+}^{\alpha,\rho}f(b)+I_{b^-}^{\alpha,\rho}f(a)\right] \le (b-a)^{\alpha-1}\sum_{n=0}^{\infty}\left(\frac{\rho-1}{\rho}(b-a)\right)^n$$
$$\frac{1}{(n+\alpha)n!}(f(a)+f(b))$$

the proof is completed.

Theorem 5 *Let* $f:[a,b]\to\mathbb{R}$ *be positive function with* $0\le a<b, 0\le c<d$ *and* $f\in L_1[a,b]$. *If* f *is a convex functionon* $[a,b]$ *then the following inequalities hold:*

$$
\begin{aligned}
& f\left(\frac{a+b}{2},\frac{c+d}{2}\right)\\
& \leq \frac{\Gamma(\alpha)\rho^{\alpha}\left[I_{a+}^{\alpha,\rho}f\left(b,\frac{c+d}{2}\right)+I_{b-}^{\alpha,\rho}f\left(a,\frac{c+d}{2}\right)\right]}{4(b-a)^{\alpha}\Omega_1(m)}+\frac{\Gamma(\beta)\rho^{\beta}\left[I_{c+}^{\beta,\rho}f\left(\frac{a+b}{2},d\right)+I_{d-}^{\beta,\rho}f\left(\frac{a+b}{2},c\right)\right]}{4(d-c)^{\beta}\Omega_2(n)}\\
& \leq \frac{\left[I_{a^+,c^+}^{\alpha,\beta,\rho}f(b,d)+I_{a^+,d^-}^{\alpha,\beta,\rho}f(b,c)+I_{c^+,b^-}^{\alpha,\beta,\rho}f(a,d)+I_{b^-,d^-}^{\alpha,\beta,\rho}f(a,c)\right]}{(\Gamma(\alpha)\Gamma(\beta))^{-1}\rho^{-(\alpha+\beta)}4(b-a)^{\alpha}(d-c)^{\beta}\Omega_1(m)\Omega_2(n)}\\
& \leq \frac{\Gamma(\alpha)\rho^{\alpha}\left[I_{a+}^{\alpha,\rho}f(b,c)+I_{a+}^{\alpha,\rho}f(b,d)+I_{b-}^{\alpha,\rho}f(a,c)+I_{b-}^{\alpha,\rho}f(a,d)\right]}{4(b-a)^{\alpha}\Omega_1(m)}\\
& +\frac{\Gamma(\beta)\rho^{\beta}\left[I_{c+}^{\beta,\rho}f(a,d)+I_{c+}^{\beta,\rho}f(b,d)+I_{d-}^{\beta,\rho}f(a,c)+I_{d-}^{\beta,\rho}f(b,c)\right]}{4(d-c)^{\beta}\Omega_2(n)}\\
& \leq \frac{f(a,c)+f(a,d)+f(b,c)+f(b,d)}{4}
\end{aligned}
\tag{31}
$$

where,

$$
\Omega_1(m)=\sum_{m=0}^{\infty}\left(\frac{\rho-1}{\rho}(b-a)\right)^m\frac{1}{(m+\alpha)m!},
$$

and

$$
\Omega_2(n)=\sum_{n=0}^{\infty}\left(\frac{\rho-1}{\rho}(d-c)\right)^n\frac{1}{(n+\beta)n!}.
$$

Proof Since $f : \Delta \to \mathbb{R}$ is convex on the co-ordinates, it follows that the mapping $g_x : [c,d] \to R$, $g_x(y)=f(x,y)$, is convex on $[c,d]$ for all $x \in [a,b]$, then by using inequalities (16) we can write.

$$
\begin{aligned}
f\left(x,\frac{c+d}{2}\right) & \leq \frac{\Gamma(\beta)\rho^{\beta}}{2(d-c)^{\beta}\sum_{n=0}^{\infty}\left(\frac{\rho-1}{\rho}(d-c)\right)^n\frac{1}{(n+\beta)n!}}\\
& \times\left[\int_c^d(d-y)^{\alpha-1}e^{\frac{\rho-1}{\rho}(d-y)}f(x,y)dy+\int_c^d(y-c)^{\alpha-1}e^{\frac{\rho-1}{\rho}(y-c)}f(x,y)dy\right]\\
& \leq \frac{f(x,c)+f(x,d)}{2}
\end{aligned}
\tag{32}
$$

for all $x \in [a,b]$. Then multiplying both side of (32) by

$$\left[(b-x)^{\alpha-1}\frac{1}{2(b-a)^{\alpha}\sum_{m=0}^{\infty}\left(\frac{\rho-1}{\rho}(b-a)\right)^{m}\frac{1}{(m+\alpha)m!}}\right]e^{\frac{\rho-1}{\rho}(b-x)}$$

and

$$\left[(x-a)^{\alpha-1}\frac{1}{2(b-a)^{\alpha}\sum_{m=0}^{\infty}\left(\frac{\rho-1}{\rho}(b-a)\right)^{m}\frac{1}{(m+\alpha)m!}}\right]e^{\frac{\rho-1}{\rho}(x-a)}$$

and integrating with respect to x over $[a, b]$, respectively, we have

$$\frac{1}{2(b-a)^{\alpha}\sum_{m=0}^{\infty}\left(\frac{\rho-1}{\rho}(b-a)\right)^{m}\frac{1}{(m+\alpha)m!}}\int_{a}^{b}(b-x)^{\alpha-1}e^{\frac{\rho-1}{\rho}(b-x)}f\left(x,\frac{c+d}{2}\right)dx$$

$$\leq\frac{1}{4(b-a)^{\alpha}\sum_{m=0}^{\infty}\left(\frac{\rho-1}{\rho}(b-a)\right)^{m}\frac{1}{(m+\alpha)m!}}\frac{1}{(d-c)^{\beta}\sum_{n=0}^{\infty}\left(\frac{\rho-1}{\rho}(d-c)\right)^{n}\frac{1}{(n+\beta)n!}}$$

$$\times\left[\int_{a}^{b}\int_{c}^{d}(b-x)^{\alpha-1}(d-y)^{\beta-1}e^{\frac{\rho-1}{\rho}(b-x)}e^{\frac{\rho-1}{\rho}(d-y)}f(x,y)dxdy\right.$$

$$\left.+\int_{a}^{b}\int_{c}^{d}(b-x)^{\alpha-1}(y-c)^{\beta-1}e^{\frac{\rho-1}{\rho}(b-x)}e^{\frac{\rho-1}{\rho}(y-c)}f(x,y)dxdy\right]$$

$$\leq\frac{1}{4(b-a)^{\alpha}\sum_{m=0}^{\infty}\left(\frac{\rho-1}{\rho}(b-a)\right)^{m}\frac{1}{(m+\alpha)m!}}$$

$$\left[\int_{a}^{b}(b-x)^{\alpha-1}e^{\frac{\rho-1}{\rho}(b-x)}f(x,c)dx+\int_{a}^{b}(b-x)^{\alpha-1}e^{\frac{\rho-1}{\rho}(b-x)}f(x,d)dx\right] \tag{33}$$

Thus, similarly, we get

$$\frac{1}{2(b-a)^{\alpha}\sum_{m=0}^{\infty}\left(\frac{\rho-1}{\rho}(b-a)\right)^{m}\frac{1}{(m+\alpha)m!}}\int_{a}^{b}(x-a)^{\alpha-1}e^{\frac{\rho-1}{\rho}(x-a)}f\left(x,\frac{c+d}{2}\right)dx$$

$$\leq\frac{1}{4(b-a)^{\alpha}\sum_{m=0}^{\infty}\left(\frac{\rho-1}{\rho}(b-a)\right)^{m}\frac{1}{(m+\alpha)m!}}\frac{1}{(d-c)^{\beta}\sum_{n=0}^{\infty}\left(\frac{\rho-1}{\rho}(d-c)\right)^{n}\frac{1}{(n+\beta)n!}}$$

$$\times\left[\int_{a}^{b}\int_{c}^{d}(x-a)^{\alpha-1}(d-y)^{\beta-1}e^{\frac{\rho-1}{\rho}(x-a)}e^{\frac{\rho-1}{\rho}(d-y)}f(x,y)dxdy\right.$$

$$\left.+\int_{a}^{b}\int_{c}^{d}(x-a)^{\alpha-1}(y-c)^{\beta-1}e^{\frac{\rho-1}{\rho}(x-a)}e^{\frac{\rho-1}{\rho}(y-c)}f(x,y)dxdy\right]$$

$$\leq\frac{1}{4(b-a)^{\alpha}\sum_{m=0}^{\infty}\left(\frac{\rho-1}{\rho}(b-a)\right)^{m}\frac{1}{(m+\alpha)m!}}$$

$$\times\left[\int_{a}^{b}(x-a)^{\alpha-1}e^{\frac{\rho-1}{\rho}(x-a)}f(x,c)dx+\int_{a}^{b}(x-a)^{\alpha-1}e^{\frac{\rho-1}{\rho}(x-a)}f(x,d)dx\right] \tag{34}$$

$$\frac{1}{2(d-c)^{\beta}\sum_{n=0}^{\infty}\left(\frac{\rho-1}{\rho}(d-c)\right)^{n}\frac{1}{(n+\beta)n!}}\int_{c}^{d}(d-y)^{\beta-1}e^{\frac{\rho-1}{\rho}(d-y)}f\left(\frac{a+b}{2},y\right)dy$$

$$\leq\frac{1}{4(b-a)^{\alpha}\sum_{m=0}^{\infty}\left(\frac{\rho-1}{\rho}(b-a)\right)^{m}\frac{1}{(m+\alpha)m!}}\frac{1}{(d-c)^{\beta}\sum_{n=0}^{\infty}\left(\frac{\rho-1}{\rho}(d-c)\right)^{n}\frac{1}{(n+\beta)n!}}$$

$$\times\left[\int_{a}^{b}\int_{c}^{d}(b-x)^{\alpha-1}(d-y)^{\beta-1}e^{\frac{\rho-1}{\rho}(b-x)}e^{\frac{\rho-1}{\rho}(d-y)}f(x,y)dxdy\right.$$

$$\left.+\int_{a}^{b}\int_{c}^{d}(x-a)^{\alpha-1}(d-y)^{\beta-1}e^{\frac{\rho-1}{\rho}(x-a)}e^{\frac{\rho-1}{\rho}(d-y)}f(x,y)dxdy\right]$$

$$\leq\frac{1}{4(d-c)^{\beta}\sum_{n=0}^{\infty}\left(\frac{\rho-1}{\rho}(d-c)\right)^{n}\frac{1}{(n+\beta)n!}}$$

$$\times\left[\int_{c}^{d}(d-y)^{\beta-1}e^{\frac{\rho-1}{\rho}(d-y)}f(a,y)dy+\int_{c}^{d}(d-y)^{\beta-1}e^{\frac{\rho-1}{\rho}(d-y)}f(b,y)dy\right] \tag{35}$$

and

$$\frac{1}{2(d-c)^{\beta}\sum_{n=0}^{\infty}\left(\frac{\rho-1}{\rho}(d-c)\right)^{n}\frac{1}{(n+\beta)n!}}\int_{c}^{d}(y-c)^{\beta-1}e^{\frac{\rho-1}{\rho}(y-c)}f\left(\frac{a+b}{2},y\right)dy$$

$$\leq\frac{1}{4(b-a)^{\alpha}\sum_{m=0}^{\infty}\left(\frac{\rho-1}{\rho}(b-a)\right)^{m}\frac{1}{(m+\alpha)m!}}\frac{1}{(d-c)^{\beta}\sum_{n=0}^{\infty}\left(\frac{\rho-1}{\rho}(d-c)\right)^{n}\frac{1}{(n+\beta)n!}}$$

$$\times\left[\int_{a}^{b}\int_{c}^{d}(b-x)^{\alpha-1}(y-c)^{\beta-1}e^{\frac{\rho-1}{\rho}(b-x)}e^{\frac{\rho-1}{\rho}(y-c)}f(x,y)dxdy\right.$$

$$\left.+\int_{a}^{b}\int_{c}^{d}(x-a)^{\alpha-1}(y-c)^{\beta-1}e^{\frac{\rho-1}{\rho}(x-a)}e^{\frac{\rho-1}{\rho}(y-c)}f(x,y)dxdy\right]$$

$$\leq\frac{1}{4(d-c)^{\beta}\sum_{n=0}^{\infty}\left(\frac{\rho-1}{\rho}(d-c)\right)^{n}\frac{1}{(n+\beta)n!}}$$

$$\left[\int_{c}^{d}(y-c)^{\beta-1}e^{\frac{\rho-1}{\rho}(y-c)}f(a,y)dy+\int_{c}^{d}(y-c)^{\beta-1}e^{\frac{\rho-1}{\rho}(y-c)}f(b,y)dy\right] \tag{36}$$

adding the inequalities (33–36), we have

$$\frac{\Gamma(\alpha)\rho^{\alpha}}{4(b-a)^{\alpha}\sum_{m=0}^{\infty}\left(\frac{\rho-1}{\rho}(b-a)\right)^{m}\frac{1}{(m+\alpha)m!}}\left[I_{a^{+}}^{\alpha,\rho}f\left(b,\frac{c+d}{2}\right)+I_{b-}^{\alpha,\rho}f\left(a,\frac{c+d}{2}\right)\right]$$

$$+\frac{\Gamma(\beta)\rho^{\beta}}{4(d-c)^{\beta}\sum_{n=0}^{\infty}\left(\frac{\rho-1}{\rho}(d-c)\right)^{n}\frac{1}{(n+\beta)n!}}\left[I_{a^{+}}^{\beta,\rho}f\left(\frac{a+b}{2},d\right)+I_{d-}^{\beta,\rho}f\left(\frac{a+b}{2},c\right)\right]$$

$$\leq\frac{\Gamma(\alpha)\rho^{\alpha}}{4(b-a)^{\alpha}\sum_{m=0}^{\infty}\left(\frac{\rho-1}{\rho}(b-a)\right)^{m}\frac{1}{(m+\alpha)m!}}\frac{\Gamma(\beta)\rho^{\beta}}{(d-c)^{\beta}\sum_{n=0}^{\infty}\left(\frac{\rho-1}{\rho}(d-c)\right)^{n}\frac{1}{(n+\beta)n!}}$$

$$\times\left[I_{a^{+},c^{+}}^{\alpha,\beta,\rho}f(b,d)+I_{a^{+},d^{-}}^{\alpha,\beta,\rho}f(b,c)+I_{c^{+},b^{-}}^{\alpha,\beta,\rho}f(a,d)+I_{b^{-},d^{-}}^{\alpha,\beta,\rho}f(a,c)\right]$$

which give the second and thired inequlities in (31).

Now, by using the first inequality in (16), we also have

$$f\left(\frac{a+b}{2},\frac{c+d}{2}\right)\leq\frac{1}{2(b-a)^{\alpha}\sum_{m=0}^{\infty}\left(\frac{\rho-1}{\rho}(b-a)\right)^{m}\frac{1}{(m+\alpha)m!}}$$

$$\times\left[\int_{a}^{b}(b-x)^{\alpha-1}e^{\frac{\rho-1}{\rho}(b-x)}f\left(x,\frac{c+d}{2}\right)dx\right.$$

$$+ \int_a^b (x-a)^{\alpha-1} e^{\frac{\rho-1}{\rho}(x-a)} f\left(x, \frac{c+d}{2}\right) dx\Bigg]$$

and

$$f\left(\frac{a+b}{2}, \frac{c+d}{2}\right) \leq \frac{1}{2(d-c)^{\beta} \sum_{n=0}^{\infty} \left(\frac{\rho-1}{\rho}(d-c)\right)^{n} \frac{1}{(n+\beta)n!}}$$
$$\times \left[\int_c^d (d-y)^{\beta-1} e^{\frac{\rho-1}{\rho}(d-y)} f\left(\frac{a+b}{2}, y\right) dy \right.$$
$$\left. + \int_c^d (y-c)^{\beta-1} e^{\frac{\rho-1}{\rho}(y-c)} f\left(\frac{a+b}{2}, y\right) dy\right]$$

by adding we have

$$f\left(\frac{a+b}{2}, \frac{c+d}{2}\right) \leq \frac{\Gamma(\alpha)\rho^{\alpha}}{4(b-a)^{\alpha} \sum_{m=0}^{\infty} \left(\frac{\rho-1}{\rho}(b-a)\right)^{m} \frac{1}{(m+\alpha)m!}}$$
$$\left[I_{a^+}^{\alpha,\rho} f\left(b, \frac{c+d}{2}\right) + I_{b-}^{\alpha,\rho} f\left(a, \frac{c+d}{2}\right)\right]$$
$$+ \frac{\Gamma(\beta)\rho^{\beta}}{4(d-c)^{\beta} \sum_{n=0}^{\infty} \left(\frac{\rho-1}{\rho}(d-c)\right)^{n} \frac{1}{(n+\beta)n!}}$$
$$\left[I_{c^+}^{\alpha,\rho} f\left(\frac{a+b}{2}, d\right) + I_{d-}^{\alpha,\rho} f\left(\frac{a+b}{2}, c\right)\right]$$

which give the first inewuality in (31).

Finally, by using the second inequality in (16) we can also state

$$\frac{1}{2(b-a)^{\alpha} \sum_{m=0}^{\infty} \left(\frac{\rho-1}{\rho}(b-a)\right)^{m} \frac{1}{(m+\alpha)m!}} \left[\int_a^b (b-x)^{\alpha-1} e^{\frac{\rho-1}{\rho}(b-x)} f(x, c) dx \right.$$
$$\left. + \int_a^b (x-a)^{\alpha-1} e^{\frac{\rho-1}{\rho}(x-a)} f(x, c) dx\right] \leq \frac{f(a, c) + f(b, c)}{2}$$

$$\frac{1}{2(b-a)^{\alpha} \sum_{m=0}^{\infty} \left(\frac{\rho-1}{\rho}(b-a)\right)^{m} \frac{1}{(m+\alpha)m!}} \left[\int_a^b (b-x)^{\alpha-1} e^{\frac{\rho-1}{\rho}(b-x)} f(x, d) dx \right.$$
$$\left. + \int_a^b (x-a)^{\alpha-1} e^{\frac{\rho-1}{\rho}(x-a)} f(x, d) dx\right] \leq \frac{f(a, d) + f(b, d)}{2}$$

$$\frac{1}{2(d-c)^{\beta}\sum_{n=0}^{\infty}\left(\frac{\rho-1}{\rho}(d-c)\right)^{n}\frac{1}{(n+\beta)n!}}\left[\int_{c}^{d}(d-y)^{\beta-1}e^{\frac{\rho-1}{\rho}(d-y)}f(a,y)dy\right.$$
$$\left.+\int_{c}^{d}(y-c)^{\beta-1}e^{\frac{\rho-1}{\rho}(y-c)}f(a,y)dy\right]\leq\frac{f(a,c)+f(a,d)}{2}$$

and

$$\frac{1}{2(d-c)^{\beta}\sum_{n=0}^{\infty}\left(\frac{\rho-1}{\rho}(d-c)\right)^{n}\frac{1}{(n+\beta)n!}}\left[\int_{c}^{d}(d-y)^{\beta-1}e^{\frac{\rho-1}{\rho}(d-y)}f(b,y)dy\right.$$
$$\left.+\int_{c}^{d}(y-c)^{\beta-1}e^{\frac{\rho-1}{\rho}(y-c)}f(b,y)dy\right]\leq\frac{f(b,c)+f(b,d)}{2}$$

which give, by addition the last inequality in (31) the proof completed.

Lemma 1 Let $f : \Delta \subset \mathbb{R}^2 \to \mathbb{R}$ *be a partial differentiable mapping on* $\Delta := [a,b]\times[c,d]$ *in* $\mathbb{R}^2$ *with* $0 \leq a < b,\ 0 \leq c < d$. *If* $\frac{\partial^2 f}{\partial t \partial s} \in L_1(\Delta)$, *then the following equality holds:*

$$\frac{(b-a)(d-c)}{4}\left\{\int_0^1\int_0^1 s^{\beta}e^{\frac{\rho-1}{\rho}(d-c)s}t^{\alpha}e^{\frac{\rho-1}{\rho}(b-a)t}\right.$$
$$\frac{\partial^2 f(ta+(1-t)b,sc+(1-s)d)}{\partial t\partial s}dsdt$$
$$-\int_0^1\int_0^1 s^{\beta}e^{\frac{\rho-1}{\rho}(d-c)s}(1-t)^{\alpha}e^{\frac{\rho-1}{\rho}(b-a)t}$$
$$\frac{\partial^2 f(ta+(1-t)b,sc+(1-s)d)}{\partial t\partial s}dsdt$$
$$-\int_0^1\int_0^1 (1-s)^{\beta}e^{\frac{\rho-1}{\rho}(d-c)s}t^{\alpha}e^{\frac{\rho-1}{\rho}(b-a)t}$$
$$\frac{\partial^2 f(ta+(1-t)b,sc+(1-s)d)}{\partial t\partial s}dsdt$$
$$=\frac{e^{\frac{\rho-1}{\rho}(d-c+b-a)}}{4}\big[f(a,c)+f(a,d)+f(b,c)+f(b,d)\big]$$
$$+\frac{\Gamma(\alpha+1)\Gamma(\beta+1)\rho^{\alpha+\beta}}{4(b-a)^{\alpha}(d-c)^{\beta}}$$

$$\times\left[I^{\alpha,\beta,\rho}_{a^+,c^+}f(b,d)+I^{\alpha,\beta,\rho}_{a^+,d^-}f(b,c)\right.$$
$$\left.+I^{\alpha,\beta,\rho}_{b^-,c^+}f(a,d)+I^{\alpha,\beta,\rho}_{b^-,d^-}f(a,c)\right]$$
$$+(\rho-1)\frac{\Gamma(\alpha+1)\Gamma(\beta+1)\rho^{\alpha+\beta}}{4(b-a)^{\alpha}(d-c)^{\beta}}$$
$$\left[I^{\alpha,\beta+1,\rho}_{a^+,c^+}f(b,d)+I^{\alpha,\beta+1,\rho}_{a^+,d^-}f(b,c)+I^{\alpha,\beta+1,\rho}_{b^-,c^+}f(a,d)\right.$$
$$+I^{\alpha,\beta+1,\rho}_{b^-,d^-}f(a,c)+I^{\alpha+1,\beta,\rho}_{a^+,c^+}f(b,d)+I^{\alpha+1,\beta,\rho}_{a^+,d^-}f(b,c)$$
$$\left.+I^{\alpha+1,\beta,\rho}_{b^-,c^+}f(a,d)+I^{\alpha+1,\beta,\rho}_{b^-,d^-}f(a,c)\right]$$
$$+(\rho-1)^2\frac{\Gamma(\alpha+1)\Gamma(\beta+1)\rho^{\alpha+\beta}}{4(b-a)^{\alpha}(d-c)^{\beta}}$$
$$\left[I^{\alpha+1,\beta+1,\rho}_{a^+,c^+}f(b,d)+I^{\alpha+1,\beta+1,\rho}_{a^+,d^-}f(b,c)\right.$$
$$\left.+I^{\alpha+1,\beta+1,\rho}_{b^-,c^+}f(a,d)+I^{\alpha+1,\beta+1,\rho}_{b^-,d^-}f(a,c)\right]-A \tag{37}$$

where

$$A=\left[\frac{\Gamma(\beta+1)\rho^{\beta}e^{\frac{\rho-1}{\rho}(b-a)}}{4(d-c)^{\beta}}\left[I^{\beta,\rho}_{c^+}f(a,d)+I^{\beta,\rho}_{c^+}f(b,d)+I^{\beta,\rho}_{d-}f(a,c)+I^{\beta,\rho}_{d-}f(b,c)\right]\right.$$
$$+\frac{\Gamma(\alpha+1)\rho^{\alpha}e^{\frac{\rho-1}{\rho}(d-c)}}{4(b-a)^{\alpha}}\left[I^{\alpha,\rho}_{a^+}f(b,c)+I^{\alpha,\rho}_{a^+}f(b,d)+I^{\alpha,\rho}_{b-}f(a,c)+I^{\alpha,\rho}_{b-}f(a,d)\right]$$
$$+(\rho-1)\frac{\Gamma(\beta+1)\rho^{\beta}e^{\frac{\rho-1}{\rho}(b-a)}}{4(d-c)^{\beta}}$$
$$\left[I^{\beta+1,\rho}_{c^+}f(a,d)+I^{\beta+1,\rho}_{c^+}f(b,d)+I^{\beta+1,\rho}_{d-}f(a,c)+I^{\beta+1,\rho}_{d-}f(b,c)\right]$$
$$+(\rho-1)\frac{\Gamma(\alpha+1)\rho^{\alpha}e^{\frac{\rho-1}{\rho}(d-c)}}{4(b-a)^{\alpha}}$$
$$\left.\left[I^{\alpha+1,\rho}_{a^+}f(b,c)+I^{\alpha+1,\rho}_{a^+}f(b,d)+I^{\alpha+1,\rho}_{b-}f(a,c)+I^{\alpha+1,\rho}_{b-}f(a,d)\right]\right]$$

Proof By integration by parts, we get

$$I_1=\int_0^1\int_0^1 s^{\beta}e^{\frac{\rho-1}{\rho}(d-c)s}t^{\alpha}e^{\frac{\rho-1}{\rho}(b-a)t}\frac{\partial^2 f(ta+(1-t)b,sc+(1-s)d)}{\partial t\partial s}dsdt$$
$$-\frac{e^{\frac{\rho-1}{\rho}(b-a)}}{(b-a)}\frac{\rho-1}{\rho}\int_0^1 s^{\beta-1}e^{\frac{\rho-1}{\rho}(d-c)s}f(a,sc+(1-s)d)ds$$
$$-\frac{\alpha e^{\frac{\rho-1}{\rho}(d=c)}}{(b-a)(d-c)}\int_0^1 t^{\alpha-1}e^{\frac{\rho-1}{\rho}(b-a)t}f(ta+(1-t)b,c)dt$$

$$
\begin{aligned}
&+\frac{\alpha\beta}{(b-a)(d-c)}\int_0^1\int_0^1 s^{\beta-1}e^{\frac{\rho-1}{\rho}(d-c)s}t^{\alpha-1}e^{\frac{\rho-1}{\rho}(b-a)t}f(ta+(1-t)b,sc\\
&+(1-s)d)dsdt\\
&+\frac{\rho-1}{\rho}\frac{\alpha}{(b-a)}\int_0^1\int_0^1 s^{\beta}e^{\frac{\rho-1}{\rho}(d-c)s}t^{\alpha-1}e^{\frac{\rho-1}{\rho}(b-a)t}f(ta+(1-t)b,sc\\
&+(1-s)d)dsdt\\
&-\frac{\rho-1}{\rho}\frac{e^{\frac{\rho-1}{\rho}(d=c)}}{(d-c)}\int_0^1 t^{\alpha}e^{\frac{\rho-1}{\rho}(b-a)t}f(ta+(1-t)b,c)dt\\
&+\frac{\rho-1}{\rho}\frac{\beta}{(d-c)}\int_0^1\int_0^1 s^{\beta-1}e^{\frac{\rho-1}{\rho}(d-c)s}t^{\alpha}e^{\frac{\rho-1}{\rho}(b-a)t}f(ta+(1-t)b,sc\\
&+(1-s)d)dsdt\\
&+\left(\frac{\rho-1}{\rho}\right)^2\int_0^1\int_0^1 s^{\beta}e^{\frac{\rho-1}{\rho}(d-c)s}t^{\alpha}e^{\frac{\rho-1}{\rho}(b-a)t}f(ta+(1-t)b,sc\\
&+(1-s)d)dsdt,
\end{aligned}
$$

$$
\begin{aligned}
I_2=&\int_0^1\int_0^1 s^{\beta}e^{\frac{\rho-1}{\rho}(d-c)s}(1-t)^{\alpha}e^{\frac{\rho-1}{\rho}(b-a)(1-t)}\\
&\frac{\partial^2 f(ta+(1-t)b,sc+(1-s)d)}{\partial t\partial s}dsdt\\
=&-\frac{e^{\frac{\rho-1}{\rho}(d-c+b-a)}}{(b-a)(d-c)}f(b,c)\\
&+\frac{\beta e^{\frac{\rho-1}{\rho}(b-a)}}{(b-a)(d-c)}\int_0^1 s^{\beta-1}e^{\frac{\rho-1}{\rho}(d-c)s}f(b,sc+(1-s)d)ds\\
&+\frac{e^{\frac{\rho-1}{\rho}(b-a)}}{(b-a)}\frac{\rho-1}{\rho}\int_0^1 s^{\beta-1}e^{\frac{\rho-1}{\rho}(d-c)s}f(b,sc+(1-s)d)ds\\
&+\frac{\alpha e^{\frac{\rho-1}{\rho}(d-c)}}{(b-a)(d-c)}\int_0^1(1-t)^{\alpha-1}e^{\frac{\rho-1}{\rho}(b-a)(1-t)}f(ta+(1-t)b,c)dt\\
&-\frac{\alpha\beta}{(b-a)(d-c)}\int_0^1\int_0^1 s^{\beta-1}e^{\frac{\rho-1}{\rho}(d-c)s}(1-t)^{\alpha-1}\\
&e^{\frac{\rho-1}{\rho}(b-a)(1-t)}f(ta+(1-t)b,sc+(1-s)d)dsdt\\
&-\frac{\rho-1}{\rho}\frac{\alpha}{(b-a)}\int_0^1\int_0^1 s^{\beta}e^{\frac{\rho-1}{\rho}(d-c)s}(1-t)^{\alpha-1}\\
&e^{\frac{\rho-1}{\rho}(b-a)(1-t)}f(ta+(1-t)b,sc+(1-s)d)dsdt\\
&+\frac{\rho-1}{\rho}\frac{e^{\frac{\rho-1}{\rho}(d=c)}}{(d-c)}\int_0^1(1-t)^{\alpha}e^{\frac{\rho-1}{\rho}(b-a)(1-t)}f(ta+(1-t)b,c)dt
\end{aligned}
$$

$$-\frac{\rho-1}{\rho}\frac{\beta}{(d-c)}\int_0^1\int_0^1 s^{\beta-1}e^{\frac{\rho-1}{\rho}(d-c)s}(1-t)^{\alpha}$$
$$e^{\frac{\rho-1}{\rho}(b-a)(1-t)}f(ta+(1-t)b,sc+(1-s)d)dsdt$$
$$-\left(\frac{\rho-1}{\rho}\right)^2\int_0^1\int_0^1 s^{\beta}e^{\frac{\rho-1}{\rho}(d-c)s}(1-t)^{\alpha}$$
$$e^{\frac{\rho-1}{\rho}(b-a)(1-t)}f(ta+(1-t)b,sc+(1-s)d)dsdt$$

$$I_3=\int_0^1\int_0^1(1-s)^{\beta}e^{\frac{\rho-1}{\rho}(d-c)(1-s)}t^{\alpha}$$
$$e^{\frac{\rho-1}{\rho}(b-a)t}\frac{\partial^2 f(ta+(1-t)b,sc+(1-s)d)}{\partial t\partial s}dsdt$$
$$=-\frac{e^{\frac{\rho-1}{\rho}(d-c+b-a)}}{(b-a)(d-c)}f(a,d)$$
$$+\frac{\beta e^{\frac{\rho-1}{\rho}(b-a)}}{(b-a)(d-c)}\int_0^1(1-s)^{\beta-1}e^{\frac{\rho-1}{\rho}(d-c)(1-s)}f(a,sc+(1-s)d)ds$$
$$+\frac{e^{\frac{\rho-1}{\rho}(b-a)}}{(b-a)}\frac{\rho-1}{\rho}\int_0^1(1-s)e^{\frac{\rho-1}{\rho}(d-c)(1-s)}f(a,sc+(1-s)d)ds$$
$$+\frac{\alpha e^{\frac{\rho-1}{\rho}(d-c)}}{(b-a)(d-c)}\int_0^1 t^{\alpha-1}e^{\frac{\rho-1}{\rho}(b-a)t}f(ta+(1-t)b,d)dt$$
$$-\frac{\alpha\beta}{(b-a)(d-c)}\int_0^1\int_0^1(1-s)^{\beta-1}e^{\frac{\rho-1}{\rho}(d-c)(1-s)}t^{\alpha-1}$$
$$e^{\frac{\rho-1}{\rho}(b-a)t}f(ta+(1-t)b,sc+(1-s)d)dsdt$$
$$-\frac{\rho-1}{\rho}\frac{\alpha}{(b-a)}\int_0^1\int_0^1(1-s)^{\beta}e^{\frac{\rho-1}{\rho}(d-c)(1-s)}t^{\alpha-1}$$
$$e^{\frac{\rho-1}{\rho}(b-a)t}f(ta+(1-t)b,sc+(1-s)d)dsdt$$
$$+\frac{\rho-1}{\rho}\frac{e^{\frac{\rho-1}{\rho}(d-c)}}{(d-c)}\int_0^1 t^{\alpha}e^{\frac{\rho-1}{\rho}(b-a)t}f(ta+(1-t)b,d)dt$$
$$-\frac{\rho-1}{\rho}\frac{\beta}{(d-c)}\int_0^1\int_0^1(1-s)^{\beta-1}e^{\frac{\rho-1}{\rho}(d-c)(1-s)}t^{\alpha}$$

$$e^{\frac{\rho-1}{\rho}(b-a)t} f(ta+(1-t)b, sc+(1-s)d)dsdt$$

$$-\left(\frac{\rho-1}{\rho}\right)^2 \int_0^1 \int_0^1 (1-s)^\beta e^{\frac{\rho-1}{\rho}(d-c)(1-s)} t^\alpha$$

$$e^{\frac{\rho-1}{\rho}(b-a)t} f(ta+(1-t)b, sc+(1-s)d)dsdt$$

$$I_4 = \int_0^1 \int_0^1 (1-s)^\beta e^{\frac{\rho-1}{\rho}(d-c)(1-s)} (1-t)^\alpha$$

$$e^{\frac{\rho-1}{\rho}(b-a)(1-t)} \frac{\partial^2 f(ta+(1-t)b, sc+(1-s)d)}{\partial t \partial s} dsdt$$

$$= \frac{e^{\frac{\rho-1}{\rho}(d-c+b-a)}}{(b-a)(d-c)} f(b,d)$$

$$- \frac{\beta e^{\frac{\rho-1}{\rho}(b-a)}}{(b-a)(d-c)} \int_0^1 (1-s)^{\beta-1} e^{\frac{\rho-1}{\rho}(d-c)(1-s)} f(b, sc+(1-s)d)ds$$

$$- \frac{e^{\frac{\rho-1}{\rho}(b-a)}}{(b-a)} \frac{\rho-1}{\rho} \int_0^1 (1-s) e^{\frac{\rho-1}{\rho}(d-c)(1-s)} f(b, sc+(1-s)d)ds$$

$$- \frac{\alpha e^{\frac{\rho-1}{\rho}(d-c)}}{(b-a)(d-c)} \int_0^1 (1-t)^{\alpha-1} e^{\frac{\rho-1}{\rho}(b-a)(1-t)} f(ta+(1-t)b, d)dt$$

$$+ \frac{\alpha\beta}{(b-a)(d-c)} \int_0^1 \int_0^1 (1-s)^{\beta-1} e^{\frac{\rho-1}{\rho}(d-c)(1-s)} (1-t)^{\alpha-1} e^{\frac{\rho-1}{\rho}(b-a)(1-t)}$$

$$\times f(ta+(1-t)b, sc+(1-s)d)dsdt$$

$$+ \frac{\rho-1}{\rho} \frac{\alpha}{(b-a)} \int_0^1 \int_0^1 (1-s)^\beta e^{\frac{\rho-1}{\rho}(d-c)(1-s)} (1-t)^{\alpha-1} e^{\frac{\rho-1}{\rho}(b-a)(1-t)}$$

$$\times f(ta+(1-t)b, sc+(1-s)d)dsdt$$

$$- \frac{\rho-1}{\rho} \frac{e^{\frac{\rho-1}{\rho}(d-c)}}{(d-c)} \int_0^1 (1-t)^\alpha e^{\frac{\rho-1}{\rho}(b-a)(1-t)} f(ta+(1-t)b, d)dt$$

$$+ \frac{\rho-1}{\rho} \frac{\beta}{(d-c)} \int_0^1 \int_0^1 (1-s)^{\beta-1} e^{\frac{\rho-1}{\rho}(d-c)(1-s)} (1-t)^\alpha e^{\frac{\rho-1}{\rho}(b-a)(1-t)}$$

$$\times f(ta+(1-t)b, sc+(1-s)d)dsdt$$

$$+\left(\frac{\rho-1}{\rho}\right)^2\int_0^1\int_0^1(1-s)^{\beta}e^{\frac{\rho-1}{\rho}(d-c)(1-s)}(1-t)^{\alpha}e^{\frac{\rho-1}{\rho}(b-a)(1-t)}$$
$$\times f(ta+(1-t)b, sc+(1-s)d)dsdt$$

From $I_1, I_2, I_3 and I_4$, using the change of variables $x = at + (1-t)b$, and $y = sc + (1-s)d$ for $t, s \in [0, 1]^2$, we can write

$$\begin{aligned}
[I_1 - I_2 - I_3 + I_4] &= \frac{e^{\frac{\rho-1}{\rho}(d-c+b-a)}}{(b-a)(d-c)}\big[f(a,c)+f(a,d)+f(b,c)+f(b,d)\big] \\
&+ \frac{\Gamma(\alpha+1)\Gamma(\beta+1)\rho^{\alpha+\beta}}{(b-a)^{\alpha+1}(d-c)^{\beta+1}} \\
&\Big[I^{\alpha,\beta,\rho}_{a^+,c^+}f(b,d) + I^{\alpha,\beta,\rho}_{a^+,d^-}f(b,c) + I^{\alpha,\beta,\rho}_{b^-,c^+}f(a,d) + I^{\alpha,\beta,\rho}_{b^-,d^-}f(a,c)\Big] \\
&+ (\rho-1)\frac{\Gamma(\alpha+1)\Gamma(\beta+1)\rho^{\alpha+\beta}}{(b-a)^{\alpha+1}(d-c)^{\beta+1}} \\
&\Big[I^{\alpha,\beta+1,\rho}_{a^+,c^+}f(b,d) + I^{\alpha,\beta+1,\rho}_{a^+,d^-}f(b,c) + I^{\alpha,\beta+1,\rho}_{b^-,c^+}f(a,d) \\
&+ I^{\alpha,\beta+1,\rho}_{b^-,d^-}f(a,c) + I^{\alpha+1,\beta,\rho}_{a^+,c^+}f(b,d) + I^{\alpha+1,\beta,\rho}_{a^+,d^-}f(b,c) \\
&+ I^{\alpha+1,\beta,\rho}_{b^-,c^+}f(a,d) + I^{\alpha+1,\beta,\rho}_{b^-,d^-}f(a,c)\Big] \\
&+ (\rho-1)^2\frac{\Gamma(\alpha+1)\Gamma(\beta+1)\rho^{\alpha+\beta}}{(b-a)^{\alpha+1}(d-c)^{\beta+1}} \\
&\Big[I^{\alpha+1,\beta+1,\rho}_{a^+,c^+}f(b,d) + I^{\alpha+1,\beta+1,\rho}_{a^+,d^-}f(b,c) \\
&+ I^{\alpha+1,\beta+1,\rho}_{b^-,c^+}f(a,d) + I^{\alpha+1,\beta+1,\rho}_{b^-,d^-}f(a,c)\Big] \\
&- \frac{4}{(b-a)(d-c)} \times A
\end{aligned} \tag{38}$$

multiplying the both sides of (38) by $\frac{(b-a)(d-c)}{4}$, we obtain (37), which the completes the proof.

Theorem 6 *Let* $f : \Delta \subset \mathbb{R}^2 \to \mathbb{R}$ *be partial differentiable mapping on* $\Delta := [a,b] \times [c,d]$ *in* $\mathbb{R}^2$ *with,* $0 \le a < b$, $0 \le c < d$. *If* $\left|\frac{\partial^2 f}{\partial t \partial s}\right|$ *is a convex on co-ordinated on* Δ, *then one has the inequalities.*

$$\begin{aligned}
&\Bigg|\frac{e^{\frac{\rho-1}{\rho}(d-c+b-a)}}{4}\left[f(a,c)+f(a,d)+f(b,c)+f(b,d)\right] \\
&+\Phi\left[I^{\alpha,\beta,\rho}_{a^+,c^+}f(b,d)+I^{\alpha,\beta,\rho}_{a^+,d^-}f(b,c)+I^{\alpha,\beta,\rho}_{b^-,c^+}f(a,d)+I^{\alpha,\beta,\rho}_{b^-,d^-}f(a,c)\right] \\
&+(\rho-1)\Phi\Big[I^{\alpha,\beta+1,\rho}_{a^+,c^+}f(b,d)+I^{\alpha,\beta+1,\rho}_{a^+,d^-}f(b,c)+I^{\alpha,\beta+1,\rho}_{b^-,c^+}f(a,d) \\
&+I^{\alpha,\beta+1,\rho}_{b^-,d^-}f(a,c)+I^{\alpha+1,\beta,\rho}_{a^+,c^+}f(b,d)+I^{\alpha+1,\beta,\rho}_{a^+,d^-}f(b,c) \\
&+I^{\alpha+1,\beta,\rho}_{b^-,c^+}f(a,d)+I^{\alpha+1,\beta,\rho}_{b^-,d^-}f(a,c)\Big] \\
&+(\rho-1)^2\Phi\Big[I^{\alpha+1,\beta+1,\rho}_{a^+,c^+}f(b,d)+I^{\alpha+1,\beta+1,\rho}_{a^+,d^-}f(b,c) \\
&+I^{\alpha+1,\beta+1,\rho}_{b^-,c^+}f(a,d)+I^{\alpha+1,\beta,\rho}_{b^-,d^-}f(a,c)\Big]-A \\
&\le \frac{(b-a)(d-c)}{4}\Omega_1\Omega_2 \\
&\left(\left|\frac{\partial^2 f(a,c)}{\partial t \partial s}\right|+\left|\frac{\partial^2 f(a,d)}{\partial t \partial s}\right|+\left|\frac{\partial^2 f(b,c)}{\partial t \partial s}\right|+\left|\frac{\partial^2 f(b,d)}{\partial t \partial s}\right|\right)
\end{aligned} \tag{39}$$

where, $\Phi = \frac{\Gamma(\alpha+1)\Gamma(\beta+1)\rho^{\alpha+\beta}}{4(b-a)^{\alpha}(d-c)^{\beta}}$, $\Omega_1 = \sum_{m=0}^{\infty}\left(\frac{\rho-1}{\rho}(b-a)\right)^m \frac{1}{(m+\alpha+1)m!}$, $\Omega_2 = \sum_{n=0}^{\infty}\left(\frac{\rho-1}{\rho}(d-c)\right)^n \frac{1}{(n+\beta+1)n!}$.

Proof From Lemma 1, we have

$$\begin{aligned}
&\Bigg|\frac{e^{\frac{\rho-1}{\rho}(d-c+b-a)}}{4}\left[f(a,c)+f(a,d)+f(b,c)+f(b,d)\right] \\
&+\frac{\Gamma(\alpha+1)\Gamma(\beta+1)\rho^{\alpha+\beta}}{4(b-a)^{\alpha}(d-c)^{\beta}} \\
&\left[I^{\alpha,\beta,\rho}_{a^+,c^+}f(b,d)+I^{\alpha,\beta,\rho}_{a^+,d^-}f(b,c)+I^{\alpha,\beta,\rho}_{b^-,c^+}f(a,d)+I^{\alpha,\beta,\rho}_{b^-,d^-}f(a,c)\right] \\
&+(\rho-1)\frac{\Gamma(\alpha+1)\Gamma(\beta+1)\rho^{\alpha+\beta}}{4(b-a)^{\alpha}(d-c)^{\beta}} \\
&\Big[I^{\alpha,\beta+1,\rho}_{a^+,c^+}f(b,d)+I^{\alpha,\beta+1,\rho}_{a^+,d^-}f(b,c)+I^{\alpha,\beta+1,\rho}_{b^-,c^+}f(a,d) \\
&+I^{\alpha,\beta+1,\rho}_{b^-,d^-}f(a,c)+I^{\alpha+1,\beta,\rho}_{a^+,c^+}f(b,d)+I^{\alpha+1,\beta,\rho}_{a^+,d^-}f(b,c)
\end{aligned}$$

$$+ I_{b^-,c^+}^{\alpha+1,\beta,\rho} f(a,d) + I_{b^-,d^-}^{\alpha+1,\beta,\rho} f(a,c)]$$

$$+ (\rho-1)^2 \frac{\Gamma(\alpha+1)\Gamma(\beta+1)\rho^{\alpha+\beta}}{4(b-a)^\alpha (d-c)^\beta}$$

$$\Big[I_{a^+,c^+}^{\alpha+1,\beta+1,\rho} f(b,d) + I_{a^+,d^-}^{\alpha+1,\beta+1,\rho} f(b,c)$$

$$+ I_{b^-,c^+}^{\alpha+1,\beta+1,\rho} f(a,d) + I_{b^-,d^-}^{\alpha+1,\beta+1,\rho} f(a,c) \Big] - A \Big|$$

$$\le \frac{(b-a)(d-c)}{4}$$

$$\left\{ \int_0^1 \int_0^1 s^\beta t^\alpha \left| e^{\frac{\rho-1}{\rho}(d-c)s} e^{\frac{\rho-1}{\rho}(b-a)t} \frac{\partial^2 f(ta+(1-t)b, sc+(1-s)d)}{\partial t \partial s} \right| dsdt \right.$$

$$+ \int_0^1 \int_0^1 s^\beta (1-t)^\alpha \left| e^{\frac{\rho-1}{\rho}(d-c)s} e^{\frac{\rho-1}{\rho}(b-a)t} \frac{\partial^2 f(ta+(1-t)b, sc+(1-s)d)}{\partial t \partial s} \right| dsdt$$

$$+ \int_0^1 \int_0^1 (1-s)^\beta \left| t^\alpha e^{\frac{\rho-1}{\rho}(d-c)s} e^{\frac{\rho-1}{\rho}(b-a)t} \frac{\partial^2 f(ta+(1-t)b, sc+(1-s)d)}{\partial t \partial s} \right| dsdt$$

$$+ \int_0^1 \int_0^1 (1-s)^\beta (1-t)^\alpha$$

$$\left. \left| e^{\frac{\rho-1}{\rho}(d-c)s} e^{\frac{\rho-1}{\rho}(b-a)t} \frac{\partial^2 f(ta+(1-t)b, sc+(1-s)d)}{\partial t \partial s} \right| dsdt \right\} \tag{40}$$

Since $\frac{\partial^2 f}{\partial t \partial s}$ is convex function on the co-ordinates on Δ, then one has:

$$\Big| \frac{e^{\frac{\rho-1}{\rho}(d-c+b-a)}}{4} [f(a,c)+f(a,d)+f(b,c)+f(b,d)]$$

$$+ \frac{\Gamma(\alpha+1)\Gamma(\beta+1)\rho^{\alpha+\beta}}{4(b-a)^\alpha (d-c)^\beta}$$

$$\Big[I_{a^+,c^+}^{\alpha,\beta,\rho} f(b,d) + I_{a^+,d^-}^{\alpha,\beta,\rho} f(b,c) + I_{b^-,c^+}^{\alpha,\beta,\rho} f(a,d) + I_{b^-,d^-}^{\alpha,\beta,\rho} f(a,c) \Big]$$

$$+ (\rho-1) \frac{\Gamma(\alpha+1)\Gamma(\beta+1)\rho^{\alpha+\beta}}{4(b-a)^\alpha (d-c)^\beta}$$

$$\Big[I_{a^+,c^+}^{\alpha,\beta+1,\rho} f(b,d) + I_{a^+,d^-}^{\alpha,\beta+1,\rho} f(b,c) + I_{b^-,c^+}^{\alpha,\beta+1,\rho} f(a,d)$$

$$+ I_{b^-,d^-}^{\alpha,\beta+1,\rho} f(a,c) + I_{a^+,c^+}^{\alpha+1,\beta,\rho} f(b,d) + I_{a^+,d^-}^{\alpha+1,\beta,\rho} f(b,c)$$

$$+ I_{b^-,c^+}^{\alpha+1,\beta,\rho} f(a,d) + I_{b^-,d^-}^{\alpha+1,\beta,\rho} f(a,c)]$$

$$+(\rho-1)^2\frac{\Gamma(\alpha+1)\Gamma(\beta+1)\rho^{\alpha+\beta}}{4(b-a)^\alpha(d-c)^\beta}$$

$$\Big[I_{a^+,c^+}^{\alpha+1,\beta+1,\rho}f(b,d)+I_{a^+,d^-}^{\alpha+1,\beta+1,\rho}f(b,c)$$

$$+I_{b^-,c^+}^{\alpha+1,\beta+1,\rho}f(a,d)+I_{b^-,d^-}^{\alpha+1,\beta+1,\rho}f(a,c)\Big]-A\Big|$$

$$\leq\frac{(b-a)(d-c)}{4}$$

$$\Bigg\{\int_0^1\int_0^1 s^\beta t^\alpha\left|e^{\frac{\rho-1}{\rho}(d-c)s}e^{\frac{\rho-1}{\rho}(b-a)t}\frac{\partial^2 f(ta+(1-t)b,sc+(1-s)d)}{\partial t\partial s}\right|dsdt$$

$$+\int_0^1\int_0^1 s^\beta(1-t)^\alpha\left|e^{\frac{\rho-1}{\rho}(d-c)s}e^{\frac{\rho-1}{\rho}(b-a)t}\frac{\partial^2 f(ta+(1-t)b,sc+(1-s)d)}{\partial t\partial s}\right|dsdt$$

$$+\int_0^1\int_0^1 (1-s)^\beta\left|t^\alpha e^{\frac{\rho-1}{\rho}(d-c)s}e^{\frac{\rho-1}{\rho}(b-a)t}\frac{\partial^2 f(ta+(1-t)b,sc+(1-s)d)}{\partial t\partial s}\right|dsdt$$

$$+\int_0^1\int_0^1 (1-s)^\beta(1-t)^\alpha$$

$$\left|e^{\frac{\rho-1}{\rho}(d-c)s}e^{\frac{\rho-1}{\rho}(b-a)t}\frac{\partial^2 f(ta+(1-t)b,sc+(1-s)d)}{\partial t\partial s}\right|dsdt\Bigg\}$$

By calculating the integral in above inequality, we have the inequality (39).

Theorem 7 *Let* $f:\Delta\subset\mathbb{R}^2\to\mathbb{R}$ *be partial differentiable mapping on* $\Delta := [a,b]\times[c,d]$ *in* $\mathbb{R}^2$ *with,* $0\leq a<b, 0\leq c<d$. *If* $\left|\frac{\partial^2 f}{\partial t\partial s}\right|^q, q>1$, *is a convex on co-ordinated on* Δ, then one has the inequalities.

$$\Bigg|\frac{e^{\frac{\rho-1}{\rho}(d-c+b-a)}}{4}\big[f(a,c)+f(a,d)+f(b,c)+f(b,d)\big]$$

$$+\frac{\Gamma(\alpha+1)\Gamma(\beta+1)\rho^{\alpha+\beta}}{4(b-a)^\alpha(d-c)^\beta}$$

$$\Big[I_{a^+,c^+}^{\alpha,\beta,\rho}f(b,d)+I_{a^+,d^-}^{\alpha,\beta,\rho}f(b,c)+I_{b^-,c^+}^{\alpha,\beta,\rho}f(a,d)+I_{b^-,d^-}^{\alpha,\beta,\rho}f(a,c)\Big]$$

$$+(\rho-1)\frac{\Gamma(\alpha+1)\Gamma(\beta+1)\rho^{\alpha+\beta}}{4(b-a)^\alpha(d-c)^\beta}$$

$$\Big[I_{a^+,c^+}^{\alpha,\beta+1,\rho}f(b,d)+I_{a^+,d^-}^{\alpha,\beta+1,\rho}f(b,c)+I_{b^-,c^+}^{\alpha,\beta+1,\rho}f(a,d)$$

$$+I_{b^-,d^-}^{\alpha,\beta+1,\rho}f(a,c)+I_{a^+,c^+}^{\alpha+1,\beta,\rho}f(b,d)+I_{a^+,d^-}^{\alpha+1,\beta,\rho}f(b,c)$$

$$+I_{b^-,c^+}^{\alpha+1,\beta,\rho}f(a,d)+I_{b^-,d^-}^{\alpha+1,\beta,\rho}f(a,c)\Big]$$

$$+(\rho-1)^2\frac{\Gamma(\alpha+1)\Gamma(\beta+1)\rho^{\alpha+\beta}}{4(b-a)^\alpha(d-c)^\beta}$$

$$\Big[I_{a^+,c^+}^{\alpha+1,\beta+1,\rho}f(b,d)+I_{a^+,d^-}^{\alpha+1,\beta+1,\rho}f(b,c)$$

$$+I_{b^-,c^+}^{\alpha+1,\beta+1,\rho}f(a,d)+I_{b^-,d^-}^{\alpha+1,\beta+1,\rho}f(a,c)\Big]-A\Bigg|$$

$$\leq(b-a)(d-c)$$

$$\left(\sum_{m=0}^{\infty}\left(\frac{\rho-1}{\rho}(b-a)\right)^m\frac{1}{(m+\alpha p+1)m!}\sum_{n=0}^{\infty}\left(\frac{\rho-1}{\rho}(d-c)\right)^n\frac{1}{(n+\beta p+1)n!}\right)^{\frac{1}{p}}$$

$$\times\left(\frac{\left|\frac{\partial^2 f(a,c)}{\partial t\partial s}\right|^q+\left|\frac{\partial^2 f(a,d)}{\partial t\partial s}\right|^q+\left|\frac{\partial^2 f(b,c)}{\partial t\partial s}\right|^q+\left|\frac{\partial^2 f(b,d)}{\partial t\partial s}\right|^q}{4}\right)^{\frac{1}{q}}$$

Proof From Lemma 1, we have inequality (40). By using the well known Hölder's inequality for double integral, we get

$$\Bigg|\frac{e^{\frac{\rho-1}{\rho}(d-c+b-a)}}{4}\big[f(a,c)+f(a,d)+f(b,c)+f(b,d)\big]$$

$$+\frac{\Gamma(\alpha+1)\Gamma(\beta+1)\rho^{\alpha+\beta}}{4(b-a)^\alpha(d-c)^\beta}$$

$$\Big[I_{a^+,c^+}^{\alpha,\beta,\rho}f(b,d)+I_{a^+,d^-}^{\alpha,\beta,\rho}f(b,c)+I_{b^-,c^+}^{\alpha,\beta,\rho}f(a,d)+I_{b^-,d^-}^{\alpha,\beta,\rho}f(a,c)\Big]$$

$$+(\rho-1)\frac{\Gamma(\alpha+1)\Gamma(\beta+1)\rho^{\alpha+\beta}}{4(b-a)^\alpha(d-c)^\beta}$$

$$\Big[I_{a^+,c^+}^{\alpha,\beta+1,\rho}f(b,d)+I_{a^+,d^-}^{\alpha,\beta+1,\rho}f(b,c)+I_{b^-,c^+}^{\alpha,\beta+1,\rho}f(a,d)$$

$$+I_{b^-,d^-}^{\alpha,\beta+1,\rho}f(a,c)+I_{a^+,c^+}^{\alpha+1,\beta,\rho}f(b,d)+I_{a^+,d^-}^{\alpha+1,\beta,\rho}f(b,c)$$

$$+I_{b^-,c^+}^{\alpha+1,\beta,\rho}f(a,d)+I_{b^-,d^-}^{\alpha+1,\beta,\rho}f(a,c)\Big]$$

$$+(\rho-1)^2\frac{\Gamma(\alpha+1)\Gamma(\beta+1)\rho^{\alpha+\beta}}{4(b-a)^\alpha(d-c)^\beta}$$

$$\Big[I_{a^+,c^+}^{\alpha+1,\beta+1,\rho}f(b,d)+I_{a^+,d^-}^{\alpha+1,\beta+1,\rho}f(b,c)$$

$$+I_{b^-,c^+}^{\alpha+1,\beta+1,\rho}f(a,d)+I_{b^-,d^-}^{\alpha+1,\beta+1,\rho}f(a,c)\Big]-A\Bigg|$$

$$\leq\frac{(b-a)(d-c)}{4}\left\{\int_0^1\int_0^1\left(s^{\beta p}t^{\alpha p}e^{\frac{\rho-1}{\rho}(d-c)sp}e^{\frac{\rho-1}{\rho}(b-a)tp}dsdt\right)^{\frac{1}{p}}\right.$$

$$\times \int_0^1\int_0^1 \left(s^{\beta p}(1-t)^{\alpha p} e^{\frac{\rho-1}{\rho}(d-c)sp} e^{\frac{\rho-1}{\rho}(b-a)(1-t)p} dsdt\right)^{\frac{1}{p}}$$

$$\times \int_0^1\int_0^1 \left((1-s)^{\beta p} t^{\alpha p} e^{\frac{\rho-1}{\rho}(d-c)(1-s)p} e^{\frac{\rho-1}{\rho}(b-a)tp} dsdt\right)^{\frac{1}{p}}$$

$$\left.\times \int_0^1\int_0^1 \left((1-s)^{\beta p}(1-t)^{\alpha p} e^{\frac{\rho-1}{\rho}(d-c)(1-s)p} e^{\frac{\rho-1}{\rho}(b-a)(1-t)p} dsdt\right)^{\frac{1}{p}}\right\}$$

$$\times \left(\int_0^1\int_0^1 \left|\frac{\partial^2 f(ta+(1-t)b, sc+(1-s)d)}{\partial t\partial s}\right|^q dsdt\right)^{\frac{1}{q}}$$

since $\left|\frac{\partial^2 f}{\partial t\partial s}\right|^q$ is convex on the co-ordinated on Δ, we know that for $t \in [0, 1]$

$$\left|\frac{\partial^2 f(ta+(1-t)b, sc+(1-s)d)}{\partial t\partial s}\right|^q \le t\left|\frac{\partial^2 f(a, sc+(1-s)d)}{\partial t\partial s}\right|^q + (1-t)\left|\frac{\partial^2 f(b, sc+(1-s)d)}{\partial t\partial s}\right|^q$$

and

$$\left|\frac{\partial^2 f(ta+(1-t)b, sc+(1-s)d)}{\partial t\partial s}\right|^q \le ts\left|\frac{\partial^2 f(a,c)}{\partial t\partial s}\right|^q + t(1-s)\left|\frac{\partial^2 f(a,d)}{\partial t\partial s}\right|^q + (1-t)s\left|\frac{\partial^2 f(b,c)}{\partial t\partial s}\right|^q + (1-t)(1-s)\left|\frac{\partial^2 f(b,d)}{\partial t\partial s}\right|^q$$

hence, it follows that

$$\left|\frac{e^{\frac{\rho-1}{\rho}(d-c+b-a)}}{4}[f(a,c)+f(a,d)+f(b,c)+f(b,d)]\right.$$

$$+\frac{\Gamma(\alpha+1)\Gamma(\beta+1)\rho^{\alpha+\beta}}{4(b-a)^{\alpha}(d-c)^{\beta}}$$

$$\left[I_{a^+,c^+}^{\alpha,\beta,\rho}f(b,d)+I_{a^+,d^-}^{\alpha,\beta,\rho}f(b,c)+I_{b^-,c^+}^{\alpha,\beta,\rho}f(a,d)+I_{b^-,d^-}^{\alpha,\beta,\rho}f(a,c)\right]$$

$$+(\rho-1)\frac{\Gamma(\alpha+1)\Gamma(\beta+1)\rho^{\alpha+\beta}}{4(b-a)^{\alpha}(d-c)^{\beta}}$$

$$\left[I_{a^+,c^+}^{\alpha,\beta+1,\rho}f(b,d)+I_{a^+,d^-}^{\alpha,\beta+1,\rho}f(b,c)+I_{b^-,c^+}^{\alpha,\beta+1,\rho}f(a,d)\right.$$

$$+I_{b^-,d^-}^{\alpha,\beta+1,\rho}f(a,c)+I_{a^+,c^+}^{\alpha+1,\beta,\rho}f(b,d)+I_{a^+,d^-}^{\alpha+1,\beta,\rho}f(b,c)$$

$$\left.+I_{b^-,c^+}^{\alpha+1,\beta,\rho}f(a,d)+I_{b^-,d^-}^{\alpha+1,\beta,\rho}f(a,c)\right]$$

$$+(\rho-1)^2\frac{\Gamma(\alpha+1)\Gamma(\beta+1)\rho^{\alpha+\beta}}{4(b-a)^{\alpha}(d-c)^{\beta}}$$

$$\left[I_{a^+,c^+}^{\alpha+1,\beta+1,\rho}f(b,d)+I_{a^+,d^-}^{\alpha+1,\beta+1,\rho}f(b,c)\right.$$

$$\left.\left.+I_{b^-,c^+}^{\alpha+1,\beta+1,\rho}f(a,d)+I_{b^-,d^-}^{\alpha+1,\beta+1,\rho}f(a,c)\right]-A\right|$$

$$\leq (b-a)(d-c)$$

$$\left(\sum_{m=0}^{\infty}\left(\frac{\rho-1}{\rho}(b-a)\right)^m\frac{1}{(m+\alpha p+1)m!}\sum_{n=0}^{\infty}\left(\frac{\rho-1}{\rho}(d-c)\right)^n\frac{1}{(n+\beta p+1)n!}\right)^{\frac{1}{p}}$$

$$\times\left\{\int_0^1\int_0^1\left(ts\left|\frac{\partial^2 f(a,c)}{\partial t\partial s}\right|^q+t(1-s)\left|\frac{\partial^2 f(a,d)}{\partial t\partial s}\right|^q\right.\right.$$

$$\left.\left.+(1-t)s\left|\frac{\partial^2 f(b,c)}{\partial t\partial s}\right|^q+(1-t)(1-s)\left|\frac{\partial^2 f(b,d)}{\partial t\partial s}\right|^q\right)dtds\right\}^{\frac{1}{q}}$$

$$=(b-a)(d-c)$$

$$\left(\sum_{m=0}^{\infty}\left(\frac{\rho-1}{\rho}(b-a)\right)^m\frac{1}{(m+\alpha p+1)m!}\sum_{n=0}^{\infty}\left(\frac{\rho-1}{\rho}(d-c)\right)^n\frac{1}{(n+\beta p+1)n!}\right)^{\frac{1}{p}}$$

$$\times\left(\frac{\left|\frac{\partial^2 f(a,c)}{\partial t\partial s}\right|^q+\left|\frac{\partial^2 f(a,d)}{\partial t\partial s}\right|^q+\left|\frac{\partial^2 f(b,c)}{\partial t\partial s}\right|^q+\left|\frac{\partial^2 f(b,d)}{\partial t\partial s}\right|^q}{4}\right)^{\frac{1}{q}}$$

which the completes the proof.

References

1. Azpeitia, A.G.: Convex functions and the Hadamard inequality. Rev. Colomb. Math. **28**, 7–12 (1994)
2. Bakula, M.K., Pecaric, J.: Note on some Hadamard-type inequalities. J. Inequal. Pure Appl. Math. **5**(3), 74 (2004)
3. Dragomir, S.S.: On Hadamard's inequality for convex functions on the co-ordinates in a rectangle from the plane. Taiwanese J. Math. **4**, 775–788 (2001)
4. Dragomir, S.S., Pearce, C.E.M.: Selected Topics on Hermite-Hadamard Inequalities and Applications. Victoria University, RGMIA Monographs (2000)
5. Gorenflo, R., Mainardi, F.: Fractional Calculus: Integral and Differential Equations of Fractional Order, pp. 223–276. Springer Verlag, Wien (1997)
6. Kilbas, A.A., Srivastava, H.M., Trujillo, J.J.: Theory and Applications of Fractional Differential Equations, North-Holland Mathematics Studies, 204. Elsevier Sci. B.V., Amsterdam (2006)
7. Kırmacı, U.S., Dikici, R.: On some Hermite-Hadamard type inequalities for twice differentiable mappings and applications. Tamkang J. Math. **44**(1), 41–51 (2013)
8. Latif, M.A., Dragomir, S.S.: On some new inequalities for differentiable co-ordinated convex functions. J. Inequal. Appl. **1**, 1–13 (2012)
9. Özdemir, M.E., Set, E., Sarikaya, M.Z.: Some new Hadamard's type inequalities for co-ordinated m-convex and (α, m) -convex functions. Hacettepe J. Math. Stat. **40**, 219–229 (2011)
10. Samko, S.G., Kilbas, A.A., Marichev, O.I.: Fractional Integrals and Derivatives, Theory and Applications. Gordon and Breach, Yverdon et alibi (1993)
11. Sarikaya, M.Z.: On the Hermite-Hadamard-type inequalities for co-ordinated convex function via fractional integrals. Integr. Transf. Spec. Funct. **25**, 134–147 (2014)
12. Sarikaya, M.Z.: Some inequalities for differentiable co-ordinated convex mappings. Asian-Eur. J. Math. **8** (2015)
13. Sarikaya, M.Z., Set, E., Ozdemir, M.E., Dragomir, S.S.: New some Hadamard's type inequalities for co-ordinated convex functions. Tamsui Oxford J. Inf. Math. Sci. **28**, 137–152 (2012)
14. Sarikaya, M.Z., Set, E., Yaldiz, H., Basak, N.: Hermite-Hadamard's inequalities for fractional integrals and related fractional inequalities. Math. Comput. Model. **57**, 2403–2407 (2012)
15. Akkurt, A., Sarıkaya, M., Budak, H., Yıldırım, H.: On the Hadamard's type inequalities for co-ordinated convex functions via fractional integrals. J. King Saud Univ. Sci. **29**(3) (2016). https://doi.org/10.1016/j.jksus.2016.06.003
16. Akkurt, A., Sarıkaya, M., Budak, H., Yıldırım, H.: On the Hermite-Hadamard type inequalities for co-ordinated convex functions. Appl. Comput. Math. Int. J. **20**(3), 408–420 (2021)
17. Jarad, F., Abdeljawad, T., Alzabut, J.: Generalized fractional derivatives generated by a class of local proportional derivatives. Eur. Phys. J. Spec. Top. **226**, 3457–3471 (2017). https://doi.org/10.1140/epjst/e2018-00021-7

Ruled Surfaces According to the Direction Curve

Aykut Has and Beyhan Yılmaz

Abstract In this study, ruled surfaces in 3-dimensional Euclidean space are investigated based on their direction curves. Firstly, the ruled surface, whose direction curve is a constant curve, is characterized and the obtained results are given. Afterwards, base curve is expressed as $\gamma(u) = c_1T + c_2N + c_3B$ from the type of Frenet elements at a point $s \in I$ of the direction curve and characterized depending on the c_1, c_2 and c_3 coefficients. Finally, necessary examples are given to make the obtained results better understood and drawings are made with the help of the Mathematica program.

Keywords Ruled surface · Frenet frame · Rectifying curve · Normal curve · Oskulating curve

1 Introduction

Differential geometry, a branch of mathematics, explores the geometric characteristics of mathematical objects utilizing analytical and differential calculus. Within this field, ruled surfaces emerge as a significant area of study. Ruled surfaces are characterized by possessing at least one directional line through each point, typically perceived as combinations of curves. The basic properties of ruled surfaces are closely related to the properties of the curves on them. The behavior of these curves on a ruled surface profoundly influences the surface's overall shape. These surfaces have found widespread applications across various domains in mathematics and physics, particularly in physical theories like differential geometry and the theory of general relativity. Ruled surfaces play a crucial role in describing the curved structures inherent in both space and time. G. Monge introduced ruled surfaces to the literature, after which the theory of ruled surfaces was further developed using the E. Study

A. Has (✉) · B. Yılmaz
Department of Mathematics, Kahramanmaraş Sütçü İmam University, Kahramanmaraş, Turkey
e-mail: ahas@ksu.edu.tr

B. Yılmaz
e-mail: beyhanyilmaz@ksu.edu.tr

O. Castillo et al. (eds.), *Advanced Mathematics for the Modeling and Solution of Challenging Problems in Engineering*, Studies in Systems, Decision and Control 582,
https://doi.org/10.1007/978-3-031-82342-8_3

transformation. This transformation enables the examination of surface theory using a single real parameter. These surfaces are formed through the continuous movement of a line in three-dimensional Euclidean space along a space curve known as the direction curve. Key studies on ruled surfaces encompass their classification, properties of base curves, exploration of geodesics, examination of the surface's shape operator, developable of surfaces, and investigation of regular surfaces that cannot be developable into simpler structures known as developable surfaces.

In recent times, the study of ruled surfaces has garnered substantial interest among researchers. Various scholars have endeavored to characterize ruled surfaces using different direction curves and distinct base curves. This pursuit involves exploring and understanding the diverse properties and characteristics of ruled surfaces through the utilization of different directional and base curves. Researchers aim to uncover new insights into the behavior, properties, and possible applications of ruled surfaces by investigating their relationships with these varied curves. [1–4]. M. Ergüt investigated the curvatures of both three-dimensional and n-dimensional ruled surfaces [5]. I. Kamenaroviç ise Galile uzayında regle yüzeyleri incelemiş ve karakterizasyonlarını sunmuştur [6]. B. Yılmaz et al. they obtained characterizations of developable surfaces according to the alternative moving frame [7]. Furthermore, general rectifying ruled surfaces in Euclidean 3-space are defined [8].

In this study, ruled surfaces are investigated according to states of the direction curve in 3-dimensional Euclidean space. Various properties are obtained on the ruled surface by choosing the direction curve as constant or different from the constant. The direction curve is first selected as a constant and the ruled surface is characterized by the type of curvature of the surface. Then, by obtaining the direction curve of the base curve under different conditions from the Frenet frame elements, the surface curvatures are obtained and the properties of the surface are given with such a rule. Finally, this study is supported and visualized with the help of an example.

2 Preliminaries

Let $\alpha = \alpha(s)$ be a regular unit speed curve in the Euclidean 3− space where s measures its arc length. Also, let $T = \alpha\prime$ be its unit tangent vector, $N = \frac{T\prime}{T\prime}$, be its principal normal vector and $B = T \times N$ be its binormal vector. The triple.

$\{T, N, B\}$ be the Frenet frame of the curve α. Then the Frenet formula of the curve is given by

$$\begin{pmatrix} T\prime(s) \\ N\prime(s) \\ B\prime(s) \end{pmatrix} = \begin{pmatrix} 0 & \kappa(s) & 0 \\ -\kappa(s) & 0 & \tau(s) \\ 0 & -\tau(s) & 0 \end{pmatrix} \begin{pmatrix} T(s) \\ N(s) \\ B(s) \end{pmatrix} \tag{1}$$

where $\kappa(s) = \frac{d^2\alpha}{ds^2}$ and $\tau(s) = \frac{dN}{ds}$, B are curvature and torsion of α, respectively [9].

Definition 1 *Surfaces formed by the movement of a line along a curve in space are called ruled surfaces. The parameterization of the ruled surface for any two differentiable curves α and γ is as follows.*

$$\phi(u, v) = \alpha(u) + v\gamma(u) \tag{2}$$

where $\alpha(u)$ is called base curve of the ruled surface and $\gamma(u)$ is a unit direction vector of an oriented line in E^3 whose various positions are called rulings. In addition, if the direction curve is not constant, with, $\gamma(u) \neq 0$, the surface is called a non-cylindrical ruled surface. Also the surfaces with a constant direction curve are called the generalized cone surface, [10].

The standart unit normal vector field n on surface ϕ can be defined by

$$n = \frac{\phi_u \times \phi_v}{\phi_u \times \phi_v} \tag{3}$$

where $\phi_u = \frac{d\phi(u,v)}{du}$ and $\phi_v = \frac{d\phi(u,v)}{dv}$, [11]. Also, the first and second fundamental forms of the ϕ surface are as follows

$$I = Edu^2 + 2Fdudv + Gdv^2$$
$$II = edu^2 + 2fdudv + gdv^2$$

where the E, F and G components are called the coefficients of the first fundemental form of the surface, and the e, f and g components are called the coefficients of the second fundemental form, respectively. The following equations are given for the first and second fundamental form coefficients of the surface, [12]

$$E = \phi_u, \phi_u, F = \phi_u, \phi_v, G = \phi_v, \phi_v \tag{4}$$

and

$$e = \phi_{uu}, n, f = \phi_{uv}, n, g = \phi_{vv}, n. \tag{5}$$

On the other hand, the Gaussian curvature K and the mean curvature H of the surface are as follows, [12]

$$K = \frac{eg - f^2}{EG - F^2} \tag{6}$$

and

$$H = \frac{Eg + Ge - 2Ff}{2(EG - F^2)}. \tag{7}$$

Finally, the geodesic curvature, normal curvature and geodesic torsion of the ϕ surface are given below, respectively, [11]

$$\kappa_g = n \times T, T\prime, \kappa_n = \alpha\prime\prime, n, \tau_g = n \times n\prime, T\prime. \tag{8}$$

Thorem 1 *Let $\alpha : I \subset R \to E^3$ be arc length parametrized curve. Let M be a regular ruled surface, with the parametrization $\phi(u, v) = \alpha(u) + v\gamma(u)$. If the Gaussian curvature of the surface M is zero, the surface is called the developable surface,* [11].

Thorem 2 *Let $\alpha : I \subset R \to E^3$ be arc length parametrized curve. Let M be a regular ruled surface, with the parametrization $\phi(u, v) = \alpha(u) + v\gamma(u)$. If the mean curvature of the M surface is zero, the surface M is called the minimal surface,* [11].

Thorem 3 *Let $\alpha : I \subset R \to E^3$ be arc length parametrized curve. For a curve $\alpha(s)$ lying on a surface, the following are well-known* [11],

i. *If the curve $\alpha(s)$ is a geodesic curve, the geodesic curvature κ_g is zero.*
ii. *If the curve $\alpha(s)$ is a normal section curve, the geodesic curvature κ_g is zero.*
iii. *If the curve $\alpha(s)$ is a asymptotic curve, the normal curvature κ_n is zero.*
iv. *If the curve $\alpha(s)$ is a line of curve, the geodesic torsion τ_g is zero.*

Definition 2 *Let $\alpha : I \subset R \to E^3$ be arc length parametrized curve. The curve $\alpha(s)$ is called rectifying curve for all $s \in I$, if the orthogonal complement of $N(s)$ contains a fixed point. Since the orthogonal complement of N is $N^{\perp} = \{v \in T_\alpha E^3 | \langle v, N = 0\rangle\}$, the position vector of rectifying curve α in.*

E^3 can be written as,

$$\alpha(s) = \lambda T(s) + \mu B(s)$$

where

λ and μ are differentiable function, [13].

Definition 3 *Let $\alpha : I \subset R \to E^3$ be arc length parametrized curve. It is said that α is an osculating curve if its position vector is in the orthogonal complement of the binormal vector for all $s \in I$ and consequently,*

$$\alpha(s) = \lambda T(s) + \mu N(s)$$

where λ, μ are differentiable function, [14].

Definition 4 *Let $\alpha : I \subset R \to E^3$ be arc length parametrized curve. It is said that α is a normal curve if its position vector is in the orthogonal complement of the tangent vector for all $s \in I$ and consequently,*

$$\alpha(s) = \lambda N(s) + \mu B(s)$$

where λ, μ *are differentiable function,* [14].

3 Main Results

In this section, ruled surfaces in 3-dimensional Euclidean space are examined in different situations according to the direction curve.

Let $\alpha : I \subset R \rightarrow E^3$ be arc length parametrized curve. The ruled surface $\phi(u, v) = \alpha(u) + v\gamma(u)$ which is composed of the unit speed vector $\gamma(u) = c_1T + c_2N + c_3B$ that is $c_1^2 + c_2^2 + c_3^3 = 1$. The naturel frame $\{\phi_u, \phi_v\}$ of Eq. (2)and Frenet formulas are obtained as

$$\phi_u = (1 - c_2 v\kappa)T + (c_1 v\kappa - c_3 v\tau)N + c_2 v\tau B,$$

$$\phi_v = c_1 T + c_2 N + c_3 B,$$

and

$$\begin{aligned}
&\phi_{uu} = \left(c_3 v\kappa\tau - c_1 v\kappa^2 - c_2 v\kappa\prime\right)T + \left(\kappa - c_2 v\kappa^2 + c_1 v\kappa\prime - c_3 v\tau\prime - c_2 v\tau^2\right)N \\
&+\left(c_1 v\kappa\tau - c_3 v\tau^2 + c_2 v\tau\prime\right)B, \\
&\phi_{uv} = -c_2\kappa T + (c_1\kappa - c_3\tau)N + c_2\tau B, \\
&\phi_{vv} = 0.
\end{aligned}$$

On the other hand, as can be easily seen from Eq. (3), the normal of the surface can be written is as follows

$$n = \frac{1}{\chi}\left(c_1c_3v\kappa - c_3^2 v\tau - c_2^2 v\tau, -c_3 + c_2c_3v\kappa + c_1c_2v\tau, c_2 - c_2^2 v\kappa - c_1^2 v\kappa + c_1c_3v\tau\right) \quad (9)$$

where
$$\chi = \sqrt{\begin{array}{c}\left(c_1c_3v\kappa - c_3^2 v\tau - c_2^2 v\tau\right)^2 + (-c_3 + c_2c_3v\kappa + c_1c_2v\tau)^2 \\ +\left(c_2 - c_2^2 v\kappa - c_1^2 v\kappa + c_1c_3v\tau\right)^2.\end{array}}$$
The coefficients of the first and second fundamental form from the above equations

$$E = (c_2 v\kappa - 1)^2 + (c_1 v\kappa - c_3 v\tau)^2 + c_2^2 v^2\tau^2, \quad (10)$$

$$F = c_1, \quad (11)$$

$$G = 1, \quad (12)$$

and

$$e = \frac{1}{\chi}\begin{bmatrix} \left(c_2 c_3^2 + c_2^3\right)(\tau\kappa\prime - \kappa\tau\prime) - c_1\left(c_2^2 + c_3^2\right)\tau^3 - c_3\left(1 - 3c_1^2\right)\kappa\tau^2 \\ -c_1\left(1 - 3c_3^2\right)\kappa^2\tau - c_3\left(c_1^2 + c_2^2\right)\kappa^3 v^2 + \\ \left[2c_2(c_3\kappa + c_1\tau)\kappa - c_1 c_3\kappa\prime + (^{c_2}\tau\prime\right]v - c_3\kappa \end{bmatrix} \tag{13}$$

$$f = \frac{1}{\chi}\left[\left(c_2^2 + c_3^2\right)\tau - c_1 c_3\kappa\right], \tag{14}$$

$$g = 0, \tag{15}$$

are obtained.

3.1 Ruled Surface with Constant Direction Curve

Let $\gamma(u)$, which is the direction curve of the ruled surface $\phi(u, v)$ be constant. Also, let the constant vectors $\{T, N, B\}$ be equivalent to the vectors $\{e_1, e_2, e_3\}$. Parametric curves of a ruled surface with a constant direction curve are obtained below

$$\phi_u = T,\ \phi_{uu} = \kappa N,$$
$$\phi_v = c_1 e_1 + c_2 e_2 + c_3 e_3,\ \phi_{vv} = 0,\ \phi_{uv} = 0.$$

As can be seen from Eq. (12), the coefficients of the first fundamental form are as follows

$$E = 1,\ F = c_1 T,\ G = 1$$

The normal of the surface is as follows

$$n = \frac{1}{\sqrt{c_3^2 + c_2^2}}(0, -c_3, c_2)$$

On the other hand, the coefficients of the second fundamental form, similar to Eq. (13), are as follows

$$e = \alpha\prime\prime, n = \frac{-c_3\kappa}{\sqrt{c_3^2 + c_2^2}}$$

$$f = 0, g = 0.$$

From the above equations, Gaussian curvature and mean curvature are obtained as

$$K = 0, \tag{16}$$

$$H = \frac{-c_3\kappa}{2(1 - c_{1^2})\left(c_3^2 + c_2^2\right)}. \tag{17}$$

In addition, the following equation is available from Eq. (16) and Frenet formulas

$$\begin{aligned} \kappa_n &= \phi_{uu}, n = \kappa N, n, \\ \kappa_g &= n \times T, \phi_{uu} = \kappa n \times T, N, \\ \tau_g &= n \times n\prime, \phi_{uu} = \kappa n \times n\prime, N. \end{aligned}$$

Conclusion 1. *Let $\alpha : I \subset R \to E^3$ be arc length parametrized curve. Considering 1 and in Preliminaries section, the following equations are available from Eqs.*(16) (17)

i. A ruled surface with a constant direction curve is an developable surface as can be seen from Eq. (16).
ii. As can be seen from Eq. (17), if a ruled surface with a constant direction curve is a minimal surface, the following equation exists

$$c_3 = 0 \text{ or } \kappa = 0.$$

Conclusion 2. *Let $\alpha : I \subset R \to E^3$ be arc length parametrized curve. Considering the* Eq. (16) (17) *and Frenet formulas, the following results can be given.*

i. In the case of $n//T$, the normal curvature, geodesic curvature and geodesic torsion of the ruled surface are as follows

$$\kappa_n = 0, \kappa_g = 0, \tau_g = 0.$$

As can be easily seen from the above results, the curve α is an asymptotic curve since the normal curvature is vanishes, similarly the curve α is an geodesic curve since the geodesic curvature is vanishes, and finally the curve α is a line of curve since the geodesic torsion is vanishes.

ii. In the case of $n//N$, the normal curvature, geodesic curvature and geodesic torsion of the ruled surface are as follows

$$\kappa_n = \lambda\kappa, \kappa_g = 0, \tau_g = 0$$

where λ is constant. As can be easily seen from the above results, the curve α is an geodesic curve since the geodesic curvature is vanishes and similarly the curve α is a line of curve since the geodesic torsion is vanishes. On the other hand, the curve α is an asymptotic curve if and only if $\lambda = 0$ or $\kappa = 0$.

iii. In the case of $n//B$, the normal curvature, geodesic curvature and geodesic torsion of the ruled surface are as follows

$$\kappa_n = 0, \kappa_g = \lambda\kappa, \tau_g = 0$$

where λ is constant. As can be easily seen from the above results, the curve α is an asymptotic curve since the normal curvature is vanishes, similarly the curve α is a line of curve since the geodesic torsion is vanishes. On the other hand, the curve α is an geodesic curve if and only if $\lambda = 0$ or $\kappa = 0$.

iv. In the case of $n = aT + bN + cB$, the normal curvature, geodesic curvature and geodesic torsion of the ruled surface are as follows

$$\kappa_n = b\kappa, \kappa_g = c\kappa, \tau_g = -ab\kappa\tau - cb\kappa^2$$

where a, b and c are constant numbers. As can be easily seen from the above results, the curve α is an asymptotic curve if and only if $b = 0$ or $\kappa = 0$, similarly the curve α is an geodesic curve if and only if $c = 0$ or $\kappa = 0$, finally curve α is an a line of curve if and only if $\frac{\tau}{\kappa} = \frac{c}{a}$.

3.2 *Ruled Surfaces with Non-Constant Direction Curve*

In this section, we will examine ruled surfaces with nonzero constants under 3 different situations.

Case 1 Let $\alpha : I \subset R \to E^3$ be arc length parametrized curve. The ruled surface $\phi(u, v) = \alpha(u) + v\gamma(u)$ which is composed of the unit speed non-constant vector $\gamma(u) = c_1T + c_2N + c_3B$ with $c_1 = 1$,

$$c_2 = 0,$$

$c_3 = 0$. As can be seen from Eq. (*12*), *the coefficients of the first fundamental form as follows*

$$E = 1 + v^2\kappa^2, \ F = 1, \ G = 1$$

On the other hand, the normal to the surface is

$$n = (0, 0, -1). \tag{18}$$

The coefficients of the second fundamental form of the ruled surface from Eq. (13) Are as follows

$$e = -v\kappa\tau, f = 0, \ g = 0.$$

Considering the above equations, gaussian curvature and mean curvature of the ruled surface from Eq. (14) and (15) are obtaiined as

$$K = 0 \tag{19}$$

$$H = -\frac{\tau}{2v\kappa}. \tag{20}$$

On the other hand, the normal curvature, geodesic curvature and geodesic torsion of the ruled surface can be written using Eq. (9) *as follows*

$$\kappa_n = 0, \ \kappa_g = -\kappa, \ \tau_g = 0. \tag{21}$$

Conclusion 3. *Let $\alpha : I \subset R \to E^3$ be arc length parametrized curve. The following results are available from Eqs. (19) and (20)*

i. A ruled surface with a non-constant direction curve according to *Case 2* is an developable surface.
ii. If a ruled surface with a non-constant direction curve for *Case 2* is a minimal surface, the following equation exists

$$\tau = 0.$$

Conclusion 4. *Let $\alpha : I \subset R \to E^3$ be arc length parametrized curve. Considering Theorem 3 the following equations are available from Eq. (21).*

i. Since $\kappa_n = 0$, it is clearly seen that the base curve of ruled surface whose direction curve is α is also an asymptotic curve.
ii. Since $\kappa_g = 0$, it is clearly seen that the base curve of ruled surface whose direction curve is α is also an geodesic curve and normal section curve.
iii. Since $\tau_g = 0$, it is clearly seen that the base curve of ruled surface whose direction curve is α is also an line of curve.

Case 2 Let $\alpha : I \subset R \to E^3$ be arc length parametrized curve. The ruled surface $\phi(u, v) = \alpha(u) + v\gamma(u)$ which is composed of the unit speed non-constant vector.

$\gamma(u) = c_1T + c_2N + c_3B$ with $c_1 = 0, \ c_2 = 1, \ c_3 = 0$. As can be seen from Eq. *(4), the coefficients of the first fundamental form are as follows*

$$E = (v\kappa - 1)^2 + v^2\tau^2, \ F = 0, \ G = 1$$

On the other hand, the normal to the surface is

$$n = \frac{1}{\chi}(-v\tau, 0, 1 - v\kappa), \chi = \sqrt{(v\kappa - 1)^2 + v^2\tau^2}.$$

The coefficients of the second fundamental form of the ruled surface from Eq. (5) are as follows

$$e = \frac{1}{\chi}\left(v^2\kappa\prime\tau + v\tau'^2\kappa\tau\prime\right), f = \frac{\tau}{\sqrt{(v\kappa - 1)^2 + v^2\tau^2}}, g = 0.$$

Considering the above equations, gaussian curvature and mean curvature of the ruled surface

$$K = -\frac{\tau^2}{\chi^2} \tag{22}$$

$$H = \frac{v^2\left(\tau\kappa' - \kappa\tau'\right) + v\tau'}{2\chi^{\frac{3}{2}}} \tag{23}$$

are obtained. On the other hand, the normal curvature, geodesic curvature and geodesic torsion of the ruled surface are obtained as follows

$$\kappa_n = 0,\ \kappa_g = \frac{\kappa(1 - v\kappa)}{\chi},\ \tau_g = \frac{v^2\kappa\left(\tau\kappa' - \kappa\tau'\right) + v\kappa\tau'}{\chi^2}. \tag{24}$$

Conclusion 5. *Let $\alpha : I \subset R \to E^3$ be arc length parametrized curve. The following equations are available from Eqs. (22) and (23).*

i. If a ruled surface with a non-constant direction curve, according to *Case 2* is a developable surface, the following equation exists

$$\tau = 0.$$

ii. If a ruled surface with a non-constant direction curve for *Case 2* is a minimal surface, the following equation exists

$$\left(\frac{\tau}{\kappa}\right)\prime = \frac{\tau'}{v\kappa}.$$

Conclusion 6. *Let $\alpha : I \subset R \to E^3$ be arc length parametrized curve. The following conclusions are available from Eq. (24).*

i. Since $\kappa_n = 0$, it is clearly seen that the base curve of ruled surfaces whose α is also an asymptotic curve.
ii. The following equation should be provided so that α, which is the direction curve of the ruled surface, geodesic curve and normal section curve

$$\kappa = \frac{1}{v}.$$

iii. The following equation should be provided so that α, which is the direction curve of the ruled surface, line of curve

$$\left(\frac{\tau}{\kappa}\right)\prime = \frac{\tau'}{v\kappa^2}.$$

Case 3 Let $\alpha : I \subset R \to E^3$ be arc length parametrized curve. The ruled surface $\phi(u, v) = \alpha(u) + v\gamma(u)$ which is composed of the unit speed non-constant vector $\gamma(u) = c_1 T + c_2 N + c_3 B$ with $c_1 = 0,\ c_2 = 0,\ c_3 = 1$. The coefficients of the first fundamental form are as follows

$$E = 1 + v^2\tau^2,\ F = 0,\ G = 1$$

On the other hand, the normal to the surface is

$$n = \frac{1}{\chi}(-v\tau, -1, 0),\ \chi = \sqrt{1 + v^2\tau^2}.$$

The coefficients of the second fundamental form of the ruled surface are as follows

$$e = \frac{-v^2\tau^2\kappa + v\tau\prime - \kappa}{\sqrt{1 + v^2\tau^2}}, f = \frac{\tau}{\sqrt{1 + v^2\tau^2}}, g = 0.$$

Considering the above equations, gaussian curvature and mean curvature of the ruled surface can be written as

$$K = -\frac{\tau^2}{\chi^4} \tag{25}$$

$$H = \frac{v\tau' - v^2\tau^2\kappa - \kappa}{2\chi^3} \tag{26}$$

On the other hand, the normal curvature, geodesic curvature and geodesic torsion of the ruled surface are obtained as follows

$$\kappa_n = \frac{-\kappa}{\chi},\ \kappa_g = 0,\ \tau_g = 0. \tag{27}$$

Conclusion 7. *Let $\alpha : I \subset R \to E^3$ be arc length parametrized curve. The following conclusions are available from Eqs. (17) and (18).*

i. If a ruled surface with a constant direction curve for is a developable surface, the following equation exists

$$\tau = 0.$$

ii. If a ruled surface with a non-constant direction curve, for is a minimal surface, the following equation exists

$$\frac{\tau'}{\kappa} = \frac{\chi^2}{v}.$$

Conclusion 8. *Let $\alpha : I \subset R \to E^3$ be arc length parametrized curve. The following equations are available from Eq. (27).*

i. The following equation should be provided so that α, which is the direction curve of the ruled surface, asymptotic curve.

$$\kappa = 0.$$

ii. Since $\kappa_g = 0$, it is clearly seen that the base curve of ruled surfaces whose direction curve is α is also an geodesic curve and normal section curve.
iii. Since $\tau_g = 0$, it is clearly seen that the base curve of ruled surfaces whose direction curve is α is also an line of curve.

Case 4 Let $\alpha : I \subset R \to E^3$ be arc length parametrized curve. The ruled surface $\phi(u, v) = \alpha(u) + v\gamma(u)$ which is composed of the unit speed non-constant vector $\gamma(u) = c_1T + c_2N + c_3B$ with $c_1 = 0$, $c_2^2 + c_3^2 = 1$. That is, the direction curve of the ruled surface lies in the normal plane. The coefficients of the first fundamental form are as follows

$$E = (c_2v\kappa - 1)^2 + v^2\kappa^2,\ F = 0,\ G = 1$$

On the other hand, the normal to the surface is

$$n = \frac{1}{\chi}\left(-v\tau, c_2c_3v\kappa - c_3, c_3 - c_2^2\kappa\right), \chi = \sqrt{(c_2v\kappa - 1)^2 + v^2\kappa^2}.$$

The coefficients of the second fundamental form of the ruled surface are as follows

$$e = \frac{\left[c_2(\tau\kappa\prime - \kappa\tau\prime) - c_3\kappa\tau^2 - c_2^2c_3\kappa^3\right]\prime v^2 + \left(2c_2c_3\kappa^2 + c_3^2\tau\prime\right)v - c_3\kappa}{\chi},$$

$$f = \frac{\tau}{\chi}, g = 0.$$

Considering the above equations, Gaussian curvature and mean curvature of the ruled surface

$$K = \frac{\tau^2}{\chi^4} \tag{28}$$

$$H = \frac{\left[c_2\left(\tau\kappa' - \kappa\tau'\right) - c_3\kappa\tau^2 - c_2^2 c_3 \kappa^3\right] v^2 + \left(2c_2c_3\kappa^2 + c_3^2\tau'\right)v - c_3\kappa}{2\chi^3} \tag{29}$$

are obtained. On the other hand, the normal curvature, geodesic curvature and geodesic torsion of the ruled surface can be written as follows

$$\kappa_n = \frac{c_2c_3v\kappa^2 - c_3\kappa}{\chi}, \ \kappa_g = \frac{c_2\kappa - c_2^2 v\kappa^2}{\chi}, \ \tau_g = \frac{c_2 v\kappa\tau'}{\chi^2}. \tag{30}$$

Conclusion 9. *Let $\alpha : I \subset R \to E^3$ be arc length parametrized curve. The following equations are available from Eqs.* (28) *and* (29).

i. If a ruled surface with a non-constant direction curve for *Case 4* is a developable surface, the following equation exists

$$\tau = 0.$$

ii. If a ruled surface with a non-constant direction curve is a minimal surface, the following equation exists

$$\left[c_2\left(\tau\kappa' - \kappa\tau'\right) - c_3\kappa\tau^2 - c_2^2 c_3\kappa^3\right] v^2 + \left(2c_2c_3\kappa^2 + c_3^2\tau'\right)v - c_3\kappa = 0.$$

Conclusion 10. *Let $\alpha : I \subset R \to E^3$ be arc length parametrized curve. The following equations are available from Eq. (30).*

i. The following equation should be provided so that α, which is the direction curve for *Case 4* of the ruled surface, asymptotic curve

$$\kappa = \frac{1}{c_2 v}.$$

ii. The following equation should be provided so that α, which is the base curve for *Case 4* of the ruled surface, geodesic curve and normal section curve

$$\kappa = \frac{1}{c_2 v}.$$

iii. The following equation should be provided so that α, which is the base curve for *Case 4* of the ruled surface, line of curve

$$\kappa = 0 \text{ or } \tau = constant.$$

Case 5 Let $\alpha : I \subset R \to E^3$ be arc length parametrized curve. The ruled surface $\phi(u, v) = \alpha(u) + v\gamma(u)$ which is composed of the unit speed non-constant vector $\gamma(u) = c_1T + c_2N + c_3B$ with $c_2 = 0$, $c_1^2 + c_3^2 = 1$. That is, the direction curve of the ruled surface is the rectifying curve. The coefficients of the first fundamental form are as follows

$$E = 1 + (c_1 v\kappa - c_3 v\tau)^2,\ F = c_1,\ G = 1$$

On the Other Hand, the Normal to the Surface is

$$n = \frac{1}{\chi}\left(c_1 c_3 v\kappa - c_3^2 v\tau, -c_3, c_1 c_3 v\tau - c_1^2 v\kappa\right), \chi = \sqrt{c_3^2 + (c_1 v\kappa - c_3 v\tau)^2}.$$

The coefficients of the second fundamental form of the ruled surface are as follows

$$e = \frac{\left[-c_1 c_3^2\left(\tau^3 - 3\kappa^2\tau\right) - c_1^2 c_3\left(\kappa^3 - 3\kappa\tau^2\right) - \kappa\tau(c_3\tau + c_1\kappa)\right] v^2 + \left(-c_1 c_3\kappa' + c_3^2\tau'\right)v - c_3\kappa}{\chi},$$

$$f = \frac{c_3^2\tau - c_1 c_3\kappa}{\chi},\ g = 0.$$

Considering the above equations, Gaussian curvature and mean curvature of the ruled surface can be written

$$K = -\frac{\left(c_3^2\tau - c_1 c_3\kappa\right)^2}{\chi^2\left[1 + (c_1 v\kappa - c_3 v\tau)^2 - c_1^2\right]} \tag{31}$$

$$H = \frac{\begin{array}{c}\left[c_1 c_3^2\left(\tau^3 - 3\kappa^2\tau\right) + c_1^2 c_3\left(\kappa^3 - 3\kappa\tau^2\right) + \kappa\tau(c_3\tau + c_1\kappa)\right] \\ v^2\left(-c_1 c_3\kappa' + c_3^2\tau'\right)v + c_3\kappa + 2c_1 c_3^2\tau - 2c_1^2 c_3\kappa\end{array}}{2\chi\left[1 + (c_1 v\kappa - c_3 v\tau)^2 - c_1^2\right]} \tag{32}$$

On the other hand, the normal curvature, geodesic curvature and geodesic torsion of the ruled surface are obtained as follows

$$\kappa_n = -\frac{c_3\kappa}{\chi},\ \kappa_g = \frac{c_1 c_3 v\kappa\tau - c_1^2 v\kappa^2}{\chi},\ \tau_g = \frac{c_1^2 c_3^2 v^2\kappa(\kappa\tau)\prime - v^2 c_1 c_3\left(c_1^2\kappa\kappa' - c_3\tau\tau'\right)}{\chi^2}. \tag{33}$$

Conclusion 11. *Let $\alpha : I \subset R \to E^3$ be arc length parametrized curve. The following equations are available from Eqs. (31) and (32).*

i. If a ruled surface with a non-constant direction curve for this *Case 5* is a developable surface, the following equation exists

$$\frac{\tau}{\kappa} = \frac{c_1}{c_3}.$$

ii. If a ruled surface with a non-constant direction curve for this is a minimal surface, the following equation exists

$$\left[c_1c_3^2\left(\tau^3-3\kappa^2\tau\right)+c_1^2c_3\left(\kappa^3-3\kappa\tau^2\right)+\kappa\tau(c_3\tau+c_1\kappa)\right]$$
$$v^2-\left(-c_1c_3\kappa'+c_3^2\tau'\right)v+c_3\kappa+2c_1c_3^2\tau-2c_1^2c_3\kappa=0.$$

Conclusion 12. *Let $\alpha : I \subset R \to E^3$ be arc length parametrized curve. The following equations are available from Eq. (33).*

i. The following equation should be provided so that α, which is the base curve of the ruled surface according to *Case 5* , asymptotic curve

$$\kappa=0.$$

ii. The following equation should be provided so that α, which is the base curve of the ruled surface according to *Case 5* , geodesic curve and normal section curve

$$\frac{\tau}{\kappa}=\frac{c_1}{c_3}.$$

iii. The following equation should be provided so that α, which is the base curve of the ruled surface according *Case 5* , line of curve

$$\frac{c_1^2(\kappa^2)\prime-c_3(\tau^2)\prime}{(\kappa\tau)\prime}=2c_1c_3.$$

Case 6 Let $\alpha : I \subset R \to E^3$ be arc length parametrized curve. The ruled surface $\phi(u, v) = \alpha(u) + v\gamma(u)$ which is composed of the unit speed non-constant vector $\gamma(u) = c_1T + c_2N + c_3B$ with $c_3 = 0$, $c_1^2 + c_2^2 = 1$. That is, the direction curve of the ruled surface is the osculating curve. So, the coefficients of the first fundamental form are as follows

$$E=(c_2v\kappa-1)^2+v^2\left(c_1^2\kappa^2+c_2^2\tau^2\right),\ F=c_1,\ G=1$$

On the Other Hand, the Normal to the Surface is

$$n=\frac{1}{\chi}\left(-c_2^2v\tau, c_1c_2v\tau, c_2-v\kappa\right), \chi=\sqrt{c_2^2v^2\tau^2+(c_2-v\kappa)^2}.$$

The coefficients of the second fundamental form of the ruled surface are as follows

$$e=\frac{\left[c_2\left(\tau\kappa'-\kappa\tau'\right)-c_1c_2^2\tau^3-c_1\kappa^2\tau\right]v^2+\left(2c_1c_2\kappa\tau+c_2^2\tau'\right)v}{\chi},$$
$$f=\frac{c_2^2\tau}{\chi},\ g=0.$$

Considering the above equations, gaussian curvature and mean curvature of the ruled surface

$$K = -\frac{c_2^4\tau^2}{\chi^2\left[(c_2v\kappa - 1)^2 + v^2\left(c_1^2\kappa^2 + c_2^2\tau^2\right) - c_1^2\right]} \tag{34}$$

$$H = \frac{\left[c_2\left(\tau\kappa' - \kappa\tau'\right) - c_1c_2^2\tau^3 - c_1\kappa^2\tau\right]v^2 + \left(2c_1c_2\kappa\tau + c_2^2\tau'\right)v - 2c_1c_2^2\tau}{2\chi\left[1 + (c_1v\kappa - c_3v\tau)^2 - c_1^2\right]} \tag{35}$$

are obtained. On the other hand, the normal curvature, geodesic curvature and geodesic torsion of the ruled surface caan be easily see that

$$\kappa_n = -\frac{c_1c_2v\kappa\tau}{\chi}, \ \kappa_g = \frac{c_2\kappa - v\kappa^2}{\chi}, \ \tau_g = \frac{c_2^2v^2\left(\kappa\tau' - \tau\kappa'\right) - c_2^3\tau'}{\chi^2}. \tag{36}$$

Conclusion 13. *Let $\alpha : I \subset R \to E^3$ be arc length parametrized curve. The following equations are available from Eq.* (34) and (35).

i. If a ruled surface with a non-constant direction curve is a developable surface, the following equation exists

$$\tau = 0.$$

ii. If a ruled surface with a non-constant direction curve is a minimal surface, the following equation exists

$$\left[c_2\left(\tau\kappa' - \kappa\tau'\right) - c_1c_2^2\tau^3 - c_1\kappa^2\tau\right]v^2 + \left(2c_1c_2\kappa\tau + c_2^2\tau'\right)v - 2c_1c_2^2\tau = 0.$$

Conclusion 14. *Let $\alpha : I \subset R \to E^3$ be arc length parametrized curve. The following equations are available.*

i. The following equation should be provided so that α, which is the base curve of the ruled surface according to , asymptotic curve.

$$\kappa = 0 \text{ or } \tau = 0.$$

ii. The following equation should be provided so that α, which is the base curve of the ruled surface according to , geodesic curve and normal section curve

$$\kappa = \frac{c_2}{v}.$$

iii. The following equation should be provided so that α, which is the base curve of the ruled surface according to , line of curve

$$\left(\frac{\tau}{\kappa}\right)' = \frac{c_2^3\tau'}{c_2^2v\kappa^2}.$$

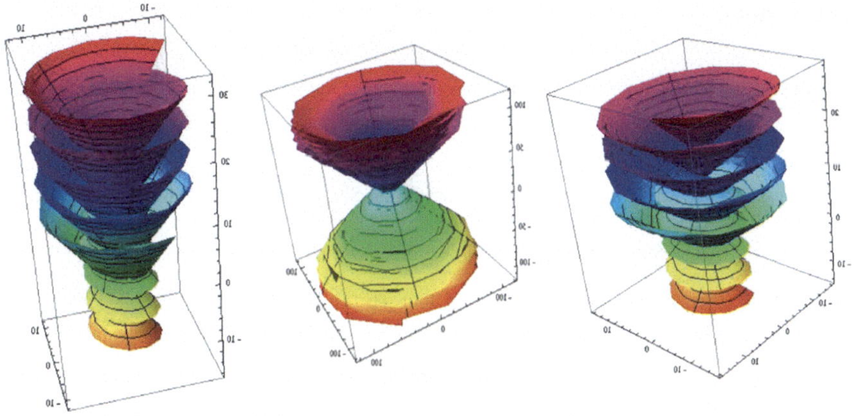

Fig. 1 *ϕ ruled surface for* $(c_1 = 1,\ c_2 = 0,\ c_3 = 0)$, $(c_1 = 0,\ c_2 = 1,\ c_3 = 0)$ *and* $(c_1 = 0,\ c_2 = 0,\ c_3 = 1)$, *respectively.*

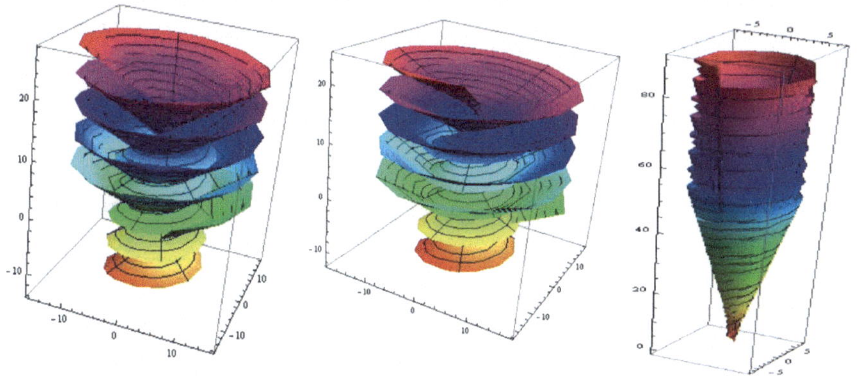

Fig. 2 *ϕ ruled surface for* $\left(c_1 = 0,\ c_2 = \frac{1}{\sqrt{2}},\ c_3 = \frac{1}{\sqrt{2}}\right)$, $\left(c_1 = \frac{1}{\sqrt{2}},\ c_2 = 0,\ c_3 = \frac{1}{\sqrt{2}}\right)$ *and* $\left(c_1 = \frac{1}{\sqrt{2}},\ c_2 = \frac{1}{\sqrt{2}},\ c_3 = 0\right)$, *respectively*

Example *Let* $\alpha(u) = \left(\frac{3}{5}\cos u,\ \frac{3}{5}\sin u,\ \frac{4}{5}u\right)$ *be a curve lying on the ruled surface M given by following parametrization:*

$$\phi(u,\ v) = \left(\frac{3}{5}\cos u,\ \frac{3}{5}\sin u,\ \frac{4}{5}u\right) + v\beta(u)$$

where $\beta(u) = c_1 T + c_2 N + c_3 B$ *and* $c_1^2 + c_2^2 + c_3^2 = 1$. *In the figures below, different states of the ruled surface ϕ are plotted according to the coefficients c_1, c_2 and c_3 of the direction curve* Figs. 1 and 2.

References

1. Divjak, B., Milin-Sipus, Z.: Special curves on the ruled surfaces in Galilean and pseudo-Galilean spaces. Acta Math. Hungar. **98**(3), 203–215 (2003)
2. Guggenheimer, H.N.: *Differential Geometry,* College of Liberal Arts University of Minnesota (1963)
3. Carmo, M.P.: Differential Geometry of Curves and Surfaces. Prentice-Hall, Inc., Englewood Cliffs, New Jersey (1976)
4. Milin Sipus, Z.: Ruled Weingarten surfaces in the Galilean Space. *Periodica Mathematica Hungerica* **45**(2), 213–225 (2008)
5. Ergüt, M.: *On Generalized Ruled Surfaces,* Phd. Thesis, Firat University, Elaziğ (1983)
6. Kamenarovic, I.: Existence theorems for ruled surfaces in the Galilean Space G3. Rad HAZU. Mat. **456**(10), 183–196 (1991)
7. Yilmaz, B., Ramis, Ç., Yayli, Y.: On developable ruled surface of the principal-direction curve. Konuralp J. Math. **5**(2), 172–178 (2017)
8. Önder, M., Kahraman, T.: On rectifying ruled surfaces. Kuwait J. Sci. **47**(4), 1–11 (2020)
9. Struik, D.J.: Lectures on Classical Differential Geometry. Dover Publications, New York (1988)
10. Kreyszig, E.: Differential Geometry. Dover Publications, New York (1991)
11. Özdemir, M.: Differential Geometry. Alti Nokta Press, İzmir (2020)
12. O'Neill, B.: Semi-Riemannian Geometry. Academic press, New York (1983)
13. Chen, B.Y.: When does the position vector of a space curve always lie in its rectifying plane. Am. Math. Mon. **110**, 147–152 (2003)
14. Ilarslan, K., Nesovic, E.: Some characterizations of osculating curves in the Euclidean spaces. Demonstratio Math. **41**(4), 931–939 (2008)

Thermal Oxidation of Ti-6Al-4 V Alloy for Engine Valve-Seat Insert Applications

Aysenur Egercioglu Yazici and Harun Mindivan

Abstract Engine valves could use titanium alloys instead of stainless steel to achieve a reduction in weight. However, they need to be more resistant to wear. To address this difficulty, researchers have devised various surface modification methods. Thermal oxidation (TO) of Ti-6Al-4V at 600 °C for 60 h resulted in the formation of a surface rutile, TiO_2, layer with a hardened oxygen diffusion zone. The TO treatment increased the surface hardness of the Ti-6Al-4V alloy from 290 ± 2 $HV_{0.025}$ to 890 ± 10 $HV_{0.025}$. We conducted a laboratory-scale impact-sliding wear test to compare the performance of untreated Ti-6Al-4V and TO-Ti-6Al-4V against SAE 52100-grade bearing steel balls. Before testing, we set the maximum spring and pneumatic cylinder loads at 40 N and 240 N, respectively, and ran 4297 loading cycles to simulate engine valve operation on TO-Ti-6Al-4V. Thermal oxidation reduced wear rate at the worn track's impact and sliding zones, reducing plastic deformation.

Keywords Ti-6Al-4 V alloy · Thermal oxidation · Wear resistance

1 Introduction

Titanium (Ti) alloys are ideal for use in the aircraft sector due to their low density, good mechanical properties, and high corrosion resistance. Automobiles utilize Ti alloys due to their high strength-to-weight ratio. Ti alloys' low hardness and wear resistance, on the other hand, limit their usage in sliding contact components and service life [1]. The valve-seat insert assembly's periodic impact causes localized wear through short-distance sliding contact. Wear scars that develop on the valve seating face can lead to valve recession and a decline in engine performance. It is believed that the formation of tribolayers reduces the intensity of wear. Using Ti alloy

A. Egercioglu Yazici · H. Mindivan (✉)
Bilecik Seyh Edebali University, Bilecik, Turkey
e-mail: harun.mindivan@bilecik.edu.tr

A. Egercioglu Yazici
e-mail: aysenur.egercioglu@bilecik.edu.tr

O. Castillo et al. (eds.), *Advanced Mathematics for the Modeling and Solution of Challenging Problems in Engineering*, Studies in Systems, Decision and Control 582,
https://doi.org/10.1007/978-3-031-82342-8_4

valves results in a 45% reduction in mass compared to conventional stainless-steel valves, which enables an increase in engine speed. Additionally, researchers have observed minimal vibration and noise as other benefits [2].

One can increase the wear performance of Ti alloys by using powder metallurgy, altering the composition of the alloy, or surface modification. Surface modification, as compared to other approaches, can extend the wear resistance of Ti alloys by improving surface properties while keeping the performance advantages of Ti alloys [3]. Ti alloys were treated with surface engineering techniques to enhance their wear resistance. These techniques include (i) physical vapor deposition (PVD) [4], (ii) boriding [5], and iii) thermal oxidation (TO) [6]. In recent years, the TO process has become the most cost-effective and, as such, may become the preferred procedure for mass production [7]. At temperatures of 600 °C [7] and 750 °C [8], exposing Ti alloys to atmospheres containing oxygen would create a hard oxide layer on the surface and an oxygen diffusion zone next to it. Research has demonstrated that Ti alloys oxidize following a parabolic law between 600 and 1000 °C [9].

Researchers [10] have investigated the wear behavior of TO-Ti alloys at room temperature. To test the wear resistance of TO-Ti-6Al-4V (700 °C for 4 h), Wang et al. [11] used a point-to-flat contact model tribometer with a load of 90 N and an average sliding speed of 0.05 m/s. It was found that TO-Ti-6Al-4V exhibited superior wear resistance when compared to Ti-6Al-4V that had not been treated. Yazdanian et al. [1] used a ball-on-disk tribometer with a load of 2 N and a sliding speed of 0.05 m/s to test the wear performance of untreated and TO-Ti-6Al-4V (600 °C for 60 h) in two different testing environments: normal (40% RH) and vacuum (7×10^{-6} torr). The presence of a TiO_2 (rutile) layer on the surface of Ti-6Al-4V shielded it from oxidative degradation in the air and plastic deformation-induced degradation in the vacuum. Lou and Alpas [2] found that TO-Ti-6Al-4V had improved sliding wear resistance compared to untreated Ti-6Al-4 V in high-temperature impact wear tests. When tested against a cast iron pin at 550 °C, the TO-Ti-6Al-4V disk made the subsurface less likely to deform and made it more resistant to wear by creating an oxygen diffusion zone [2].

Although numerous researchers have shown that thermal oxidation improves the wear resistance of Ti alloys, few studies have evaluated their wear behavior under impact and sliding wear tests. Valve-seat inserts in engine valve applications frequently experience combined impact and sliding loads, necessitating the evaluation of wear behavior for these lightweight automotive components under periodic impact and sliding motions. Nie and Zhang [12] developed the ball-on-plate inclined impact-sliding test method to simulate the wear behavior of materials subjected to combined impact and sliding loads.

We studied the impact-sliding wear behavior of the untreated Ti-6Al-4V and TO-Ti-6Al-4V, which are commonly used due to their excellent specific strength and high corrosion resistance, to determine their performance during the engine valve process.

2 Experimental Procedures

A surface profilometer (Mitutoyo Surftest SJ-400) was used to measure the surface roughness (R_a) of 20 mm diameter and 5 mm thick disks made of Ti-6Al-4V alloy. The disks were then metallographically ground and polished until the R_a value was 60 nm. After oxidizing the disks in an air furnace at 600 °C for 60 h [1], we cooled the furnace to 25 °C. The R_a value of the thermally oxidized sample surfaces was 85 nm.

We performed microscopic examinations and microhardness tests to characterize the structure of the untreated Ti-6Al-4V and TO-Ti-6Al-4V, respectively. In order to evaluate the impact-sliding wear performance of untreated and TO-Ti-6Al-4V samples, a home-made impact sliding tester designed at Bilecik Seyh Edebali University, as shown in Fig. 1, which was inspired by the study of [2] and intended to simulate engine valve-seat insert assembly, was used with a 10 mm steel ball. For one loading cycle, the impact loading was 40 N and the maximal sliding loading was 240 N, corresponding to 1.25 and 2.28 GPa Hertzian contact pressures, respectively. The impact and sliding cycles were repeated 4297 times, resulting in a 60 m sliding distance. At ambient temperature (22 °C and 55% RH), we conducted an impact-sliding wear test on untreated and TO-Ti-6Al-4V samples. A 2-D surface profilometer examined two distinct regions of wear scar (impact and sliding) to calculate the quantity of wear loss following the impact-sliding wear test. We calculated the wear rates by dividing the volumetric wear loss by the number of loading cycles in mm^3/cycle. Additionally, we performed light optical microscope (LOM) examinations on the worn surfaces of untreated Ti-6Al-4V and TO-Ti-6Al-4V to ascertain the wear mechanism. We performed hardness measurements on the surfaces of the untreated Ti-6Al-4V and TO-Ti-6Al-4V using conventional microhardness testers with Vickers pyramid indenters at a load of 0.245 N [13].

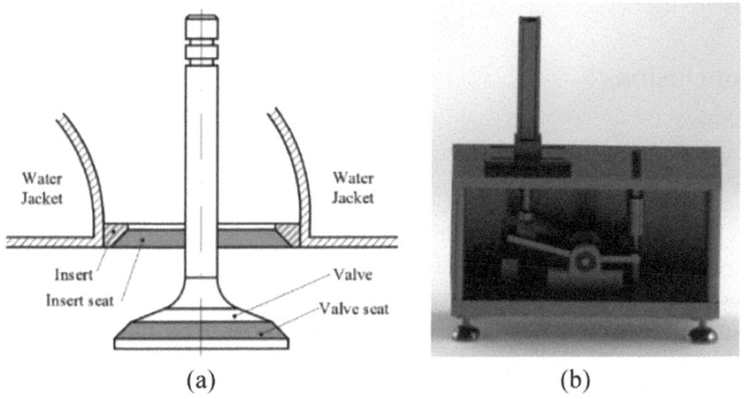

Fig. 1 **a** Schematic representation of a valve-seat insert assembly illustrating the contact surfaces; **b** illustration of the inclined impact-sliding tester used in this investigation

3 Results and Discussion

Figure 2a shows a simplified picture of a wear track. After impact-sliding wear testing, untreated Ti-6Al-4V and TO-Ti-6Al-4V samples display typical two-dimensional wear tracks in Fig. 2b and c. For untreated Ti-6Al-4V, the impact zones have greater depth and width than the sliding zones. However, thermal oxidation imposed a sliding zone that was broader and deeper than the impact zone. Shear-lips were observed along the boundaries of the areas affected by impact and sliding. This showed that plastic deformation helped the wear progress for untreated Ti-6Al-4V, but thermal oxidation did not help shear-lips.

The impact and sliding zone wear rates of the samples are shown in Fig. 3. The impact zone's wear rate was decreased by 66% by thermal oxidation, however the sliding zone's wear rate was only lowered by 14%. On the other hand, thermal oxidation resulted in a 20% decrease in the overall wear rate.

LOM photographs of the wear tracks generated throughout the tests are presented in Fig. 4. The untreated Ti-6Al-4V sample exhibited severe plastic deformation on its worn surfaces, characterized by the presence of shear lips along the borders of the impact and sliding zones (Fig. 2). These lip shapes were stronger in the impact zones than in the sliding zones. Plastic deformation under an impact load could induce surface layer fracture and exfoliation in the Ti-6Al-4V core, as it was composed of hard surface layers on a soft matrix. However, the rutile-based TiO_2 layer on the surface did not come off (Fig. 4) and showed ring cracks that formed around the indentation imprint (Fig. 4), which is similar to what other research has found [2]. Thermal oxidation resulted in the production of hard layers, consisting of the surface TiO_2 layer and the oxygen diffusion zone beneath it, which had a significant hardening impact, as anticipated. The surface TiO_2 layer exhibited hardness values of 890 ± 10 $HV_{0.025}$, whereas untreated Ti-6Al-4V exhibited softer hardness values of 290 ± 2 $HV_{0.025}$.

4 Conclusions

Below are the impact-sliding wear test results for untreated Ti-6Al-4V and TO-Ti-6Al-4V samples.

1. The TO-Ti-6Al-4V sample exhibited a wear rate that was 66% lower than the untreated Ti-6Al-4 V sample in the impact zone and 14% lower in the sliding zone, under a contact pressure of 1.25/2.28 GPa.
2. Thermally oxidized Ti-6Al-4V showed a lower total wear rate, which includes both impact and sliding wear rates, compared to untreated Ti-6Al-4V. This suggests that TO-Ti-6Al-4V has the potential to be used for engine valve applications.

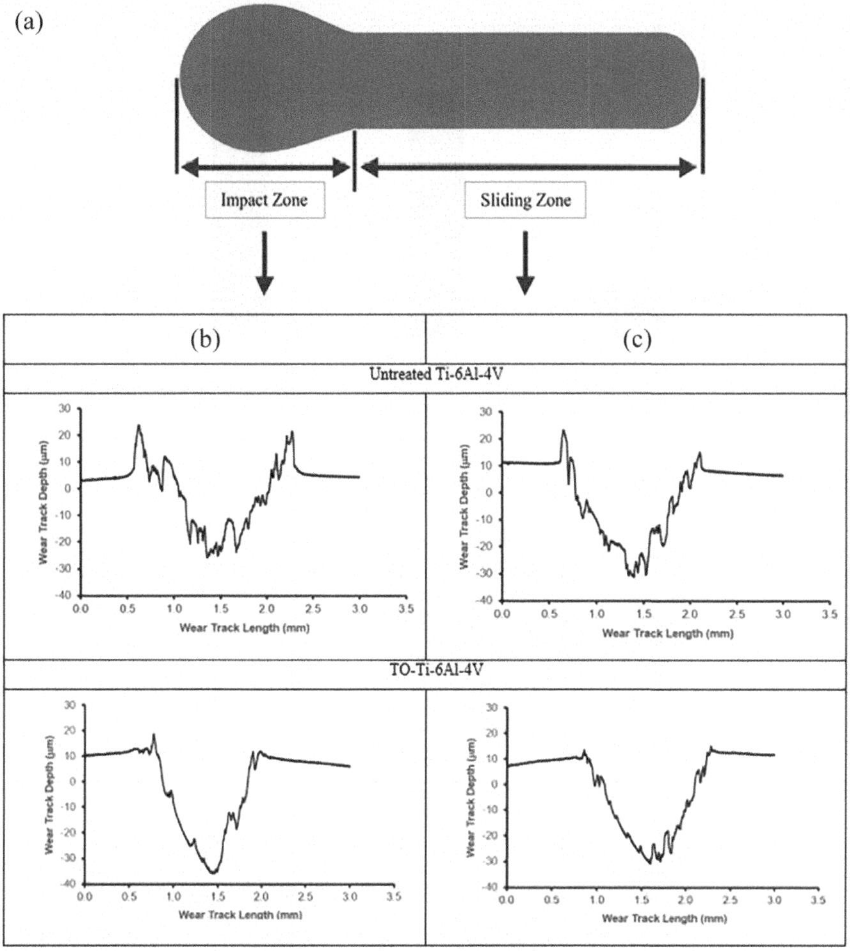

Fig. 2 **a** An illustration of the wear track that occurs on the surfaces of the samples. Following impact-sliding wear testing, typical two-dimensional profiles of **b** impact and **c** sliding zones obtained from the samples [13]

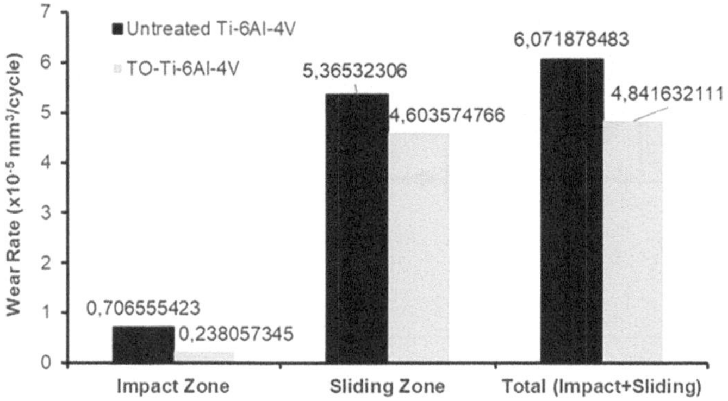

Fig. 3 Wear rates of the wear tracks formed on the samples' surfaces [13]

	Impact Zone	Sliding Zone
Untreated Ti-6Al-4V		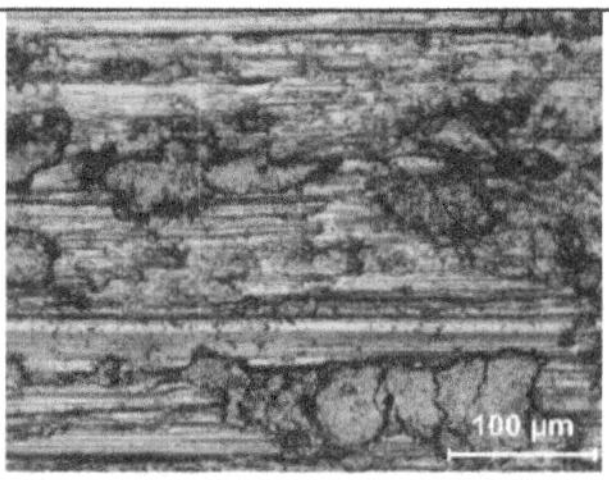
TO-Ti-6Al-4V		

Fig. 4 LOM images taken from the worn surface of the samples

Acknowledgements We express our gratitude for the financial assistance provided by the research foundation of Bilecik Seyh Edebali University [grant number TEZ-Y-2024-509].

References

1. Yazdanian, M.M., Edrisy, A., Alpas, A.T.: Vacuum sliding behaviour of thermally oxidized Ti–6Al–4V alloy. Surf. Coat. Technol. **202**(4–7), 1182–1188 (2007)

2. Lou, M., Alpas, A.T.: High temperature wear mechanisms in thermally oxidized titanium alloys for engine valve applications. Wear **426**, 443–453 (2019)
3. Ibrahim, M.K., Kaba, M., Muhaffel, F., Agaogullari, D., Cimenoglu, H.: Thermal oxidation of a porous Ti–23Nb alloy for wear related biomedical applications: effect of oxidation duration. Surf. Coat. Technol. **439**, 1–10 (2022)
4. Marin, E., Offoiach, R., Regis, M., Fusi, S., Lanzutti, A., Fedrizzi, L.: Diffusive thermal treatments combined with PVD coatings for tribological protection of titanium alloys. Mater. Des. **89**, 314–322 (2016)
5. Duan, Y., Li, P., Chen, Z., Shi, J., Ma, L.: Surface evolution and growth kinetics of Ti6Al4V alloy in pack boriding. J. Alloy. Compd. **742**, 690–701 (2018)
6. Cetiner, D., Atar, E., Derin, B., Cimenoglu, H.: Thermal oxidation of cold sprayed Ti-5Al-XZn coatings for tribological applications. Mater. Lett. **274**, 1–5 (2020)
7. Aniołek, K., Kupka, M.: Mechanical, tribological and adhesive properties of oxide layers obtained on the surface of the Ti–6Al–7Nb alloy in the thermal oxidation process. Wear **432–433**, 1–14 (2019)
8. García-Rueda, A.K., Guzman-Castillo, D., García-Gonzalez, L., Zamora-Peredo, L., Hernandez-Torres, J.: Surface modification of a Ti6Al4V alloy by thermal oxidation to improve its tribological properties. Mater. Lett. **317**, 1–4 (2022)
9. El Bakali, A., Gilblas, R., Pottier, T., Lieurey, A., Le Maoult, Y.: Effect of oxidation on spectral and integrated emissivity of Ti-6Al-4V alloy at high temperatures. J. Alloy. Compd. **889**, 1–11 (2021)
10. Aniołek, K., Kupka, M., Barylski, A.: Sliding wear resistance of oxide layers formed on a titanium surface during thermal oxidation. Wear **356–357**, 23–29 (2016)
11. Wang, S., Liao, Z., Liu, Y., Liu, W.: Influence of thermal oxidation temperature on the microstructural and tribological behavior of Ti6Al4V alloy. Surf. Coat. Technol. **240**, 470–477 (2014)
12. Nie, Y., Zhang, J.: Cyclic impact-sliding fatigue wear testing instrument, Patent No: US 8402,811 B2 (2013)
13. Egercioglu Yazici, A.: Investigation of impact-sliding wear behavior and development of wear maps under dry sliding conditions of surface treated Ti-6Al-4V alloy. M.Sc. Thesis, Bilecik Seyh Edebali University, Institute of Graduate Education, (2014) (in Turkish)

Solvability and GUH Stability Results of Fuzzy Nonlinear ABC-Fractional Coupled System

Aziz El Ghazouani, M'Hamed Elomari, and Said Melliani

Abstract In this investigation, we focus on a fuzzy nonlinear Atangana Baleau Caputo fractional coupled equation. Following a discussion of numerous significant concepts on the fuzzy ABC operator, various novel and readily demonstrable adequate solvability conditions are found. We even more show that our main problem is stable under generalized Ulam Hyers (GUH). Lastly, a case study is used to show the accessibility of our primary findings.

Keywords Solvability · Atangana Baleau Caputo fractional derivative · Banach's contraction mapping principle · Generalized Ulam Hyers stability

1 Introduction

The use of fractional differential equations in mathematical modelling of physical and engineering subjects has developed significantly in recent years [1–6]. Many major investments, such as books and articles, have been produced on connected themes. The earliest textbook on the fuzzy notion is [7], and another book is [8] by Kilbas et al. Furthermore, numerous articles on fractional differential equation results have been released [1, 9]. a while ago, Agrawal et al. [10] suggested the notion of approaches to fractional differential Equations via uncertainty, and they investigated the Riemann–Liouville's differentiation to find a solution to problems, which is a mixture of Hukuhara difference and Riemann–Liouville derivatives.

A. El Ghazouani (✉) · M. Elomari · S. Melliani
Laboratory of Applied Mathematics and Scientific Computing, Sultan Moulay Slimane University, Beni Mellal, Morocco
e-mail: aziz.elghazouani@usms.ac.ma

M. Elomari
e-mail: m.elomari@usms.ma

S. Melliani
e-mail: s.melliani@usms.ma

O. Castillo et al. (eds.), *Advanced Mathematics for the Modeling and Solution of Challenging Problems in Engineering*, Studies in Systems, Decision and Control 582,
https://doi.org/10.1007/978-3-031-82342-8_5

Stability analysis is a key part of mathematically applied science which is useful in an assortment of industrial and scientific disciplines. Ulam's stable can be seen of as a subset of materials elastic that began with Ulam [11], who defined the stability of a functional equation, which was subsequently handled by Hyers [12] via the additive function specified on the Banach space.

Several investigators are currently focusing on the idea of Hukuhara difference, which was first proposed by Bede et al. [13, 14]. As a result, in this work, we will apply an extension of extensively generalised differentiation to the fractional setting in the nonlinear $ABC-$fuzzy fractional coupled systems shown below:

$$\begin{cases} {}^{\mathrm{ABC}}\mathcal{D}_0^{p_1}[\hat{u}_1(s)] = K_1(s, \hat{u}_1(s), \hat{u}_2(s)), s \in J \\ {}^{\mathrm{ABC}}\mathcal{D}_0^{p_2}[\hat{u}_2(s)] = K_2(s, \hat{u}_1(s), \hat{u}_2(s)), s \in J \\ \hat{u}_1(0) = a_1, \hat{u}_2(0) = a_2 \end{cases} \tag{1}$$

where $\hat{u}_1(s), \hat{u}_2(s) \in \Lambda = C^F(J) \cap L^F(J), 0 < p_1, p_2 < 1, J = [0, T] \subseteq \mathbb{R}$

and a_1, a_2 represent two fuzzy values, and $C^F(J)$ and $L^F(J)$, correspondingly, are the space of all continuously fuzzy value functions and the area of all Lebesgue integrated fuzzy value functions on $J \cdot K_i \in C(J \times \Lambda \times \Lambda, \Lambda), i = 1, 2$ is a fuzzy operator.

As far as I am concerned, there are no works on the ABC fuzzy fractional coupled equations, which I believe is a fascinating and tough subject.

This paper examines the solvability and stability of the (1). In Sect. 2, we will go over some basic $ ABC $-fractional calculus knowledge. We use the Banach fixed point theorem in Sect. 3 to show that system (1). has an individual response. In Sect. 4 we confirmed the GUH stability of the system (1). more thoroughly. Section 5 gives an example to demonstrate the accuracy of our primary conclusions. Section 6 contains a summary.

2 Preliminaries

In this part, we explain the notion of the ABC fractional derivative and introduce several key ideas.

Definition 1 A fuzzy element $\hat{u}$ is a map [15].
$\hat{u} : \mathbb{R} \to [0, 1]$ that has the next features:

- $\hat{u}$ is upper semicontinuous, normal, and convex.
- $\mathrm{supp}(\hat{u})$ has a compact closure.

Definition 2 Letting $\hat{u}$ a fuzzy element [15]. So, the parametric interval shape of $\hat{u}$ is provided by.

$$\hat{u} = \left[\underline{u}(\gamma), \overline{u}(\gamma)\right], \gamma \in \left[0, 1\right]$$

and

- $\underline{u}(\gamma)$ is left continuous and non-decreasing according to γ.
- $\overline{u}(\gamma)$ is right continuous and non-increasing according to γ.
- Every
- $\gamma \in [0, 1]$, we have $\overline{u}(\gamma) \geq \underline{u}(\gamma)$.

Definition 3 Allow $\hat{u}$ and [15].

$\hat{v}$ to be two fuzzy elements. Thus, we have

$$(\hat{u} \oplus \hat{v}) = \left[\underline{u}(\gamma) + \overline{u}(\gamma), \underline{v}(\gamma) + \overline{v}(\gamma)\right],$$
$$(\lambda \odot \hat{u}) = \begin{cases} \left[\lambda \underline{u}(\gamma), \lambda \overline{u}(\gamma)\right], \lambda \geq 0 \\ \left[\lambda \overline{u}(\gamma), \lambda \underline{u}(\gamma)\right], \lambda < 0 \end{cases}$$

where $\gamma \in [0, 1]$.

The parameterization interval shape of $\hat{u}(s, \gamma)$ is calculated by.

$$\hat{u}(s, \gamma) = \left[\underline{u}(s; \gamma), \overline{u}(s; \gamma)\right], \gamma \in \left[0, 1\right]$$

For $\hat{\rho}, \hat{\varrho} \in E^n$, where E^n is the space of all fuzzy numbers. The gH difference [13] of $\hat{\rho}$ and $\hat{\varrho}$, as shown by,$\hat{\rho} \ominus_{\mathrm{gH}} \hat{\varrho}$ is defined as the number $\hat{\sigma} \in \mathbb{R}_{\mathrm{F}}$ such that.

$$\hat{\rho} \ominus_{gH} \hat{\varrho} = \hat{\sigma} \Leftrightarrow \begin{cases} (i)\hat{\rho} = \hat{\varrho} + \hat{\sigma}\, or \\ (ii)\hat{\varrho} = \hat{\rho} + (-1)\hat{\sigma} \end{cases}.$$

Regarding γ-cuts, we obtain.

$$\left(\hat{\rho} \ominus_{\mathrm{gH}} \underline{\varrho}\right)^{\gamma} = [\min\left\{\underline{\rho}(\gamma) - \underline{\underline{\rho}}(\gamma), \bar{\rho}(\gamma) - \hat{\varrho}(\gamma)\right\},$$
$$\max\left\{\underline{\rho}(\gamma) - \underline{\rho}(\gamma), \bar{\rho}(\gamma) - \bar{\rho}(\gamma)\right\}]$$

and the prerequisites for the existence of $\sigma = \rho \ominus_{g} \varrho \in \mathbb{R}_{\mathcal{F}}$ are

$$\text{case}(i) \begin{cases} \underline{\sigma}(\gamma) = \underline{\rho}(\gamma) - \underline{\underline{g}}(\gamma) \text{and } \delta(\gamma) = \overline{p}\gamma) - \underline{\underline{g}}(\gamma) \\ \text{with} \underline{\sigma}(\gamma) \text{increasing}, \overline{\sigma}(\gamma) \text{decreasing}, \underline{\sigma}(\gamma) \leq \overline{\sigma}(\gamma) \end{cases}$$
$$\text{case}(ii) \begin{cases} \underline{\sigma}(\gamma) = \overline{\rho}(\gamma) - \underline{\overline{\rho}}(\gamma) \text{and} \hat{\sigma}(\gamma) = \underline{\rho}(\gamma) - \underline{\underline{g}}(\gamma) \\ \text{with } \underline{\sigma}(\gamma) \text{increasing}, \hat{\sigma}(\gamma) \text{decreasing}, \underline{\sigma}(\gamma) \leq \hat{\sigma}(\gamma) \end{cases}$$

for all $\gamma \in [0, 1]$.

Definition 4 The Hausdorff distance is calculated as follows:

$$d : \mathbb{R}_{\mathcal{F}} \times \mathbb{R}_{F} \to \mathbb{R} \cup \{0\}$$
$$d(\hat{y}, \hat{z}) = \sup_{r \in [0,1]} \max\left\{\left|\underline{y}(r) - \underline{z}(r)\right|, \left|\overline{y}(r) - \overline{z}(r)\right|\right\}$$

where R_F is the set of all fuzzy numbers and $(R_{\mathcal{F}}, d)$ is a complete metric space and the traits listed below are highly recognized [16]

(1). $d(\hat{y} \oplus \hat{w}, \hat{z} \oplus \hat{w}) = d(\hat{y}, \hat{z}), \forall \hat{y}, \hat{z}, \hat{w} \in \mathbb{R}_F$
(2). $d(\hat{y} \oplus \hat{z}, 0) = d(\hat{y}, 0) + d(\hat{z}, 0), \forall \hat{y}, \hat{z} \in \mathbb{R}_F$
(3). $d(\hat{y} \oplus \hat{z}, \hat{y} \oplus \hat{w}) = d(\hat{z}, \hat{w}), \forall \hat{y}, \hat{z}, \hat{w} \in \mathbb{R}_F$
(4). $d(\hat{y} \oplus \hat{z}, \hat{w} \oplus \hat{s}) \leq d(\hat{y}, \hat{w}) + d(\hat{z}, \hat{s}), \forall \hat{y}, \hat{z}, \hat{w}, \hat{s} \in \mathbb{R}_F$
(5). $d(\hat{y} \ominus \hat{z}, \hat{w} \ominus \hat{s}) \leq d(\hat{y}, \hat{w}) + d(\hat{z}, \hat{s}), \forall \hat{y}, \hat{z}, \hat{w}, \hat{s} \in \mathbb{R}_\mathrm{F}$, $\hat{y} \ominus \hat{z}$, $\hat{w} \ominus \hat{s}$ exist
(6). $d(\lambda \odot \hat{y}, \lambda \odot \hat{z}) = |\lambda| d(\hat{y}, \hat{z}), \forall \hat{y}, \hat{z} \in \mathbb{R}F, \lambda \in \mathbb{R}$

Definition 5 The generalized Hukuhara derivative of $\hat{u}$ may be characterized as [15].

$$\hat{u}'_{gH}(s) = \lim_{h \to 0} \frac{\hat{u}(s+h) \ominus_{\mathrm{gH}} \hat{\mathrm{u}}(\mathrm{s})}{h},$$

if $f'_{gH}(s) \in \mathbb{R}_{\mathcal{F}}$, we declare that $\hat{u}$ is gH-differentiable,In addition we claim that $\hat{u}$ is $\left[(i) - gH\right]$-differentiable if $Case(2.1)$, $\left(\tilde{u}'_{gH}\right)_\alpha(s) = \left[\underline{u}'(s, \alpha), u\prime(s, \alpha)\right], 0 \leq \gamma \leq 1$ and that $\hat{u}$ is $\left[(ii) - gH\right]$-differentiable, if $Case(2.2)$, $\left(\tilde{u}'_{gH}\right)_\alpha(s) = \left[\overline{u}'(s, \alpha), \underline{u}'(s, \alpha)\right], 0 \leq \gamma \leq 1$.

Definition 6 The ABC fractional derivative of $\hat{u}$ of order $p \in (0, 1)$ can be obtained as [17].

$${}^{ABC}D_0^p \hat{u}(s) = \frac{\mathcal{N}(p)}{1-p} \int_0^s E_p\left(\frac{-p}{p-1}(s-\zeta)^p\right) d\zeta,$$

where $\mathcal{N}(p) = 1 - p + \frac{p}{r(p)}$ and E_p is the Mittag Leffler function characterized by

$$E_p(\zeta) = \sum_{n=0}^{\infty} \frac{\zeta^n}{\Gamma(np+1)}, \mathrm{Re}(p) > 0, \zeta \in \mathbb{C}.$$

Definition 7 The AB fractional integral for $\hat{u}$ of rank $p \in (0, 1]$ is stated in the following manner [17]:

$${}^{AB}I_0^p \hat{u}(s) = \frac{1-p}{\mathcal{N}(p)} \hat{u}(s) + \frac{p}{\mathcal{N}(p)} \frac{1}{\Gamma(p)} \int_0^s (s-\zeta)^{p-1} \hat{u}(\zeta) d\zeta \cdot \tag{2}$$

In the following definitions, we will present the fuzzy fractional derivative and fractional integral in the context of the ABC operator.

Definition 8 The ABC fractional derivative is offered in two forms [18]:

$$ {}_{0}^{ABC}D_{s}^{i,p}\hat{u}(s;\gamma) = \left[{}_{0}^{ABC}D_{s}^{i,p}\underline{u}(s;\gamma), {}_{0}^{ABC}D_{s}^{i,p}\overline{u}(s;\gamma)\right], Case(2.1), $$
$$ {}_{0}^{ABC}D_{s}^{ii,p}\hat{u}(s;\gamma) = \left[{}_{0}^{ABC}D_{s}^{ii,p}\overline{u}(s;\gamma) {}_{0}^{ABC}D_{s}^{ii,p}\underline{u}(s;\gamma)\right], Case(2.1), $$

where

$$ {}_{0}^{ABC}D_{s}^{i,p}\hat{u}(s;\gamma) = \begin{cases} \frac{N(p)}{1-p}\int_{0}^{s}((i) - gH^{T}(\zeta;\gamma))E_{p}\left(-\frac{p}{1-p}(s-\zeta)^{p}\right)\Big|_{E_{p}\geq 0} d\zeta \\ +\frac{N(p)}{1-p}\int_{0}^{s}((ii) - gH^{2}\hat{u}\prime(\zeta;\gamma))E_{p}\left(-\frac{p}{1-p}(s-\zeta)^{p}\right)\Big|_{E_{p}<0} d\zeta, \end{cases} $$

and

$$ {}_{0}^{ABC}D_{s}^{ii,p}\hat{u}(s;\gamma) = \begin{cases} \frac{N(p)}{1-p}\int_{0}^{s}((ii) - gH^{\hat{u}}(\zeta;\gamma))E_{p}\left(-\frac{p}{1-p}(s-\zeta)^{p}\right)\Big|_{E_{p}\geq 0} d\zeta \\ +\frac{N(p)}{1-p}\int_{0}^{s}((i) - gH^{2}\hat{u}\prime(\zeta;\gamma))E_{p}\left(-\frac{p}{1-p}(s-\zeta)^{p}\right)\Big|_{E_{p}<0} d\zeta \end{cases} $$

Following that, the *ABC* fractional derivative's end points are determined:

$$ {}_{0}^{ABC}D_{s}^{*,p}\underline{u}(s;\gamma) = \begin{cases} \frac{N(p)}{1-q}\int_{0}^{s}\underline{\underline{u}}'(s;\gamma)E_{p}\left(-\frac{p}{1-p}(s-\zeta)^{p}\right)\Big|_{E_{p}\geq 0} d\zeta \\ +\frac{N(p)}{1-p}\int_{0}^{s}\overline{u}'(s;\gamma)E_{p}\left(-\frac{p}{1-p}(s-\zeta)^{p}\right)\Big|_{E_{p}<0} d\zeta \end{cases} \tag{3} $$

and

$$ {}_{0}^{ABC}D_{s}^{*,p}\overline{u}(s;\gamma) = \begin{cases} \frac{N(p)}{1-p}\int_{0}^{s}\overline{u}'(s;\gamma)E_{p}\left(-\frac{p}{1-p}(s-\zeta)^{p}\right)\Big|_{E_{p}\geq 0} d\zeta \\ +\frac{N(p)}{1-p}\int_{0}^{s}\underline{u}'(s;\gamma)E_{p}\left(-\frac{p}{1-p}(s-\zeta)^{p}\right)\Big|_{E_{p}<0} d\zeta \end{cases} \tag{4} $$

where $* \in \{i, ii\}$.

Definition 9 On fuzzy value interval parameterized functions, the fractional integral *AB* is connected to the fractional derivative *ABC* by [18].

$$ \left[{}_{0}^{ABC}I_{s}^{p}u(s)\right]_{\gamma} = \left[{}_{0}^{AB}I_{s}^{p}\underline{u}(s,\gamma), {}_{0}^{AB}I_{s}^{p}\overline{u}(s,\gamma)\right], $$

and it's defined as,

$$
{}_{0}^{AB}I_{s}^{p}\underline{u}(s,\gamma)=\frac{1-p}{N(p)}\underline{u}(s,\gamma)+\frac{p}{\mathcal{N}(p)}\frac{1}{\Gamma(p)}\int_{0}^{s}(s-\zeta)^{p-1}\underline{u}(\zeta,\gamma)d\zeta,
$$

$$
{}_{0}^{AB}I_{s}^{p}\overline{u}(s,\gamma)=\frac{1-p}{N(p)}\overline{u}(s,\gamma)+\frac{p}{\mathcal{N}(p)}\frac{1}{\Gamma(p)}\int_{0}^{s}(s-\zeta)^{p-1}\overline{u}(\zeta,\gamma)d\zeta.
$$

Because the fractional integral is a summation, its γ-cut is the total of γ-cuts. Regarding (3, 4) and the influence of this fractional integral on the ABC fraction derivative, the following may be clarified:

Where

$$
\begin{aligned}
&{}_{0}^{AB}I_{s}^{p}\left({}_{0}^{ABC}D_{s}^{i,p}\hat{u}(s;\gamma)\right)=\left[{}_{0}^{AB}I_{s}^{p}\left({}_{0}^{ABC}D_{s}^{i,p}u(s;\gamma)\right),{}_{0}^{AB}I_{s}^{p}\left({}_{0}^{ABC}D_{s}^{i,p}\overline{u}((s;\gamma)\right)\right], Case(2.1)\\
&{}_{0}^{AB}I_{s}^{p}\left({}_{0}^{ABC}D_{s}^{ii,p}\hat{u}(s;\gamma)=\left[{}_{0}^{AB}I_{s}^{p}\left({}_{0}^{ABC}D_{s}^{ii,p}\overline{u}(s;\gamma)\right),{}_{0}^{AB}I_{s}^{q}\left({}_{0}^{ABC}D_{s}^{ii,p}\underline{u}(s;\gamma)\right)\right], Case(2.2)
\end{aligned}
\tag{5}
$$

$$
\begin{aligned}
{}_{0}^{AB}I_{s}^{p}\left[{}_{0}^{ABC}D_{s}^{*,p}\underline{u}(s;\gamma)\right]&=\begin{cases}{}_{0}^{AB}I_{s}^{p}\left(\frac{N(p)}{1-q}\int_{0}^{s}\underline{u}'(s;\gamma)E_{p}\left(-\frac{p}{1-p}(s-\zeta)^{p}\right)\Big|_{E_{p}\geq 0}d\zeta\right)\\ +{}_{0}^{AE}I_{s}^{p}\left(\frac{N(p)}{1-p}\int_{0}^{*}\overline{u}'(s;\gamma)E_{p}\left(-\frac{p}{1-p}(s-\zeta)^{p}\right)\Big|_{E_{p}<0}d\zeta\right)\end{cases}\\
&={}_{0}^{AB}I_{s}^{p}\left({}_{0}^{ABC}D_{s}^{*,p}\underline{u}(s;\gamma)\right)_{E_{p}\geq 0}+{}_{0}^{AE}I_{t}^{\alpha}\left({}_{0}^{ABC}D_{s}^{*,p}\overline{u}(s;\gamma)\right)_{E_{p}<0}\\
&\left(\underline{u}(s;\gamma)-\underline{u}(0;\gamma)\right)_{E_{p}\geq 0}+(\overline{u}(s;\gamma)-\overline{u}(0;\gamma))_{E_{p}<0}
\end{aligned}
$$

and

$$
\begin{aligned}
{}_{0}^{AE}I_{s}^{p}\left[{}_{0}^{ABC}D_{s}^{*,p}\overline{u}(s;\gamma)\right]&=\begin{cases}{}_{0}^{AB}I_{s}^{p}\left(\frac{N(p)}{1-q}\int_{0}^{s}\overline{u}'(s;\gamma)E_{p}\left(-\frac{p}{1-p}(s-\zeta)^{p}\right)\Big|_{E_{p}\geq 0}d\zeta\right)\\ +{}_{0}^{AB}I_{s}^{p}\left(\frac{N(p)}{1-p}\int_{0}^{*}\underline{u}'(s;\gamma)E_{p}\left(-\frac{p}{1-p}(s-\zeta)^{p}\right)\Big|_{E_{p}<0}d\zeta\right)\end{cases}\\
&={}_{0}^{AB}I_{s}^{p}\left({}_{0}^{ABC}D_{s}^{*},\overline{u}(s;\gamma)\right)_{E_{s}\geq 0}+{}_{0}^{AB}I_{s}^{p}\left({}_{0}^{ABC}D_{s}^{*,p}\underline{u}(s;\gamma)\right)_{E_{p}<0}\\
&=(\overline{u}(s;\gamma)-\overline{u}(0;\gamma))_{E_{p}\geq 0}+\left(\underline{u}(s;\gamma)-\underline{u}(0;\gamma)\right)_{E_{p}<0}
\end{aligned}
$$

Thus, for $0\leq\gamma\leq 1$ we have

$$
\begin{aligned}
\left[{}_{0}^{AB}I_{s}^{p}\left({}_{0}^{ABC}D_{s}^{p}\hat{u}(s)\right)\right]_{\gamma}=\Big[&\left(\underline{u}(s;\gamma)-\underline{u}(0;\gamma)\right)_{E_{p}\geq 0}+(\bar{u}(s;\gamma)-\bar{u}(0;\gamma))_{E_{p}<0},\\
&(\bar{u}(s;\gamma)-\bar{u}(0;\gamma))_{E_{p}\geq 0}+\left(\underline{u}(s;\gamma)-\underline{u}(0;\gamma)\right)_{E_{\gamma}<0}\Big],
\end{aligned}
$$

and

$$
{}_{0}^{AB}I_{s}^{p}\left({}_{0}^{ABC}D_{s}^{p}\hat{u}(s)\right)=(\hat{u}(s)\ominus\hat{u}(0))_{E_{p}\geq 0}\ominus(1)(\hat{u}(s)\ominus\hat{u}(0))_{E_{p}<0}.
$$

or in general,

$$ {}_{0}^{AB}I_{s}^{p}\left({}_{0}^{ABC}D_{s}^{p}\hat{u}(s)\right) = \left(\hat{u}(s)\ominus_{\mathrm{gH}}\hat{\mathrm{u}}(0)\right)_{E_p\geq 0} \oplus \left(\hat{u}(s)\ominus_{\mathrm{g}}\mathrm{H}\hat{\mathrm{u}}(0)\right)_{E_p<0}. \tag{6} $$

3 Existence and Uniqueness of Fuzzy Solution

We first need to turn the problem (1) into an analogous integral expression to shift it to a fixed-point dilemma. We offer the next Lemma, which is the linear version of the (1) boundaries values problem.

Lemma 1 *(Existence) Let* $a_1, a_2 \in \mathbb{R}_F, 0 < p_1, p_2 \leq 1, K_i \in C\big(J \times \Lambda^2, \Lambda\big), i =$ 1,2. *Then the fuzzy nonlinear ABC-fractional coupled* Eq. (1) *is equivalent to the following integral equations.*

$$ \begin{aligned} &\left(\hat{u}_1(s)\ominus_{\mathrm{gH}}\mathrm{a}_1\right)_{E_{p_1\geq 0}} \oplus \left(\hat{u}_1(s)\ominus_{\mathrm{gH}}\mathrm{a}_1\right)_{E_{p_1<0}} = \frac{1-p_1}{\mathcal{N}(\mathrm{p}_1)}K_1(s,\hat{u}_1(s),\hat{u}_2(s)) \\ &\oplus\frac{p_1}{\mathcal{N}(\mathrm{p}_1)\Gamma(\mathrm{p}_1)}\int_0^s (s-\zeta)^{p_1-1}K_1(\zeta,\hat{u}_1(\zeta),\hat{u}_2(\zeta))d\zeta \\ &\left(\hat{u}_2(s)\ominus_{\mathrm{gH}}\mathrm{a}_2\right)_{E_{p_2\geq 0}} \oplus \left(\hat{u}_2(s)\ominus_{\mathrm{gH}}\mathrm{a}_2\right)_{E_{p_2<0}} = \frac{1-p_2}{\mathcal{N}(\mathrm{p}_2)}K_2(s,\hat{u}_1(s),\hat{u}_2(s)) \\ &\oplus\frac{p_2}{\mathcal{N}(\mathrm{p}_2)\Gamma(\mathrm{p})}\int_0^s (s-\zeta)^{p_2-1}K_2(\zeta,\hat{u}_1(\zeta),\hat{u}_2(\zeta))d\zeta. \end{aligned} \tag{7} $$

Proof. Assume that $(\hat{u}_1(s), \hat{u}_2(s)) \in \Lambda^2$ satisfies the Eq. (1). Then, by applying ${}_{0}^{AB}I_{y}^{P_1}$ on both side of the first equation of (1) we obtain

$$ {}_{0}^{AB}I_{s}^{p_1}\left[{}^{ABC}\mathcal{D}_{0}^{p_1}\hat{u}_1(s)\right] = {}_{0}^{AB}I_{s}^{p_1}[K_1(s,\hat{u}_1(s),\hat{u}_2(s))] \tag{8} $$

It follows from (2) and (5) that

$$ \begin{aligned} &\left(\hat{u}_1(s)\ominus_{\mathrm{gH}}\hat{\mathrm{u}}_1(0)\right)_{E_{p_1}\geq 0} \oplus \left(\hat{u}_1(s)\ominus_{\mathrm{gH}}\hat{\mathrm{u}}_1(0)\right)_{E_{p_1}<0} = \frac{1-p_1}{\mathcal{N}(\mathrm{p}_1)}K_1(s,\hat{u}_1(s),\hat{u}_2(s)) \\ &\oplus\frac{p_1}{\mathcal{N}(\mathrm{p}_1)\Gamma(\mathrm{p}_1)}\int_0^{\star}(s-\zeta)^{p_1-1}K_1(\zeta,\hat{u}_1(\zeta),\hat{u}_2(\zeta))d\zeta \end{aligned} \tag{9} $$

$$ \begin{aligned} &\left(\hat{u}_2(s)\ominus_{\mathrm{gH}}\hat{\mathrm{u}}_2(0)\right)_{E_{p_2}\geq 0} \oplus \left(\hat{u}_2(s)\ominus_{\mathrm{gH}}\hat{\mathrm{u}}_2(0)\right)_{E_{p2}<0} = \frac{1-p_2}{N(p_2)}K_2(s,\hat{u}_1(s),\hat{u}_2(s)) \\ &\oplus\frac{p_2}{\mathcal{N}(\mathrm{p}_2)\Gamma(\mathrm{p}_2)}\int_0^{*}(s-\zeta)^{p_2-1}K_2(\zeta,\hat{u}_1(\zeta),\hat{u}_2(\zeta))d\zeta \end{aligned} \tag{10} $$

Substituting the initial value conditions $\hat{u}_1(0) = a_1, \hat{u}_2(0) = a_2$, one easily gets (6). that is, $(\hat{u}_1(s), \hat{u}_2(s)) \in \Lambda^2$ is a solution of the integral Eqs. (6). one knows that the above derivation is completely reversible. Vice versa, if $(\hat{u}_1(s), \hat{u}_2(s))$ is the solution of (6), then it is also a solution of (1). The evidence is finished.

Remark 1 In case the kernels E_{p_i} for every $s, p_i, i = 1, 2$ is completely nonnegative then by considering the initials values we have:

In **case (2.1),**

$$\begin{cases} \hat{u}_1(s) = a_1 \oplus \dfrac{1-p_1}{N(p_1)} K_1(s, \hat{u}_1(s), \hat{u}_2(s)) \\ \qquad \oplus \dfrac{p_1}{N(p_1)\Gamma(p_1)} \displaystyle\int_0^s (s-\zeta)^{p_1-1} K_1(\zeta, \hat{u}_1(\zeta), \hat{u}_2(\zeta)) d\zeta \\ \hat{u}_2(s) = a_2 \oplus \dfrac{1-p_2}{N(p_2)} K_2(s, \hat{u}_1(s), \hat{u}_2(s)) \\ \qquad \oplus \dfrac{p_2}{N(p_2)\Gamma(p)} \displaystyle\int_0^s (s-\zeta)^{p_2-1} K_2(\zeta, \hat{u}_1(\zeta), \hat{u}_2(\zeta)) d\zeta. \end{cases}$$

In **case (2.2)**,

$$\begin{cases} \hat{u}_1(s) = a_1 \ominus (-1) \dfrac{1-p_1}{N(p_1)} K_1(s, \hat{u}_1(s), \hat{u}_2(s)) \\ \qquad \ominus \dfrac{p_1}{N(p_1)\Gamma(p_1)} \displaystyle\int_0^s (s-\zeta)^{p_1-1} K_1(\zeta, \hat{u}_1(\zeta), \hat{u}_2(\zeta)) d\zeta \\ \hat{u}_2(s) = a_2 \ominus \dfrac{1-p_2}{N(p_2)} K_2(s, \hat{u}_1(s), \hat{u}_2(s)) \\ \qquad \ominus (-1) \dfrac{p_2}{N(p_2)\Gamma(p)} \displaystyle\int_0^s (s-\zeta)^{p_2-1} K_2(\zeta, \hat{u}_1(\zeta), \hat{u}_2(\zeta)) d\zeta. \end{cases}$$

Remark 2 In case the kernels E_{p_i} for every $s, p_i, i = 1, 2$ is negatives the mentioned above relations are trues conversely.

Theorem 1 (*Uniqueness*) *Suppose that the functions* $K_i(., \hat{u}_1(), .\hat{u}_2().)$ *in* (1) *are fuzzy values continuous functions defined on* Λ, *and meeting the Lipschitz requirements in.*

Λ *regarding* $\hat{u}_1$ *and* $\hat{u}_2$,

$$\begin{aligned} & d(K_i(s, \hat{u}_1(s), \hat{u}_2(s)), K_i(s, \hat{v}_1(s), \hat{v}_2(s))) \\ & \leq L_{i1} d(\hat{u}_1(s), \hat{v}_1(s)) + L_{i2} d(\hat{u}_2(s), \hat{v}_2(s)), \forall \hat{u}_1, \hat{u}_2 \in \Lambda. \\ & M = \max\{L_{i1}, L_{i2}\}, i = 1, 2 \end{aligned} \tag{11}$$

Once the subsequent criterion is met, the issue (1) *has a singular fuzzy solution* $\hat{u} = (\hat{u}_1(s), \hat{u}_2(s))$,

$$\max\{\Delta_1, \Delta_2\} M < 1$$

where

$$\Delta_1 = \left(\frac{1-p_1}{\mathcal{N}(p_1)} + \frac{T^{p_1}}{\mathcal{N}(p_1)\Gamma(p_1)}\right)$$

and

$$\Delta_2 = \left(\frac{1-p_2}{\mathcal{N}(p_2)} + \frac{T^{p_2}}{\mathcal{N}(p_2)\Gamma(p_2)}\right)$$

Proof. To demonstrate, we'll use the operator $\mathcal{B} : \Lambda \times \Lambda \to \Lambda$ on non-negative and negative kernel solutions independently as:

$$\mathcal{B}(\hat{u}_1, \hat{u}_2)(s) = (\mathcal{B}_1(\hat{u}_1, \hat{u}_2)(s), \mathcal{B}_2(\hat{u}_1, \hat{u}_2)(s))$$

where, In **case (2.1),**

$$\begin{aligned}\mathcal{B}_1(\hat{u}_1, \hat{u}_2)(s) &= a_1 \oplus \frac{1-p_1}{N(p_1)} K_1(s, \hat{u}_1(s), \hat{u}_2(s)) \\ &\oplus \frac{p_1}{N(p_1)\Gamma(p_1)} \int_0^s (s-\zeta)^{p_1-1} K_1(\zeta, \hat{u}_1(\zeta), \hat{u}_2(\zeta)) d\zeta \\ \mathcal{B}_2(\hat{u}_1, \hat{u}_2)(s) &= a_2 \oplus \frac{1-p_2}{\mathcal{N}(p_2)} K_2(s, \hat{u}_1(s), \hat{u}_2(s)) \\ &\oplus \frac{p_2}{\mathcal{N}(p_2)\Gamma(p)} \int_0^* (s-\zeta)^{p_2-1} K_2(\zeta, \hat{u}_1(\zeta), \hat{u}_2(\zeta)) d\zeta\end{aligned}$$

and, in **case (2.2),**

$$\begin{aligned}\mathcal{B}_1(\hat{\mathrm{u}}_1, \hat{\mathrm{u}}_2)(\mathrm{s}) &= a_1 \ominus (-1) \frac{1-\mathrm{p}_1}{\mathcal{N}(\mathrm{p}_1)} \mathrm{K}_1(\mathrm{s}, \hat{\mathrm{u}}_1(\mathrm{s}), \hat{\mathrm{u}}_2(\mathrm{s})) \\ &\ominus (-1) \frac{\mathrm{p}_1}{\mathcal{N}(\mathrm{p}_1)\Gamma(\mathrm{p}_1)} \int_0^{\mathrm{s}} (\mathrm{s}-\zeta)^{\mathrm{p}_1-1} \mathrm{K}_1(\zeta, \hat{\mathrm{u}}_1(\zeta), \hat{\mathrm{u}}_2(\zeta)) \mathrm{d}\zeta \\ \mathcal{B}_2(\hat{\mathrm{u}}_1, \hat{\mathrm{u}}_2)(\mathrm{s}) &= a_2(-1)) \frac{1-p_2}{\mathcal{N}(\mathrm{p}_2)} K_2(s, \hat{u}_1(s), \hat{u}_2(s)) \\ &\ominus (-1) \frac{\mathrm{p}_2}{\mathcal{N}(\mathrm{p}_2)\Gamma(\mathrm{p})} \int_0^{\mathrm{s}} (\mathrm{s}-\zeta)^{\mathrm{p}_2-1} \mathrm{K}_2(\zeta, \hat{\mathrm{u}}_1(\zeta), \hat{\mathrm{u}}_2(\zeta)) \mathrm{d}\zeta\end{aligned}$$

In **case (1),** we have

$$d(\mathcal{B}_1(\hat{u}_1,\hat{u}_2)(s),\mathcal{B}_1(\hat{v}_1,\hat{v}_2)(s))$$
$$= d\left(\begin{array}{l} a_1 \oplus \frac{1-p_1}{\mathcal{N}(p_1)} K_1(s,\hat{u}_1(s),\hat{u}_2(s)) \\ \oplus \frac{p_1}{\mathcal{N}(p_1)\Gamma(p_1)} \int_0^s (s-\zeta)^{p_1-1} K_1(\zeta,\hat{u}_1(\zeta),\hat{u}_2(\zeta))d\zeta, \\ a_1 \oplus \frac{1-p_1}{\mathcal{N}(p_1)} K_1(s,\hat{v}_1(s),\hat{v}_2(s)) \\ \oplus \frac{p_1}{\mathcal{N}(p_1)\Gamma(p_1)} \int_0^* (s-\zeta)^{p_1-1} K_1(\zeta,\hat{v}_1(\zeta),\hat{v}_2(\zeta))d\zeta \end{array}\right)$$

Using the properties of Definition 4, we get

$$\begin{aligned} d(\mathcal{B}_1(\hat{u}_1,\hat{u}_2)(s),\mathcal{B}_1(\hat{v}_1,\hat{v}_2)(s)) &= d\Bigg(\frac{1-p_1}{\mathcal{N}(p_1)} K_1(s,\hat{u}_1(s),\hat{u}_2(s)) \\ &\oplus \frac{p_1}{\mathcal{N}(p_1)\Gamma(p_1)} \int_0^s (s-\zeta)^{p_1-1} K_1(\zeta,\hat{u}_1(\zeta),\hat{u}_2(\zeta))d\zeta n \\ &\frac{1-p_1}{\mathcal{N}(p_1)} K_1(s,\hat{v}_1(s),\hat{v}_2(s)) \\ &\oplus \frac{p_1}{\mathcal{N}(p_1)\Gamma(p_1)} \int_0^s (s-\zeta)^{p_1-1} K_1(\zeta,\hat{v}_1(\zeta),\hat{v}_2(\zeta)d\zeta\Bigg) \\ &\leq \frac{1-p_1}{\mathcal{N}(p_1)} d(K_1(s,\hat{u}_1(s),\hat{u}_2(s)),K_1(s,\hat{v}_1(s),\hat{v}_2(s)))n \\ &\oplus \frac{p_1}{\mathcal{N}(p_1)\Gamma(p_1)} \int_0^* (s-\zeta)^{p_1-1} d(K_1(s,\hat{u}_1(s),\hat{u}_2(s)),K_1(s,\hat{v}_1(s),\hat{v}_2(s))) \end{aligned} \tag{12}$$

By (10), we obtain

$$\begin{aligned} d(\mathcal{B}_1(\hat{\mathrm{u}}_1,\hat{\mathrm{u}}_2)(\mathrm{s}),\mathcal{B}_1(\hat{\mathrm{v}}_1,\hat{\mathrm{v}}_2)(\mathrm{s})) &\leq \frac{1-p_1}{N(p_1)}(L_{11}d(\hat{u}_1(s),\hat{v}_1(s))+L_{12}d(\hat{u}_2(s),\hat{v}_2(s))) \\ &\oplus \frac{p_1}{\mathcal{N}(\mathrm{p}_1)\Gamma(\mathrm{p}_1)} \int_0^s (s-\zeta)^{p_1-1}(L_{11}d(\hat{u}_1(s),\hat{v}_1(s))+L_{12}d(\hat{u}_2(s),\hat{v}_2(s)))d\zeta \\ &\leq \frac{1-|\mathrm{p}_1}{N(p_1)}(L_{11}d(\hat{u}_1(s),\hat{v}_1(s))+L_{12}d(\hat{u}_2(s),\hat{v}_2(s))) \\ &\oplus \frac{s^{p_1}}{\mathcal{N}(\mathrm{p}_1)\Gamma(\mathrm{p}_1)}(L_{11}d(\hat{u}_1(s),\hat{v}_1(s))+L_{12}d(\hat{u}_2(s),\hat{v}_2(s))) \\ &\leq \left(\frac{1-p_1}{\mathcal{N}(\mathrm{p}_1)}+\frac{s^{p1}}{\mathcal{N}(\mathrm{p}_1)\Gamma(\mathrm{p}_1)}\right)(L_{11}d(\hat{u}_1(s),\hat{v}_1(s))+L_{12}d(\hat{u}_2(s),\hat{v}_2(s))) \\ &\leq \left(\frac{1-p_1}{\mathcal{N}(\mathrm{p}_1)}+\frac{T^{p_1}}{\mathcal{N}(\mathrm{p}_1)\Gamma(\mathrm{p}_1)}\right)Md(\hat{u},\hat{v}) = \Delta_1 Md(\hat{u},\hat{v}). \end{aligned}$$

By similar calculus, we obtain

$$d(\mathcal{B}_2(\hat{u})(s), \mathcal{B}_2(\hat{v})(s)) \leq \left(\frac{1-p_2}{\mathcal{N}(p_2)} + \frac{T^{p_2}}{\mathcal{N}(p_2)\Gamma(p_2)}\right) Md(\hat{u}, \hat{v}) = \Delta_2 Md(\hat{u}, \hat{v})$$

Therefore,

$$d(\mathcal{B}(\hat{u})(s), \mathcal{B}(\tilde{v})(s)) \leq \max\{\Delta_1, \Delta_2\} Md(\hat{u}, \hat{v})$$

Since

$$\max\{\Delta_1, \Delta_2\}M < 1.$$

Thus, the operator B is a contraction on Λ. As a result, our issue (1) has a singular solution relying on the Banach fixed point theorem. The similar strategy may be employed in the **case (2.2).**

4 Generalized Ulam Hyers Stability Results

We primarily analyze the GUH stability of (1) using straightforward analysis.

techniques in this section. We begin by defining Ulam-Hyers and Generalized Ulam Hyers stability in relation to the issue (1), as below.

Allow $\hat{u} = (\hat{u}_1(s), \hat{u}_2(s)) \in \Lambda$ and $\varepsilon > 0$. Given the subsequent inequality.

$$\begin{cases} {}^{ABC}\mathcal{D}_0^{p_1}[\hat{u}_1(s)] - K_1(s, \hat{u}_1(s), \hat{u}_2(s)) < \varepsilon, s \in J, \\ {}^{ABC}\mathcal{D}_0^{p_2}[\hat{u}_2(s)] - K_2(s, \hat{u}_1(s), \hat{u}_2(s)) < \varepsilon, s \in J, \\ \hat{u}_1(0) = a_1, \hat{u}_2(0) = a_2, \end{cases} \tag{13}$$

Definition 10 Assume that $\forall\varepsilon > 0$ and $\forall\hat{u} = (\hat{u}_1, \hat{u}_2) \in \Lambda^2$ satisfying (12), there exist a unique $\hat{v} = (\hat{v}_1, \hat{v}_2) \in \Lambda$ satisfying (1) and a constant $\Omega_1 > 0$ such that.

$$\|\hat{u}(s) - \hat{v}(s)\| \leq \Omega_1\varepsilon$$

then problem (1.1) is called Ulam-Hyers (UH) stable.

Definition 11 Assume that $\forall\varepsilon > 0$ and $\forall\hat{u} = (\hat{u}_1, \hat{u}_2) \in \Lambda^2$ satisfying (12), there exist a unique $\hat{v} = (\hat{v}_1, \hat{v}_2) \in \Lambda$ satisfying (1) and $\kappa \in C(\mathbb{R}, \mathbb{R}^+)$ with $\kappa(0) = 0$ such that.

$$\|\hat{u}(s) - \hat{v}(s)\| \leq \kappa(\varepsilon)$$

then problem (1) is called generalized Ula

m-Hyers (GUH) stable.

Remark 3 $\hat{u} = (\hat{u}_1, \hat{u}_2) \in \Lambda^2$ is a solution of inequality (12) if there exists $\eta = (\eta_1, \eta_2) \in \Lambda$ such that.

(1) $d\big(\eta_1(s), \hat{0}\big) \leq \varepsilon$, and $d\big(\eta_2(s), \hat{0}\big) \leq \varepsilon$, $s \in J$;
(2) ${}^{\text{ABC}}\mathcal{D}_0^{p_1}[\hat{u}_1(s)] = K_1(s, \hat{u}_1(s), \hat{u}_2(s)) + \eta_1(t), s \in J$;
(3) ${}^{\text{ABC}}\mathcal{D}_0^{p_2}[\hat{u}_2(s)] = K_2(s, \hat{u}_1(s), \hat{u}_2(s)) + \eta_2(t), s \in J$;
(4) $\hat{u}_1(0) = a_1, \hat{u}_2(0) = a_2\cdot$

Theorem 2 *If conditions* (10) *hold, then the problem* (1) *is GUH-stable.*
Proof. Based on Lemma 1 and Remark 1, the inequality (12) is solved by.
In **case (2.1),**

$$\begin{cases} \hat{u}_1(s) = a_1 \oplus \dfrac{1-p_1}{\mathcal{N}(p_1)}[K_1(s, \hat{u}_1(s), \hat{u}_2(s)) + \eta_1(s)] \\ \oplus \dfrac{p_1}{\mathcal{N}(p_1)\Gamma(p_1)} \displaystyle\int_0^s (s-\zeta)^{p_1-1}[K_1(\zeta, \hat{u}_1(\zeta), \hat{u}_2(\zeta)) + \eta_1(\zeta)]d\zeta \hat{u}_2(s) \\ = a_2 \oplus \dfrac{1-p_2}{\mathcal{N}(p_2)}[K_2(s, \hat{u}_1(s), \hat{u}_2(s)) + \eta_2(s)] \\ \oplus \dfrac{p_2}{\mathcal{N}(p_2)\Gamma(p)} \displaystyle\int_0^s (s-\zeta)^{p_2-1}[K_2(\zeta, \hat{u}_1(\zeta), \hat{u}_2(\zeta)) + \eta_2(\zeta)]d\zeta. \end{cases}$$

In **case (2.2),**

$$\begin{cases} \hat{u}_1(s) = a_1 \ominus (-1)\dfrac{1-\mathrm{p}_1}{\mathcal{N}(\mathrm{p}_1)}[\mathrm{K}_2(\mathrm{s}, \hat{\mathrm{u}}_1(\mathrm{s}), \hat{\mathrm{u}}_2(\mathrm{s})) + \eta_2(\mathrm{s})] \\ \quad \ominus (-1)\dfrac{\mathrm{p}_1}{\mathcal{N}(\mathrm{p}_1)\Gamma(\mathrm{p}_1)} \displaystyle\int_0^{\mathrm{s}} (\mathrm{s}-\zeta)^{\mathrm{p}_1-1}[\mathrm{K}_2(\zeta, \hat{\mathrm{u}}_1(\zeta), \hat{\mathrm{u}}_2(\zeta)) + \eta_2(\zeta)]\mathrm{d}\zeta \\ \hat{u}_2(s) = a_2 \ominus (-1)\dfrac{1-\mathrm{p}_2}{\mathcal{N}(\mathrm{p}_2)}[\mathrm{K}_2(\mathrm{s}, \hat{\mathrm{u}}_1(\mathrm{s}), \hat{\mathrm{u}}(\mathrm{s})) + \eta(\mathrm{s})] \\ \quad \ominus (-1)\dfrac{\mathrm{p}_2}{\mathcal{N}(\mathrm{p}_2)\Gamma(\mathrm{p})} \displaystyle\int_0^{\mathrm{s}} (\mathrm{s}-\zeta)^{\mathrm{p}_2-1}\big[\mathrm{K}_2(\zeta, \hat{\mathrm{u}}_1(\zeta), \hat{\mathrm{u}}_2(\zeta)) + \eta_{29\zeta}\big]\mathrm{d}\zeta. \end{cases}$$

According to Theorem 1 and Lemma 1 the unique solution $\hat{v} = (\hat{v}_1(s), \hat{v}_2(s)) \in \Lambda^2$ of (1) satisfies (6). For all $\varepsilon > 0$, from (10), and (1) of 4.3 and using the same calculus as (11), In case (2.1), we get

$$d(\hat{u}_1, \hat{v}_1) \leq \frac{1-p_1}{\mathcal{N}(p_1)} d(K_1(s, \hat{u}_1(s), \hat{u}_2(s)) + \eta_1(s), K_1(s, \hat{v}_1(s), \hat{v}_2(s)))$$

$$\oplus \frac{p_1}{\mathcal{N}(p_1)\Gamma(p_1)} \int_0^s (s-\zeta)^{p_1-1} d(K_1(\zeta, \hat{u}_1(\zeta), \hat{u}_2(\zeta)) + \eta_1(\zeta), K_1(\zeta, \hat{v}_1(\zeta), \hat{v}_2(\zeta)))d\zeta$$

$$\leq \frac{1-p_1}{\mathcal{N}(p_1)} d\big(\eta_1(s), \hat{0}\big) \oplus \frac{p_1}{\mathcal{N}(p_1)\Gamma(p_1)} \int_0^s (s-\zeta)^{p_1-1} d\big(\eta_1(\zeta), \hat{0}\big)d\zeta$$

$$\leq \left(\frac{1-p_1}{\mathcal{N}(p_1)} + \frac{T^{p_1}}{\mathcal{N}(p_1)\Gamma(p_1)}\right)\varepsilon = \Delta_1 \varepsilon.$$

Analogy we get

$$d(\hat{u}_2, \hat{v}_2) \leq \left(\frac{1-p_2}{\mathcal{N}(p_2)} + \frac{T^{p_2}}{\mathcal{N}(p_2)\Gamma(p_2)}\right)\varepsilon = \Delta_2\varepsilon.$$

Thus,

$$d(\hat{u}, \hat{v}) \leq \max\{\Delta_1, \Delta_2\}\varepsilon\cdot \tag{14}$$

Therefore, we know from (13) and Definition 11 that problem (1) is GUH-stable. The proof is completed.

5 An Illustrative Example

The goal of this part is to demonstrate the validity and application of our primary findings using an illustrated case (14).

To carry out so, look at the nonlinear ABC-fractional coupled system shown below (15)

$$\begin{cases} {}^{ABC}\mathcal{D}^{0,2}_{0+}\hat{u}_1(s) = \frac{2+\cos(\hat{u}_1(s))}{100} + \frac{1}{50}\frac{\hat{u}_2(s)}{1+\hat{u}_2^2(s)}, \; s \in [0,1] \\ {}^{ABC}\mathcal{D}^{0.6}_{0+}\hat{u}_2(s) = \frac{1}{50}\left[\frac{3\pi}{4} + \arctan(\hat{u}_1(s) + \hat{u}_2(s))\right], \; s \in [0,1] \\ \hat{u}_1(0) = \hat{0}, \; \hat{u}_2(0) = (1,2,3), \end{cases} \tag{15}$$

Here, we have

$$K_1(s, \hat{u}_1(s), \hat{u}_2(s)) = \frac{2+\cos(\hat{u}_1(s))}{100} + \frac{1}{50}\frac{\hat{u}_2(s)}{1+\hat{u}_2^2(s)}$$
$$K_2(s, \hat{u}_1(s), \hat{u}_2(s)) = \frac{1}{50}\left[\frac{3\pi}{4} + \arctan(\hat{u}_1(s) + \hat{u}_2(s))\right]$$

Therefore,

$$d(K_1(s, \hat{u}_1(s), \hat{u}_2(s)), K_1(s, \hat{v}_1(s), \hat{v}_2(s))) \leq \frac{1}{100}d(\hat{u}_1(s), \hat{v}_1(s)) + \frac{1}{100}d(\hat{u}_2(s), \hat{v}_2(s))$$

$$d(K_2(s, \hat{u}_1(s), \hat{u}_2(s)), K_2(s, \hat{v}_1(s), \hat{v}_2(s))) \leq \frac{1}{50}d(\hat{u}_1(s), \hat{v}_1(s)) + \frac{1}{50}d(\hat{u}_2(s), \hat{v}_2(s)).$$

Thus, the conditions (10) are true. Consequently, $L_{11} = L_{12} = \frac{1}{100}$, $L_{21} = L_{22} = \frac{1}{50}\frac{1}{50}$.

In addition, we get $M = \frac{1}{50}$.

and $\Delta_1 = 1, 20$, $\Delta_2 = 1, 33$ So, $\max\{\Delta_1, \Delta_2\} = 1, 33$.

Then,

$$\max\{\Delta_1, \Delta_2\}M = \frac{1,33}{50} = 0,02 < 1.$$

As a consequence, the conditions of Theorem 1 are met. We assert that system (14) has a singular solution and is GUH-stable based on Theorems 1 and 2.

6 Conclusion

Fuzzy fractional differential equations have piqued the curiosity of researchers due to the rapid growth of fuzzy fractional calculus and its potential uses. Many scientists have also studied Riemann–Liouville, Caputo, Caputo-Fabrizio, Hilfer, Hadamard and Grünwald-Letnikov fuzzy fractional differential equations. Nevertheless, no research on ABC-fuzzy fractional differential coupled equations has been published so far. As a result, selecting the system (1) as the subject of study is new and fascinating to us. We use the Banach's contraction mapping principle and the straightforward analysis approach to show the existence, uniqueness, and GUH-stability of the answer for problem (1). The manuscript's methodology and processes can be summed as followed: (1) Convert the differential system (1) to the integral structure (6); (2) Construct the operator $\mathcal{B}$ based on the integral scheme (6); (3) Demonstrate that the operator $\mathcal{B}$ is contractive. The methodology and procedures outlined previously may be employed to the analysis of additional forms of fuzzy fractional differential equations. In the years to come, we plan to use fractional calculus theory and fuzzy sets theory to explore some new fuzzy systems.

References

1. El Ghazouani, A., et al.: On the existence and uniqueness of fuzzy mild solution of fractional evolution equations. Kragujevac J. Math. **49**(6), 949–966 (2025)
2. Allahviranloo, T., Salahshour, S., Abbasbandy, S.: Explicit solutions of fractional differential equations with uncertainty. Soft Comput **16**, 297 (2012)
3. Arara, A., Benchohra, M., Hamidi, N., Nieto, J.J.: Fractional order differential equations on an unbounded domain. Nonlinear Anal. **72**, 580–586 (2010)
4. Babenko, Y.I.: Heat, and mass transfer. Chemia, Leningrad (1986)
5. El Ghazouani, A., Amir, F.I.A., Elomari, M.H., Melliani, S.: Existence and asymptotic behavior of non-linear hybrid functional equations including the fuzzy Caputo Nabla fractional difference. J.I MESA **14**(3), 737–750 (2023)
6. El Ghazouani, A., Talhaoui, A., M'hamed Elomari, S.M.: Existence and uniqueness results for a semilinear fuzzy fractional equation. Filomat **37**(27), 9315–9326 (2023)
7. Zimmermann, H.J.: *Fuzzy Set Theory-and its Applications*. Kiuwer Academic Publishers (1985)
8. Kilbas, A.A., Srivastava, H.M., Trujillo, J.J.: Theory and Applications of Fractional Differential Equations. Elsevier Science B.V, Amesterdam (2006)

9. Lakshmikantham, V., Leela, S., Devi, J.V.: *Theory of Fractional Dynamic Systems*. Cambridge, UK: Cambridge Scientific Pub (2009)
10. Agrawal, R.P., Lakshmikantham, V., Nieto, J.J.: On the concept of solution for fractional differential equations with uncertainty. Nonlinear Anal. **72**, 2859–2862 (2010)
11. Ulam, S.M.: *A Collection of Mathematical Problems*. Interscience Publishers (1960)
12. Hyers, D.H.: On the stability of the linear functional equation. P. Natl. Acad. Sci. USA **27**, 222–224 (1941)
13. Bede, B., Gal, S.G.: Generalizations of the differentiability of fuzzy-number-valued functions with applications to fuzzy differential equations. Fuzzy Sets Syst. **151**(3), 581–599 (2005)
14. Bede, B., Rudas, I.J., Bencsik, A.L.: First order linear fuzzy differential equations under generalized differentiability. Inf. Sci. **177**, 1648–1662 (2007)
15. Bede, B., Stefanini, L.: Generalized differentiability of fuzzy-valued functions. Fuzzy Sets Syst. **230**, 119–141 (2013)
16. Allahviranloo, T., Armand, A., Gouyandeh, Z.: Explicit solutions of fractional differential equations with uncertainty. J. Intell. Fuzzy Syst. **26**(3), 1481–1490 (2014)
17. Atangana, A., Baleanu, D.: New fractional derivatives with nonlocal and non-singular kernel: theory and application to heat transfer model. Therm. Sci. **20**, 763–769 (2016)
18. Allahviranloo, T., Ghanbari, B.: On the fuzzy fractional differential equation with interval Atangana-Baleanu fractional derivative approach. Chaos Soliton. Fract. **130**, 109–397 (2020)

Effect of Electrochemical Boriding Treatment on the Tribological Behavior of an AISI 316L Coating Sprayed with HVOF

Bülent Ermis and Harun Mindivan

Abstract Advanced coating technologies and surface techniques are needed to apply high-performance materials in hostile environments. It is possible to get a unique mix of bulk and surface properties by using thermal spraying to apply metallic coatings and then diffusion-based boriding to finish. Boriding enhances the surface hardness of materials and protects against wear efficiently. Several studies have shown that crystal lattice defects and residual stresses primarily affect boron diffusion in the boriding process. Thermally sprayed coatings exhibit a significant degree of deformation due to particle impact, resulting in the presence of defects and stress. This study examined the effect of thermal diffusion-based boriding on HVOF-applied 316L coating. Due to the high coating adherence, no delamination occurred during the thermal diffusion-based boriding. Through the application of the boriding coating, the hardness, wear resistance, and coefficient of friction were all enhanced. There is a possibility that surface boriding will increase the durability of the coating.

Keywords AISI 316L · Boriding · High Velocity Oxygen Fuel Spraying (HVOF) · Thermal spraying · Wear

1 Introduction

Industries such as petrochemical, chemical, paper, and medical extensively utilize austenitic stainless steels due to their corrosion resistance. Nevertheless, their tribological properties, characterized by low hardness and inadequate wear resistance, may impose limitations on their applications in certain cases [1–3].

Thermochemical treatments provide the opportunity to improve the property profile. Two primary categories of thermochemical processes exist. Nitrogen and

B. Ermis · H. Mindivan (✉)
Bilecik Şeyh Edebali University, Bilecik, Turkey
e-mail: harun.mindivan@bilecik.edu.tr

O. Castillo et al. (eds.), *Advanced Mathematics for the Modeling and Solution of Challenging Problems in Engineering*, Studies in Systems, Decision and Control 582,
https://doi.org/10.1007/978-3-031-82342-8_6

carbon enrichment are capable of producing interstitial hardening. Precipitation hardening procedures are used for applications with higher demands and at higher temperatures. The process of boriding enables the creation of thermally stable precipitates and maximum hardness values. For ferrous materials, pack, paste, liquid, and molten salt boriding are available. Pack boriding is widely employed in industry, despite its long processing time, solid waste, and gaseous emissions. These disadvantages make conventional boriding costly commercially and environmentally. Electrochemical-based boriding has been developed to address the issues with the current favored method. Researchers have demonstrated that electrochemical boriding can rapidly produce a thick and dense layer of desired boride phases [4]. However, thermochemical procedures primarily restrict the use of austenitic stainless steel in bulk material states. Functional coatings and steel substrates offer a cost-effective alternative to the comparatively expensive austenitic stainless steels.

Thermal spray methods enable the deposition of coatings in a wide variety of thicknesses, making this type of deposition economically viable. The feedstock material can preserve its chemical makeup due to the absence of metallurgical bonding and minimal temperature input. The formation of characteristic structural defects, such as pores and oxides, is a natural consequence of coating operations carried out under typical atmospheric circumstances. Through the utilization of high-kinetic procedures, it is possible to mitigate the severity of these flaws. High-velocity oxygen fuel (HVOF) thermal spraying is the most common way that metal coatings and cemented carbides are put on in industrial settings [5].

Previous research [6–9] have examined the nitriding process of thermally sprayed stainless-steel coatings. To our knowledge, this is the first attempt at electrochemical boriding of thermally applied stainless steel coating, as we were unable to locate any published papers or reports in the open literature. In order to implement a combined surface technology that includes a coating process and a thermochemical surface hardening process, it is necessary to select and develop appropriate coating and surface treatment concepts. This study examines the viability and tribological behavior of faster and more efficient boriding of HVOF-sprayed AISI 316L coating using an electrochemical process.

2 Experimental Procedures

Plates of low-carbon steel were cut to dimensions of $20 \times 50 \times 4$ mm. The steel plates were initially roughened by sandblasting and then coated with an intermediate Ni–Cr with a particle size in the range of $53 + 11$ μm that enhanced the coating's adhesion. As the feedstock material, a powder of austenitic stainless steel (AISI 316L) with a particle size in the range of $53 + 20$ μm was used. The HVOF thermal spray process was utilized to apply a 316L stainless steel coating to the steel plates. The thickness of the final coating was 330 ± 4 μm.

Before electrochemical boriding, we ground the surfaces of the HVOF-sprayed AISI 316 stainless steel coating with 1200 grit emery paper and then polished them with 1 μm alumina paste to achieve a satisfactory surface preparation.

After that, an ultrasonic bath in acetone was used to clean them, and then they were dried. The boriding process was conducted on the samples using a borax-based electrolyte of 90% borax ($Na_2B_4O_7$) and 10% sodium carbonate (Na_2CO_3). The process lasted for 15 min at a temperature of 950 °C, with a constant current density of 200 mA/cm^2. The steel sample served as the cathode, while the graphite crucible was the anode. After the electrochemical boriding, the applied current was turned off, and a post-processing or soaking treatment was initiated in which the sample was left in the molten bath for 45 min without polarization. Mindivan [10] stated that the constant parameters employed in this study were the initial optimum settings. After completing the boriding procedure, we removed the borided sample from the graphite crucible, cooled it in air, and rinsed it in hot water to prepare it for further examination.

For further characterization, cross-sections were then prepared according to a standard procedure and polished to a mirror finish. After HVOF coating and thermochemical surface hardening, the thickness and morphology of the coating and the boron-enriched surface layer were examined using a Nikon Eclipse LV150 Light Optic Microscope (LOM). In addition, the surface hardness distribution of the coatings in the as-sprayed (AS) and AS+ borided states was determined using a Shimadzu HVM microhardness tester with 100 g of load and 15 s of dwell time. Sliding wear tests were performed using a specially designed ball-on-flat setup in dry conditions. The tests were conducted on a linear reciprocating wear tester at a room temperature of 30 °C and a relative humidity of 50 ± 5%. The testing procedure followed the guidelines outlined in ASTM G133-05. The test parameters included a 10 mm Al_2O_3 ball, a sliding speed of 1.9 cm s^{-1}, a sliding distance of 57.5 m, a track length of 11.5 mm, and a normal load of 20 N. We used a surface profilometer (Mitutoyo Surftest SJ-400) to determine the depth and width of wear tracks. Additionally, we performed light optical microscope (LOM) examinations on the worn surfaces of the coatings in the AS and AS+ borided states to determine the wear mechanism.

3 Results and Discussion

Figure 1 displays the cross-sectional views of the coating in the AS state. The coating in the AS form often exhibits a characteristic structure. The high kinetic energy of the HVOF coating process flattens the single spray particles, which can be observed. The bonding between these particles is strong. As a result of the interaction with the environment during the coating process, oxides form at the boundaries. The formation of porosity is minimal. Furthermore, the LOM images clearly illustrate that the HVOF-316L coating consisted of two clearly distinguishable phases: a pale gray phase and a deep gray phase. The deep gray phase may contain oxides generated by the HVOF process.

Fig. 1 LOM images of the coating in the AS state (low and high magnifications)

Figure 2 depicts representative images of the coating in its borided state. In the AS+ borided state, the coating surface exhibited a tendency towards a smooth and flat morphology. The thick oxides operating as a barrier at the splat boundaries and the high amount of alloying elements in the AISI 316L coating in the AS state slow down the diffusion process, according to the literature [1, 11]. In conjunction with electrochemical boriding at 950 °C, the precipitate of borides near the surface dissolves the splat boundaries. The precipitation layer's oxides and pores did not induce cracking. In addition, there was no delamination of the coating systems. We thus confirm the overall applicability of the process combination under consideration.

Figure 3 displays the surface hardness measurements of the coatings in both the AS and AS+ borided states. The coating in the AS condition exhibited an average microhardness of 289 ± 36 $HV_{0.1}$. Adachi and Ueda found that the plasma-sprayed 316L coating had a surface hardness of approximately 336 HV. This level of hardness renders them unsuitable as wear-resistant coatings in situations involving intense friction [7]. Boriding increased the surface microhardness of the coating to above 1419 ± 378 $HV_{0.1}$, five times higher than the coating without boriding.

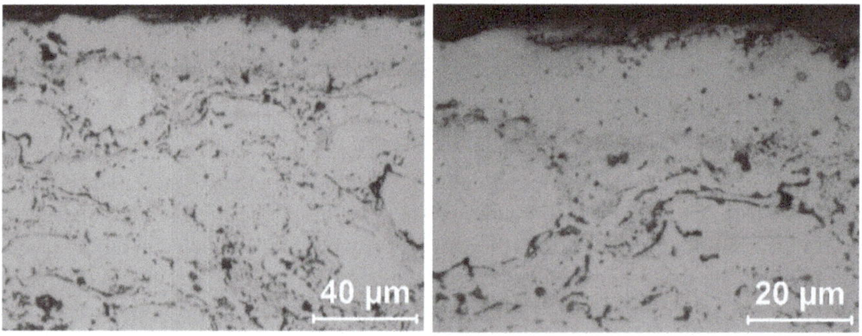

Fig. 2 Cross-sections of the coating in the AS+ borided state (low and high magnifications)

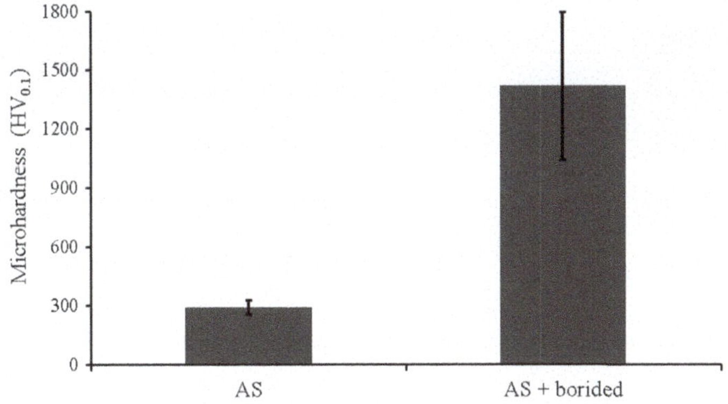

Fig. 3 Surface microhardnesses of the coatings in the AS and AS+ borided states

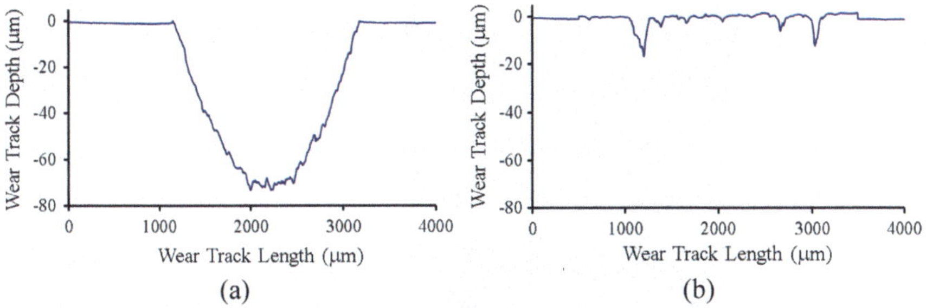

Fig. 4 2-D profiles of worn tracks on coated surfaces in **a** AS and **b** AS+ borided states

When 2-D profiles of the wear tracks were shown (Fig. 4), they showed that after electrochemical boriding, the wear depth and width were greatly reduced.

Reciprocating wear investigations are summarized in Fig. 5. Boriding leads to a decrease in the depth of wear (Fig. 4b) and consequently a lower rate of wear (Fig. 5a). Figure 5b illustrates the variation in the coefficient of friction (CoF) for the coatings in both the AS and AS+ borided states as the sliding distance progressively rises. The CoF of the coating in the AS+ borided state stabilizes at an approximately constant value (~0.17 to 0.18), achieving a stable state of sliding with minimal fluctuations. In the wear track (Fig. 6), the coating in the AS state exhibited grooving and plastic deformation. This suggests that the debris particles were flattened due to the plasticity effect caused by the sliding process in the tribosystem. Conversely, the coating in the AS+ borided state has polished surfaces that are typically smooth or shiny, with minimal material loss and no obvious scratching or plastic deformation (Fig. 6). The increased wear rate and CoF values of the coating in the AS state (Fig. 5) were due to significant plastic deformation and severe detachment of abrasive particles that had become entangled in the wear track (Fig. 6). On the other hand, a slight transfer of material was observed in the Al_2O_3 ball for the coating in the AS+ borided state, which is related to a lesser removal of material compared to that non-borided sample

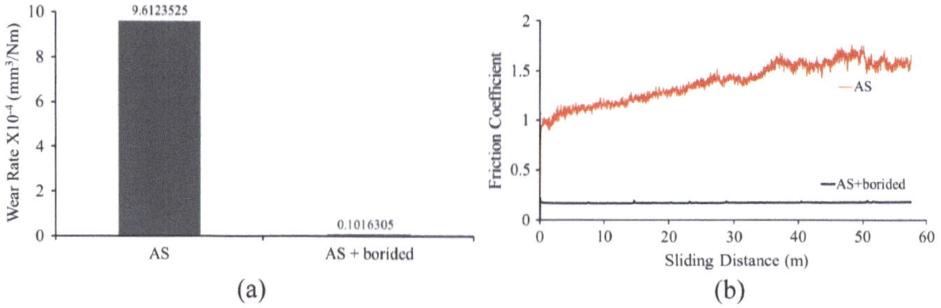

Fig. 5 **a** Wear rate and **b** friction coefficient plots of worn-out samples

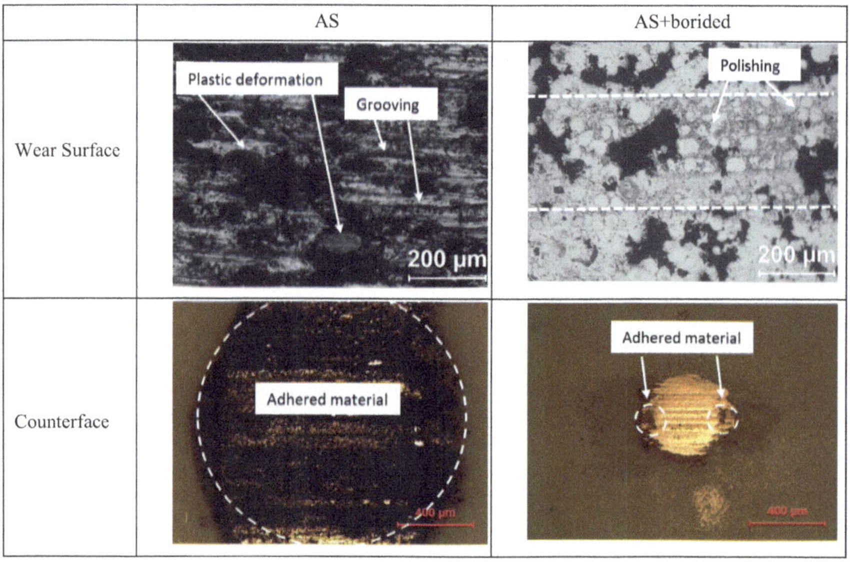

Fig. 6 LOM images of tested samples' wear tracks and Al_2O_3 ball surface following reciprocating wear test

(Fig. 6). The lower wear rate can be attributed to both an increase in hardness value and enhanced internal structural bonding as a result of the boriding process.

4 Conclusions

The hardness and wear rate of the coating in the AS state were improved considerably by the combined electrochemical boriding process. The distribution of hardness across the cross-section of the sprayed 316L coating treated at 950 °C became one that would be expected to increase durability.

Acknowledgements We express our gratitude for the financial assistance provided by Bilecik Şeyh Edebali University Scientific Research Projects Coordination Unit (Project No.: 2023-01.BŞEÜ.03-01). We express our gratitude to the Eti Maden Operations General Directorate Technology Development Department for providing samples for the purpose of conducting electrochemical boriding research.

References

1. Kayali, Y., Büyüksagis, A., Yalçin, Y.: Corrosion and wear behaviors of boronized AISI 316L stainless steel. Met. Mater. Int. **19**(5), 1053–1061 (2013)
2. Kheyrodin, M., Habibolahzadeh, A., Babak Mousavi, S.: Wear and corrosion behaviors of duplex surface treated 316L austenitic stainless steel via combination of boriding and chromizing. Prot. Met. Phys. Chem. Surf. **53**(1), 105–111 (2017)
3. Arteaga-Hernandez, L.A., Cuao-Moreu, C.A., Gonzalez-Rivera, C.E., Alvarez-Vera, M., Ortega-Saenz, J.A., Hernandez-Rodriguez, M.A.L.: Study of boriding surface treatment in the tribological behavior of an AISI 316L stainless steel. Wear **477**, 1–10 (2021)
4. Arslan, M., Karimzadehkhoei, M., Kartal Sireli, G., Coskun, O.K., Sert, M., Timur, S.: Investigating growth of iron borides with the formation of monolithic Fe_2B Layer on AISI 304 stainless steel via cathodic reduction and thermal diffusion-based boriding. J. Mater. Eng. Perform. **3**, 3274–3286 (2022)
5. Padmavathi, G., Sarada, B.N., Shanmuganathan, S.P., Padmini, B.V., Mohan, N.: Effects of high velocity oxy fuel thermal spray coating on mechanical and tribological properties of materials–a review. Mater. Today: Proceed. **27**, 2152–2157 (2020)
6. Park, G., Bae, G., Moon, K., Lee, C.: Effect of plasma nitriding and nitrocarburizing on hvof-sprayed stainless steel coatings. J. Therm. Spray Technol. **22**, 1366–1373 (2013)
7. Adachi, S., Ueda, N.: Surface hardness improvement of plasma-sprayed AISI 316L stainless steel coating by low-temperature plasma carburizing. Adv. Powder Technol. **24**, 818–823 (2013)
8. Adachi, S., Ueda, N.: Combined plasma carburizing and nitriding of sprayed AISI 316L steel coating for improved wear resistance. Surf. Coat. Technol. **259**, 44–49 (2014)
9. Mindivan, F., Mindivan, H.: Surface properties and tribocorrosion behaviour of a thermal sprayed martensitic stainless steel coating after pulsed plasma nitriding process. Adv. Mater. Process. Technol. **2**(4), 514–526 (2016)
10. Mindivan, H.: High-temparature wear and oxidation behaviour of electrochemically borided low carbon steel. J. Facul. Eng. Architect. Gazi Univ. **38**(2), 937–945 (2023)
11. Mindivan, H.: Pulsed plasma nitriding of high velocity oxy-fuel sprayed Inconel 625 coatings. Proceed. Instit. Mech. Eng. Part J: J. Eng. Tribol. **236**(10), 1950–1961 (2020)

An Encoding–Decoding Algorithm Based on Narayana Numbers

Engin Eser, Bahar Kuloğlu, and Engin Özkan

Abstract In this study we present an encoding/decoding algorithm using Narayana numbers. We use the Narayana Q-matrices applying them into matrices which are blocks. By using this method, we get different keys and massages. This process aims to not only increase the reliability of information security technology, but also to provide the ability to verify information at a high rate.

Keywords Coding/decoding algorithm · Matrix · Narayana numbers

1 Introduction

Communication for people has shown its importance throughout history. Encryption depends on the coding/decoding theory. This theory was studied different ways. One of the people who work on coding/decoding theory is Lester Hill. He created Hill algorithm. It is included matrix multiplying and matrix inverses. This algorithm is the first step of the modern mathematical theory. However, this method was not secure nowadays. It is relatively break it using computer. After Alan Turing, a computer scientist, invented an encryption machine. He broke the encrypted messages which were made by Germans Enigma machine, helping to turn the tide of World War II.

Many researchers have introduced coding/decoding algorithm using Fibonacci and Lucas numbers. Uçar et al. introduced Fibonacci Q-matrices and R-matrices in

E. Eser
Erzincan Binali Yıldırım University, Erzincan, Turkey
e-mail: engineser1978@gmail.com

B. Kuloğlu (✉)
Department of Engineering Basic Sciences, Sivas Science and Technology University, Sivas, Türkiye
e-mail: bahar_kuloglu@hotmail.com; bkuloglu@sivas.edu.tr

E. Özkan
Department of Mathematis, Marmara University, İstanbul, Türkiye
e-mail: engin.ozkan@marmara.edu.tr

O. Castillo et al. (eds.), *Advanced Mathematics for the Modeling and Solution of Challenging Problems in Engineering*, Studies in Systems, Decision and Control 582,
https://doi.org/10.1007/978-3-031-82342-8_7

[8]. Prasad introduced a new method for coding/decoding algorithm using Lucas p-numbers [6]. Diskaya et al. presented a new encryption algorithm based on Fibonacci polynomials and matrices in [4]. Shtayat et al. presents an encoding/decoding algorithm using Padovan numbers in [7]. In our study we present a new encoding/decoding algorithm using Narayana numbers. Narayana sequence N_n is defined by the third order recurrence relation as follow

$$N_n = N_{n-1} + N_{n-3} \quad n \geq 3, \quad N_0 = 0, N_1 = 1, N_2 = 1$$

with initial conditions. Indian mathematician Narayana Pandit introduced Narayana numbers in fourteenth century. Narayana numbers are the solution of the cows and calf problem. In this problem, we suppose that there is a one cow in the beginning and that every cow produces one calf each year. Every calf which reaches the age four gives birth to a calf. This problem depends on the counts the number of calves [1, 5]. In fact, this problem is similar to Fibonacci rabbit problem. There are many applications in coding theory, cryptography MIMO communication system and graph theory [3]. In [2], Bala and Mishra calculated the matrices of Narayana numbers. These matrices are defined by as follows

$$Q^0 = \begin{bmatrix} 1\,0\,0 \\ 0\,1\,0 \\ 0\,0\,1 \end{bmatrix}, Q^1 = \begin{bmatrix} 1\,0\,1 \\ 1\,0\,0 \\ 0\,1\,0 \end{bmatrix}, Q^2 = \begin{bmatrix} 1\,1\,1 \\ 1\,0\,1 \\ 1\,0\,0 \end{bmatrix}$$

$$\{Q^i\}_{i=0}^{\infty} = \left\{ Q^0 = \begin{bmatrix} 1\,0\,0 \\ 0\,1\,0 \\ 0\,0\,1 \end{bmatrix}, Q^1 = \begin{bmatrix} 1\,0\,1 \\ 1\,0\,0 \\ 0\,1\,0 \end{bmatrix}, \right.$$
$$\left. Q^2 = \begin{bmatrix} 1\,1\,1 \\ 1\,0\,1 \\ 1\,0\,0 \end{bmatrix}, Q^3 = \begin{bmatrix} 2\,1\,1 \\ 1\,1\,1 \\ 1\,0\,1 \end{bmatrix}, \ldots \right\}$$

As you see above det $(Q^n) = 1$. The general form of this matrix is defined by

$$N^n = \begin{bmatrix} N_{n+1} & N_{n-1} & N_n \\ N_n & N_{n-2} & N_{n-1} \\ N_{n-1} & N_{n-3} & N_{n-2} \end{bmatrix}$$

In this study we present a new coding–decoding system using Narayana Q-matrix. The main idea of our study is that diving the message matrix into block matrices which is size 3×3. By using the mix type algorithm and different coded letters of alphabets we increase the security of data transfer. We determine the alphabet using the number of the block matrices of message matrix. Using all these methods we have more safely coding/decoding algorithm.

This study consists of two parts. In first part we give the matrices which we use in coding/decoding algorithm. In second part we give the coding algorithm and decoding algorithm.

2 The Matrices in Coding–Decoding Algorithm

In this part first we give character table which we use in coding/decoding. First character is matched with n. This table is according to mod27.

The size of the message matrix is $3m \times 3m$. We put "·" between each word. The block matrices of M are shown by T_i, $1 \leq i \leq m^2$. The size of each block is 3×3. Assume that matrices T_i, A_i, Q_n are of the following forms:

$$T_i = \begin{bmatrix} t_1^i & t_2^i & t_3^i \\ t_4^i & t_5^i & t_6^i \\ t_7^i & t_8^i & t_9^i \end{bmatrix}$$

$$A_i = \begin{bmatrix} a_1^i & a_2^i & a_3^i \\ a_4^i & a_5^i & a_6^i \\ a_7^i & a_8^i & a_9^i \end{bmatrix}$$

$$Q_n = \begin{bmatrix} q_1 & q_2 & q_3 \\ q_4 & q_5 & q_6 \\ q_7 & q_8 & q_9 \end{bmatrix}$$

2.1 Coding Algorithm

First, in case of any minor of $(T_i)_{22}$ is zero we have to add "·" of the beginning of the T_i matrix. This process repeats until the minors get rid of zero. Our coding algorithm is based on four steps. We give the steps as follows:

1. We choose n as follows,

$$n = \begin{cases} 4 & m = 1 \\ m^2 & otherwise \end{cases}$$

2. Using the table as follow we change M matrix into T_i.
3. We compute the each $w_i = \det(T_i)$.
4. And we find the coded matrix $P = \left[w_i, t_k^i\right]_{k=1,2,3,4,6,7,8,9}$.

2.2 Decoding Algorithm

To decode the message, we take the coded matrix and applying inversing process to it and we get the M matrix. Our decoding algorithm is based on six steps. We give these steps are as follows:

First we know that det $(Q^n T_i) = w_i$

1. Compute

a_1^i	$q_1 t_1^i + q_2 t_4^i + q_3 t_7^i$	a_4^i	$q_4 t_1^i + q_5 t_4^i + q_6 t_7^i$	a_7^i	$q_7 t_1^i + q_8 t_4^i + q_9 t_7^i$
a_2^i	$q_1 t_2^i + q_1 t_1^i$	a_5^i	$q_4 t_2^i + q_6 t_8^i$	a_8^i	$q_7 t_1^i + q_9 t_8^i$
a_3^i	$q_1 t_3^i + q_2 t_6^i + q_3 t_9^i$	a_6^i	$q_4 t_3^i + q_5 t_6^i + q_6 t_9^i$	a_9^i	$q_7 t_3^i + q_8 t_6^i + q_9 t_9^i$

2. Solve the equation as follows

$$w_i = \left(a_2^i + q_2 x_i\right) \begin{vmatrix} a_6^i & a_4^i \\ a_9^i & a_7^i \end{vmatrix} - \left(a_5^i + q_5 x_i\right) \begin{vmatrix} a_7^i & a_1^i \\ a_9^i & a_3^i \end{vmatrix} + \left(a_8^i + q_8 x_i\right) \begin{vmatrix} a_3^i & a_1^i \\ a_6^i & a_4^i \end{vmatrix}$$

3. Calculate the elements x_i
4. Substitute for $x_i = t_5^i$
5. Construct T_i
6. And finally, we obtain M.

Let's end the situation with an example so that the above-mentioned algorithms can be better understood.

Considering the massage "SMOOTH SEA NEVER MADE SKILLED SAILOR"

$$M = \begin{bmatrix} S & M & O & O & T & H \\ . & S & E & A & . & N \\ E & V & E & R & . & M \\ A & D & E & . & S & K \\ I & L & L & E & D & . \\ S & A & I & L & O & R \end{bmatrix}$$

Using Table 1, we convert M into L matrix as follows

$$L = \begin{bmatrix} 22 & 16 & 18 & 18 & 23 & 11 \\ 3 & 22 & 8 & 4 & 3 & 17 \\ 18 & 25 & 18 & 21 & 3 & 16 \\ 4 & 7 & 8 & 3 & 22 & 14 \\ 12 & 15 & 15 & 18 & 17 & 3 \\ 22 & 4 & 12 & 15 & 18 & 21 \end{bmatrix}$$

Table 1 Number values corresponding to letters

A	B	C	D	E	F	G	H	I	J
n	$n+1$	$n+2$	$n+3$	$n+4$	$n+5$	$n+6$	$n+7$	$n+8$	$n+9$
K	L	M	N	O	P	Q	R	S	T
$n+10$	$n+11$	$n+12$	$n+13$	$n+14$	$n+15$	$n+16$	$n+17$	$n+18$	$n+19$
U	V	W	X	Y	Z	.			
$n+20$	$n+21$	$n+22$	$n+23$	$n+24$	$n+25$	$n+26$			

Dividing the L matrix of size 6×6 to the matrices T_i, matrices of size 3×3 since $m = 2$, we have four blocks.

$$T_1 = \begin{bmatrix} 22 & 16 & 18 \\ 18 & 23 & 11 \\ 3 & 22 & 8 \end{bmatrix}, T_2 = \begin{bmatrix} 4 & 3 & 17 \\ 18 & 25 & 18 \\ 21 & 3 & 16 \end{bmatrix}, T_3 = \begin{bmatrix} 4 & 7 & 8 \\ 3 & 22 & 14 \\ 12 & 15 & 15 \end{bmatrix},$$

$$T_4 = \begin{bmatrix} 18 & 17 & 3 \\ 22 & 4 & 12 \\ 15 & 18 & 21 \end{bmatrix}$$

$$w_1 = 2834, w_2 = -6353, w_3 = -456, w_4 = -6162$$

At the end we have encoded matrix P as follows

$$P = \begin{bmatrix} 2834 & 22 & 16 & 18 & 18 & 11 & 3 & 22 & 8 \\ -6353 & 4 & 3 & 17 & 18 & 18 & 21 & 3 & 16 \\ -411 & 4 & 7 & 8 & 3 & 14 & 12 & 15 & 15 \\ -6162 & 18 & 17 & 3 & 22 & 12 & 15 & 18 & 21 \end{bmatrix}$$

For decoding we process as follows
For T_1

a_1^1	90	a_4^1	65	a_7^1	43
a_2^1	92	a_5^1	54	a_8^1	38
a_3^1	81	a_6^1	55	a_9^1	37

Solve $2834 = 28 + 122x_1$
We get $x_1 = 23$ as desired.
For T_2

a_1^2	72	a_4^2	47	a_7^2	43
a_2^2	15	a_5^2	9	a_8^2	6
a_3^2	101	a_6^2	68	a_9^2	51

Solve $-6353 = 792 - 293x_2$
We get $x_2 = 25$ as desired.
For T_3

a_1^3	39	a_4^3	23	a_7^3	19
a_2^3	51	a_5^3	29	a_8^3	22
a_3^3	68	a_6^3	45	a_9^3	37

Solve $-411 = 381 - 36x_3$
We get $x_3 = 22$ as desired.
For T_4

a_1^4	106	a_4^4	73	a_7^4	55
a_2^4	87	a_5^4	52	a_8^4	35
a_3^4	63	a_6^4	39	a_9^4	36

Solve $-6162 = -7494 + 333x_3$
We get $x_4 = 4$ as desired.
Substitute for $x_1 = t_5^1 = 23, x_2 = t_5^2 = 25, x_3 = t_5^3 = 22$ and $x_4 = t_5^4 = 4$.
Now we construct the block matrices T_i.
We get L matrix as follows

$$L = \begin{bmatrix} 22 & 16 & 18 & 18 & 23 & 11 \\ 3 & 22 & 8 & 4 & 3 & 17 \\ 18 & 25 & 18 & 21 & 3 & 16 \\ 4 & 7 & 8 & 3 & 22 & 14 \\ 12 & 15 & 15 & 18 & 17 & 3 \\ 22 & 4 & 12 & 15 & 18 & 21 \end{bmatrix}$$

So, we have the message matrix M.

3 Conclusions

In this study, we give coding algorithm associated with Narayana numbers and their matrices. Then we constructed coding and decoding matrices algorithm. Then we calculated the determinants of each block matrix which we found. Then we applied decoding process, and we obtain the entries of L. Then, when the letter values corresponding to the L matrix are replaced, the M matrix is obtained. By using Narayana sequence we found more secure keys. With the use of the Narayana number sequence, many coding studies previously done with two-recurring number sequences have been studied in three-recurrence equations to create a more reliable coding and decoding algorithm. In addition, by combining the Narayana numbers with different numbers with coding operations called "Minesweeper", an algorithm can be created that systematically increases the reliability more.

References

1. Allouche, J.P., Johnson, T.: Narayana's cows and delayed morphisms. J. d'Inf. Music. (1996)
2. Bala, R., Mishra, V.: Narayan matrix sequence. Proc. Jangjeon Math. Soc. **25**(4), 427–434 (2022)
3. Balamurugan, B.J., Thirusangu, K., Murali, B.J., Venkateswara Rao, J.: Computation of Narayana prime cordial labeling of book graphs. In: Applied Mathematics and Scientific Computing: International Conference on Advances in Mathematical Sciences, Vellore, India, 2017 Dec 2019, vol. II, pp. 547–553. Springer International Publishing (2019)
4. Dişkaya, O., Avaroğlu, E., Menken, H., Emsal, A.: A new encryption algorithm based on Fibonacci polynomials and matrices. Trait. Signal **39**(5), 1453 (2022)
5. Lin, X.: On the recurrence properties of Narayana's cows sequence. Symmetry **13**(1), 149 (2021)
6. Prasad, B.: Coding theory on Lucas p numbers. Discrete Math. Algorithms Appl. **8**(04), 1650074 (2016)
7. Shtayat, J., Al-Kateeb, A.: An encoding-decoding algorithm based on Padovan numbers. arXiv preprint arXiv:1907.02007 (2019)
8. Uçar, S., Taş, N., Özgür, N.Y.: A new application to coding theory via Fibonacci and Lucas numbers. Math. Sci. Appl. E-Notes **7**(1), 62–70 (2019)

Tzitzeica Curves According to Alternative Moving Frame

Ilkay Guven, Beyhan Yilmaz, and Aykut Has

Abstract In this present paper, we discuss the Tzitzeica curves from a different point in Euclidean 3-space. These curves have been studied before in many different dimensions in many different spaces. The difference of this study is to describe these curves with the help of an alternative framework. We called these special curves, with which we are defined with the help of this alternative moving frame, alternative Tzitzeica curves. We have obtained many characterizations for these curves.

Keywords Tzitzeica curve · Alternative moving frame · Harmonic curvature · Slant helix · Spherical curve

1 Introduction

Gheorghe Tzitzeica introduced a new type of spatial curves in 1911 with the property

$$\frac{\tau(s)}{(d(s))^2} = a$$

where $\tau(s)$ is the second curvature of the curve and $d^2(s)$ is both the square of the distance from the origin to the osculating plane at an arbitrary point of the curve and a nonzero constant [1]. These spatial curves are called as Tzitzeica curves. Tzitzeica curves represent early examples of affine invariant geometrical objects.

I. Guven
Department of Mathematics, Gaziantep University, Gaziantep, Turkey
e-mail: iarslan@gantep.edu.tr

B. Yilmaz (✉) · A. Has
Department of Mathematics, Kahramanmaraş Sütçü İmam University, Kahramanmaraş, Turkey
e-mail: beyhanyilmaz@ksu.edu.tr

A. Has
e-mail: ahas@ksu.edu.tr

O. Castillo et al. (eds.), *Advanced Mathematics for the Modeling and Solution of Challenging Problems in Engineering*, Studies in Systems, Decision and Control 582,
https://doi.org/10.1007/978-3-031-82342-8_8

These curves have attracted a lot of attention since the day they were defined. Therefore, these curves have been studied by many authors in various spaces [2–5]. Crasmareanu in 2002 expressed elliptic and hyperbolic cylindrical curves satisfying the Tzitzeica condition in E^3, depending on the solution of the harmonic equation [6]. Karacan and Bükcü obtained elliptic cylindrical Tzitzeica curves in E_1^3 space depending on the solution of a harmonic equation in their study in 2009. Also, the conditions for a curve to be a spacelike, timelike or null elliptic cylindrical Tzitzeica curve are given [7]. Constantinescu and Crasmareanu obtained all three parametric, open and closed equations of Tzitzeica hypersurfaces in R^n in their study in 2011 [8]. Bobe et al. examined Tzitzeica curves and surfaces in terms of central affine invariants in E_1^3 in their study in 2012 [9]. Agnew et al., in their study in 2010, stated that Tzitzeica curves and surfaces represent examples of affine invariant geometric objects [10]. In [11], authors characterize Tzitzeica curves according to their curvatures in Euclidean 3-space.

In this present paper, we introduce Tzitzeica curves from a different perspective. For this, we use an alternative moving frame, different from the Frenet frame. The existing Tzitzeica conditions in the literature are re-characterized with this alternative frame and useful results are obtained. Alternative frame curvatures and harmonic curvatures for the sake of brevity are used while obtaining these characterizations. Finally, Tzitzeica conditions for tangent indicatrix, principal normal indicatrix and binormal indicatrix of the curve are obtained according to both the Frenet frame and the alternative moving frame.

2 Preliminaries

In Euclidean 3-space, it is well known that to each unit speed curve with at least four continuous derivatives, one can associate three mutually orthogonal unit vector fields T, N and B are the tangent, the principal normal and the binormal vector fields, respectively [12]. We consider the usual metric in E^3, that is

$$\langle , \rangle = dx_1^2 + dx_2^2 + dx_3^2,$$

where (x_1, x_2, x_3) is a rectangular coordinate system of E^3. In particular, the norm of an arbitrary vector $x \in R^n$ is given by $\|x\|^2 = \langle x, x\rangle$. If $X = (x_1, x_2, x_3)$ and $Y = (y_1, y_2, y_3)$ are arbitrary vectors in E^3, we define the vector product of X and Y as the following,

$$X \wedge Y = (x_2y_3 - x_3y_2, x_3y_1 - x_1y_3, x_1y_2 - x_2y_1).$$

Let $\alpha = \alpha(s)\colon I \subset R \to E^3$ be an arbitrary curve in E^3. The curve α is said to be a unit speed (or parameterized by the arc-length parameter s) if $\langle \alpha'(s), \alpha'(s)\rangle = 1$ for any $s \in I$. Let $\{T(s), N(s), B(s)\}$ be the moving frame of the curve α which satisfies the Frenet equations

$$\frac{d}{ds}\begin{bmatrix} T(s) \\ N(s) \\ B(s) \end{bmatrix} = \begin{bmatrix} 0 & \kappa(s) & 0 \\ -\kappa(s) & 0 & \tau(s) \\ 0 & \tau(s) & 0 \end{bmatrix}\begin{bmatrix} T(s) \\ N(s) \\ B(s) \end{bmatrix} \tag{2.1}$$

where $\kappa(s)$ and $\tau(s)$ are the curvature and torsion of $\alpha(s)$, respectively [13–16].

Definition 1 Let α: $I \subset R \to E^3$ be a unit speed curve in Euclidean 3-space. The curve α is a slant helix if and only if the geodesic curvature of spherical image of principal normal indicatrix (α_N) of a space curve α

$$\sigma = \frac{\kappa^2}{\left(\kappa^2 + \tau^2\right)^{3/2}}\left(\frac{\tau}{\kappa}\right)' \tag{2.2}$$

is a constant function [17].

Denote by $\{N, C, N \wedge C = W\}$ the alternative moving frame along the curve α in Euclidean 3-space. Note that N, $C = \frac{N'}{\|N'\|}$ and $W = \frac{\tau T + \kappa B}{\sqrt{\kappa^2+\tau^2}}$ are the unit principal normal vector, the derivative of principal normal vector and the darboux vector, respectively. For the derivatives of the alternative moving frame, we have

$$\frac{d}{ds}\begin{bmatrix} N(s) \\ C(s) \\ W(s) \end{bmatrix} = \begin{bmatrix} 0 & f(s) & 0 \\ -f(s) & 0 & g(s) \\ 0 & -g(s) & 0 \end{bmatrix}\begin{bmatrix} N(s) \\ C(s) \\ W(s) \end{bmatrix} \tag{2.3}$$

where $f = \sqrt{\kappa^2 + \tau^2}$ and $g = \sigma f$ are curvatures of the curve α in terms of alternative moving frame (κ and τ are the curvature and the torsion of the curve α in terms of Frenet frame, respectively) [18].

Lemma 1 *Let* α: $I \subset R \to E^3$ *be a unit speed curve in Euclidean* 3-*space in terms of the alternative moving frame apparatus* $\{N, C, W, f, g\}$. *Then* $\frac{g(s)}{f(s)} = \sigma$ *is a constant function if and only if* α *is a slant helix* [18].

Theorem 1 *Let* α: $I \subset R \to E^3$ *be a unit speed curve in Euclidean* 3-*space according to Frenet frame.*

i. *If the Frenet frame apparatus of the tangent indicatrix* (α_T) *of a space curve* α *is* {T, N, B, κ_T, τ_T}, *then we have the following*

$$\mathrm{T} = N, \mathrm{N} = \frac{\kappa}{f}(-T + HB), B = \frac{\kappa}{f}(HT + B)$$

and

$$s_T = \int \kappa(s)ds, \;\; \kappa_T = \frac{f}{\kappa}, \;\; \tau_T = \frac{g}{\kappa} \tag{2.4}$$

s_T *is natural representation of the tangent indicatrix of the curve* α *and equal to the total curvature of the curve* α.

ii. *If the Frenet frame apparatus of the normal indicatrix* (α_N) *of a space curve* α *is* $\{\mathcal{T}, \mathcal{N}, \mathcal{B}, \kappa_N, \tau_N\}$, *then we have the following*

$$\begin{aligned} \mathcal{T} &= \frac{\kappa}{f}(-T + HB), \\ \mathcal{N} &= \frac{\sigma}{\sqrt{1+\sigma^2}}\left[\frac{\kappa}{f}(HT + B) - \frac{N}{\sigma}\right], \\ \mathcal{B} &= \frac{1}{\sqrt{1+\sigma^2}}\left[\frac{\kappa}{f}(HT + B) + \sigma N\right], \end{aligned} \tag{2.5}$$

and

$$s_N = \int f(s)ds,\ \kappa_N = \sqrt{1+\sigma^2},\ \tau_N = \Gamma\sqrt{1+\sigma^2},\ \Gamma = \frac{\sigma'(s)}{f(s)\left(1+\sigma^2(s)\right)^{3/2}}. \tag{2.6}$$

s_N *is natural representation of the principal normal indicatrix of the curve* α.

iii. *If the Frenet frame apparatus of the binormal indicatrix* (α_B) *of a space curve* α *is* $\{\boldsymbol{T}, \boldsymbol{N}, \boldsymbol{B}, \kappa_B, \tau_B\}$, *then we have the following*

$$\boldsymbol{T} = -N,\ \boldsymbol{N} = \frac{\kappa}{f}(T - HB),\ \boldsymbol{B} = \frac{\kappa}{f}(HT + B) \tag{2.7}$$

and

$$s_B = \int \tau(s)ds,\ \kappa_B = \frac{f}{\tau},\ \tau_B = -\frac{g}{\tau}. \tag{2.8}$$

s_B *is natural representation of the binormal indicatrix of the curve* α [19].

Definition 2 Let $\alpha = \alpha(s)\colon I \subset R \to E^3$ be an arbitrary curve in E^3. Harmonic curvatures of α is defined by

$$H(s) = \frac{\tau(s)}{\kappa(s)}$$

in the paper [20].

Definition 3 Let $\alpha\colon I \subset \mathrm{R} \to E^3$ be a unit speed curve with curvatures $\kappa(s) > 0$ and $\tau(s) > 0$ in Euclidean 3-space. Then α is a spherical curve if and only if

$$\frac{\tau(s)}{\kappa(s)} = \left(\frac{\kappa'(s)}{\tau(s)\kappa^2(s)}\right)' \tag{2.9}$$

holds [21].

3 Tzitzeica Curves According to Alternative Moving Frame

In this section, we define a new type Tzitzeica curve according to alternative moving frame $\{N, C, W\}$ in Euclidean 3-space. Due to its definition with the help of alternative frame, this new curve is called the alternative Tzitzeica curve. We give some characterizations about this curve.

Definition 4 Let $\alpha = \alpha(s): I \subset \mathrm{R} \to E^3$ be a unit speed curve according to alternative moving frame $\{N(s), C(s), W(s)\}$ in Euclidean 3-space where s measures its arc length. The plane containing the N and C vectors of the curve at a given point on the curve α is called the NC plane.

Definition 5 Let $\alpha = \alpha(s): I \subset \mathrm{R} \to E^3$ be a unit speed curve according to alternative moving frame $\{N(s), C(s), W(s)\}$ in Euclidean 3-space. A curve is called an alternative Tzitzeica curve in which the ratio of the second curvature of the curve α at the point $\alpha(s)$ to the square of the distance of the NC plane to the origin is constant. Let g be the second curvature of the curve α and d_{NC} be the distance of the curve α from the origin of the plane NC at an arbitrary point $\alpha(s)$. In this case, the condition of being an alternative Tzitzeica curve is expressed as follows

$$\frac{g}{(d_{NC})^2} = c \tag{3.1}$$

where c is a nonzero real constant.

Lemma 2 *Let $\alpha = \alpha(s): I \subset \mathrm{R} \to E^3$ be a unit speed curve according to alternative moving frame $\{N, C, W\}$ in Euclidean 3-space. If d_{NC} is the distance of the curve α from the origin of the plane NC at an arbitrary point $\alpha(s)$, we can say*

$$d_{NC} = \frac{\langle W, \alpha \rangle}{\|W\|} = \langle W, \alpha \rangle.$$

So, Eq. (3.1) *can be expressed again with distance of the curve α from the origin of the NC plane as follows*

$$\frac{g}{\langle W, \alpha \rangle^2} = c \tag{3.2}$$

where c is a nonzero real constant.

Theorem 2 *Let $\alpha: I \subset \mathrm{R} \to E^3$ be a unit speed curve according to alternative moving frame $\{N, C, W\}$ in Euclidean 3-space. α is an alternative Tzitzeica curve if and only if*

$$g'\langle W, \alpha \rangle + 2g^2\langle C, \alpha \rangle - \frac{2gH}{\sqrt{1 + H^2}} = 0 \tag{3.3}$$

where H is the harmonic curvature and g is the torsion of the curve α.

Proof Let α be an alternative Tzitzeica curve in Euclidean 3-space. If the derivative of Eq. (3.2) is taken, we get

$$\frac{g'}{\langle W,\alpha\rangle^2}-\frac{2g\left(-g\langle C,\alpha\rangle+\frac{\tau}{\sqrt{\kappa^2+\tau^2}}\right)}{\langle W,\alpha\rangle^3}=0.$$

Since $\frac{\tau}{\sqrt{\kappa^2+\tau^2}}=\frac{\tau}{\kappa\sqrt{1+\frac{\tau^2}{\kappa^2}}}$ and $\frac{\tau}{\kappa}=H$, we can write the above equation again as

$$\frac{g'\langle W,\alpha\rangle+2g^2\langle C,\alpha\rangle-2g\frac{H}{\sqrt{1+H^2}}}{\langle W,\alpha\rangle^3}=0.$$

So, we can say that α is an alternative Tzitzeica curve if and only if

$$g'\langle W,\alpha\rangle+2g^2\langle C,\alpha\rangle-\frac{2gH}{\sqrt{1+H^2}}=0.$$

Theorem 3 *Let $\alpha\colon I\subset \mathbb{R}\to E^3$ be a unit speed curve according to alternative moving frame $\{N,C,W\}$ in Euclidean 3-space. Parametric equation of α is given by*

$$\alpha(s)=a_1(s)N(s)+a_2(s)C(s)+a_3(s)W(s) \tag{3.4}$$

where $a_i\colon I\subset\mathbb{R}\to\mathbb{R}$ are differentiable functions for $1\le i\le 3$. If α is an alternative Tzitzeica curve, then following equations holds:

1. $\frac{(a_2g+a_3')H+a_3g-a_1f-a_2'}{\sqrt{1+H^2}}=1,$
2. $a_1'-a_2f=0,$
3. $(a_1f+a_2'-a_3g)H+a_2g+a_3'=0,$
4. $g'a_3+2g^2a_2-\frac{2gH}{\sqrt{1+H^2}}=0.$

Proof Let α be the alternative Tzitzeica curve according to alternative moving frame with the parametric equation as

$$\alpha(s)=a_1(s)N(s)+a_2(s)C(s)+a_3(s)W(s).$$

If the derivative of Eq. (3.4) is taken, we obtain

$$\alpha'(s)=\left(a_1'-a_2f\right)N+\left(a_1f+a_2'-a_3g\right)C+\left(a_2g+a_3'\right)W.$$

If the following equations

$$C=\frac{N'}{\|N'\|}=\frac{-\kappa T+\tau B}{\sqrt{\kappa^2+\tau^2}}\text{ and }W=\frac{\tau T+\kappa B}{\sqrt{\kappa^2+\tau^2}}$$

are used here and the selection $X = a_1 f + a_2' - a_3 g$ and $Y = a_2 g + a_3'$, we get

$$\alpha'(s) = (a_1' - a_2 f)N + X\frac{(-\kappa T + \tau B)}{\sqrt{\kappa^2+\tau^2}} + Y\frac{(\tau T + \kappa B)}{\sqrt{\kappa^2+\tau^2}}$$

and

$$\begin{aligned}\alpha'(s) = &\left(-X\frac{\kappa}{\sqrt{\kappa^2+\tau^2}} + Y\frac{\tau}{\sqrt{\kappa^2+\tau^2}}\right)T + (a_1' - a_2 f)N \\ &+ \left(X\frac{\tau}{\sqrt{\kappa^2+\tau^2}} + Y\frac{\kappa}{\sqrt{\kappa^2+\tau^2}}\right)B.\end{aligned}$$

So, we obtain the following equation more simply using the harmonic curvature function as

$$\begin{aligned}\alpha'(s) = &\left(\frac{-X}{\sqrt{1+H^2}} + \frac{YH}{\sqrt{1+H^2}}\right)T + (a_1' - a_2 f)N \\ &+ \left(\frac{XH}{\sqrt{1+H^2}} + \frac{Y}{\sqrt{1+H^2}}\right)B.\end{aligned}$$

Using reciprocal equations, we can easily write that

$$\begin{aligned}\frac{-X + YH}{\sqrt{1+H^2}} &= 1, \\ a_1' - a_2 f &= 0, \\ \frac{XH + Y}{\sqrt{1+H^2}} &= 0.\end{aligned}$$

If the functions X and Y are used in above equations respectively, we have

$$\frac{a_2 g + a_3' - a_1 f - a_2' + a_3 g}{\sqrt{1+H^2}} = 1,$$

$$a_1' - a_2 f = 0,$$

$$\frac{(a_1 f + a_2' - a_3 g)H + a_2 g + a_3'}{\sqrt{1+H^2}} = 0.$$

Finally, from Eq. (3.3), we get the fourth item of the theorem

$$g' a_3 + 2g^2 a_2 - \frac{2gH}{\sqrt{1+H^2}} = 0.$$

So, the proof is complete.

Theorem 4 *Let $\alpha\colon I \subset \mathrm{R} \to E^3$ be a unit speed curve according to alternative moving frame $\{N, C, W\}$ in Euclidean 3-space. Then α is spherical curve if and only if*

$$H = \left(\frac{f'}{Hf^3} + \frac{f'H}{f^3} - \frac{H'}{f^2}\right)'. \tag{3.5}$$

Proof Let $\alpha\colon I \subset \mathrm{R} \to E^3$ be arc length parametrized curve according to alternative moving frame $\{N, C, W\}$. Let the curvature and torsion of the curve α be f and g, respectively. Using Definition 3, we give the definition of spherical curve through alternative frame elements. The expressions of alternative curvatures in terms of κ and τ are given below

$$H = \frac{\tau}{\kappa},\ f = \kappa\sqrt{1+H^2},\ g = \sigma f.$$

For the curvature

if we take the derivative of the above curvature, we obtain

$$\kappa' = \frac{f'(1+H^2) - fHH'}{(1+H^2)^{\frac{3}{2}}}.$$

On the other hand, from Eq. (2.9) we get the following

$$H = \left(\frac{f'(1+H^2) - fHH'}{(1+H^2)^{\frac{3}{2}}}.\frac{1}{H\kappa^3}\right)', \quad \kappa^3 = \frac{f^3}{(1+H^2)^{\frac{3}{2}}}.$$

If necessary algebraic arrangements are made, we can see that

$$H = \left(\frac{f'}{Hf^3} + \frac{f'H}{f^3} - \frac{H'}{f^2}\right)'.$$

Theorem 5 *Let $\alpha\colon I \subset \mathrm{R} \to E^3$ be a unit speed spherical curve according to alternative moving frame $\{N, C, W\}$ in Euclidean 3-space. If α is an alternative Tzitzeica curve, then the following equation is holds.*

$$g'\left(\left(\frac{f'(1+H^2) - fHH'}{f^3\sqrt{1+H^2}}\right)'\frac{1}{g} + \frac{1-g(1+H^2)}{g\sqrt{1+H^2}}\right)$$
$$+\, 2g^2\left(\frac{f'(1+H^2) - fHH'}{f^3\sqrt{1+H^2}}\right) - \frac{2gH}{\sqrt{1+H^2}} = 0$$

Proof Let $\alpha \colon I \subset \mathrm{R} \to E^3$ be a unit speed spherical curve according to alternative moving frame. Let's give the condition that this spherical curve α is an alternative Tzitzeica curve. Since α is a spherical curve, we can write that

$$\|\alpha(s)\| = r \tag{3.6}$$

where r is the radius of the sphere. Differentiating the Eq. (3.6) with respect to s, we obtain

$$\langle \alpha(s), T(s) \rangle = 0.$$

If we take the derivative of above equation, we get

$$\langle \alpha, N \rangle = -\frac{1}{\kappa}. \tag{3.7}$$

Differentiating Eq. (3.7), we obtain

$$\langle \alpha, C \rangle = \frac{\kappa'}{\kappa^2 f}. \tag{3.8}$$

Again differentiating Eq. (3.8) with respect to s, we can see

$$\frac{1}{\|N'\|}\langle T, -\kappa T + \tau B \rangle - f\langle \alpha, N \rangle + g\langle \alpha, W \rangle = \left(\frac{\kappa'}{\kappa^2 f}\right)'.$$

If necessary adjustments are made in the above equation, we have

$$-\frac{\kappa}{f} + \frac{f}{\kappa} + g\langle \alpha, W \rangle = \left(\frac{\kappa'}{\kappa^2 f}\right)'$$

and

$$\langle \alpha, W \rangle = \left(\frac{\kappa'}{\kappa^2 f}\right)' \frac{1}{g} + \frac{1}{g\sqrt{1+H^2}} - \sqrt{1+H^2}. \tag{3.9}$$

Hence, if Eqs. (3.8) and (3.9) are substituted in the alternative Tzitzeica condition in Eq. (3.3), we can write as

$$g'\left(\left(\frac{\kappa'}{\kappa^2 f}\right)' \frac{1}{g} + \frac{1}{g\sqrt{1+H^2}} - \sqrt{1+H^2}\right) + 2g^2 \frac{\kappa'}{\kappa^2 f} - \frac{2gH}{\sqrt{1+H^2}} = 0.$$

If necessary calculations are made, we get the following

$$g'\left(\left(\frac{f'(1+H^2)-fHH'}{f^3\sqrt{1+H^2}}\right)'\frac{1}{g}+\frac{1-g(1+H^2)}{g\sqrt{1+H^2}}\right)$$
$$+2g^2\left(\frac{f'(1+H^2)-fHH'}{f^3\sqrt{1+H^2}}\right)-\frac{2gH}{\sqrt{1+H^2}}=0.$$

Theorem 6 *Let $\alpha\colon I\subset \mathrm{R}\to E^3$ be a unit speed curve according to alternative moving frame $\{N, C, W\}$ in Euclidean 3-space. Tzitzeica and alternative Tzitzeica condition for the curve α are as follows*

$$\frac{fH\sqrt{1+H^2}}{(a_2H+a_3)^2}=\text{constant (Tzitzeica condition)},$$

and

$$\frac{g}{a_3^2}=\text{constant (alternative Tzitzeica condition)}$$

where $a_i\colon I\subset \mathrm{R}\to \mathrm{R}$ are differentiable functions for $1\le i\le 3$.

Proof Let $\alpha\colon I\subset \mathrm{R}\to E^3$ be arc length parametrized curve whose parametric equation is given by

$$\alpha(s)=a_1(s)N(s)+a_2(s)C(s)+a_3(s)W(s) \tag{3.10}$$

where $a_i\colon I\subset \mathrm{R}\to \mathrm{R}$ are differentiable functions for $1\le i\le 3$ and $\{N, C, W, f, g\}$ be the alternative frame apparatus. The Tzitzeica condition with respect to the alternative frame apparatus can also be considered as follows. Using the following

$$f=\kappa\sqrt{1+H^2},\quad g=\frac{H'}{(1+H^2)} \tag{3.11}$$

and

$$f=\frac{\tau\sqrt{1+H^2}}{H},\quad \tau=\frac{fH}{\sqrt{1+H^2}}, \tag{3.12}$$

we obtain that

$$\langle B,\alpha\rangle=\frac{a_2}{\sqrt{\kappa^2+\tau^2}}\langle B,-\kappa T+\tau B\rangle+\frac{a_3}{\sqrt{\kappa^2+\tau^2}}\langle B,\tau T+\kappa B\rangle,$$
$$=\frac{a_2}{\sqrt{\kappa^2+\tau^2}}\tau+\frac{a_3}{\sqrt{\kappa^2+\tau^2}}\kappa.$$

From Eqs. (3.11) and (3.12), we get the

$$\langle B, \alpha \rangle = \frac{a_2 H + a_3}{\sqrt{1 + H^2}}.$$

Let's substitute this equation in the equation of being the Tzitzeica condition as

$$\frac{\tau}{(d_{osc})^2} = \frac{fH\sqrt{1 + H^2}}{(a_2 H + a_3)^2}.$$

Now let's get a new characterization for the alternative Tzitzeica curve according to alternative moving frame.

$$\begin{aligned}
\langle W, \alpha \rangle &= \left\langle \frac{\tau T + \kappa B}{\sqrt{\kappa^2 + \tau^2}}, a_1(s)N(s) + a_2(s)C(s) + a_3(s)W(s) \right\rangle, \\
&= \frac{a_2}{f} \left\langle \tau T + \kappa B, \frac{-\kappa T + \tau B}{\sqrt{\kappa^2 + \tau^2}} \right\rangle + a_3, \\
&= a_3.
\end{aligned}$$

If we substitute this obtained equation in the alternative Tzitzeica curve condition, we can see that

$$\begin{aligned}
\frac{g}{(d_{NC})^2} &= \frac{H'}{(1 + H^2)\langle W, \alpha \rangle^2} \\
&= \frac{H'}{(1 + H^2) a_3^2}
\end{aligned}$$

or

$$\frac{g}{(d_{NC})^2} = \frac{g}{a_3^2}.$$

4 Tzitzeica Curve Conditions for the Indicatrix of Tangent, Principal Normal and Binormal

In this section, we give the condition that the tangents indicatrix, principal normal indicatrix and binormal indicatrix of the curve α to be Tzitzeica curve in Euclidean 3-space.

Theorem 7 *Let $\alpha \colon I \subset \mathbb{R} \to E^3$ be a unit speed curve and α_T be the tangent indicatrix of the curve α in Euclidean 3-space. The condition of the Tzitzeica curve that the tangent indicatrix α_T of the curve α are as follows with respect to the Frenet frame and the alternative moving frame, respectively,*

$$\frac{\sigma\left(1+H^2\right)^{\frac{3}{2}}}{H^2}=\text{constant},$$

and

$$\frac{1+H^2}{H^2}\left(\frac{g}{f}\right)'=\text{constant}.$$

Proof Let α_T be the tangent indicatrix of the curve α. For the curve α_T, we know that

$$\tau_T=\frac{g}{\kappa},\ B_T=\frac{\kappa}{f}(HT+B)\text{ and }\alpha_T=T.$$

The necessary and sufficient condition for the curve α_T to be the Tzitzeica curve is as follows

$$\frac{\tau_T}{\left(d_{osc_T}\right)^2}=\frac{\tau_T}{\langle B_T,\alpha_T\rangle^2}=\text{constant}.$$

If necessary arrangements are made, we have

$$\begin{aligned}\frac{\tau_T}{\left(d_{osc_T}\right)^2}&=\frac{\frac{g}{\kappa}}{\left\langle\frac{\kappa}{f}(HT+B),T\right\rangle^2},\\&=\frac{\sigma\left(1+H^2\right)^{\frac{3}{2}}}{H^2}.\end{aligned}$$

Considering the alternative Tzitzeica condition for α_T according to alternative moving frame, we get

$$\frac{g_T}{(d_{NC})_T^2}=\text{constant}.$$

Using the following equalities,

$$\begin{aligned}\sigma_T&=\frac{\kappa_T^2}{\left(\kappa_T^2+\tau_T^2\right)^{\frac{3}{2}}}\left(\frac{\tau_T}{\kappa_T}\right)',\\&=\frac{f^2}{\kappa^2}\frac{\kappa^3}{\left(f^2+g^2\right)^{\frac{3}{2}}}\left(\frac{g}{f}\right)',\end{aligned}$$

and

$$f_T=\kappa_T\sqrt{1+H_T^2},$$

$$= \frac{\sqrt{f^2 + g^2}}{\kappa},$$

we can write the second curvatures of the tangent indicatrix of the curve α as

$$\begin{aligned} g_T &= \sigma_T f_T \\ &= \frac{f^2}{f^2 + g^2}\left(\frac{g}{f}\right)'. \end{aligned}$$

On the other hand, for the vector

$$W_T = \frac{\kappa}{\sqrt{f^2 + g^2}}\left(\frac{g}{\kappa}N + HT + B\right),$$

we obtain $(d_{NC})_T$ as

$$(d_{NC})_T = \langle W_T, \alpha_T \rangle = \frac{\kappa}{\sqrt{f^2 + g^2}}\left\langle \frac{g}{\kappa}N + HT + B, T \right\rangle.$$

Hence, we can write the alternative Tzitzeica conditon as

$$\frac{g_T}{(d_{NC})_T^2} = \frac{\kappa^2(1 + H^2)}{\kappa^2 H^2}$$

and

$$\frac{1 + H^2}{H^2}\left(\frac{g}{f}\right)' = \text{constant}.$$

Considering this theorem, we can give the following result.

Corollary 1 *The curve α is slant helix if and only if there is not exist an alternative Tzitzeica curve of tangent indicatrix curve of α.*

Theorem 8 *Let $\alpha \colon I \subset \mathrm{R} \to E^3$ be a unit speed curve and α_N be the principal normal indicatrix of the curve α in Euclidean 3-space. The condition of the Tzitzeica curve that the principal normal indicatrix α_N of the curve α are as follows with respect to the Frenet frame and the alternative moving frame, respectively,*

$$\frac{\sigma'}{f\sigma^2} = \text{constant},$$

and

$$\frac{\Gamma'(1 + \sigma^2)}{\sigma^2} = \text{constant}, \quad \text{where } \Gamma = \frac{\sigma'}{f(1 + \sigma^2)^{\frac{3}{2}}}.$$

Proof Let α_N be the principal normal indicatrix of the curve α. For the curve α_N,

$$\tau_N = \Gamma\sqrt{1+\sigma^2}, \text{ where } \Gamma = \frac{\sigma'}{f\left(1+\sigma^2\right)^{\frac{3}{2}}}.$$

The necessary and sufficient condition for the curve α_N to be the Tzitzeica curve is as follows

$$\frac{\tau_N}{\left(d_{osc_N}\right)^2} = \frac{\tau_N}{\langle B_N, \alpha_N\rangle^2} = \text{constant}.$$

Since

$$B_N = \frac{1}{\sqrt{1+\sigma^2}}\left(\frac{\kappa}{f}(HT+B)+\sigma N\right) \text{ and } \alpha_N = N,$$

we can say that

$$\langle B_N, \alpha_N\rangle^2 = \frac{\sigma^2}{1+\sigma^2}.$$

So, we get the condition as

$$\frac{\tau_N}{\langle B_N, \alpha_N\rangle^2} = \frac{\sigma'}{f\left(1+\sigma^2\right)}\frac{\left(1+\sigma^2\right)}{\sigma^2}$$

and

$$\frac{\sigma'}{f\sigma^2} = \text{constant}.$$

On the other hand, alternative Tzitzeica condition of α_N is

$$\frac{g_N}{(d_{NC})^2_N} = \text{constant}.$$

Let's get these equations data one by one.

$$\begin{aligned}\sigma_N &= \frac{\kappa_N^2}{\left(\kappa_N^2+\tau_N^2\right)^{\frac{3}{2}}}\left(\frac{\tau_N}{\kappa_N}\right)', \\ &= \frac{\Gamma'}{\sqrt{1+\sigma^2}\left(1+\Gamma^2\right)^{\frac{3}{2}}},\end{aligned}$$

and

$$f_N = \kappa_N\sqrt{1+H_N^2},$$
$$= \sqrt{1+\sigma^2}\sqrt{1+\Gamma^2}.$$

So, we obtain the second curvature of the curve α_N

$$g_N = \sigma_N f_N,$$
$$= \frac{\Gamma'}{1+\Gamma^2}.$$

If the expression in the denominator of the Tzitzeica condition is calculated at this time, we obtain

$$W_N = \frac{1}{\sqrt{1+\sigma^2}\sqrt{1+\Gamma^2}}\left(\frac{\kappa\Gamma\sqrt{1+\sigma^2}}{f}(-T+HB) + \left(\frac{\kappa}{f}(HT+B)+\sigma N\right)\right),$$

and

$$(d_{NC})_N = \langle W_N, \alpha_N\rangle = \frac{1}{\sqrt{1+\sigma^2}\sqrt{1+\Gamma^2}}\sigma.$$

Hence, we can write as

$$\frac{g_N}{(d_{NC})_N^2} = \frac{\Gamma'}{1+\Gamma^2}\frac{\left(1+\sigma^2\right)\left(1+\Gamma^2\right)}{\sigma^2}$$

and

$$\frac{\Gamma'\left(1+\sigma^2\right)}{\sigma^2} = \text{constant}.$$

Theorem 9 *Let α: $I \subset \mathbb{R} \to E^3$ be a unit speed curve and α_B be the binormal indicatrix of the curve α in Euclidean 3-space. The condition of the Tzitzeica curve that the binormal indicatrix α_B of the curve α are as follows with respect to the Frenet frame and the alternative moving frame, respectively,*

$$-\frac{\sigma\left(1+H^2\right)^{\frac{3}{2}}}{H} = \text{constant},$$

and

$$\left(1+H^2\right)\left(-\frac{g}{f}\right)' = \text{constant}.$$

Proof Let α_B be the binormal indicatrix of the curve α. For the curve α_B,

$$\tau_B = -\frac{g}{\tau},$$

$$B_B = \frac{\kappa}{f}(HT + B) \text{ and } \alpha_B = B.$$

$$\frac{\tau_B}{(d_{osc})_B^2} = \frac{\tau_B}{\langle B_B, \alpha_B\rangle^2}$$

$$\frac{\tau_B}{\langle B_B, \alpha_B\rangle^2} = -\frac{\sigma f\left(1 + H^2\right)}{\tau}.$$

If necessary algebraic operations are done, we obtain

$$\frac{\tau_B}{\langle B_B, \alpha_B\rangle^2} = -\frac{\sigma\kappa\sqrt{1 + H^2}\left(1 + H^2\right)}{\tau}$$

and

$$-\frac{\sigma\left(1 + H^2\right)^{\frac{3}{2}}}{H} = \text{constant}.$$

For the alternative Tzitzeica condition of the curve α_B, we have the following

$$\sigma_B = \frac{f^2\tau}{\left(f^2 + g^2\right)^{\frac{3}{2}}}\left(-\frac{g}{f}\right)',$$

and

$$f_B = \frac{\sqrt{f^2 + g^2}}{\tau}.$$

So, we can easily write the curvature as

$$g_B = \frac{f^2}{f^2 + g^2}\left(-\frac{g}{f}\right)'.$$

Also, we can calculate the following as

and

$$(d_{NC})_B = \langle W_B, \alpha_B\rangle = \frac{\tau}{H\sqrt{f^2 + g^2}}.$$

If the found data is substituted in the condition of Tzitzeica, we obtain

$$\frac{g_B}{(d_{NC})_B^2} = \frac{\kappa^2(1+H^2)H^2}{\tau^2}\left(-\frac{g}{f}\right)'$$

and

$$\left(1+H^2\right)\left(-\frac{g}{f}\right)' = \text{constant}$$

References

1. Tzitzeica, G.: Sur certaines courbes gauches. Ann. Sci. Ecole Norm. Sup. **28**, 9–32 (1911)
2. Bila, N.V., Eni, M.: Particular solutions to the Tzitzeica curve equation. Math Comput. Sci. Fac. Work. Pap. **15**, 1–10 (2012)
3. Aydin, M.E., Ergüt, M.: Non-null curves of Tzitzeica type in Minkowski 3-space. Rom. J. Math. Comput. Sci. **4**, 81–90 (2014)
4. Hathout, F., Yayli, Y.: Darboux vectors and the spherical indicatrix curves satisfying the Tzitzeica condition in Minkowski space. Rom. J. Math. Comput. Sci. **8**(1), 7–16 (2018)
5. Karacan, M.K., Bükcü, B.: On the hyperbolic cylindrical Tzitzeica curves in Minkowski 3-space. Balkesir Üniv. Fen Bilim. Enst. Derg. **10**(1), 46–51 (2008)
6. Crasmareanu, M.: Cylindrical Tzitzeica curves implies forced harmonic oscillators. Balkan J. Geom. Appl. **7**(1), 37–42 (2002)
7. Karacan, M.K., Bükcü, B.: On the elliptic cylindrical Tzitzeica curves in Minkowski 3-space. Sci. Manga **5**, 44–48 (2009)
8. Constantinescu, O., Crasmareanu, M.: A new Tzitzeica hypersurface and cubic Finslerian metrics of Berwald type. Balkan J. Geom. Appl. **16**(2), 27–34 (2011)
9. Bobe, A., Boskoff, W.G., Ciuca, M.G.: Tzitzeica type centro-affine invariants in Minkowski spaces. An. St. Univ. Ovidius Constanta **20**(2), 27–34 (2012)
10. Agnew, A.F., Bobe, A., Boskoff, W.G., Suceava, B.D.: Tzitzeica curves and surfaces. Math. J. **12**, 1–18 (2010)
11. Bayram, B., Tunç, E., Arslan, K., Öztürk, G.: On Tzitzeica curves in Euclidean 3-space E^3. Facta Univ. Ser. Math. Inform. **33**(3), 409–416 (2018)
12. Struik, D.J.: Lectures on Classical Differential Geometry. Addison-Wesley Publishing Company, Inc., Massachusetts (1961)
13. Hacisalihoğlu, H.H.: Diferensiyel Geometri. Inönü Üniversitesi Fen Edebiyat Fakültesi Yaynlari (1983)
14. Sabuncuoğlu, A.: Diferensiyel Geometri. Nobel Basmevi, Ankara (2004)
15. Do Carmo, M.P.: Differential Geometry of Curves and Surfaces, p. 503. Prentice-Hall, Englewood Cliffs, New Jersey (1976)
16. Gluck, H.: Higher curvatures of curves in Euclidean space. Am. Math. Mon. **73**, 699–704 (1966)
17. Izumiya, S., Takeuchi, N.: New special curves and developable surfaces. Turk. J. Math. **28**, 153–163 (2004)
18. Uzunoğlu, B., Gök, İ, Yayli, Y.: A new approach on curves of constant precession. Appl. Math. Comput. **275**, 317–323 (2016)
19. Ali, A.T.: New special curves and their spherical indicatrices. arXiv preprint arXiv:0909.2390 (2009)
20. Özdamar, E., Hacisalihoğlu, H.H.: A characterization of inclined curves in Euclidean *n*-space. Commun. Fac. Sci. Univ. Ankara **24**, 15–22 (1975)
21. Gray, A.: Modern Differential Geometry of Curves and Surface. CRS Press, Inc. (1993)

Thermal Efficacy of a Li-Ion Battery Pack: The Effects of Different BTMS Setups

Soner Birinci, Mehmet Saglam, Bugra Sarper, M. Yusuf Yazici, and Orhan Aydin

Abstract The aim of this work is to conduct a numerical analysis on a battery thermal management system (BTMS) that utilizes liquid cooling and is equipped with a cold plate. The system is specifically designed for a 2 $\times$ 6 format of a 18650 cylindrical Li-ion battery pack (BP). In order to analyze cooling performance of the BTMS, three distinct cold plate setups of the BTMS and varying volume flow rates (0, 0.1, 0.25, 0.5, 0.75, and 1 l/min) are utilized at a fixed discharge rate of 2C. The initial setup, denoted as Setup I, pertains to the implementation of symmetrical cooling. The subsequent setup, referred as Setup II, involves the utilization of asymmetrical cooling. Lastly, the third setup, designated as Setup III, encompasses both asymmetrical cooling and the incorporation of an aluminum block. Numerical analyses are performed with the ANSYS Fluent software. The study involves comparing of these setups in terms of the pressure drop throughout the cold plate, the greatest temperature difference between the cells, the hot spot temperature within the BP, the cooling efficacy as well as the energy density of the BP. The findings indicate that there is a negative correlation between flow rate and both hot spot temperature and the greatest temperature difference within the BP. Nevertheless, once the flow rate reaches 0.5 l/min, any subsequent increase in flow rate does not appear to significantly impact the cooling performance. Furthermore, it is observed that the pressure drop exhibits

S. Birinci · M. Saglam · O. Aydin
Department of Mechanical Engineering, Karadeniz Technical University, Trabzon, Türkiye
e-mail: sonerbirinci@ktu.edu.tr

M. Saglam
e-mail: mehmetsaglam@ktu.edu.tr

O. Aydin
e-mail: oaydin@ktu.edu.tr

B. Sarper (✉)
Department of Mechanical Engineering, Tarsus University, Mersin, Türkiye
e-mail: bugrasarper@tarsus.edu.tr

M. Yusuf Yazici
Department of Mechanical Engineering, Samsun University, Samsun, Türkiye
e-mail: myusuf.yazici@samsun.edu.tr

O. Castillo et al. (eds.), *Advanced Mathematics for the Modeling and Solution of Challenging Problems in Engineering*, Studies in Systems, Decision and Control 582,
https://doi.org/10.1007/978-3-031-82342-8_9

a positive correlation with the flow rate. Hence, it can be concluded that flow rates ranging from 0.25 to 0.5 l/min are seen to be the most favorable for these BTMS setups. When comparing the BTMS setups, it can be concluded that Setup III is the optimal setup, particularly when considering factors such as the hot spot temperature and the greatest temperature difference within the BP.

Keywords Li-ion batteries · Liquid-cooling · Thermal behavior · Computational fluid dynamics (CFD) · Battery thermal management systems (BTMS)

1 Introduction

In the present day, there exists a persistent increase in the requirement for energy resources, particularly fossil fuels such as oil, natural gas, and coal. These resources currently account for approximately 80% of the world energy demand [1]. According to statistical data from the BP World Energy Review [2], it is evident that there has been a notable rise in worldwide oil consumption, with an increase of 0.9 million barrels per day. Notably, the transportation sector accounts for about two-thirds of this overall increase [3]. Hence, the utilization of internal combustion engines powered by fossil fuels leads to an increase in environmental pollutants and the release of greenhouse gas emissions.

There has been a growing emphasis on the advancement of environmentally sustainable transportation options, particularly in relation to pollutants and greenhouse gas emissions. This has led to increased attention towards the development of pure electric vehicles, hybrid electric vehicles, and plug-in electric vehicles (EVs), which are considered to be more efficient compared to traditional internal combustion engine systems [4]. The European Commission's study titled "Sustainable and Smart Transport Strategy" outlines the objective of achieving a 90% reduction in greenhouse gas emissions associated with transportation sectors in Europe by the year 2050 [5]. The primary objective for this purpose is that at least 30 million vehicles will be comprised of zero-emission vehicles by the year 2030, with the aim of achieving complete adoption of zero-emission vehicles for all cars, minibuses, buses, and heavy-duty vehicles in Europe by the year 2050.

The main challenge in the development of electric vehicles is the design of an appropriate energy storage system that supports long range, fast charging, reliability, and high performance. With their high energy density, fast charging, lightness, low self-discharge rates, higher recyclability, long cycle life and low memory effect, Li-ion batteries (LIBs) have an important potential for electric vehicles.

Temperature is one of the most important parameters affecting the performance, cycle life and safety of LIBs. LIBs release heat due to environmental conditions and chemical reactions during charging/discharging, which causes an increase in temperature between batteries. Besides the absolute temperature of a BP, the non-uniform temperature distribution inside of the BP also needs to be considered. The presence of a non-uniform temperature distribution among the Li-ion cells gives rise to different charge and discharge characteristics for each module. Consequently, this leads to electrical instability in the modules/packages and a subsequent decline in performance. During the process of charging or discharging, if the temperature within the BP increases and the system is unable to dissipate the heat, a condition known as thermal runaway occurs when the temperature exceeds 80 °C [6]. The occurrence of thermal runaway in BPs can lead to the release of toxic gases, the production of smoke, the initiation of fires, and even the possibility of explosions [7, 8]. Furthermore, it has been shown that an increase in temperature within the range of 30–40 °C is associated with a reduction in the lifespan of Li-ion cells by a duration of two months [9]. Li-ion battery performance and reliability are significantly reduced except within the recommended optimal operating temperature range. In order to mitigate temperature-related problems in LIBs, it is recommended to maintain an operational temperature range from 15 to 35 °C, and the maximum temperature difference (ΔT_{max}) between the cells should not exceed 5 °C [10]. Hence, effective operation of LIBs necessitates an effective BTMS.

The BTMSs may be categorized into several cooling techniques including, air cooling [11–16], liquid cooling [17–23], cooling with phase change materials (PCMs) [24–28] and hybrid cooling [28–38]. Air-cooled BTMSs have several benefits in terms of their structural design, lightweight nature, safety features, application, and accessibility. On the other hand, phase change materials (PCMs) give an advantage by effectively storing energy in the form of latent heat throughout the process of phase transition. Liquid cooled BTMSs are often utilized and are increasingly becoming the prevailing choice in comparison to alternative cooling techniques. This is primarily attributed to their notable versatility in adapting to various environmental and operational conditions, as well as their exceptional performance and high heat capacity [39]. There are numerous studies in the literature concerning the liquid cooling of Li-ion battery modules. Some of these studies have focused on cooling of cylindrical LIBs employing cold plates. The heat transfer rate in liquid-cooled BTMSs is restricted by the geometric constraints of cylindrical LIBs, making it challenging to maximize the surface area for heat transfer between the cold plates and the batteries. However, it is seen that the studies on the liquid cooling of cylindrical LIBs employing cold plates are limited [40]. Qi et al. [41] performed numerical simulations and experimental analyses, to ascertain the optimal structural characteristics of the Swiss roll BTMS cooling belt for EVs. This study examined the impact of coolant type, flow rate, and inlet temperature on the heat transfer performance. Wiriyasart et al. [42] proposed the utilization of a mini-channel cold plate-based BTMS with the incorporation of nanofluids for cooling a cylindrical Li-ion BP. This study examined the impact of coolant flow direction, mass flow rate, and coolant types on cooling performance. In their study, Liao et al. [43] conducted a numerical investigation on the thermal performance of a Li-ion BP. The cooling plate utilized in their research was a wavy flat tube, and they examined the impact of different factors on the cooling process. It was determined that the utilization of nanofluids as a coolant exhibits superior efficacy compared to water in terms of cooling capabilities. Xie et al. [44] introduced a micro-channel based BTMS with a baffle design. This study aimed to examine the impact of height and number of baffles on heat transfer performance. In their work, Zhao et al. [45] conducted a numerical investigation to examine the thermal characteristics of cylindrical LIBs cooled via a cooling channel. This study investigated the impact of many factors, including C-rate, heat transfer surface area between batteries, liquid flow rate, and interface area of the battery and channel outside wall, on the overall performance of the cooling system. Liquid based BTMS with wavy tubes was proposed by Guo et al. [46]. The effects of separated dual tube structure, the coolant mass flow rate, series and parallel cooling modes, and flow direction were considered. Ke et al. [47] conducted an experimental investigation with the objective of mitigating the occurrence of thermal runaway in a Li-ion BP. This was achieved by the implementation of a liquid-cooling system utilizing serpentine channels. According to their statement, the suggested BTMS effectively mitigated the propagation of thermal runaway.

This study investigates a liquid-cooled active BTMS with a cold plate for cooling of cylindrical Li-ion BP configured in a 2×6 format. The focus is given on three distinct setups of the BTMS, along with varying volume flow rates ranging from 0 to 1 l/min, while maintaining a discharge rate of 2C. The initial setup, denoted as Setup I, pertains to the implementation of symmetrical cooling. The subsequent setup, referred to as Setup II, involves the utilization of asymmetrical cooling. Lastly, the third setup, designated as Setup III, encompasses both asymmetrical cooling and utilization of aluminum block. The numerical investigations are carried out using ANSYS Fluent software to analyze several aspects of the BP. Specifically, the study focuses on comparing of these setups in terms of the pressure drop throughout the cold plate, the greatest temperature difference between the cells, the hot spot temperature within the BP, the cooling efficacy as well as the energy density of the BP.

2 Numerical Study

The thermal characteristics of a cylindrical Li-ion BP, namely the 18650 model, placed in a 2×6 format, are examined using numerical simulations conducted with ANSYS Fluent software [48]. A commercial Panasonic NCR18650GB Li-on battery with a nominal voltage of 4.2 V and a capacity of 3.5 Ah is used in this study, and thermophysical properties of battery are given in Table 1 [49].

In this study, a concave-shaped unique cold plate is employed for the purpose of cooling the BP. In all setups, the height (h), width (w) and wall thickness (t) of

Table 1 The thermophysical properties of the aluminum, Li-ion battery, and water

Material	ρ (kg m^{-3})	c_p (J kg^{-1} K^{-1})	k (W m^{-1} K^{-1})	μ (kg m^{-1} s^{-1})
Liquid water	996.58	4072.7	0.61229	0.0008684
Battery	2939	2400	$k_{axial} = 28.05$ $k_{radial} = 3$	–
Aluminum	2719	871	202.4	–

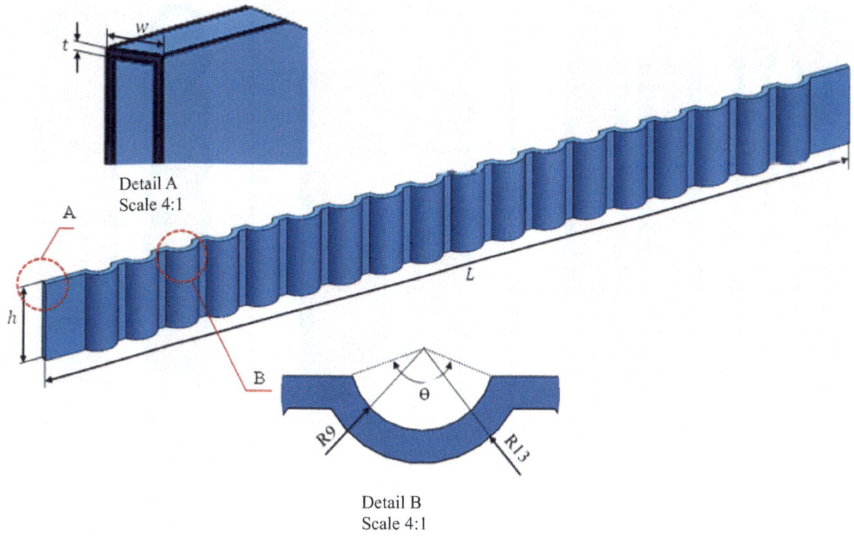

Fig. 1 Schematic view of cold plate

the cold plate are 30 mm, 3 mm and 0.5 mm, respectively, while the contact angle (θ) between the cold plate and the batteries is 120°. The heat generated by LIBs is delivered to the coolant by conduction inside the plate and convection within the coolant. The schematic depiction of the cold plate is presented in Fig. 1, providing a full overview. Aluminum is used as the material for the cold plate, while water is chosen as the coolant. Table 1 also presents the thermophysical characteristics of aluminum and water. As previously stated, this study examines three distinct BTMS setups, which are seen in Fig. 2.

Simulations are performed at different volume flow rates (0, 0.1, 0.25, 0.5, 0.75, 1 l/min), resulting in laminar flow. The temperature at the inlet is equal to the ambient temperature and equals to 25 °C. In this study, a fixed discharge rate of 2C is chosen to determine the optimal setups of different BTMSs. Three-dimensional numerical simulations are performed in transient regime because of transient nature of heat generation of the LIBs, and time step is chosen as 1s in all simulation.

The equations for continuity, momentum, energy, and energy transport for the battery are given as in Eqs. (1)–(4) [49]. The equation for the heat production source term in Eq. (4), as established by Bernardi et al. [50], is presented in Eq. (5).

$$\nabla \cdot V_f = 0 \tag{1}$$

$$\rho \frac{DV_f}{Dt} = \rho_f g - \nabla \cdot P_f + \mu_f \left(\nabla^2 V_f\right) \tag{2}$$

$$\rho_f c_{p,f} \frac{\partial T_f}{\partial t} + \rho_f c_{p,f} \nabla \cdot \left(T_f V_f\right) = \nabla \cdot \left(k_f \nabla T_f\right) \tag{3}$$

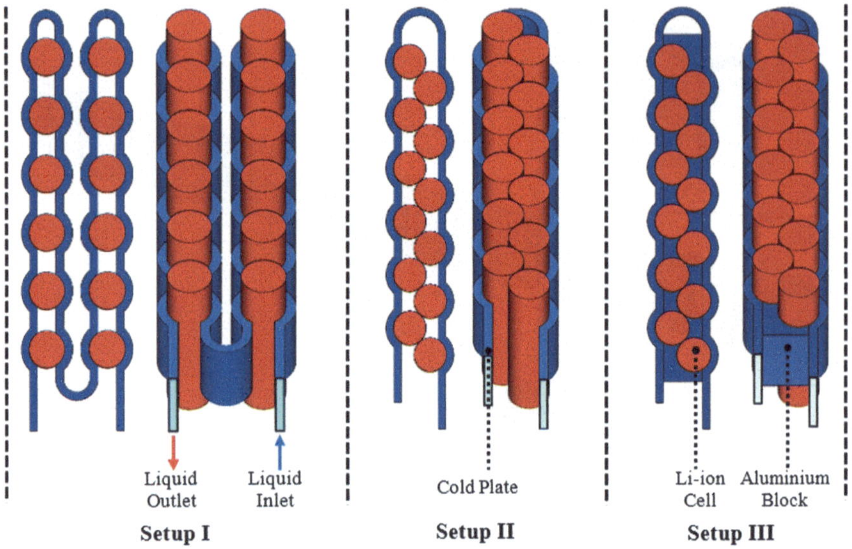

Fig. 2 Three different setups of the BTMS

$$\rho_b c_{p,b} \frac{\partial T_b}{\partial t} = k_b \nabla \cdot (\nabla T_b) + \dot{q} \tag{4}$$

$$\dot{q} = I\left[(OCV - V) - T_p \frac{dOCV}{dT}\right] \tag{5}$$

Here, the velocity vector is denoted as V, the density as ρ, the dynamic viscosity of the fluid as μ, gravitational acceleration as g, the pressure as P, the conductivity as k, the specific heat as c_p, and the temperature as T. The subscripts f and b are used to denote fluid and battery, respectively.

For pressure–velocity coupling and pressure interpolation, respectively, the SIMPLE algorithm and PRESTO scheme are utilized. In simulations, the 2nd order upwind scheme discretizes momentum and energy equations.

The pressure drop throughout the cold plate, the greatest temperature difference between the cells, the hot spot temperature within the BP, the cooling efficacy, and the energy density of the BP are obtained to determine the optimal setup of the BTMS. Highest temperature difference between the cells is calculated as follows:

$$\Delta T_{\max} = T_{\max} - T_{\min} \tag{6}$$

where $T_{\max}$ and $T_{\min}$ stand for the highest and lowest temperatures on the surface of the battery cell, respectively.

The power consumption required to provide water to the cold plate is as follows:

$$\dot{W} = \left(\Delta P \times \dot{Q}_{water}\right) \quad (7)$$

where the power consumption, pressure drop throughout the cold plate, and volume flow rate of the coolant are denoted as $\dot{W}$, ΔP, and $\dot{Q}_{water}$, respectively.

The following describes the heat removed from the battery back by water:

$$\dot{Q} = \dot{m} \times c_p \times \Delta T \quad (8)$$

The cooling efficacy [51] is as follows:

$$\beta = \frac{\dot{Q}}{\dot{W}} \quad (9)$$

This equation demonstrates that a greater β value is associated with increased energy efficacy.

The energy density of the BP is as follows:

$$\rho_{energy} = \frac{Total\ nominal\ battery\ voltage\ (V) \times Total\ battery\ capacity\ (Ah)}{Volume\ of\ battery\ unit\ \left(m^3\right)} \quad (10)$$

At channel entrance and exit, velocity inlet and pressure outlet boundary conditions are applied, respectively. A user defined function (UDF) is utilized to express the heat generation rate of Li-ion battery utilizing the Bernardi's heat generation equation. The energy balance is applied at the contact surfaces of the battery and cold plate, and velocity of the coolant is zero on the channel walls. The convective boundary condition is implemented at the interfaces of the air-battery and air-cold plates, with a heat transfer coefficient of 5 W/m^2 K.

The simulations employ a Polyhedral mesh structure, as seen in Fig. 3a. The grid independency study examines the cell numbers of 187,767, 792,318, 1,699,150, 2,392,600, 2,741,182, and 2,967,520. In the final simulations, a grid size of 2,392,600 is chosen because the difference between the battery's surface temperature and the pressure drop in the cold plate is less than 1% (see Fig. 3b).

To ensure the validity of the thermal model for LIBs, a single battery is employed, and a series of experimental tests are performed based on the Bernardi's equation [50]. Subsequently, the volumetric heat generation profiles obtained at various discharge rates are included into the energy equation by the utilization of a UDF. Following this, numerical validation tests are conducted as stated in Ref. [49]. To determine the accuracy of the UDF code, the surface temperatures of the battery at discharge rates of 0.5C, 1C, and 2C are compared to experimental data. The experimental and numerical results are in particularly acceptable agreement as shown in Fig. 4.

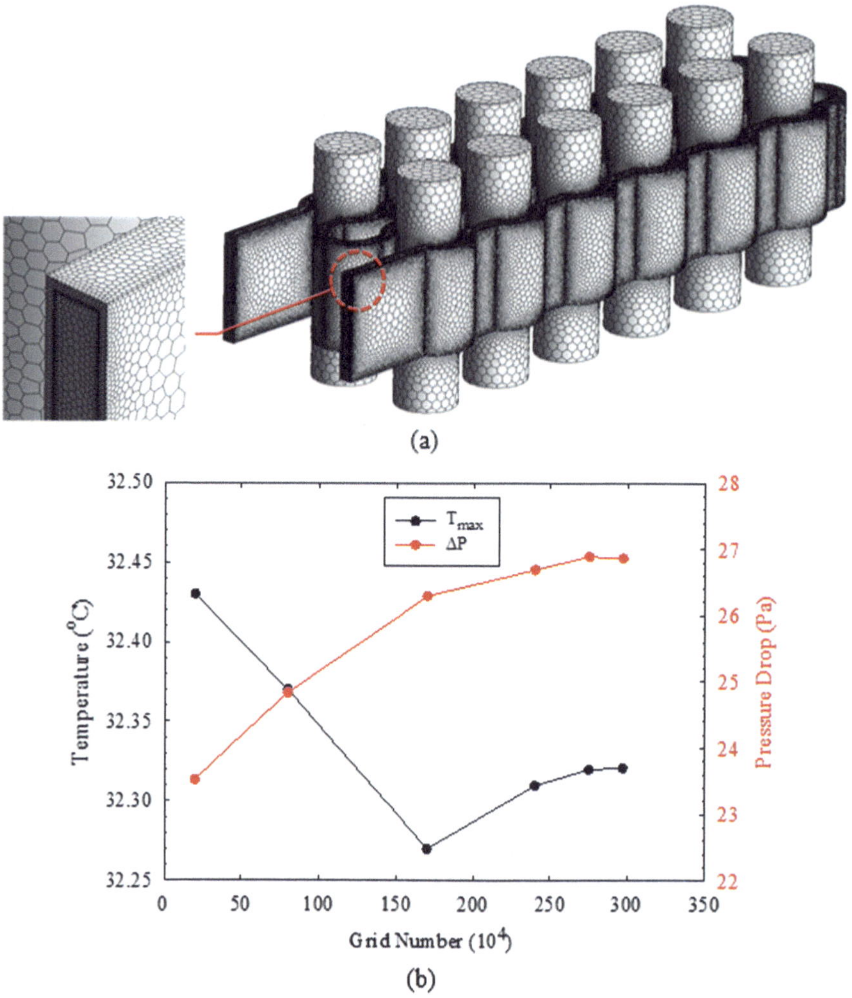

Fig. 3 **a** Grid structure used in the simulations, **b** results of the grid independency test

3 Results

This research focuses on liquid-cooled active BTMS with a cold plate for cooling of cylindrical Li-ion BP configured in 2×6 format. At fixed discharge rate of 2C, the effects of cold plate setups and various volume flow rates are investigated. The optimal BTMS is determined by comparing the pressure drop throughout the cold plate, greatest temperature difference between the cells, the hot spot temperature within the BP, cooling efficacy, and energy density of the BP for various setups.

Variation of the hot spot temperature and the greatest temperature difference within the BP with volume flow rate are depicted in Fig. 5 for different BTMS

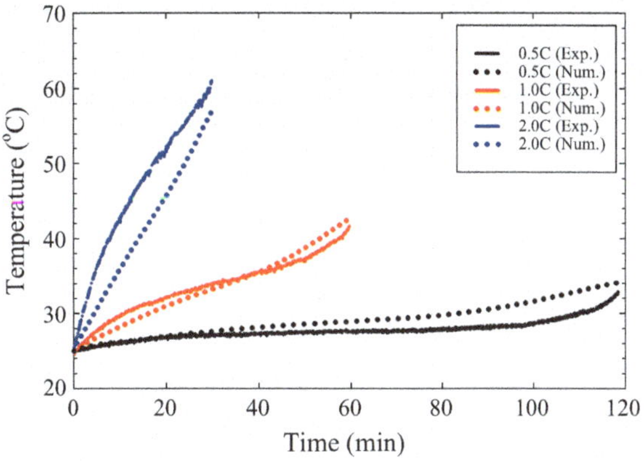

Fig. 4 Validation results against experimental and numerical studies

setups. As shown in Fig. 5a, the hot spot temperature within the BP decreases as the flow rate increases for all BTMS setups. In almost all cases, the hot spot temperature decreased by 20 °C when compared to the condition without cooling. After 0.5 l/min, however, increasing the flow rate has no discernible effect on cooling efficacy. Examining the variation of hot spot temperature with flow rate reveals that the greatest temperatures of Setups I and III are very close to each other, and at all flow rates, the hot spot temperatures for Setups I and III are lower than those of Setup II. Evaluating the results obtained from Fig. 5a, it can be seen that the hot spot temperature of the BP remains below the recommended maximum operating temperature at almost all flow rates. When the greatest temperature difference between the cells is analyzed, it is found that the greatest temperature difference reduces as the flow rate increases in all BTMS setups. However, as stated above, beyond 0.5 l/min does not lead to significant changes in cooling performance. In Setup III the greatest temperature difference is less than the acceptable maximum temperature difference (5 °C) at all flow rates, whereas in Setup II, the greatest temperature difference exceeds 5 °C at all flow rates. When the temperature difference between the cells exceeds 5 °C, the resistance and capacity differences between cells grow dramatically, reducing Setup II's operational efficacy. An examination of the three BTMS setups shows that Setup III is the optimal setup because it provides a more uniform temperature distribution.

Figure 6 illustrates the temperature contours of the BP for various flow rates. As stated previously, it is evident that Setup III provides more uniform temperature distribution. When the temperature contours are examined in detail, the hot spot temperature of the BP is observed in the last cell for all BTMS setups. This is the result of the water entering the cold plate gradually heating up as it sweeps up the cells, and cooling performance decreases in the flow direction.

The cold plate pressure drop and cooling efficacy are shown in Fig. 7. As can be seen in Fig. 7a, the cold plate pressure drop increases as the flow rate increases. However, the pressure drop values for Setups II and III are lower than that of the Setup I due to the one-sided cooling. This results in a decrease in pumping power. When considering the cooling efficacy, the cooling efficacy decreases with increasing flow

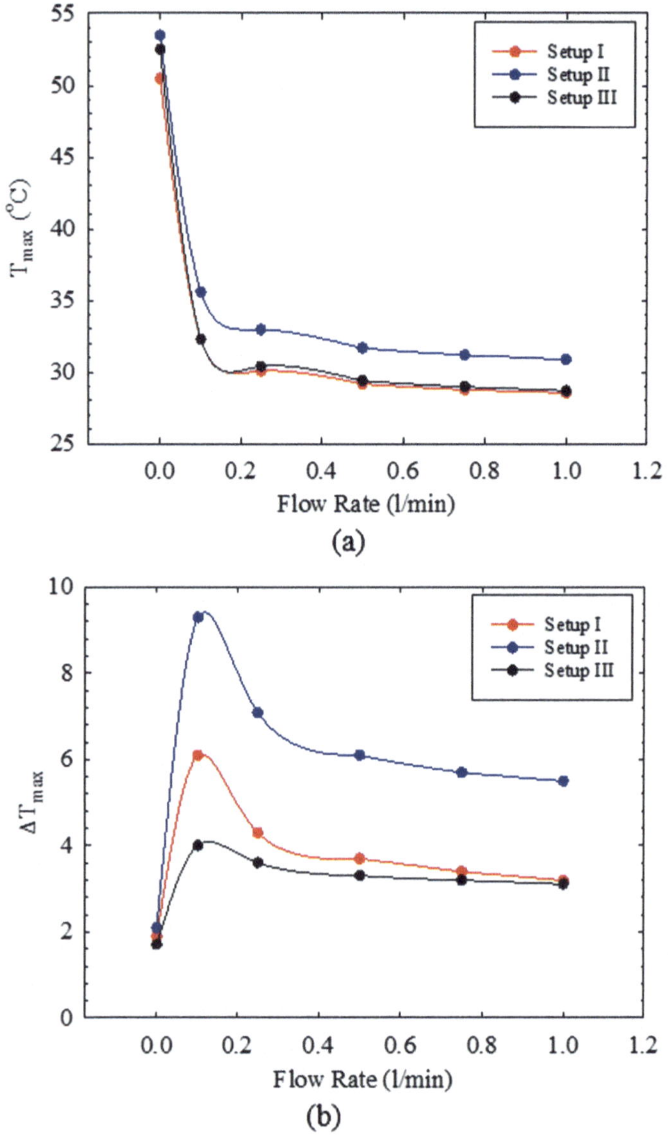

Fig. 5 **a** The hot spot temperature of BP, **b** the greatest temperature difference between the cells

rate in all BTMS setups as depicted in Fig. 7b. In addition, beyond the flow rate of 0.5 l/m, the influence of flow rate on cooling efficacy decreases significantly (Fig. 7). Furthermore, the pressure contours at various flow rates within the cold plate are illustrated in Fig. 8

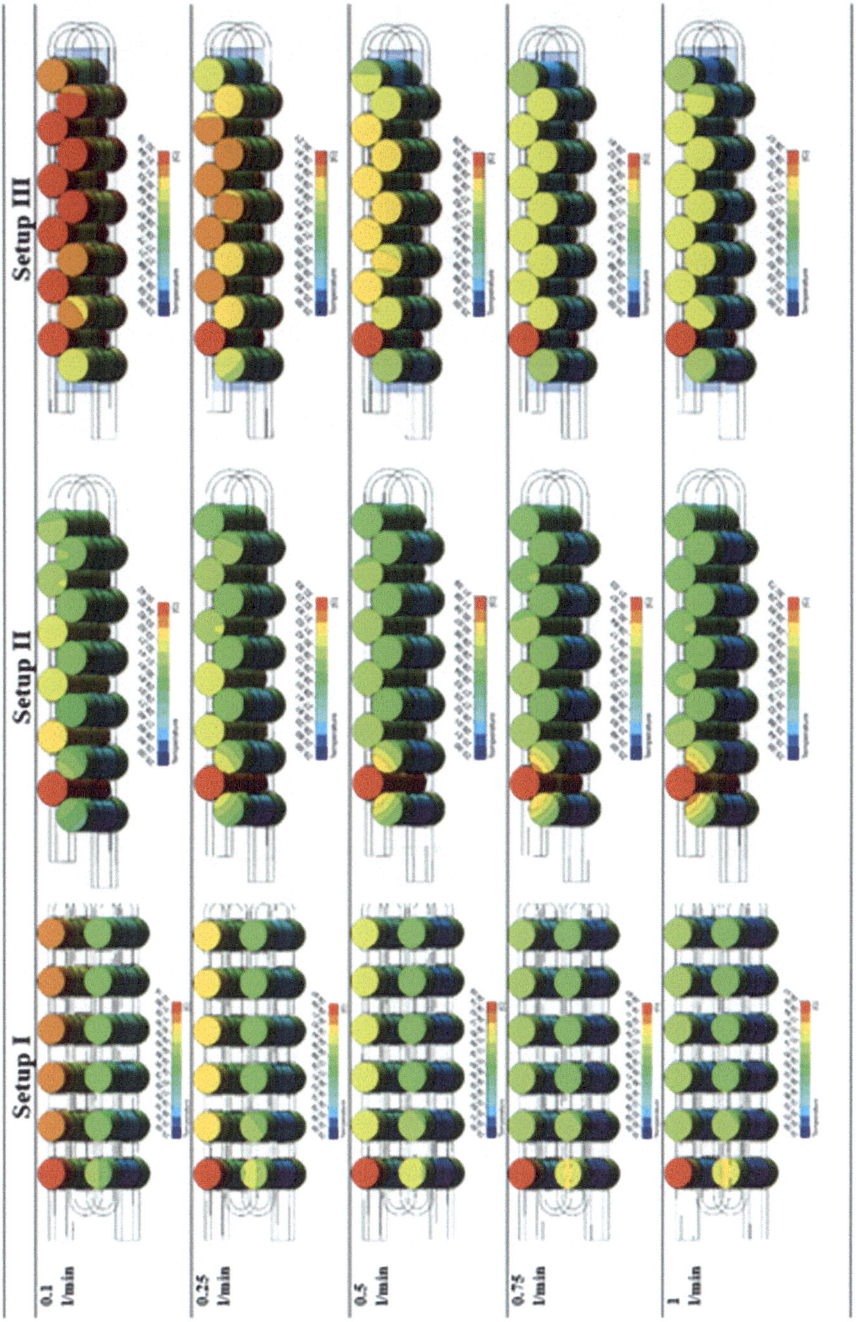

Fig. 6 Temperature contours of the BP at different flow rates

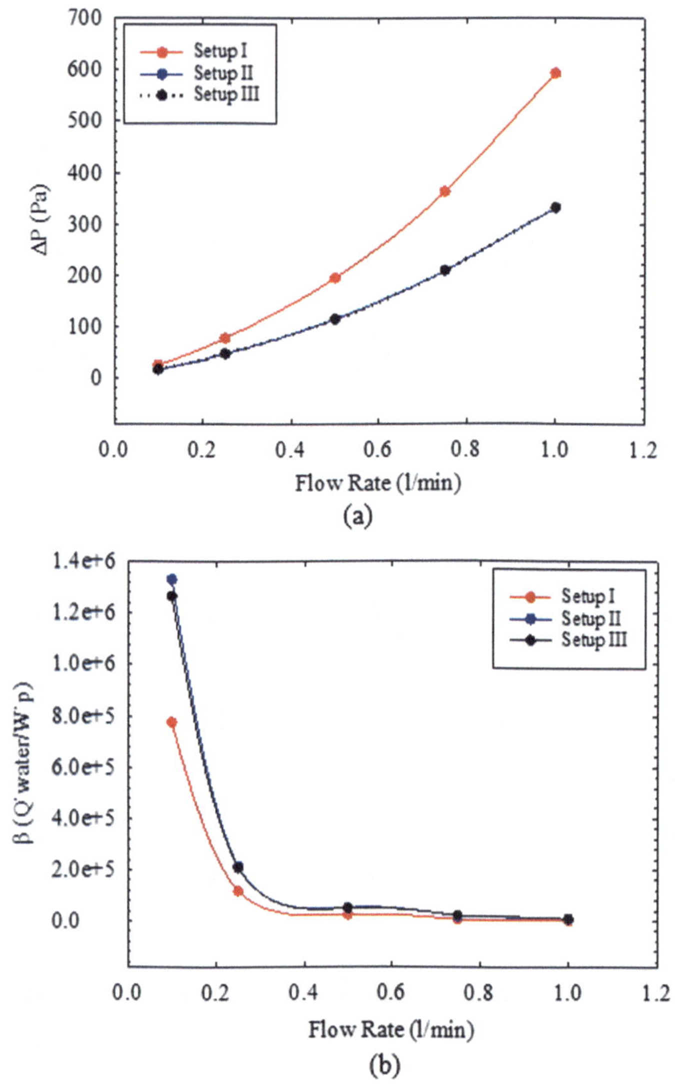

Fig. 7 **a** Pressure drop and **b** cooling efficacy of all setups

Figure 9 depicts the energy density of the BP for different BTMS setups. It can be seen that the Setup II has a greater energy density than those in Setup I and Setup III. This is because the volume of Setup II is less than in the Setups I and III.

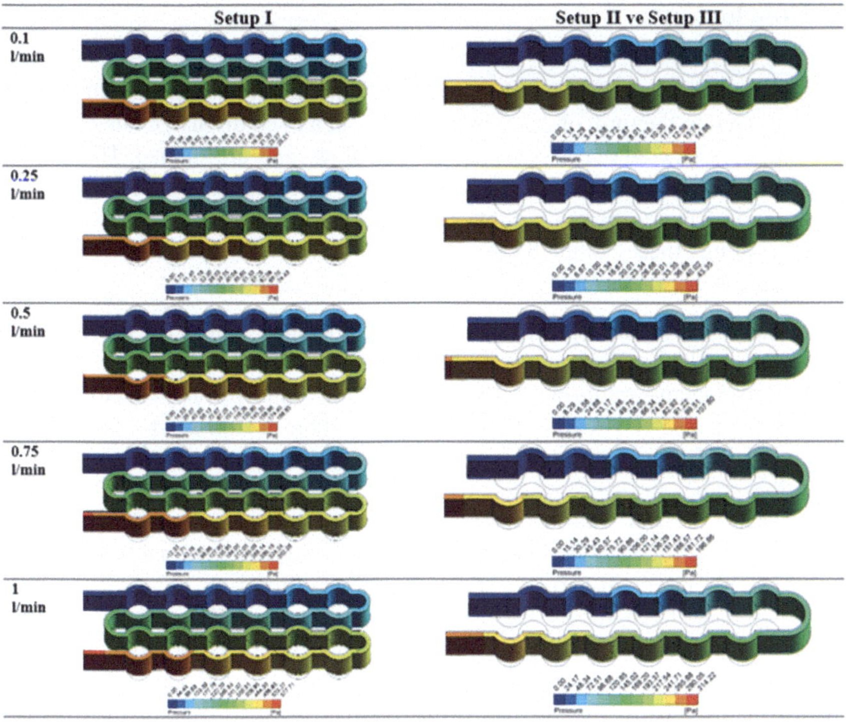

Fig. 8 Pressure contours within the cold plate at different flow rates

Fig. 9 The energy density of the BP

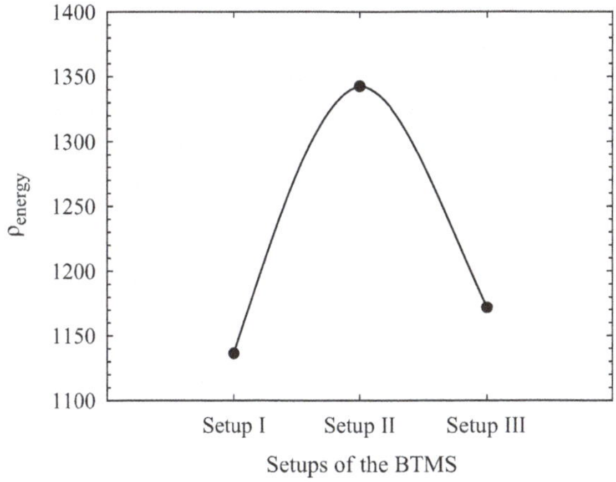

4 Conclusions

A liquid-cooled active BTMS with a cold plate is numerically analyzed for a Li-ion BP configured in the form of 2 × 6 in this study. At a fixed discharge rate of 2C, the effects of various cold plate setups and volume flow rates are studied. The optimal BTMS is determined by comparing the pressure drop throughout the cold plate, the greatest temperature difference between the cells, the hot spot temperature within the BP, cooling efficacy of the BTMS, and the energy density of the BP for multiple setups. According to the study, the following are the results:

- In almost all BTMS setups, the hot spot temperature of the BP decreased by approximately 20 °C when compared to the situation without cooling.
- The hot spot temperature of the BP is below the maximum recommended operating temperature (35 °C) at almost all flow rates in three BTMS setups, but only the Setup III is below the maximum recommended temperature difference (5 °C) at all flow rates.
- Increasing the flow rate decreases the hot spot temperature and the greatest temperature difference between the cells. Increasing the flow rate beyond 0.5 l/m, however, has no discernable effect on cooling efficacy.
- Among these three setups, the optimal flow rates are 0.25 and 0.5 l/m. When comparing three setups, Setup III is the optimal setup, especially considering the hot spot temperature within the BP and the greatest temperature difference between the cells.

References

1. Dinçer, İ., Hamut, H.S., Javani, N.: Thermal Management of Electric Vehicle Battery Systems. Wiley (2017)
2. British Petroleum: Statistical Review of World Energy (2020). https://www.bp.com/content/dam/bp/business-sites/en/global/corporate/pdfs/energy-economics/statistical-review/bp-stats-review-2020-full-report.pdf
3. Liu, H., Wei, Z., He, W., Zhao, J.: Thermal issues about Li-ion batteries and recent progress in battery thermal management systems: a review. Energy Convers. Manage. **150**, 304–330 (2017)
4. Wada, M.: Research and development of electric vehicles for clean transportation. J. Environ. Sci. **21**(6), 745–749 (2009)
5. European Commission, Communication from The Commission to The European Parliament, The Council, The European Economic and Social Committee and The Committee of the Regions. https://ec.europa.eu/transport/sites/transport/files/legislation/com20200789.pdf
6. Ling, Z., Zhang, Z., Shi, G., Fang, X., Wang, L., Gao, X., Fang, Y., Xu, T., Wang, S., Liu, X.: Review on thermal management systems using phase change materials for electronic components, Li-ion batteries and photovoltaic modules. Renew. Sustain. Energy Rev. **31**, 427–438 (2009)
7. Wang, Q., Ping, P., Zhao, X., Chu, G., Sun, J., Chen, C.: Thermal runaway caused fire and explosion of lithium ion battery. J. Power Sources **208**, 210–224 (2012)

8. Xu, J., Lan, C., Qiao, Y., Ma, Y.: Prevent thermal runaway of lithium-ion batteries with minichannel cooling. Appl. Therm. Eng. **110**, 883–890 (2017)
9. Motloch, C.G., Christophersen, J.P., Belt, J.R.: High-power battery testing procedures and analytical methodologies for HEV's. SAE Tech. Pap. Ser. 01-1950 (2002)
10. Pesaran, A., Nrel, G.K.: Addressing the impact of temperature extremes on large format Li-ion batteries for vehicle applications (presentation). In: 30th International Battery Seminar. NREL (National Renewable Energy Laboratory) (2013)
11. Lu, Z., Meng, X.Z., Wei, L.C., Hu, W.Y., Zhang, L.Y., Jin, L.W.: Thermal management of densely-packed EV battery with forced air cooling strategies. Energy Procedia **88**, 682–688 (2016)
12. Chen, K., Wang, S., Song, M., Chen, L.: Configuration optimization of battery pack in parallel air-cooled battery thermal management system using an optimization strategy. Appl. Therm. Eng. **123**, 177–186 (2017)
13. Jiaqiang, E., Yue, M., Chen, J., Zhu, H., Deng, Y., Zhu, Y.: Effects of the different air cooling strategies on cooling performance of a lithium-ion battery module with baffle. Appl. Therm. Eng. **144**, 231–241 (2018)
14. Jilte, R.D., Kumar, R., Ma, L.: Thermal performance of a novel confined flow Li-ion battery module. Appl. Therm. Eng. **146**, 1–11 (2019)
15. Sharma, A.R., Sai, C.S., Kumar, A., Jayasimha Reddy, R.V., Danyharsha, D., Jilte, R.: Three-dimensional CFD study on heat dissipation in cylindrical lithium-ion battery module. Mater. Today Proc. **46**, 10964–10968 (2021)
16. Saechan, P., Dhuchakallaya, I.: Numerical study on the air-cooled thermal management of lithium-ion battery pack for electrical vehicles. Energy Rep. **8**, 1264–1270 (2022)
17. Huo, Y., Rao, Z., Liu, X., Zhao, J.: Investigation of power battery thermal management by using mini-channel cold plate. Energy Convers. Manage. **89**, 387–395 (2015)
18. Basu, S., Hariharan, K.S., Kolake, S.M., Song, T., Sohn, D.K., Yeo, T.: Coupled electrochemical thermal modelling of a novel Li-ion battery pack thermal management system. Appl. Energy **181**, 1–13 (2016)
19. Du, X., Qian, Z., Chen, Z., Rao, Z.: Experimental investigation on mini-channel cooling-based thermal management for Li-ion battery module under different cooling schemes. Int. J. Energy Res. **42**(8), 2781–2788 (2018)
20. Tang, A., Li, J., Lou, L., Shan, C., Yuan, X.: Optimization design and numerical study on water cooling structure for power lithium battery pack. Appl. Therm. Eng. **159**, 113760 (2019)
21. Zhang, T., Gao, Q., Gu, Y., Li, Y.: Studies on thermal management of lithium-ion battery using non-metallic heat exchanger. Appl. Therm. Eng. **182** (2021)
22. Gao, R., Fan, Z., Liu, S.: A gradient channel-based novel design of liquid-cooled battery thermal management system for thermal uniformity improvement. J. Energy Storage **48** (2022)
23. Wang, J., Ruan, L.: Cell-to-cell inconsistency analysis and structure optimization for a liquid-cooled cylindrical battery module. Appl. Therm. Eng. **224**, 120021 (2023)
24. He, F., Li, X., Zhang, G., Zhong, G., He, J.: Experimental investigation of thermal management system for lithium ion batteries module with coupling effect by heat sheets and phase change materials. Int. J. Energy Res. **42**(10), 3279–3288 (2018)
25. Li, Y., Du, Y., Xu, T., Wu, H., Zhou, X., Ling, Z., Zhang, Z.: Optimization of thermal management system for Li-ion batteries using phase change material. Appl. Therm. Eng. **131**, 766–778 (2018)
26. Zou, D., Liu, X., He, R., Zhu, S.X., Bao, J., Guo, J., Hu, Z., Wang, B.: Preparation of a novel composite phase change material (PCM) and its locally enhanced heat transfer for power battery module. Energy Convers. Manage. **180**, 1196–1202 (2019)
27. Fan, R., Zheng, N., Sun, Z.: Evaluation of fin intensified phase change material systems for thermal management of Li-ion battery modules. Int. J. Heat Mass Transf. **166** (2021)
28. Choudhari, V.G., Dhoble, A.S., Panchal, S.: Numerical analysis of different fin structures in phase change material module for battery thermal management system and its optimization. Int. J. Heat Mass Transf. **163**, 120434 (2020)

29. Ling, Z., Wang, F., Fang, X., Gao, X., Zhang, X.: A hybrid thermal management system for lithium ion batteries combining phase change materials with forced-air cooling. Appl. Energy **148**, 403–409 (2015)
30. Rao, Z., Wang, Q., Huang, C.: Investigation of the thermal performance of phase change material/mini-channel coupled battery thermal management system. Appl. Energy **164**, 659–669 (2016)
31. Huang, Q., Li, X., Zhang, G., Zhang, J., He, F., Li, Y.: Experimental investigation of the thermal performance of heat pipe assisted phase change material for battery thermal management system. Appl. Therm. Eng. **141**, 1092–1100 (2018)
32. Yang, W., Zhou, F., Zhou, H., Wang, Q., Kong, J.: Thermal performance of cylindrical lithium-ion battery thermal management system integrated with mini-channel liquid cooling and air cooling. Appl. Therm. Eng. **175**, 115331 (2020)
33. Zhao, Y., Li, Q., Zou, B., Zhang, T., Jin, L., Qiao, G., Nie, B., Huang, Y., Ding, Y.: Performance of a liquid cooling-based battery thermal management system with a composite phase change material. Int. J. Energy Res. **44**(6), 4727–4742 (2020)
34. Chen, S., Wei, X., Garg, A., Gao, L.: A comprehensive flowrate optimization design for a novel air-liquid cooling coupled battery thermal management system. J. Electrochem. Energy Convers. Storage **18**(2), 1–13 (2020)
35. Cao, J., Ling, Z., Fang, X., Zhang, Z.: Delayed liquid cooling strategy with phase change material to achieve high temperature uniformity of Li-ion battery under high-rate discharge. J. Power Sources **450** (2020)
36. Bamdezh, M.A., Molaeimanesh, G.R.: Impact of system structure on the performance of a hybrid thermal management system for a Li-ion battery module. J. Power Sources **457**, 227993 (2020)
37. Safdari, M., Ahmadi, R., Sadeghzadeh, S.: Numerical investigation on PCM encapsulation shape used in the passive-active battery thermal management. Energy **193**, 116840 (2020)
38. Li, M., Liu, F., Han, B., Guo, J., Xu, Y.: Research on temperature control performance of battery thermal management system composited with multi-channel parallel liquid cooling and air cooling. Ionics **27**(6), 2685–2695 (2021)
39. Wang, H., Tao, T., Xu, J., Mei, X., Liu, X., Gou, P.: Cooling capacity of a novel modular liquid-cooled battery thermal management system for cylindrical lithium ion batteries. Appl. Therm. Eng. **178** (2020)
40. Cao, W., Zhao, C., Wang, Y., Dong, T., Jiang, F.: Thermal modeling of full-size-scale cylindrical battery pack cooled by channeled liquid flow. Int. J. Heat Mass Transf. **138**, 1178–1187 (2019)
41. Qi, W., Huang, W., Niu, J., Chen, F., Chen, B., Chen, Y.: Thermal management of power battery based on flexible Swiss roll type liquid cooling micro-channel. Appl. Therm. Eng. **209**, 119491 (2023)
42. Wiriyasart, S., Hommalee, C., Sirikasemsuk, S., Prurapark, R., Naphon, P.: Thermal management system with nanofluids for electric vehicle battery cooling modules. Case Stud. Therm. Eng. **18** (2020)
43. Liao, G., Wang, W., Zhang, F., Jiagiang, E., Chen, J., Leng, E.: Thermal performance of lithium-ion battery thermal management system based on nanofluid. Appl. Therm. Eng. **216**, 118997 (2022)
44. Xie, L., Huang, Y., Lai, H.: Coupled prediction model of liquid-cooling based thermal management system for cylindrical lithium-ion module. Appl. Therm. Eng. **178** (2020)
45. Zhao, C., Cao, W., Dong, T., Jiang, F.: Thermal behavior study of discharging/charging cylindrical lithium-ion battery module cooled by channeled liquid flow. Int. J. Heat Mass Transf. **120**, 751–762 (2018)
46. Guo, J., Liu, F., Xu, Y., Han, B., Li, M.: Optimization design and numerical study of liquid-cooling structure for cylindrical lithium-ion battery pack. J. Energy Eng. **147**(4), 04021017 (2021)
47. Ke, Q., Li, X., Guo, J., Cao, W., Wang, Y., Jiang, F.: The retarding effect of liquid-cooling thermal management on thermal runaway propagation in lithium-ion batteries. J. Energy Storage **48** (2022)

48. ANSYS: CFX-Solver Theory Guide. ANSYS, Inc., Canonsburg, PA (2021)
49. Gümüşsu, E., Ekici, Ö., Köksal, M.: 3-D CFD modeling and experimental testing of thermal behavior of a Li-ion battery. Appl. Therm. Eng. **120**, 484–495 (2017)
50. Bernardi, D.: A general energy balance for battery systems. J. Electrochem. Soc. **132**, 5 (1985)
51. Patil, M.S., Panchal, S., Kim, N., Lee, M.Y.: Cooling performance characteristics of 20 Ah lithium-ion pouch cell with cold plates along both surfaces. Energies **11**(10) (2018)

Interpretation of the Computational Application of a Susceptibility Analysis in Toxicity Studies

Burcu Yeşilbudak

Abstract Our precise knowledge of the chaotic structure of aquatic ecosystems and the toxicological input parameters in the environment is rather uncertain. In this regard, computational toxicological models provide logical approaches to include solutions to uncertainty and variability problems and to report results. In addition, computational toxicological applications have an important potential to ensure environmental safety. However, the use of computational applied methods, as well as in vitro methods, is increasing day by day in order to predict the interactions of toxic substances and in vivo susceptibility analysis in toxicity studies. It is aimed to examine and compare the toxicants in similar tissues of two different fish species taken from a freshwater ecosystem seperately with a deterministic mathematical model, in this study. Depending on the mathematical calculation, a species- and metal-specific mathematical model was determined. As a result, it has been tried to present a case study about which mathematical approaches will be suitable for which biological studies, while showing that considering different mathematical modeling approaches can provide more reliable and informative results in biological studies.

Keywords Computational · Susceptibility · Toxic substances

1 Introduction

1.1 Ecotoxicology and Computational Applications

Water pollution has always attracted attention as an environmental problem in terms of both society and the scientific community [1]. In ecotoxicological studies, it is known how aquatic ecosystems under the influence of various toxic substances and how the living things in these habitats are put under pressure. Chemical pollution can

B. Yeşilbudak (✉)
Çukurova University, Adana, Türkiye
e-mail: yesilbudak@gmail.com

O. Castillo et al. (eds.), *Advanced Mathematics for the Modeling and Solution of Challenging Problems in Engineering*, Studies in Systems, Decision and Control 582,
https://doi.org/10.1007/978-3-031-82342-8_10

somehow enter the aquatic ecosystem due to natural erosion processes and anthropogenic reasons [2]. One of these elements, such as zinc, is vital for organisms, but its excess and deficiency have negative effects for organisms [3]. Zinc, which is used in metal plating and alloys and is an important element for industrial establishments, increases its toxic effects in the environment with the corrosive effect of sewage water and acid rain [4]. And so they become toxic to aquatic organisms at unusually high concentrations [5]. Heavy metals such as cadmium are among the pollutants that most concern and harm the environment [6]. It has been stated that approximately 30,000 tons of Cd fuel–oil fuels and anthropogenic activities such as mining are released into terrestrial and aquatic environments every year [7]. Environmental pollutants such as metal toxicants cause vital problems in many aquatic organisms by changing genetic, physiological, biochemical and behavioral parameters [8]. Heavy metal toxicity varies depending on the species [9], developmental stage of the species [10], environmental conditions [11], metal concentration and duration of action of the toxicant [12]. Heavy metals can adversely affect the behavior [13], development [14], migration [15] and metabolic processes of fish even at low concentrations [16]. Regardless of the sources of various polluting factors and the paths they follow, the regions they reach are aquatic environments [17]. Above, we briefly reviewed some of the experimental studies on the effects of toxic substances on aquatic ecosystems. However, several studies have been conducted on the effects of a single toxic substance [5–7, 9, 10, 12, 14]. In a study conducted in the Arctic, while Hg concentrations were determined in some long-lived predatory arctic fish depending on mercury (Hg) screening of aquatic organisms, it was determined that Hg concentrations in shorter-lived invertebrates had no toxicological risk or were low [18].

Another study investigated the effects of microbiological contamination, total petroleum hydrocarbons (TPHs) and heavy metals (As, Cd, Hg and Pb) on 8 marine fish tissues. Accordingly, arsenic in muscle tissues was determined at the highest levels among all metals evaluated in all species [19]. The status of microplastics (Mps) and heavy metals in important environmental matrices off the southwestern coast of India has been studied. In this study, it was reported that the most dominant metals in the polyethylene of the marine coastal sediments were Cd and Pb [20]. As mentioned above, heavy metals form different combinations with many different environmental pollution materials (TPHs, Mps etc.) cause a combo effect in their toxic effect on the environment [21]. It can be noted that most of the studies reported in recent years are related to the biological effects of toxicants of different nature, but under real conditions, species invariably affect some target tissues and organs lethally by a single type of toxicant [22]. For this reason, it is important to make environmental models by examining metals one by one. In recent years, with the increase of anthropogenic effect and industrialization, studies including mathematical models of the effects on the water system and the living creatures have started [23], but these studies may still be insufficient for certain situations. A study has been carried out on the mathematical modeling of a living community consisting of one type of species under toxic stress. Accordingly, based on the hypothesis that the toxic resource capacity in the external environment is greater than the biomass of the common species and that the toxic entry into the external environment is limited, a survival threshold was obtained showing the distinction between the average survival of a single species community and the extinction of the species [24]. Although the ecotoxicological problem model is provided according to the biomass of the population of the living beings, the toxic substance in the living environment and the concentrations depending on the type of toxicant taken from outside, the total mass or volume in the habitat of the living species, it is stated that each equation starts to take shape more since it generally includes all these parameters [25].

Since fish are an important food source for humans and other organisms, any adverse effects caused by metals will also affect humans and other organisms. For this reason, it is important for the future of the ecosystem to determine the possible physiological and biochemical changes that both necessary and unnecessary metals will cause in fish [26]. In this study, it is aimed to interpret the effects of individual cadmium and zinc in different fish species (*Oreochromis niloticus* and *Cyprinus carpio*) with a mathematical application.

2 Materials and Methods

2.1 Empirical Data

For empirical datasets required to be used in mathematical calculations, datasets in which *O. niloticus* and *C. carpio* fish were exposed to 0.5 mg Cd l^{-1} cadmium and 0.5 mg Zn l^{-1} zinc for 1, 15 and 30 days were used. In the experiment, muscle tissue was subjected to mineralization processes and organically destroyed [27]. Zinc and cadmium analyzes in the tissue were determined by spectrophotometric methods (AAS-SAFAS AA181) and spectrophotometric data were obtained. Then, a series of preliminary calculations were made for the data, and the standard solution of 5 different concentrations prepared for both metals was measured. The data were calculated by placing them in the equation in the regression graph. Then it was multiplied by the dilution coefficient and the ratio of the tissue to the dry weight of 1000 was obtained [28].

2.2 The Logistic Growth

Population changes due to births and deaths. The difference ΔP in the population P per a fixed time period is defined to be births minus deaths in this period, that is

$$\Delta P(t) = B(t) - D(t) \tag{1}$$

where $B(t)$ and $D(t)$ denote the births and deaths, respectively, and the variable t stands for time.

The birth rate is defined to be the number of births per population size at a certain time period. That is, the birth rate depends upon two parameters: $P(t)$ and t. We denote by $\beta(t)$ the birth rate. Similarly, the death rate is the number of deaths per population size at a certain time period. Let us denote by $\delta(t)$ the death rate.

Under the light of above definitions, we obtain the number of births and deaths at a certain time period as follows.

$$B(t) = \beta(t)P(t)\Delta t$$

and

$$D(t) = \delta(t)P(t)\Delta t$$

Now substituting the findings in Eq. (1) we obtain the expression below.

$$\Delta P(t) \approx \beta(t)P(t)\Delta t - \delta(t)P(t)\Delta t$$

$$\Delta P \approx (\beta P - \delta P)\Delta t$$

$$\frac{\Delta P}{\Delta t} \approx (\beta - \delta)P$$

or equivalently we obtain that

$$\frac{dP}{dt} = (\beta - \delta)P \tag{2}$$

In the logistic model it is assumed that the birth rate decreases as population increases on a linear scale and the death rate is a positive real constant. Hence, one may put the notations below.

$$\beta(t) = \beta_0 - \beta_1 P$$

$$\delta(t) = \delta_0$$

for some positive real numbers $\beta_0, \beta_1, \delta_0 \in \mathbb{R}^+$. Now by (2) we have that

$$\frac{dP}{dt} = (\beta_0 - \beta_1 P - \delta_0)P$$

$$\frac{dP}{dt} = \beta_0 P - \beta_1 P^2 - \delta_0 P$$

$$\frac{dP}{dt} = (\beta_0 - \delta_0)P - \beta_1 P^2$$

or equivalently

$$\frac{dP}{dt} = \beta_1 P\left(\frac{\beta_0 - \delta_0}{\beta_1} - P\right)$$

Now denoting $r = \beta_1/k$ for $k = \frac{\beta_0 - \delta_0}{\beta_1}$ we get the logistic equation

$$\frac{dP}{dt} = rkP(k - P) \tag{3}$$

where M is the carrying capacity. One may solve this differential equation as follows. Initially, we get that

$$\frac{dP}{dt} = rP\left[1 - \frac{P}{k}\right] = Pr\frac{k - P}{k}$$

and hence

$$\frac{k}{P(k - P)}dP = rdt$$

Note that

$$\frac{k}{P(k - P)} = \frac{1}{P} + \frac{k}{k - P}$$

and thus

$$\int_{t_0}^{t} rdt = \int_{P_0}^{P} \left[\frac{1}{P} + \frac{k}{k - P}\right]dP = (\ln|P| - \ln|k - P|)_{P_0}^{P}$$

Keeping the assumptions $t_0 = 0$, $P > 0$ and $k - P > 0$ in mind, we may ignore the absolute value function. This implies that

$$rt - rt_0 = \left[\ln\frac{P}{k - P}\right]_{P_0}^{P}$$

or

$$rt = \ln\frac{P}{k - P} - \ln\frac{P_0}{k - P_0} = \ln\frac{P(k - P_0)}{P_0(k - P)}$$

Now the exponential powers of both sides give that

$$e^{rt} = \frac{P(k - P_0)}{P_0(k - P)}$$

or

$$P = (k - P)\frac{P_0}{k - P_0}e^{rt}$$

Hence

$$\left(1 + \frac{P_0}{k - P_0}e^{rt}\right)P = k\frac{P_0}{k - P_0}e^{rt}$$

$$\frac{k - P_0 + P_0 e^{rt}}{k - P_0}P = k\frac{P_0}{k - P_0}e^{rt}$$

and finally

$$P = \frac{kP_0 e^{rt}}{k - P_0 + P_0 e^{rt}}$$

We suggest readers the book by Freedman [29] and the video by Leonard [30] for details. The behavior of this function carries information about the relation between death rate and population growth. Observe that the function above can be expressed as below.

$$P = \frac{kP_0}{P_0 + (k - P_0)e^{-rt}}$$

From the denominator, one may deduce two important cases.

Case I: The initial population is smaller than the carrying capacity, i.e., $P_0 < k$. In this case observe that the growth rate is decreasing as the population is increasing. We also have that under this condition, $k - P_0$ becomes positive and we obtain that

$$P = \frac{kP_0}{P_0 + (k - P_0)e^{-rt}} < \frac{kP_0}{P_0} = k$$

Therefore, the increasing population is bounded above by the carrying capacity k. In this case the graph of P as a function of t is as in Fig. 1.

Case II: The initial population is greater than the carrying capacity, that is $P_0 > k$. In this case the growth rate is increasing as the population is decreasing. We also have that under this condition, $k - P_0$ is negative and hence

$$P = \frac{kP_0}{P_0 + (k - P_0)e^{-rt}} > \frac{kP_0}{P_0} = k$$

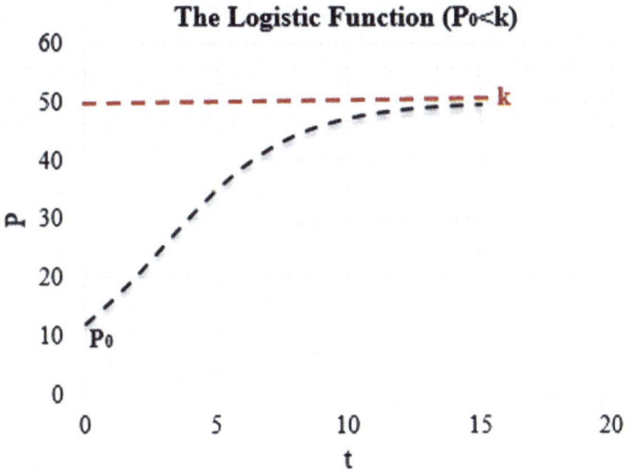

Fig. 1 The graph of the logistic function when the initial population is smaller than the carrying capacity

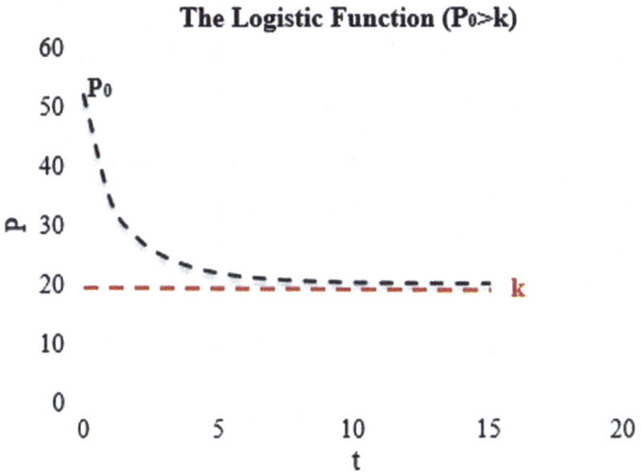

Fig. 2 The graph of the logistic function when the initial population is greater than the carrying capacity

Thus, the decreasing population is bounded below by the carrying capacity k. In this case the graph of P as a function of t is as in Fig. 2.

3 Conclusions

It was suggested that the logistic model can be utilized for amounts of toxic accumulations in tissues of living beings [25]. In the following lines, as a consequence of our experimental results, we provide logistic formulas for the effect of two toxic substances, zinc and cadmium, on two fish species *C. carpio* and *O. niloticus*. The formula for the accumulation of zinc in tissues of *C. carpio* was found to be

$$Z(t) = \frac{1.0116}{0.36 + 2.450e^{-0.3106t}}$$

and for *O. niloticus* we obtained

$$Z(t) = \frac{0.1584}{0.36 + 0.08e^{-0.1504t}}$$

On the other hand the formula for the accumulation of cadmium in tissues of *C. carpio* was found to be

$$C(t) = \frac{0.162}{0.09 + 1.71e^{-0.2141t}}$$

and for *O. niloticus* we obtained

$$C(t) = \frac{0.4056}{0.39 + 0.65e^{-0.3430t}}$$

Here $Z(t)$ and $C(t)$ are functions measuring zinc and cadmium amounts in the tissues of corresponding fish. The results show that the graphs of those four functions behave as in Fig. 1.

Biomonitoring studies based on mathematical modeling is a proven mechanism to be useful in integrating toxicant, species and habitats. Depending on the dose–response formulation, the definitive threshold value in the terminal organism concentration is the main factor that determines the existence of a permanent population of the species in the habitat or what goes to extinction. In disciplines such as molecular biology, medicine, veterinary medicine, pharmacy and agriculture, which include R&D studies, human health, animal welfare studies, many empirical trials are carried out for the rational use of risky substances, from active substances to various toxicants. Computational empirical studies provide an appropriate cost in terms of knowing the threshold values.

References

1. Tang, W., Pei, Y., Zheng, H., Zhao, Y., Shu, L., Zhang, H.: Twenty years of China's water pollution control: experiences and challenges. Chemosphere **295**, 133875 (2022)
2. Akhtar, N., Syakir Ishak, M.I., Bhawani, S.A., Umar, K.: Various natural and anthropogenic factors responsible for water quality degradation: a review. Water **13**(19), 2660 (2021)
3. Soetan, K.O., Olaiya, C.O., Oyewole, O.E.: The importance of mineral elements for humans, domestic animals and plants—a review. Afr. J. Food Sci. **4**(5), 200–222 (2010)
4. Mantovi, P., Bonazzi, G., Maestri, E., Marmiroli, N.: Accumulation of copper and zinc from liquid manure in agricultural soils and crop plants. Plant Soil **250**(2), 249–257 (2003)
5. Wepener, V., Van Vuren, J.H.J., Du Preez, H.H.: Uptake and distribution of a copper, iron and zinc mixture in gill, liver and plasma of a freshwater teleost, *Tilapia sparrmanii*. Water SA **27**(1), 99–108 (2001)
6. Singer, C., Zimmermann, S., Sures, B.: Induction of heat shock proteins (hsp70) in the zebra mussel (*Dreissena polymorpha*) following exposure to platinum group metals (platinum, palladium and rhodium): comparison with lead and cadmium exposures. Aquat. Toxicol. **75**(1), 65–75 (2005)

7. Cooper, C.A., Handy, R.D., Bury, N.R.: The effects of dietary iron concentration on gastrointestinal and branchial assimilation of both iron and cadmium in zebrafish (*Danio rerio*). Aquat. Toxicol. **79**(2), 167–175 (2006)
8. Scott, G.R., Sloman, K.A.: The effects of environmental pollutants on complex fish behaviour: integrating behavioural and physiological indicators of toxicity. Aquat. Toxicol. **68**(4), 369–392 (2004)
9. Harbar, O., Harbar, D., Stadnychenko, A., Babych, Y.: Ecotoxicological responses of two *Planorbarius corneus* s. lato (Mollusca, Gastropoda) allospecies to exposure of heavy metals. Int. J. Aquat. Biol. **9**(6), 423–431 (2022)
10. Dhara, K., Chukwuka, A.V., Saha, S., Saha, N.C., Faggio, C.: Effects of short-term selenium exposure on respiratory activity and proximate body composition of early-life stages of *Catla catla*, *Labeo rohita* and *Cirrhinus mrigala*. Environ. Toxicol. Pharmacol. **90**, 103805 (2022)
11. Witeska, M., Jezierska, B.: The effects of environmental factors on metal toxicity to fish (review). Fresenius Environ. Bull. **12**(8), 824–829 (2003)
12. Sobha, K., Poornima, A., Harini, P., Veeraiah, K.: A study on biochemical changes in the fresh water fish, *Catla catla* (Hamilton) exposed to the heavy metal toxicant cadmium chloride. Kathmandu Univ. J. Sci. Eng. Technol. **3**(2), 1–11 (2007)
13. Pandey, G., Madhuri, S.: Heavy metals causing toxicity in animals and fishes. Res. J. Anim. Vet. Fishery Sci. **2**(2), 17–23 (2014)
14. Burger, J.: Food chain differences affect heavy metals in bird eggs in Barnegat Bay, New Jersey. Environ. Res. **90**(1), 33–39 (2002)
15. Burger, J., Tsipoura, N., Niles, L.J., Gochfeld, M., Dey, A., Mizrahi, D.: Mercury, lead, cadmium, arsenic, chromium and selenium in feathers of shorebirds during migrating through Delaware Bay, New Jersey: comparing the 1990s and 2011/2012. Toxics **3**(1), 63–74 (2015)
16. Handy, R.D.: Chronic effects of copper exposure versus endocrine toxicity: two sides of the same toxicological process? Comp. Biochem. Physiol. Part A Mol. Integr. Physiol. **135**(1), 25–38 (2003)
17. Wagner, A., Boman, J.: Biomonitoring of trace elements in muscle and liver tissue of freshwater fish. Spectrochim. Acta Part B At. Spectrosc. **58**(12), 2215–2226 (2003)
18. Barst, B.D., Chételat, J., Basu, N.: Toxicological risk of mercury for fish and invertebrate prey in the Arctic. Sci. Total Environ. **836**, 155702 (2022)
19. Jisr, N., Younes, G., El Omari, K., Hamze, M., Sukhn, C., El-Dakdouki, M.H.: Levels of heavy metals, total petroleum hydrocarbons, and microbial load in commercially valuable fish from the marine area of Tripoli, Lebanon. Environ. Monit. Assess. **192**(11), 1–13 (2020)
20. Robin, R.S., Karthik, R., Purvaja, R., Ganguly, D., Anandavelu, I., Mugilarasan, M., Ramesh, R.: Holistic assessment of microplastics in various coastal environmental matrices, southwest coast of India. Sci. Total Environ. **703**, 134947 (2020)
21. Wagner, M., Scherer, C., Alvarez-Muñoz, D., Brennholt, N., Bourrain, X., Buchinger, S., Fries, E., Grosbois, C., Marti, T., Rodriguez-Mozaz, S., Urbatzka, R., Vethaak, A.D., Winther-Nielsen, M., Reifferscheid, G.: Microplastics in freshwater ecosystems: what we know and what we need to know. Environ. Sci. Eur. **26**(1), 1–9 (2014)
22. Almroth, B.C., Sturve, J., Stephensen, E., Holth, T.F., Förlin, L.: Protein carbonyls and antioxidant defenses in corkwing wrasse (*Symphodus melops*) from a heavy metal polluted and a PAH polluted site. Mar. Environ. Res. **66**(2), 271–277 (2008)
23. Ruru, S., Na, W., Xiaoyan, M., Huanle, Z., Jingkun, Z.: Optimization of model selection for toxicity evaluation of mixtures. Asian J. Ecotoxicol. **1**, 203–212 (2022)
24. Zhien, M., Baojun, S., Hallam, T.G.: The threshold of survival for systems in a fluctuating environment. Bull. Math. Biol. **51**(3), 311–323 (1989)
25. Freedman, H.I., Shukla, J.B.: Models for the effect of toxicant in single-species and predator-prey systems. J. Math. Biol. **30**(1), 15–30 (1991)
26. Javed, M., Usmani, N.: An overview of the adverse effects of heavy metal contamination on fish health. Proc. Natl. Acad. Sci. India Sect. B Biol. Sci. **89**(2), 389–403 (2019)
27. Muramoto, S.: Elimination of copper from Cu-contaminated fish by long-term exposure to EDTA and freshwater. J. Environ. Sci. Health A **18**(3), 455–461 (1983)

28. Sharma, B.K.: Instrumental Methods of Chemical Analysis. Krishna Prakashan Media, GOEL Publishing House, India (1981)
29. Freedman, H.I.: Deterministic Mathematical Models in Population Ecology. HIFR Consulting, Edmonton (1987)
30. Leonard, B.: Introduction to Population Models and Logistic Equation (Differential Equations 31). https://www.youtube.com/watch?v=7SPEvSmfK0w (2019). Accessed 14 July 2020

A Mathematical Model Used to Determine the Effects of Mixture Toxicology on Aquatic Species

Burcu Yeşilbudak

Abstract Various mathematical models are used to monitor the effects of the changing and transforming environment in aquatic ecosystems on the continuity of life, distribution of organisms and the effects of periodic use on the internal parameters (molecular and biochemical markers) of the living being. In this study, nonlinear mathematical calculations were used to understand the effects of toxicants, which reach increasing levels in the environment through natural or unnatural processes, on species. Especially in recent years, studies in which the interactions of toxicants in mixtures are examined practically and theoretically in ecotoxicological studies have become important. A generalized nonlinear model was created for the evaluation of analytical results obtained from chemical exposure to a catfish species under the influence of heavy metals individually and in a mixture, and an appropriate equation was provided, in this study. In general, an exemplary study has been provided for the use of computational and biological studies together.

Keywords Nonlinear · Species · Toxicant

1 Introduction

The harmful effects of toxic substances on the environment have been the subject of various studies, both theoretically and practically, in the last half-century [20, 9]. Heavy metal pollution, which is considered as a chemical pollution, is in the first place compared to other chemical pollutants. Because it is transported from natural and unnatural sources to the aquatic system in some way, its excretion is costly, but it can easily enter the food chain and accumulate in the living body with increasing concentrations [16, 30]. Particularly, some of the heavy metals, which are considered non-essential (cadmium (Cd), lead (Pb), chromium (Cr), arsenic (As), mercury (Hg), etc.) and accumulate in living things through natural physiological mechanisms,

B. Yeşilbudak (✉)
Çukurova University, Adana, Türkiye
e-mail: yesilbudak@gmail.com

O. Castillo et al. (eds.), *Advanced Mathematics for the Modeling and Solution of Challenging Problems in Engineering*, Studies in Systems, Decision and Control 582,
https://doi.org/10.1007/978-3-031-82342-8_11

show toxic effects in cases where tolerable limits are exceeded [6]. Because of this accumulation, creatures living in the aquatic environment may die, and even the life of people who eat seafood contaminated with metals may be endangered depending on the metal limit in these creatures [1]. Heavy metals affect all living beings in the food chain according to their metabolic status, less in the simplest organism in water and more in the most complex organism [8]. Due to the harmful effects of heavy metals in aquatic ecosystems, it is important to monitor their concentrations in aquatic environments over various times [2]. For this reason, many researches have been carried out, including the analysis of heavy metals in the sea [29, 18], the discharge analysis of toxicants to fresh waters such as rivers and lakes [34, 4], pollutants that cause deterioration of water quality such as heavy metals in lagoon [10]. Heavy metals are elements with a high atomic weight and a density of at least 5 times that of water [26]. Heavy metals such as copper (Cu), zinc (Zn) are essential elements that are needed in trace amounts for normal biological functions in organisms, and when their concentrations exceed a certain level, they cause structural and functional disorders in organisms [14]. Metal–metal interaction in organisms can be affected by many chemical processes between metals, such as membrane changes, promoting the synthesis of metal-binding proteins, and competition for carriers [35]. Heavy metals such as cadmium (Cd) and lead (Pb) have no biological function and may show toxic effects at low concentrations [36]. Studies on the determination of ecotoxicology and ecophysiology of metals have generally focused on the effect of a single metal [3, 5, 17, 33]. However, it has been stated that the metal pollution in the environment does not consist of a single metal element, but mostly consists of metal mixtures [32]. It was stated that SOD activity increased in the tissues of *Catla catla* and *Cirrhina mrigala* in freshwater fish under the effect of the binary mixtures of Fe and Mn [25]. As a result, aquatic ecosystems are often affected by various combinations of heavy metals that create significant interactions between them [7, 12].

Due to the co-existence of pollutants in the environment, metal mixture studies had better reflect the determination of environmental pollution. It is contradictory to com-pare results from different studies many times owing to the use of different reference models with inherently different definitions of antagonism, synergism, and additive effects [13]. On the other hand, mathematical models provide useful information for using various computational models to compare environmental quality standards, reporting and obtaining precise solutions [31]. In experimental studies, biologists have mostly focused on mathematical models containing only three parameters (assuming the minimum value is 0, an upper asymptote, a velocity parameter, and a time constant) with logistic, Gompertz, or monomolecular functions [37]. Although the Richards equation is the most popular of the more flexible growth equations [21], it is preferred for a variety of purposes, including growth models of perennials, mammals, and birds, and comparison of the agent or therapeutic effects of the toxicant [19, 11, 24]. In a study involving mathematical modeling, the simultaneous effect of binary mixture toxicant, one of which is more toxic than the other one on a biological species was examined [23]. It seems that most ecotoxicological studies in the past half century have been empirical, and there have been fewer phenomena in which computational and empirical studies have been modelled together. To study the simultaneous effect of two toxic substances, one more toxic than the other, on a biological population, a nonlinear model was studied using the stability theory of differential equations [23]. In this study, it is aimed to propose a mathematical model of the heavy metal data obtained experimentally from catfish, a freshwater fish exposed to a mixed metal toxicant for one week, with old and new phenomena, and to reveal the importance of some new aspects that attract significant attention.

2 Ecotoxicological Practice and Its Mathematical Model

2.1 Ecotoxicological Practice

For the required biological experimental datasets to be examined in mathematical calculations, 1 mg l^{-1} of non-essential metals (cadmium and lead) and essential metals (zinc and copper) of *Clarias gariepinus* fish for 1, 2, 3, 4, 5, 6, 7 days was exposed to a concentration of 5 mg l^{-1} and datasets were used. In the experiment, a certain piece of muscle tissue was taken from the muscle tissue at the end of the experimental period, subjected to mineralization processes and organically destroyed [15]. Non-essential metals (cadmium and lead) and essential metals (zinc and copper) analyzes in tissue were determined by spectrophotometric methods (AAS-SAFAS AA181) and spectrophotometric data were obtained. Then, a series of preliminary calculations were made for the data and standard solutions of 5 different concentrations prepared for all metals were measured. The data were calculated by inserting them into the equation in the regression graph. Then, the ratio of tissue to dry weight was obtained as 1000 by multiplying the dilution coefficient [22].

2.2 Mathematical Background

The idea of the generalized logistic model is based on the logistic model. Recall that the logistic differential equation is expressed as

$$\frac{dP}{dt} = rP\left[1 - \frac{P}{k}\right]$$

with the solution (logistic function)

$$P = \frac{kP_0}{P_0 + (k - P_0)e^{-rt}}$$

where P_0 is the initial population, k is the carrying capacity and r is a positive constant related to the birth rate. The generalized logistic differential equation is defined as follows.

$$\frac{dP}{dt} = rP^{\alpha}\left[1 - \left(\frac{P}{k}\right)^{\beta}\right]^{\gamma}$$

for some positive real numbers α, β, γ. Note that $\alpha = 1, \beta = 1, \gamma = 1$ gives the ordinary logistic function. In what follows we review the aspects of this function. Tsoularis and Wallace [27] showed the following properties: The value P reaches to carrying capacity k value as time t approaches to infinity. They also provide that the relative growth rate

$$\frac{1}{P}\frac{dP}{dt}$$

reaches its maximum at

$$P^* = k\left[1 + \frac{\beta\gamma}{\alpha - 1}\right]^{-\left(\frac{1}{\beta}\right)}$$

under the assumption that the maximum value is greater than the initial value, i.e., $P^* > P_0$.

The general solution to the generalized logistic function has to be examined case by case. As an approach, we focus on a special case called Richard's growth equation.

$$\frac{dP}{dt} = rP\left[1 - \left(\frac{P}{k}\right)^{\beta}\right]$$

when $\alpha = 1, \gamma = 1$ and β is a positive real number different from 1. We include the solution to this equation for the integrity of the paper. We deduce from the equation above that

$$\frac{dP}{dt} - rP = -\frac{r}{k^{\beta}}P^{1+\beta}$$

or

$$P^{-1-\beta}\frac{dP}{dt} - rP^{-\beta} = -\frac{r}{k^{\beta}}$$

and multiplying both sides by the scalar $-\beta$ we get that

$$-\beta P^{-1-\beta}\frac{dP}{dt} + \beta r P^{-\beta} = \frac{r\beta}{k^{\beta}}$$

Now substituting $x = P^{-\beta}$ we have that

$$\frac{dx}{dt} = -\beta P^{-1-\beta}\frac{dP}{dt}$$

and hence the equation turns into

$$\frac{dx}{dt} + \beta r x = \frac{r\beta}{k^{\beta}}$$

Now we multiply both sides by the integral factor

$$e^{\int \beta r dt} = e^{\beta r t}$$

and obtain that

$$e^{\beta r t}\frac{dx}{dt} + \beta r e^{\beta r t}x = \frac{r\beta}{k^{\beta}}e^{\beta r t}$$

which yields that

$$\frac{d\left(e^{\beta r t}x\right)}{dt} = \frac{r\beta}{k^{\beta}}e^{\beta r t}$$

or

$$d\left(e^{\beta r t}P^{-\beta}\right) = \frac{r\beta}{k^{\beta}}e^{\beta r t}dt$$

Now integrating both sides for lower limits $t_0 = 0$ and P_0; upper limits t and P, we have that

$$e^{\beta r t}P^{-\beta} - P_0^{-\beta} = \frac{e^{\beta r t} - 1}{k^{\beta}}$$

Evaluating the obtained equation we get the following equivalent lines.

$$e^{\beta r t}P^{-\beta} = \frac{P_0^{-\beta}k^{\beta} - 1 + e^{\beta r t}}{k^{\beta}}$$

$$P^{-\beta} = \frac{e^{-\beta r t}\left(P_0^{-\beta}k^{\beta} - 1\right) + 1}{k^{\beta}}$$

$$P^{-\beta} = \frac{P_0^{\beta} + \left(k^{\beta} - P_0^{\beta}\right)e^{-\beta r t}}{P_0^{\beta}k^{\beta}}$$

$$P^{\beta} = \frac{P_0^{\beta}k^{\beta}}{P_0^{\beta} + \left(k^{\beta} - P_0^{\beta}\right)e^{-\beta r t}}$$

$$P(t) = \frac{P_0 k}{\left(P_0^\beta + \left(k^\beta - P_0^\beta\right)e^{-\beta rt}\right)^{1/\beta}}$$

One may also mention the case when $\alpha = 1 + \beta(1-\gamma)$, $\beta, \gamma \in \mathbb{R}^+$ provided $\gamma < 1 + 1/\beta$. That is

$$\frac{dP}{dt} = rP^{1+\beta(1-\gamma)}\left[1 - \left(\frac{P}{k}\right)^\beta\right]^\gamma$$

which has the solution [28]

$$P(t) = \frac{k}{\left[1 + \left((\gamma - 1)\beta r k^{\beta(1-\gamma)} t + \left[\left(\frac{k}{P_0}\right)^\beta - 1\right]^{1-\gamma}\right)^{1/(1-\gamma)}\right]^{1/\beta}}$$

The above equation is a modification of so called Von Bertalanffy's growth equation. In this case, $\alpha = 2/3$, $\beta = 1/3$ and $\gamma = 1$. Hence the equation is

$$\frac{dP}{dt} = rP^{2/3}\left[1 - \left(\frac{P}{k}\right)^{1/3}\right]$$

The solution to this equation can be given as in the following lines similar to the one of Richard's growth.

$$\frac{dP}{dt} = rP^{2/3} - \frac{r}{k^{1/3}}P$$

$$\frac{dP}{dt} + \frac{r}{k^{1/3}}P = rP^{2/3}$$

$$P^{-2/3}\frac{dP}{dt} + \frac{r}{k^{1/3}}P^{1/3} = r$$

Let us substitute $x = P^{1/3}$. Then

$$\frac{dx}{dt} = \frac{P^{-2/3}}{3}\frac{dP}{dt}$$

and

$$\frac{dx}{dt} + \frac{r}{3k^{1/3}}x = \frac{r}{3}$$

Now let us multiply both sides by

$$e^{\int \frac{r}{3k^{1/3}} dt} = e^{\frac{rt}{3k^{1/3}}}$$

and get

$$e^{\frac{rt}{3k^{1/3}}} \frac{dx}{dt} + \frac{r}{3k^{1/3}} e^{\frac{rt}{3k^{1/3}}} x = \frac{r}{3} e^{\frac{rt}{3k^{1/3}}}$$

$$d\left(e^{\frac{rt}{3k^{1/3}}} x\right) = \frac{r}{3} e^{\frac{rt}{3k^{1/3}}} dt$$

After the integration we have

$$e^{\frac{rt}{3k^{1/3}}} P^{1/3} - P_0^{1/3} = k^{1/3}\left(e^{\frac{rt}{3k^{1/3}}} - 1\right)$$

$$e^{\frac{rt}{3k^{1/3}}} P^{1/3} = k^{1/3} e^{\frac{rt}{3k^{1/3}}} - k^{1/3} + P_0^{1/3}$$

$$P^{1/3} = k^{1/3} - k^{1/3} e^{\frac{-rt}{3k^{1/3}}} + P_0^{1/3} e^{\frac{-rt}{3k^{1/3}}}$$

$$P^{1/3} = k^{1/3}\left(1 - \left[1 - \left(\frac{P_0}{k}\right)^{1/3}\right] e^{\frac{-rt}{3k^{1/3}}}\right)$$

and finally

$$P(t) = k\left[1 - \left(1 - \left(\frac{P_0}{k}\right)^{1/3}\right) e^{-rt/3\sqrt[3]{k}}\right]^3$$

3 Conclusions

We know from various literatures that toxic substances siphoning into the environment for many reasons are harmful to the populations living in the aquatic ecosystem and pose health problems for the human element at the top of the ecological chain. Accordingly, it is thought that it is necessary to study them with mathematical models. Because mathematical models have become important tools for analysing and predicting, the probable route of ecosystems and their parameters. In this study, a generalized logistics mathematical model was used to examine the simultaneous effects of two or more toxic substances that may have negative effects on the vitality of the aquatic ecosystem. A mathematical model has been proposed that governs a

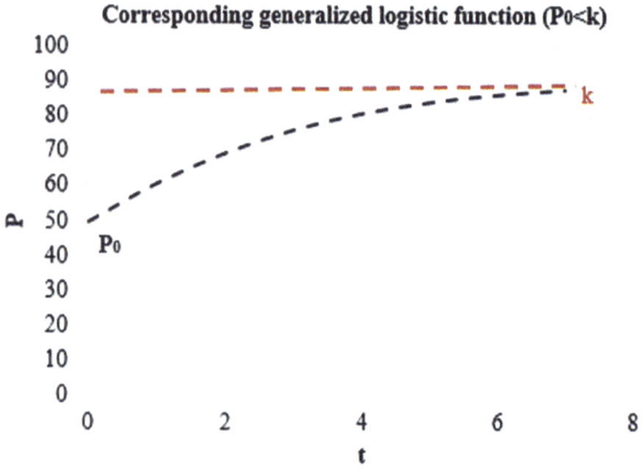

Fig. 1 The graph of the corresponding generalized logistic function

subclass of the total population that is affected by this mixed toxicant and, consequently, exhibits severe symptoms with morphological and physiological changes in the living body (such as growth arrest, deformity, changes in blood parameters, lesion in tissue, etc.). Our biologic calculation of toxic approach predicted all mixture responses accurately.

Our experimental data provide that the formula for the effect of the mixture of toxic substances copper, zinc, cadmium and lead is a special case of Richard's growth equation. The differential formula was found to be

$$\frac{dP}{dt} = 1.317P\left[1 - \frac{\sqrt[3]{P}}{4.4827}\right]$$

and the accumulation function was

$$P(t) = \frac{4469.7696}{\left(3.6746 + 0.8081e^{-0.439t}\right)^3}$$

The graph of $P(t)$ is as in Fig. 1.

References

1. Ali, H., Khan, E.: Bioaccumulation of non-essential hazardous heavy metals and metalloids in freshwater fish. Risk to human health. Environ. Chem. Lett. **16**(3), 903–917 (2018)
2. Ali, H., Khan, E., Ilahi, I.: Environmental chemistry and ecotoxicology of hazardous heavy metals: environmental persistence, toxicity, and bioaccumulation. J. Chem. **2019** (2019)
3. Alkahemal-Balawi, H.F., Ahmad, Z., Al-Akel, A.S., Al-Misned, F., Suliman, E.A.M., Al-Ghanim, K.A.: Toxicity bioassay of lead acetate and effects of its sub-lethal exposure on growth, haematological parameters and reproduction in *Clarias gariepinus*. Afr. J. Biotech. **10**(53), 11039 (2011)

4. Altındağ, A., Yiğit, S.: Assessment of heavy metal concentrations in the food web of lake Beyşehir, Turkey. Chemosphere **60**(4), 552–556 (2005)
5. Annune, P.A., Ebele, S.O., Oladimeji, A.A.: Acute toxicity of cadmium to juveniles of *Clarias gariepinus* (Teugels) and *Oreochromis niloticus* (Trewavas). J. Environ. Sci. Health Part A **29**(7), 1357–1365 (1994)
6. Birge, W.J., Price, D.J., Shaw, J.R., Spromberg, J.A., Wigginton, A.J., Hogstrand, C.: Metal body burden and biological sensors as ecological indicators. Environ. Toxicol. Chem. Int. J. **19**(4), 1199–1212 (2000)
7. Bu-Olayan, A.H., Thomas, B.V.: Trace metals toxicity and bioaccumulation in mudskipper *Periophthalmus waltoni* Koumans 1941 (Gobiidae: Perciformes). Turk. J. Fish. Aquat. Sci. **8**(2) (2008)
8. Chary, N.S., Kamala, C.T., Raj, D.S.S.: Assessing risk of heavy metals from consuming food grown on sewage irrigated soils and food chain transfer. Ecotoxicol. Environ. Saf. **69**(3), 513–524 (2008)
9. Dimić, D.S., Milenković, D.A., Avdović, E.H., Nakarada, Đ.J., Marković, J.M.D., Marković, Z.S.: Advanced oxidation processes of coumarins by hydroperoxyl radical: an experimental and theoretical study, and ecotoxicology assessment. Chem. Eng. J. **424**, 130331 (2021)
10. Dural, M., Göksu, M.L.: Çamlık Lagünü (Karataş, Adana), seston, bentoz ve sedimentinde mevsimsel ağır metal değişimi. Ege J. Fish. Aquat. Sci. **23**(1), 65–69 (2006)
11. Gaillard, J.M., Pontier, D., Allaine, D., Loison, A., Herve, J.C., Heizmann, A.: Variation in growth form and precocity at birth in eutherian mammals. Proc. R. Soc. Lond. Ser. B Biol. Sci. **264**, 859–868 (1997)
12. Hussain, S.M., Javed, M., Asghar, S., Hussain, M., Abdullah, S., Raza, S.A., Javid, A.: Studies on growth performance of metals mixture stressed *Cirrhina mrigala* in earthen ponds. Pak. J. Agri. Sci **47**(3), 263–270 (2010)
13. Ince, N.H., Dirilgen, N., Apikyan, I.G., Tezcanli, G., Üstün, B.: Assessment of toxic interactions of heavy metals in binary mixtures: a statistical approach. Arch. Environ. Contam. Toxicol. **36**(4), 365–372 (1999)
14. Łuczyńska, J., Pietrzak-Fiećko, R., Purkiewicz, A., Łuczyński, M.J.: Assessment of fish quality based on the content of heavy metals. Int. J. Environ. Res. Public Health **19**(4), 2307 (2022)
15. Muramoto, S.: Elimination of copper from Cu-contaminated fish by long-term exposure to EDTA and freshwater. J. Environ. Sci. Health A **18**(3), 455–461 (1983)
16. Nica, D.V., Bura, M., Gergen, I., Harmanescu, M., Bordean, D.M.: Bioaccumulative and conchological assessment of heavy metal transfer in a soil-plant-snail food chain. Chem. Cent. J. **6**(1), 1–15 (2012)
17. Ololade, I.A., Ogini, O.: Behavioural and hematological effects of zinc on African catfish, *Clarias gariepinus*. Int. J. Fish. Aquacult. **1**(2), 22–27 (2009)
18. Orani, A.M., Vassileva, E., Azemard, S., Thomas, O.P.: Comparative study on Hg bioaccumulation and biotransformation in Mediterranean and Atlantic sponge species. Chemosphere **260**, 127515 (2020)
19. Parresol, B.R.: Basal area growth for 15 tropical tree species in Puerto Rico. For. Ecol. Manage. **73**, 211–219 (1995)
20. Perrodin, Y., Boillot, C., Angerville, R., Donguy, G., Emmanuel, E.: Ecological risk assessment of urban and industrial systems: a review. Sci. Total Environ. **409**(24), 5162–5176 (2011)
21. Richards, F.J.: A flexible growth function for empirical use. J. Exp. Bot. **10**(2), 290–301 (1959).
22. Sharma, B.K.: Instrumental Methods of Chemical Analysis. Krishna Prakashan Media, GOEL Publishing House, India (1981)
23. Shukla, J.B., Dubey, B.: Simultaneous effect of two toxicants on biological species: a mathematical model. J. Biol. Syst. **4**(01), 109–130 (1996)
24. Sibly, R.M., Barker, D., Denham, M.C., Hone, J., Pagel, M.: On the regulation of populations of mammals, birds, fish, and insects. Science **309**(5734), 607–610 (2005)
25. Tayyab, H., Latif, F., Aziz, S., Zubair, R., Aslam, J.: Toxicological effects of a binary mixture of metals (Fe + Mn) on the oxidative stress of major carps. Pak. J. Agric. Agric. Eng. Vet. Sci. **38**(1), 37–45 (2022)

26. Tchounwou, P.B., Yedjou, C.G., Patlolla, A.K., Sutton, D.J.: Heavy metal toxicity and the environment. Mol. Clin. Environ. Toxicol. 133–164 (2012)
27. Tsoularis, A., Wallace, J.: Analysis of logistic growth models. Math. Biosci. **179**(1), 21–55 (2002)
28. Turner, M.E., Jr., Bradley, E.L., Jr., Kirk, K.A., Pruitt, K.M.: A theory of growth. Math. Biosci. **29**(3–4), 367–373 (1976)
29. Türkmen, A., Türkmen, M., Tepe, Y., Akyurt, İ: Heavy metals in three commercially valuable fish species from Iskenderun Bay, Northern East Mediterranean Sea, Turkey. Food Chem. **91**(1), 167–172 (2005)
30. Vardhan, K.H., Kumar, P.S., Panda, R.C.: A review on heavy metal pollution, toxicity and remedial measures: current trends and future perspectives. J. Mol. Liq. **290**, 111197 (2019)
31. Vezzaro, L., Eriksson, E., Ledin, A., Mikkelsen, P.S.: Quantification of uncertainty in modelled partitioning and removal of heavy metals (Cu, Zn) in a stormwater retention pond and a biofilter. Water Res. **46**(20), 6891–6903 (2012)
32. Väänänen, K., Leppänen, M.T., Chen, X., Akkanen, J.: Metal bioavailability in ecological risk assessment of freshwater ecosystems: from science to environmental management. Ecotoxicol. Environ. Saf. **147**, 430–446 (2018)
33. Wani, A.A., Malik, J., Zuber, S.M.: Determination of lethal toxicity of copper to *Clarias gariepinus*. Int. J. Adv. Res. Sci. Eng. **7**(4), 1011–1018 (2018)
34. Weber, P., Behr, E.R., Knorr, C.D.L., Vendruscolo, D.S., Flores, E.M., Dressler, V.L., Baldisserotto, B.: Metals in the water, sediment, and tissues of two fish species from different trophic levels in a subtropical Brazilian river. Microchem. J. **106**, 61–66 (2013)
35. Wicklund, A., Runn, P.: Calcium effects on cadmium uptake, redistribution, and elimination in minnows, *Phoxinus phoxinus*, acclimated to different calcium concentrations. Aquat. Toxicol. **13**(2), 109–121 (1988)
36. Witkowska, D., Słowik, J., Chilicka, K.: Heavy metals and human health: possible exposure pathways and the competition for protein binding sites. Molecules **26**(19), 6060 (2021)
37. Zeide, B.: Analysis of growth equations. For. Sci. **39**, 594–616 (1993)

Optimization of Gurney Flap Over NACA 0018 by Using Surrogate Modeling

Emre Güler, Mehmet Erdem, Şıhmehmet Yıldız, and Melike Nikbay

Abstract This research investigates the impact of the Gurney flap on the aerodynamic performance of the NACA 0018 airfoil at a Reynolds number of $Re = 1 \times 10^5$. A total of forty data sets are generated using the Halton sampling method, and computational fluid dynamics (CFD) simulations are performed on each set. The CFD outputs are then utilized to train a surrogate model based on the Kriging method. The optimization process employs the differential evolution algorithm to determine the optimal Gurney flap height, width and angle of attack values. The findings demonstrate that the optimized Gurney flap configuration leads to a significant enhancement in the lift coefficient, with an average increase of approximately 26.9%. These results underscore the effectiveness of the Gurney flap as a means to improve the aerodynamic performance of the NACA 0018 airfoil at the specified Reynolds number. Further investigations and experimental validations are recommended to validate the obtained results and explore potential applications in practical aerodynamic designs.

Keywords Aerodynamics · Flow control · Gurney flap · Optimization · Surrogate modeling

E. Güler (✉) · M. Erdem
Aerospace Engineering, Tarsus University, Mersin, Turkey
e-mail: emreguler@tarsus.edu.tr; gulere22@itu.edu.tr

M. Erdem
e-mail: mehmeterdem@tarsus.edu.tr; erdemm23@itu.edu.tr

E. Güler · M. Erdem · Ş. Yıldız · M. Nikbay
Aeronautical and Astronautical Engineering, Istanbul Technical University, Istanbul, Turkey
e-mail: yildizsih@itu.edu.tr

M. Nikbay
e-mail: nikbay@itu.edu.tr

O. Castillo et al. (eds.), *Advanced Mathematics for the Modeling and Solution of Challenging Problems in Engineering*, Studies in Systems, Decision and Control 582,
https://doi.org/10.1007/978-3-031-82342-8_12

1 Introduction

The flow characteristics around specific structures, such as micro air vehicles (MAVs) and vertical axis wind turbines (VAWTs), can exhibit low Reynolds numbers, indicating minimal airflow chord length or velocity. In particular, when the Reynolds number is below 500.000, the flow is classified as low Reynolds number flow, where viscous forces dominate over inertial forces [1]. Initially, the airflow approaches the airfoil in a laminar manner, transitioning into turbulence, and eventually becoming fully turbulent. At low Reynolds numbers, when the viscous forces predominate the inertial forces, the laminar flow separates from the airfoil, undergoes transition within the separated region, and then reattaches. This region of separation and reattachment is commonly referred to as the laminar separation bubble [2]. To mitigate the adverse effects associated with the laminar separation bubble, and Gurney flaps give a good solution for this problem which were introduced by Dan Gurney in the 1960s [3]. These flaps, affixed to the trailing edge of the airfoil, enhance aerodynamic performance without requiring complex systems. Figure 1 provides a schematic representation of the Gurney flap.

Gurney flaps have demonstrated their effectiveness in both subsonic and transonic flight regimes. Although they contribute to increased drag, they significantly improve the lift-to-drag ratio, effectively simulating an increase in the airfoil's camber [5]. Numerous studies have provided evidence of the positive impact of Gurney flaps on lift and drag forces. Liebeck's research demonstrated a drag reduction of 1–2% and a lift force increase when Gurney flaps were employed [6]. Similarly, Li et al. [7] reported a potential lift force improvement of over 25% through the application of Gurney flaps. Giguere et al. concluded that Gurney flaps were most effective when positioned on the lower surface at the trailing edge of the airfoil [8]. Traub et al. discovered that angling the Gurney flap rather than placing it perpendicular to the flow could prevent significant increases in drag [9]. Additionally, Motta found that Gurney flaps could effectively control vibrations induced by airflow [10]. Jang et al. [11] conducted research indicating that the implementation of Gurney flaps could delay separation over specific airfoils. Holst et al. [12] stated that the use of Gurney flaps on wind turbines could decrease the standard deviation of the lift force by up to 65%, thereby extending the turbine's lifespan. Lastly, Graham et al. [13] revealed

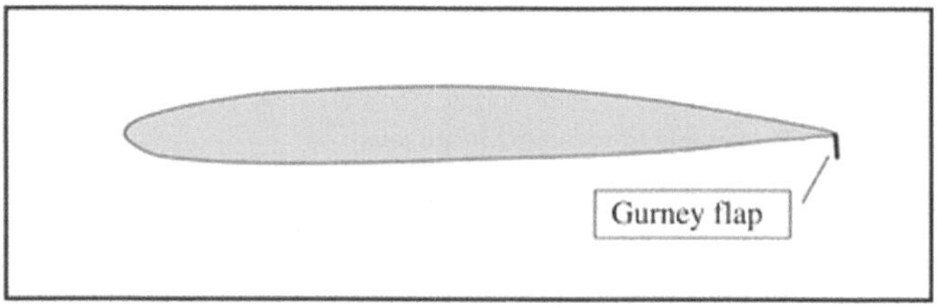

Fig. 1 Diagram of an airfoil with a Gurney flap at the trailing edge. The Gurney flap is a small, vertical tab that enhances aerodynamic performance by increasing lift [4]

that the dimensions of Gurney flaps had a significant influence on the aerodynamic performance of certain airfoils, noting that drag force directly correlated with flap height but inversely correlated with thickness at a 0° angle of attack. Surrogate modeling, also known as metamodeling or surrogate-based optimization, is a technique widely used in the field of engineering design. It provides an efficient approach to replace complex simulations with simplified models that are computationally less demanding while accurately capturing the essential phenomena [14]. These models are particularly valuable when dealing with complex problems where explicit mathematical relationships are either unknown or difficult to derive [15]. In scenarios requiring a large number of simulations, such as optimization problems, surrogate modeling offers significant advantages. It enables accurate prediction of the primary system's responses without the need for exhaustive simulations, facilitating faster exploration and optimization of design spaces [16]. Differential evolution (DE), on the other hand, is a stochastic, population-based optimization algorithm extensively employed in engineering and computer science [17]. Specifically designed for multi-dimensional, real-valued optimization problems, DE iteratively improves solutions by generating potential solutions and incorporating mutation and crossover operations [18]. The notable feature of differential evolution lies in its simple yet powerful mechanism, where trial vectors are generated by adding the weighted difference of two population vectors to a third vector [19]. This allows the algorithm to effectively explore the problem space and discover optimal or near-optimal solutions. When combined, surrogate modeling and differential evolution form a powerful toolkit for addressing complex design problems. The surrogate model significantly reduces the computational cost, while the DE method enables an efficient and thorough search of the design space [20]. This combination is particularly advantageous for computationally expensive problems, where traditional optimization methods may struggle [21]. Please note that the first paragraph of a section or subsection is not indented. The first paragraphs that follows a table, figure, equation etc. does not have an indent, either. Subsequent paragraphs, however, are indented.

2 Materials and Methods

2.1 Numerical Solution

Ansys Fluent software package is used for CFD simulations, and three geometric parameters have been defined in geometry optimization as dimensionless height to chord length h/c, dimensionless width to chord length w/c, and angle of attack, α. While creating the mesh structure after the parameterization processes, the triangular mesh structure has been used and the y^+ value has been kept close to 1 by using the structured mesh configuration on the airfoil surface. For all cases, the solutions have been carried out at constant Reynolds number (Eq. 1), $Re = 1 \times 10^5$ and it has been carried out with the k-kl-kw turbulence model [22]. The reason why this model is

Table 1 Validation of lift coefficient

h/c	w/c	α (°)	Previous study [23]	Present study	Error (%)
0.02	0.005	10	1.184	1.1787	0.45
0.01	0.005	13	1.226	1.22	0.5

preferred is that at Re $= 1 \times 10^5$ flow transitions from laminar to turbulent flow and other models are insufficient for transition flows. Durhasan and Karasu [23] performed a numerical study on the Gurney flap and showed a good result on the lift coefficient (Eq. 2). In the formula of Re and C_L, ρ is the density of the fluid, V is the free stream velocity, c is the characteristic length of the airfoil and μ is the dynamic viscosity, F_L is lift force and A indicates wing reference area. The validation study is given in Table 1 with various parametric configurations of h/c, w/c, and α.

$$\mathrm{Re} = \frac{\rho V c}{\mu} \tag{1}$$

$$C_L = \frac{F_L}{\frac{1}{2}\rho V^2 A} \tag{2}$$

2.2 *Surrogate Modeling and Optimization*

To generate the dataset, an analysis was conducted on 40 design variables determined using the Halton sampling method and the bounds of simulation are given in Table 2. Subsequently, a surrogate model was created using the Kriging method based on this dataset. The purpose of employing surrogate modeling is to decrease simulation time and mitigate the processing time required for intricate geometries and mesh structures, as well as the increased solution time associated with the chosen CFD model. By using surrogate modeling, the analysis time can be shortened, and potential errors stemming from complexity can be minimized. The selection of the Kriging method is motivated by its prediction variance and flexibility, which make it well-suited for this task.

To maximize the lift coefficient, an optimization study was conducted using the established surrogate model instead of the analysis program. The surrogate model, created using the Python library [24], was integrated with the optimization algorithm

Table 2 Bounds of simulation

Maximize	C_L
Subject to	$0.005 \leq h/c \leq 0.02$
	$0.001 \leq w/c \leq 0.02$ $-5° \leq \alpha \leq 15°$

to identify the most suitable geometric parameters. For optimization, the Differential Evolution (DE) algorithm was employed. DE is an evolutionary computation method that iteratively improves a candidate solution based on a specific quality measure [25]. It is a metaheuristic that proves effective in exploring extensive solution spaces without making excessive assumptions about the problem being solved. DE is widely applicable for searching optimal solutions across various domains. In conclusion, surrogate modeling and optimization are valuable tools for addressing problems with high computational costs. Surrogate modeling enables faster optimization by reducing the need for extensive computations. However, it is essential to consider the accuracy and alignment of the results with the true function when employing surrogate modeling techniques.

3 Results and Discussion

The main process of the data sampling, surrogate modeling and optimization are given above. By using a surrogate model, forty data sets from the CFD solution are trained and checked for optimum geometry variables due to the maximize C_L. The comparison of base and optimum geometry are depicted in Table 3 and increase in lift coefficient is found as about 26.9%.

Figure 2 illustrates the base airfoil and the optimized Gurney flapped airfoil, displaying streamwise velocity contour and streamline topology. It is evident from the figure that the addition of the optimized Gurney flap at the trailing edge of the airfoil creates a vacuum on the suction surface. This alteration in the flow pattern contributes to the observed increase in the lift coefficient. The figure provides visual evidence of the impact of the optimized Gurney flap on the airflow characteristics around the airfoil.

Table 3 Surrogate based optimization result

	Base	Optimum	Lift increase
C_L	1.04	1.32	26.9%
h/c	0	0.0190	–
w/c	0	0.0195	–
α (°)	11.79	11.79	–

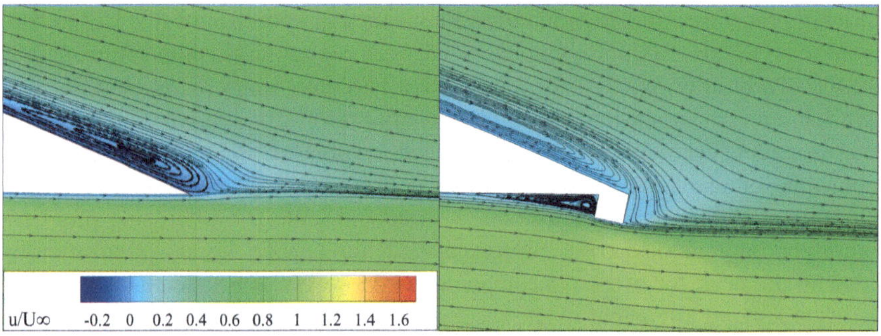

Fig. 2 A computational fluid dynamics simulation showing airflow over an airfoil. The image is divided into two sections, each depicting different flow patterns with streamlines. The left section shows base airfoil, while the right section illustrates optimized Gurney flapped airfoil. A color scale at the bottom ranges from blue to red, indicating streamwise velocity contours

4 Conclusion

In this study, the geometry of the Gurney flap was optimized using surrogate modeling. A set of forty different design parameters was generated using the Halton data sampling method, and the flow was simulated using computational fluid dynamics (CFD). The dataset obtained from the CFD simulations was used to create a Kriging surrogate model. The surrogate model was then utilized to find the optimal geometric parameters, which were subsequently compared with the CFD solutions.

The results of this study showed a significant increase in the lift coefficient, with an improvement of approximately 26.9%. This indicates that the optimized Gurney flap geometry resulted in enhanced aerodynamic performance compared to the initial design. The surrogate modeling approach proved effective in finding the optimal solution and provided valuable insights into the relationship between the geometric parameters and the lift coefficient.

References

1. Wang, J., Feng, L.: Flow Control Techniques and Applications. In: Cambridge Aerospace Series. Cambridge University Press, Cambridge (2018). https://doi.org/10.1017/9781316676448
2. Genç, M.S., Karasu, I., Açıkel, H.H., Akpolat, M.T.: Low Reynolds number flows and transition. In: Low Reynolds Number Aerodynamics and Transition. InTech (2012)
3. Jain, S., Sitaram, N., Krishnaswamy, S.: Computational investigations on the effects of Gurney flap on airfoil aerodynamics. Int. Sch. Res. Notices (2015)
4. Amini, Y., Emdad, H., Farid, M.: Adjoint shape optimization of airfoils with attached Gurney flap. Aerosp. Sci. Technol. **41**, 216–228 (2015). https://doi.org/10.1016/j.ast.2014.12.023
5. Wang, J.J., Li, Y.C., Choi, K.S.: Gurney flap—lift enhancement, mechanisms and applications. Prog. Aerosp. Sci. **44**(1), 22–47 (2008)
6. Liebeck, R.H.: Design of subsonic airfoils for high lift. J. Aircr. **15**(9), 547–561 (1978)

7. Li, Y., Wang, J., Zhang, P.: Effects of Gurney flaps on a NACA0012 airfoil. Flow Turbul. Combust. **68**(1), 27 (2002)
8. Giguere, P., Dumas, G., Lemay, J.: Gurney flap scaling for optimum lift-to-drag ratio. AIAA J. **35**(12), 1888–1890 (1997)
9. Traub, L.W., Miller, A., Rediniotis, O.: Preliminary parametric study of Gurney flap dependencies. J. Aircr. **43**(4), 1242–1244 (2006)
10. Motta, V.: Computational fluid dynamic analysis of a L-shaped Gurney flap for vibration control. Ph.D. thesis, Milano Polytechnic University, 2015
11. Jang, C.S., Ross, J.C., Cummings, R.M.: Numerical investigation of an airfoil with a Gurney flap. Aircr. Des. **1**(2), 75–88 (1998)
12. Holst, D., Bach, A.B., Nayeri, C.N., Paschereit, C.O., Pechlivanoglou, G.: Wake analysis of a finite width Gurney flap. J. Eng. Gas Turbines Power **138**(6), 062602 (2016)
13. Graham, M., Muradian, A., Traub, L.W.: Experimental study on the effect of Gurney flap thickness on airfoil performance. J. Aircr. 1–6 (2017)
14. Wang, X., Forrester, A.I.J.: Surrogate modeling review and recommendations. J. Mech. Des. **137**(4), 040301 (2015)
15. Sacks, J., Welch, W.J., Mitchell, T.J., Wynn, H.P.: Design and analysis of computer experiments. Stat. Sci. **4**(4), 409–423 (1989)
16. Hsiao, K.-L., Vincent, T., Chen, W.: Reliability-based design optimization using concurrent subspace dimension reduction. Struct. Multidiscip. Optim. **43**(2), 253–268 (2011)
17. Storn, R., Price, K.: Differential evolution—a simple and efficient heuristic for global optimization over continuous spaces. J. Glob. Optim. **11**(4), 341–359 (1997)
18. Vesterstrøm, J., Thomsen, R.: A comparative study of differential evolution, particle swarm optimization, and evolutionary algorithms on numerical benchmark problems. In: IEEE Congress on Evolutionary Computation (2004)
19. Das, S., Suganthan, P.N.: Differential evolution: a survey of the state-of-the-art. IEEE Trans. Evol. Comput. **15**(1), 4–31 (2011)
20. Müller, J., Singh, S.P.N.: Differential evolution for multiobjective portfolio optimization. Comput. Econ. **39**(2), 149–182 (2012)
21. Forrester, A.I.J., Sóbester, A., Keane, A.J.: Engineering Design Via Surrogate Modelling: A Practical Guide. Wiley (2008)
22. Walters, D.K., Leylek, J.H.: A new model for boundary layer transition using a single-point RANS approach. J. Turbomach. **126**, 193–202 (2004)
23. Durhasan, T., Karasu, İ.: Effect of Gurney flap on the aerodynamic performance of NACA 0018 airfoil at low Reynolds number. Presented at the 4th International Conference on Advances in Mechanical Engineering Istanbul 2018, Istanbul (2018)
24. Bouhlel, M.A., Hwang, J.T., Bartoli, N., Lafage, R., Morlier, J., Martins, J.R.R.A.: A Python surrogate modeling framework with derivatives. Adv. Eng. Softw. 102662 (2019). ISSN: 0965-9978. https://doi.org/10.1016/j.advengsoft.2019.03.005
25. Rocca, P., Oliveri, G., Massa, A.: Differential evolution as applied to electromagnetics. IEEE Antennas Propag. Mag. **53**(1), 38–49 (2011). https://doi.org/10.1109/MAP.2011.5773566

Examination of Recycled Papers Using an Intuitionistic Fuzzy Set-Based MCDM

Feride Tuğrul

Abstract Multi-criteria decision making (MCDM) methods are used in combination with intuitionistic fuzzy sets offer very favorable results to decision makers. Intuitionistic fuzzy set-based MCDM methods defined in this direction were utilized in many application areas. In this study, recycled papers were examined separately according to Cyan, Magenta and Yellow colors. Paper samples, which passed through a total of 4 recycling stages, were evaluated according to their L, a, and b values. The decision making mechanism established in the study was observed that provides effective results in the fields such as paper quality, print quality, recycling, etc.

Keywords Intuitionistic fuzzy set · PROMETHEE method · Multi criteria decision making · Recycling

1 Introduction

In this study, it is to examines the quality of 4 different recycling samples of papers according to colors according to L, a, b values. The main purpose is to create a decision making mechanism in which all the criteria are activated at the same time and all samples are examined at the same time. For this purpose, samples of Cyan, Magenta and Yellow colors at 4 different stages of recycling were examined. In these application stages, the intuitionistic fuzzy PROMETHEE method was used. Researchers have developed applications using PROMETHEE methods, intuitionistic fuzzy sets and fuzzy sets [1–27]. Also, there are many important studies in the literature on the print quality of papers [28–35].

Recycling of waste paper and their reusability in production gains importance at the present time [36]. But the whiteness, yellowness, brightness and $L^*a^*b^*$ values of the papers obtained as a result of recycling are other factors affecting the print

F. Tuğrul (✉)
Faculty of Engineering, Department of Computer Engineering, Munzur University, Tunceli, Türkiye
e-mail: feridetugrul@gmail.com

O. Castillo et al. (eds.), *Advanced Mathematics for the Modeling and Solution of Challenging Problems in Engineering*, Studies in Systems, Decision and Control 582,
https://doi.org/10.1007/978-3-031-82342-8_13

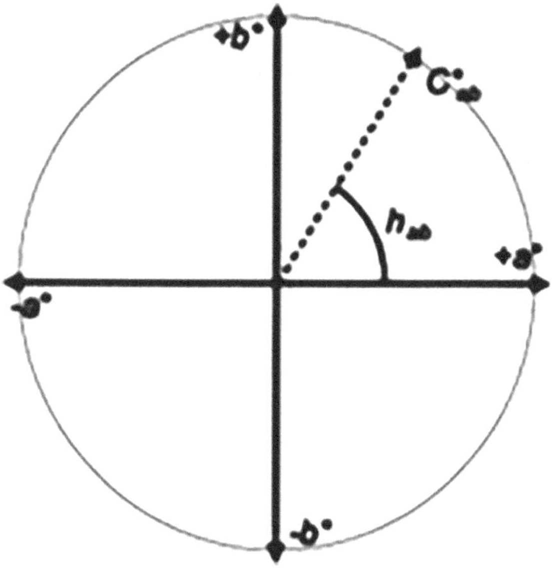

Fig. 1 CIE_{ab} chroma and hue angle

quality [37]. Another representation of the CIELab colour universe is to calculate the polar coordinates of the chroma C^*_{ab} perceptual features. High chroma indicates high colour saturation, which is an important property for good quality prints, including high colour gamut [38]. This calculation is found according to the following formula:

$$C_{ab} = \sqrt{a^{*2} + b^{*2}}$$

CIE C^*_{ab} estimates the distance from 0, that is, the achromatic centre ascending towards a chromatic centre, in other words from 0 to the L^* axis as shown in Fig. 1 [39].

2 Material and Method

Definition 1 [3, 40] Let $X \neq \emptyset$. An intuitionistic fuzzy set A in X;

$$A = \{\langle x, \mu_A(x), \nu_A(x)\rangle | x \in X\}, \tag{1}$$

$$\mu_A(x), \nu_A(x), \pi_A(x) : X \rightarrow [0, 1] \tag{2}$$

defined membership, nonmembership and hesitation degree of the element $x \in X$ respectively.

$$\mu_A(x) + \nu_A(x) + \pi_A(x) = 1 \tag{3}$$

The following is a depiction of the intuitionistic fuzzy value (IFV) defined by Xu [41]:

$$\tilde{a} = (\mu_{\tilde{a}}, \nu_{\tilde{a}}, \pi_{\tilde{a}}) \text{ and where } \mu_{\tilde{a}}, \nu_{\tilde{a}}, \pi_{\tilde{a}} \in [0, 1].$$

Definition 2 For IFVs $\tilde{a} = (\mu_{\tilde{a}}, \nu_{\tilde{a}})$ and $\tilde{b} = \left(\mu_{\tilde{b}}, \nu_{\tilde{b}}\right)$ the following operations were defined [41, 42]:

$$(1)\, \tilde{a} \oplus \tilde{b} = \left(\mu_{\tilde{a}} + \mu_{\tilde{b}} - \mu_{\tilde{a}}\mu_{\tilde{b}}, \nu_{\tilde{a}}\nu_{\tilde{b}}\right) \tag{4}$$

$$(2)\, \tilde{a} \otimes \tilde{b} = \left(\mu_{\tilde{a}}\mu_{\tilde{b}}, \nu_{\tilde{a}} + \nu_{\tilde{b}} - \nu_{\tilde{a}}\nu_{\tilde{b}}\right) \tag{5}$$

$$(3)\, \bigoplus_{j=1}^{m} \tilde{a}_j = \left(1 - \prod_{j=1}^{m} (1 - \mu_j), \prod_{j=1}^{m} \nu_j\right) \tag{6}$$

$$(4)\, \bigotimes_{j=1}^{m} \tilde{a}_j = \left(\prod_{j=1}^{m} \mu_j, \prod_{j=1}^{m} (1 - \nu_j)\right) \tag{7}$$

To compare IFVs, the following method was employed [43]:

$$\rho(\alpha) = 0.5(1 + \pi_\alpha)(1 - \mu_\alpha) \tag{8}$$

It's crucial to note that when the value $\rho(\alpha)$ decreases, the preferred value α will go up.

After determining the alternatives and criteria in multi-criteria decision-making problems, the biggest problem faced by decision-makers is determining the importance degrees of the criteria. While determining the degree of importance of the criteria, it is of great importance to express the comments and opinions of the decision-makers in an objective manner mathematically. At the same time, while determining the degree of importance; Due to the intuitionistic fuzzy set-based multi-criteria decision-making process, it is necessary to express the importance levels as intuitionistic fuzzy values. Many researchers investigated the importance levels of the criteria and defined many methods [14, 15, 44–46]. In this paper, linguistic terms were used to facilitate decision makers to express their ideas. As indicated in Table 1, the weights of the criteria will be determined by linguistic terms. The intuitionistic fuzzy numbers (IFN) equivalents of the linguistically expressed values by the decision-makers are given in Table 1.

V shape criterion type was used in this paper. Also, criteria weights may be determined using the controlled set. The basic definitions for controlled sets for decision makers who want to use a controlled set are as follows:

Table 1 Linguistic terms for rating criterion

Linguistic terms	IFNs
Very important	(0.9, 0.1)
Important	(0.75, 0.2)
Medium	(0.5, 0.45)
Unimportant	(0.35, 0.6)
Very unimportant	(0.1, 0.9)

Definition 3 [14] Let E be an universe, α is a function from E to I then E is called α- set.

Definition 4 [14] Let E be an α-set. The set E is called α-controlled set if

$$\forall x \in E, \exists y \in E \ni 1 - \alpha(x) - \alpha(y)$$

The family of α-controlled set on an universe E is represented by $E \in CS(\alpha)$.

Definition 5 [14] Let $E \in CS(\alpha)$ and $a \in E$. The following set is called control set of a,

$$\alpha = \{b \in E | 1 - \mu(a) - \mu(b)\}$$

Definition 6 [15] Let E be an α-set. We define the following mapping on E so that

$$\alpha^*(x) = \begin{cases} 1 - \alpha(x), \ x \in E_\alpha \\ sup_y \ \alpha(y), \ \ y \in E|\alpha(x) < 1 - \alpha(y) \\ 0, \ \ otherwise \end{cases}$$

where $E_\alpha = \cup_{a \in E} \overline{a}$.

Definition 7 [15] Let E be α-set. Then the set $A = \{\langle x, \alpha(x), \alpha^*(x)\rangle | x \in E\}$ is called (α, α^*)-controlled set.

Definition 8 [8] V shape criterion type;

$$P(d) = \begin{cases} 0, & d \leq q \\ \frac{d-q}{p-q}, & q < d \leq p \\ 1, & d > p \end{cases} \tag{9}$$

In which, q: indifference and p: strict preference. Decision makers are free to change these thresholds according to the desired situation. Deviations based on pairwise comparisons:

$$d_j(x, y) = c_j(x) - c_j(y) \tag{10}$$

Definition 9 [41] An intuitionistic fuzzy preference relation R on the set $X = x_1, x_2, \ldots, x_n$ is represented by a matrix $R = (r_{ik})_{n\times n}$ where $r_{ik} = \langle (x_i, x_k), \mu(x_i, x_k), \nu(x_i, x_k)\rangle$ for all $i, k = 1, 2, \ldots, n$.

$$\mu_{ik}, \nu_{ik} \in [0, 1], \quad \mu_{ik} + \nu_{ik} \leq 1, \quad \mu_{ik} = \nu_{ki}, \quad \mu_{ki} = \nu_{ik}$$

$$\pi_{ik} = 1 - \mu_{ik} - \nu_{ik}, \tag{11}$$

for $\forall i, k = 1, 2, \ldots, n$.

2.1 Algorithm of Intuitionistic Fuzzy PROMETHEE

The preference matrix is obtained [44]:

$$U^{(j)} = \left(\mu_{ik}^{(j)}\right)_{n\times n} = \begin{bmatrix} - & \mu_{12}^{(j)} & \cdots & \mu_{1n}^{(j)} \\ \mu_{21}^{(j)} & - & \cdots & \mu_{2n}^{(j)} \\ \vdots & \vdots & - & \vdots \\ \mu_{n1}^{(j)} & \mu_{n2}^{(j)} & \cdots & - \end{bmatrix} \tag{12}$$

Matrix of the intuitionistic fuzzy preference relation is obtained:

$$R^{(j)} = \left(r_{ik}^{(j)}\right)_{n\times n} = \begin{bmatrix} - & \left(\mu_{12}^{(j)}, \nu_{12}^{(j)}\right) & \cdots & \left(\mu_{1n}^{(j)}, \nu_{1n}^{(j)}\right) \\ \left(\mu_{21}^{(j)}, \nu_{21}^{(j)}\right) & - & \cdots & \left(\mu_{2n}^{(j)}, \nu_{2n}^{(j)}\right) \\ \vdots & \vdots & - & \vdots \\ \left(\mu_{n1}^{(j)}, \nu_{n1}^{(j)}\right) & \left(\mu_{n2}^{(j)}, \nu_{n2}^{(j)}\right) & \cdots & - \end{bmatrix} \tag{13}$$

Then; considering the c_j $(j = 1, 2, \ldots, m)$ criteria, we must establish the general preference index for each alternative. The fuzzy preference index for all criteria can be calculated for the alternative x_i to x_k:

$$r(x_i, x_k) = r_{ik} = \bigoplus_{j=1}^{m} \left(\tilde{w}_j \otimes r_{ik}^{(j)}\right) \tag{14}$$

$\tilde{\omega}_j = \left(\mu_{\tilde{\omega}_j}, \nu_{\tilde{\omega}_j}\right)$ then according to Eqs. (4) and (5):

$$\tilde{w}_j \otimes r_{ik}^{(j)} = \left(\mu_{ik}^{(j)} \mu_{\tilde{w}j}, \nu_{ik}^{(j)} + \nu_{\tilde{w}j} - \nu_{ik}^{(j)} \nu_{\tilde{w}j}\right) \tag{15}$$

Then;

$$r(x_i, x_k) = \overset{m}{\underset{j=1}{\oplus}}\left(\tilde{w}_j \otimes r_{ik}^{(j)}\right)$$
$$= \left(1 - \prod_{j=1}^{m}\left(1 - \mu_{ik}^{(j)}\mu_{\tilde{w}j}\right),\right.$$
$$\left.\prod_{j=1}^{m}\left(\nu_{ik}^{(j)} + \nu_{\tilde{w}j} - \nu_{ik}^{(j)}\nu_{\tilde{w}j}\right)\right) \tag{16}$$

Overall intuitionistic fuzzy preference relationship is obtained:

$$R = (r_{ik})_{n\times n} = \begin{bmatrix} - & (\mu_{12}, \nu_{12}) & \dots & (\mu_{1n}, \nu_{1n}) \\ (\mu_{21}, \nu_{21}) & - & \dots & (\mu_{2n}, \nu_{2n}) \\ \vdots & \vdots & - & \vdots \\ (\mu_{n1}, \nu_{n1}) & (\mu_{n2}\nu_{n2}) & \dots & - \end{bmatrix} \tag{17}$$

Outranking flows are achieved:

(1) The intuitionistic fuzzy positive outranking flow:

$$\tilde{\varphi}^{+}(x_i) = \frac{1}{n-1}\overset{n}{\underset{k=1,k\neq i}{\oplus}} r(x_i, x_k) = \frac{1}{n-1}\overset{n}{\underset{k=1,k\neq i}{\oplus}} r_{ik} \tag{18}$$

(2) The intuitionistic fuzzy negative outranking flow:

$$\tilde{\varphi}^{-}(x_i) = \frac{1}{n-1}\overset{n}{\underset{k=1,k\neq i}{\oplus}} r(x_k, x_i) = \frac{1}{n-1}\overset{n}{\underset{k=1,k\neq i}{\oplus}} r_{ki} \tag{19}$$

The relationship between $\tilde{\varphi}^{+}(x_i)$ and $\tilde{\varphi}^{-}(x_i)$ is explained:

$$\rho(\varphi(x_i)) = \rho\left(\tilde{\varphi}^{+}(x_i)\right) - \rho\left(\tilde{\varphi}^{-}(x_i)\right) \tag{20}$$

Three different ranking for $\tilde{\varphi}^{+}(x_i)$ and $\tilde{\varphi}^{-}(x_i)$ can be achieved [44]:

1. Partial ranking: x_i outranks x_k if $\tilde{\varphi}^{+}(x_i) > \tilde{\varphi}^{+}(x_k)$ and $\tilde{\varphi}^{-}(x_i)$ and $\tilde{\varphi}^{-}(x_k)$;
2. Equality: $\tilde{\varphi}^{+}(x_i) = \tilde{\varphi}^{+}(x_k)$ and $\tilde{\varphi}^{-}(x_i) = \tilde{\varphi}^{-}(x_k)$ hold at the same time involves indifference between two options.
3. Incomparability: This takes if $\tilde{\varphi}^{+}(x_i) > \tilde{\varphi}^{+}(x_k)$ and $\tilde{\varphi}^{-}(x_i) > \tilde{\varphi}^{-}(x_k)$ or $\tilde{\varphi}^{+}(x_i) < \tilde{\varphi}^{+}(x_k)$ ve $\tilde{\varphi}^{-}(x_i) < \tilde{\varphi}^{-}(x_k)$.

3 Examination of Recycled Papers Using the Intuitionistic Fuzzy Based-PROMETHEE

In this study, the papers recycled 4 times were evaluated according to colors and L, a, b values. The reuse of paper by recycling is a very important step in human life in terms of printing, forests, waste and recycling. In this direction, recycled papers in Cyan, Magenta and Yellow colors were examined and the L, a, b values obtained at the end of each recycling step were examined. Recycling was applied 4 times in total, and data were obtained for each of the papers. Sample 1, Sample 2, Sample 3 and Sample 4 represent each recycling step. In the decision-making mechanism, the recycling stages for each color were considered as an alternative.

Step 1: The data set in this study consisted of a total of 12 alternatives and 3 criteria. Classification of alternatives and criteria was determined.

Alternatives are as follows:

- A_1: Cyan Sample 1
- A_2: Cyan Sample 2
- A_3: Cyan Sample 3
- A_4: Cyan Sample 4
- A_5: Magenta Sample 1
- A_6: Magenta Sample 2
- A_7: Magenta Sample 3
- A_8: Magenta Sample 4
- A_9: Yellow Sample 1
- A_{10}: Yellow Sample 2
- A_{11}: Yellow Sample 3
- A_{12}: Yellow Sample 4.

Criteria are as follows:

- $C_1: L$
- $C_2: a$
- $C_3: b$.

The values obtained according to the criteria at each stage of recycling according to the colors Cyan, Magenta and Yellow were shown in Table 2.

Step 2: Decision makers were consulted to determine the degree of importance while evaluating the criteria. One of the most effective methods when determining the membership and non-membership values of the criterion weights is controlled sets. Applications using controlled sets are available in the literature. In this study, the opinions of the decision makers were expressed through linguistic terms. Decision makers consist of experts in their fields such as paper, recycling and print quality. With the joint decision of all decision makers, the importance degrees of the criteria were provided in Table 3, and the intuitionistic fuzzy value equivalents were also shown.

Table 2 Values of alternatives

	L	a	b
Cyan Sample 1	59.31	– 15.33	– 36.23
Cyan Sample 2	61.34	– 16.68	– 36.47
Cyan Sample 3	61.76	– 17.79	– 36.12
Cyan Sample 4	61.94	– 17.93	– 37.50
Magenta Sample 1	54.52	47.74	2.18
Magenta Sample 2	55.76	47.80	3.71
Magenta Sample 3	56.51	49.23	2.5
Magenta Sample 4	55.92	49.95	2.39
Yellow Sample 1	80.7	3.46	65.69
Yellow Sample 2	– 36.47	1.35	67.22
Yellow Sample 3	84.56	3.87	70.58
Yellow Sample 4	85.06	4.71	72.68

Table 3 Values of criteria

Criteria	Linguistic term	IFNs
L	Very important	(0.90, 0.1)
a	Important	(0.75, 0.2)
b	Important	(0.75, 0.2)

Step 3: Type V linear criteria type was utilized and the preference matrix $U^{(j)}$ was established.

Steps 4–5: The intuitionistic fuzzy preference relation, after that the overall intuitionistic fuzzy preference relation was established.

Step 6: The intuitionistic fuzzy positive and negative outranking flow by utilizing Eqs. (16) and (17) were created (Table 4).

Step 7: The relationship between $\tilde{\varphi}^{+}(x_i)$ and $\tilde{\varphi}^{-}(x_i)$ was calculated. The net outranking values were given in Table 5 and also the graphic of the net outranking flow was shown.

When the values of Cyan color in 4 recycling stages were examined, it was observed that the paper quality increased as the recycling stages increased. The order of quality paper according to recycling samples is as follows for Cyan color: Cyan Sample 4, Cyan Sample 3, Cyan Sample 2, Cyan Sample 1 (Fig. 2).

When the values of Magenta color in 4 recycling stages were examined, it was observed that the recycling stages and paper quality were independent. The highest quality paper was obtained in the 2nd recycling stage in Magenta color. The order of quality paper according to recycling samples is as follows for Magenta color: Magenta Sample 2, Magenta Sample 3, Magenta Sample 4, Magenta Sample 1 (Fig. 3).

Table 4 Positive and negative outranking flow of alternatives

Positive outranking flow	Negative outranking flow
$\rho(\tilde{\varphi}^+(x_1)) = 0.867539$	$\rho(\tilde{\varphi}^-(x_1)) = 0.867769$
$\rho(\tilde{\varphi}^+(x_2)) = 0.865558$	$\rho(\tilde{\varphi}^-(x_2)) = 0.867769$
$\rho(\tilde{\varphi}^+(x_3)) = 0.865153$	$\rho(\tilde{\varphi}^-(x_3)) = 0.867769$
$\rho(\tilde{\varphi}^+(x_4)) = 0.864957$	$\rho(\tilde{\varphi}^-(x_4)) = 0.867769$
$\rho(\tilde{\varphi}^+(x_5)) = 0.867769$	$\rho(\tilde{\varphi}^-(x_5)) = 0.868040$
$\rho(\tilde{\varphi}^+(x_6)) = 0.867769$	$\rho(\tilde{\varphi}^-(x_6)) = 0.868162$
$\rho(\tilde{\varphi}^+(x_7)) = 0.867769$	$\rho(\tilde{\varphi}^-(x_7)) = 0.868143$
$\rho(\tilde{\varphi}^+(x_8)) = 0.867769$	$\rho(\tilde{\varphi}^-(x_8)) = 0.868120$
$\rho(\tilde{\varphi}^+(x_9)) = 0.867769$	$\rho(\tilde{\varphi}^-(x_9)) = 0.868828$
$\rho(\tilde{\varphi}^+(x_{10})) = 0.867769$	$\rho(\tilde{\varphi}^-(x_{10})) = 0.867769$
$\rho(\tilde{\varphi}^+(x_{11})) = 0.867769$	$\rho(\tilde{\varphi}^-(x_{11})) = 0.869186$
$\rho(\tilde{\varphi}^+(x_{12})) = 0.867769$	$\rho(\tilde{\varphi}^-(x_{12})) = 0.869484$

Table 5 Net outranking flow for alternatives

$\rho(\tilde{\varphi}(x_1)) = -\ 0.000230$
$\rho(\tilde{\varphi}(x_2)) = -\ 0.002210$
$\rho(\tilde{\varphi}(x_2)) = -\ 0.002616$
$\rho(\tilde{\varphi}(x_4)) = -\ 0.002812$
$\rho(\tilde{\varphi}(x_5)) = -\ 0.000271$
$\rho(\tilde{\varphi}(x_6)) = -\ 0.000394$
$\rho(\tilde{\varphi}(x_7)) = -\ 0.000375$
$\rho(\tilde{\varphi}(x_8)) = -\ 0.000352$
$\rho(\tilde{\varphi}(x_9)) = -\ 0.001059$
$\rho(\tilde{\varphi}(x_{10})) = 0.000000$
$\rho(\tilde{\varphi}(x_{11})) = -\ 0.001417$
$\rho(\tilde{\varphi}(x_{12})) = -\ 0.001716$

When the values of Yellow color in 4 recycling stages were examined, it was observed that the recycling stages and paper quality were independent. The highest quality paper was obtained in the 4th recycling stage in Yellow color. The order of quality paper according to recycling samples is as follows for Yellow color: Yellow Sample 4, Yellow Sample 3, Yellow Sample 1, Yellow Sample 2 (Fig. 4).

Figure 5 was obtained when all recycling stages were observed. If the recycling stages were interpreted according to the colors, the print quality order of the colors in the 1st recycling stage: Yellow, Magenta, Cyan. In order of print quality when sorted by 2nd recycling stage: Cyan, Magenta, Yellow. Sorting by 3rd recycling stage: Cyan, Yellow, Magenta, and finally, according to the 4th recycling stage, the order of print quality is as follows: Cyan, Yellow, Magenta.

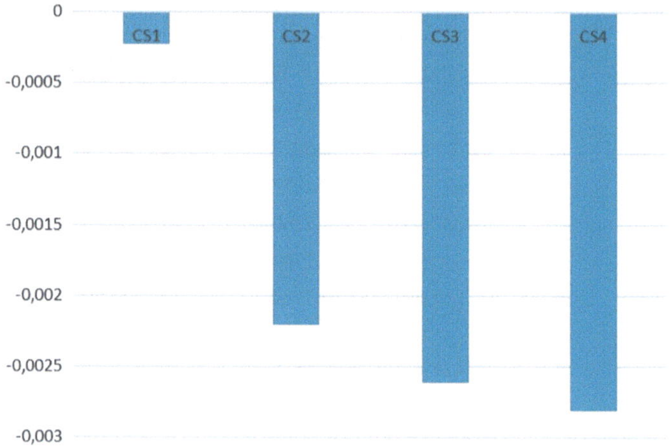

Fig. 2 Values of Cyan color by four recycling samples

Fig. 3 Values of Magenta color by four recycling samples

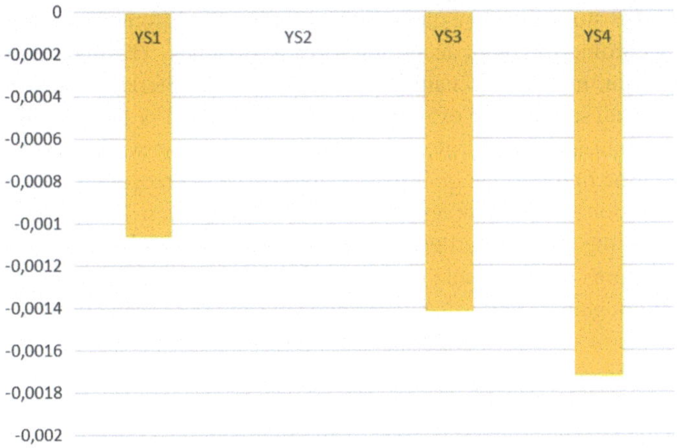

Fig. 4 Values of Yellow color by four recycling samples

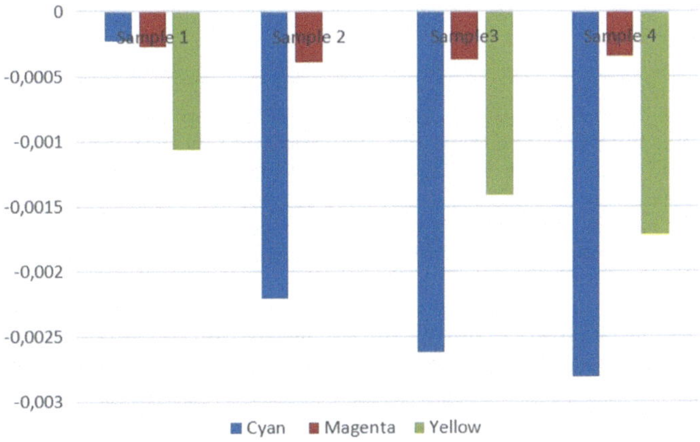

Fig. 5 Recycling samples by colors

4 Conclusion

Fuzzy sets and intuitionistic fuzzy sets have attracted the attention of researchers since the day they were defined. Intuitionistic fuzzy sets, which are used in many application areas and give favorable results, also give effective results in multi-criteria decision making actions. PROMETHEE method, which is one of the multi-criteria decision making methods defined on an intuitionistic fuzzy basis, offers the opportunity to observe positive and negative rankings as well as to obtain clear rankings. There are many methods defined for determining the weights of the criteria. In this study, linguistic terms were used to reflect the opinions of the decision makers objectively. In this way, decision makers determined the degree of importance for each criterion. In this study, it is to examines the quality of 4 different recycling samples of papers according to colors according to L, a, b values. The main purpose is to create a decision making mechanism in which all the criteria are activated at the same time and all samples are examined at the same time. For this purpose, samples of Cyan, Magenta and Yellow colors at 4 different stages of recycling were examined. According to the findings, the quality of each color at the recycling stage was interpreted. In addition, all the samples at the recycling stage were interpreted according to colors. By using the multi-criteria decision making mechanism established in this study, applications may be made on different examples. By increasing the number of criteria, point gain, gamut value, trapping value, etc. The evaluation may be made by adding criteria such as it is possible to work on different recycled papers. The main purpose here is to create a mechanism where all the alternatives and criteria are observed at the same time. Thanks to this mechanism, favorable results will be obtained in all desired application areas. This study will guide many researchers and open new ideas.

References

1. Albadvi, A.: Formulating national information technology strategies: a preference ranking model using PROMETHEE method. Eur. J. Oper. Res. **153**, 290–296 (2004)
2. Albadvi, A., Chaharsooghi, S.K., Esfahanipour, A.: Decision making in stock trading: an application of PROMETHEE. Eur. J. Oper. Res. **177**, 673–683 (2007)
3. Atanassov, K.T.: Intuitionistic fuzzy sets, VII ITKR session, Sofia, 20–23 (Deposed in Centr. Sci. Techn. Lib. Bulg. Acad. Sci., 1697/84) (in Bulgarian). Reprinted: Int. J. Bioautom. **20**(S1), 2016 (1983)
4. Atan, Ö., Kutlu, F., Castillo, O.: Intuitionistic fuzzy sliding controller for uncertain hyperchaotic synchronization. Int. J. Fuzzy Syst. **22**(5), 1430–1443 (2020)
5. Behzadian, M., Kazemzadeh, R.B., Albadvi, A., Aghdasi, M.: PROMETHEE: a comprehensive literature review on methodologies and applications. Eur. J. Oper. Res. **200**, 198–215 (2010)
6. Brans, J.P.: L'ingenierie de la decision: l'elaboration d'instruments d'aide a la decision. Colloq. d'aide a la decision. Universite Laval, Quebec, Canada (1982)
7. Brans, J.P., Mareschal, B., Vincke, P.: PROMETHEE: a new family of outranking methods in multicriteria analysis. Oper. Res. IFORS **84**, 477–490 (1984)
8. Brans, J.P., Vincke, Ph.: A preference ranking organisation method (the PROMETHEE method for multiple criteria decision making). Manage. Sci. **31**(6), 647–656 (1985)
9. Brans, J.P., Mareschal, B.: PROMETHEE methods, chap. 5. In: Multiple Criteria Decision Analysis: State of the Art Surveys. International Series in Operations Research—Management Science, vol. 78. Springer, New York, NY (2005). https://doi.org/10.1007/0-387-23081-5-5
10. Castillo, O., Kutlu, F., Atan, Ö.: Intuitionistic fuzzy control of twin rotor multiple input multiple output systems. J. Intell. Fuzzy Syst. **38**(1), 821–833 (2020)
11. Çitil, M., Tuğrul, F., Yılmaz, B.: An application of multi criteria decision making: ranking of school success. Celal Bayar Univ. J. Sci. **15**(1), 45–50 (2019)
12. Çuvalcıoğlu, G., Yılmaz, S.: Some properties of OTMOs on IFSs. Adv. Stud. Contemp. Math. **20**(4), 621–628 (2010)
13. Çuvalcıoğlu, G., Yılmaz, S., Atanassov, K.T.: Matrix representation of the second type of intuitionistic fuzzy modal operators. Notes Intuition. Fuzzy Sets **20**(5), 9–16 (2014)
14. Çuvalcıoğlu, G.: Controlled Set Theory, p. 342. Bogolyubov Readings DIF-2013, Ukraine (2013)

15. Çuvalcıoğlu, G.: Some properties of controlled set theory. Notes Intuition. Fuzzy Sets **20**(2), 37–42 (2014)
16. Çuvalcıoğlu, G., Yılmaz, S.: Some properties of intuitionistic fuzzy equivalence relations and class trees w.r.t. intuitionistic fuzzy equivalence relations. Adv. Stud. Contemp. Math. **24**(1), 77–86 (2014)
17. Çuvalcıoğlu, G., Aykut, E.: An application of some intuitionistic fuzzy modal operators to agriculture. Notes Intuition. Fuzzy Sets **21**(2), 140–149 (2015)
18. Çuvalcıoğlu, G., Tarsuslu, S., Bal, A., Artun, G.: Intuitionistic fuzzy modal operators in intelligent system for pesticide and fertilization. Ann. Fuzzy Math. Inform. **16**(1), 117–132 (2018)
19. Çuvalcıoğlu, G., Bureva, V., Michalikova, A.: Intercriteria analysis applied to university ranking system of Turkey. Notes Intuition. Fuzzy Sets **25**(4), 90–97 (2019)
20. Deniz, Ü.: Fuzzy linear mappings. J. Univers. Math. **4**(1), 81–86 (2021)
21. Kutlu, F.: On separation axioms in temporal intuitionistic fuzzy Šostak topology. Notes Intuition. Fuzzy Sets **23**(1), 21–30 (2017)
22. Kutlu, F., Tuğrul, F., Çitil, M.: Introduction to temporal intuitionistic fuzzy approximate reasoning. Commun. Fac. Sci. Univ. Ank. Ser. A1 Math. Stat. **69**(1), 232–251 (2020)
23. Tuğrul, F., Yılmaz, B., Çitil, M.: School success ranking in multi criteria decision making. Turk. J. Math. Comput. Sci. **10**, 1–6 (2018)
24. Tuğrul, F., Yılmaz, B., Çitil, M.: Application of ranking with similarity measure in multi criteria decision making. Konuralp J. Math. **7**(2), 438–441 (2019)
25. Tuğrul, F., Çitil, M., Balcı, S.: On the determination students' aptitude using intuitionistic fuzzy logic. J. Univers. Math. **2**(1), 36–41 (2019)
26. Tuğrul, F.: Application of intuitionistic fuzzy logic with a new method in multi criteria decision making process. Ph.D. thesis, Kahramanmaraş Sütcü İmam University, 2021
27. Tuğrul, F., Citil, M.: A new perspective on evaluation system in education with intuitionistic fuzzy logic and PROMETHEE algorithm. J. Univers. Math. **4**(1), 13–24 (2021)
28. Şahin, C.: Evaluation of the effect of the efficiency of operators on print quality in offset printing. Ph.D. thesis, Gazi University, Ankara, 2012
29. Şahin, C., Özomay, Z., Keskin, B.: Evaluation of weighting pressure's effect on dot gain in newsprint paper. Electron. J. Vocat. Coll. **3**(3), 121–128 (2013)
30. Yılmaz, U.: Investigation of the effect on offset print quality of physical properties of some printing papers. M.Sc. thesis, Kahramanmaraş Sütçü İmam University, Kahramanmaraş, 2016
31. Yılmaz, U., Tutuş, A., Sönmez, S.: A research of comparing of the effects on the color properties of offset, laser and inkjet print systems on uncoated recycled papers. Erciyes Univ. J. Inst. Sci. Technol. **37**(1), 130–136 (2021a)
32. Yılmaz, U., Tutuş, A., Sönmez, S.: Investigation of recycling repetition effect on deinkability efficiency and printing properties. J. Adhes. Sci. Technol. (2021b). https://doi.org/10.1080/01694243.2021.1966256
33. Yılmaz, U., Tutuş, A., Sönmez, S.: Investigation of color values of inkjet and laserjet prints on recycled papers. J. Graph. Eng. Des. **12**(3), 512 (2021c)
34. Yılmaz, U.: Analysis of the color properties of papers subjected to different recycling numbers in electrophotographic (toner) printing. Pigm. Resin Technol. (2022). https://doi.org/10.1108/PRT-09-2021-0106
35. Yılmaz, U., Tutuş, A., Sönmez, S.: Effects of using recycled paper in inkjet printing system on colour difference. Pigm. Resin Technol. **51**(3), 336–343 (2022)
36. Kasmani, E.J., Mahdavi, S., Samariha, A.: Physical properties and printability characteristics of mechanical printing paper with LWC. BioResources **8**(3), 3646–3656 (2013)
37. Hu, G., Fu, S., Chu, F., Lin, M.: Relationship between paper whiteness and color reproduction in inkjet printing. BioResources **12**(3), 4854–4866 (2017)
38. Sönmez, S.: Interactive effects of copolymers and nano-sized pigments on coated recycled paperboards in flexographic print applications. Asian J. Chem. **23**(6), 2609–2613 (2011)
39. Mclellan, M.R., Lind, L.R., Kime, R.W.: Hue angle determinations and statistical analysis for multiquadrant hunter L, a, b data. J. Food Qual. **18**(3), 235–240 (1995)

40. Atanassov, K.T.: Intuitionistic fuzzy sets. Fuzzy Sets Syst. **20**(1), 87–96 (1986)
41. Xu, Z.S.: Intuitionistic fuzzy aggregation operators. IEEE Trans. Fuzzy Syst. **15**, 1179–1187 (2007)
42. Xu, Z.S., Yager, R.R.: Some geometric aggregation operators based on intuitionistic fuzzy set. Int. J. Gen. Syst. **35**, 417–433 (2006)
43. Szmidt, E., Kacprzyk, J.: Amount of information and its reliability in the ranking of Atanassov's intuitionistic fuzzy alternatives. In: Rakus-Andersson, E., Yager, R.R., Ichalkaranje, N., Jain, L. (eds.) Recent Advances in Decision Making. Studies in Computational Intelligence, pp. 7–19. Springer, Berlin, Germany (2009)
44. Liao, H., Xu, Z.S.: Multi-criteria decision making with intuitionistic fuzzy PROMETHEE. J. Intell. Fuzzy Syst. **27**, 1703–1717 (2014a)
45. Liao, H., Xu, Z.S.: Priorities of intuitionistic fuzzy preference relation based on multiplicative consistency. IEEE Trans. Fuzzy Syst. **22**(6), 1669–1681 (2014b)
46. Wang, Z.J.: Derivation of intuitionistic fuzzy weights based on intuitionistic fuzzy preference relations. Appl. Math. Model. **37**, 6377–6388 (2013)

On the Hadamard's Type Inequalities for Exponential Type Co-ordinated Convex Functions via Fractional Integrals

Firas Cumaoğlu, Abdullah Akkurt, and Hüseyin Yıldırım

Abstract In this article, we introduce a new definition, which is called exponential type co-ordinated convexity. Using this definition, we have obtained some new fractional integral inequalities of the Hermite–Hadamard type. Our results generalize some of the existing studies in the literature.

Keywords Hermite–Hadamard inequality · Fractional calculus · Convex functions

1 Introduction

Inequality theory is one of the important topics in mathematics. In recent years, especially Hermite–Hadamard type fractional integral inequalities are one of them. The famous Hermite–Hadamard inequalities play an important role in obtaining integral inequalities. Please see the studies on this subject [1–4, 6, 8–17].

F. Cumaoğlu · A. Akkurt (✉) · H. Yıldırım
Department of Mathematics, Faculty of Science, Kahramanmaraş Sütçü İmam University, Kahramanmaraş, Turkey
e-mail: abdullahmat@gmail.com

F. Cumaoğlu
e-mail: firaselhaci1997@gmail.com

H. Yıldırım
e-mail: hyildir@ksu.edu.tr

O. Castillo et al. (eds.), *Advanced Mathematics for the Modeling and Solution of Challenging Problems in Engineering*, Studies in Systems, Decision and Control 582,
https://doi.org/10.1007/978-3-031-82342-8_14

The convexity property of a given function plays an important role in obtaining integral inequalities. Proving inequalities for convex functions has a long and rich history in mathematics. We refer the reader to a monograph written by Dragomir and Pearce [4].

Let $f : I \subset \mathbb{R} \to \mathbb{R}$ be a convex function, then,

$$f\left(\frac{a+b}{2}\right) \leq \frac{1}{b-a}\int_a^b f(x) \leq \frac{f(a)+f(b)}{2}, \tag{1.1}$$

is known in the literature as Hermite–Hadamard inequality for convex mappings. Note that some of the classical inequalities for means can be derived from (1.1) for appropriate particular selections of the mapping f.

Let us now consider a bidimensional interval $\Delta =: [a, b] \times [c, d]$ in $\mathbb{R}^2$ with $a < b$ and $c < d$. A mapping $f : \Delta \to \mathbb{R}$ is said to be convex on Δ if the following inequality:

$$f(tx + (1-t)z, ty + (1-t)w) \leq tf(x, y) + (1-t)f(z, w), \tag{1.2}$$

holds for all (x, y), $(z, w) \in \Delta$ and $t \in [0, 1]$. A function $f : \Delta \to \mathbb{R}$ is said to be convex on the co-ordinates on Δ if the partial mappings $f_y : [a, b] \to \mathbb{R}, f_y(u) = f(u, y)$ and $f_x : [c, d] \to \mathbb{R}, f_x(v) = f(x, v)$ are convex where defined for all $x \in [a, b]$ and $y \in [c, d]$ (see, [3]).

Definition 1.1 A function $f : \Delta \to \mathbb{R}$ will be called co-ordinated convex on Δ, for all $t, s \in [0, 1]$ and (x, y), $(u, w) \in \Delta$, then the following inequalty holds;

$$\begin{aligned} f(tx + (1-t)y, su + (1-s)w) \leq\ & tsf(x, u) + s(1-t)f(y, u) \\ & + t(1-s)f(x, w) + (1-t)(1-s) \\ & f(y, w). \end{aligned} \tag{1.3}$$

Earlier, Dragomir [3] established the following very useful Hermite–Hadamard type inequality.

Theorem 1.2 *Suppose that $f : \Delta \to \mathbb{R}$ is co-ordinated convex on Δ. Then one has the inequalities;*

$$\begin{aligned} f\left(\frac{a+b}{2}, \frac{c+d}{2}\right) &\leq \frac{1}{2}\left[\frac{1}{b-a}\int_a^b f\left(x, \frac{c+d}{2}\right)dx + \frac{1}{d-c}\int_c^d f\left(\frac{a+b}{2}, y\right)dy\right] \\ &\leq \frac{1}{(b-a)(d-c)}\int_a^b\int_c^d f(x, y)dydx \end{aligned}$$

$$\leq \frac{1}{4}\left[\frac{1}{b-a}\int_a^b f(x,c)dx + \frac{1}{b-a}\int_a^b f(x,d)dx\right.$$

$$\left.+\frac{1}{d-c}\int_c^d f(a,y)dy + \frac{1}{d-c}\int_c^d f(b,y)dy\right]$$

$$\leq \frac{f(a,c)+f(a,d)+f(b,c)+f(b,d)}{4}. \tag{1.4}$$

Definition 1.3 Let $f \in L_1[a,b]$. The Riemann Liouville fractional integrals $I_{a^+}^{\alpha}f$ and $I_{b^-}^{\alpha}f$ of order $\alpha > 0$, $a \geq 0$ are defined by [5, 7]

$$I_{a^+}^{\alpha}f(x) = \frac{1}{\Gamma(\alpha)}\int_a^x (x-t)^{\alpha-1}f(t)dt, \quad x > a, \tag{1.5}$$

and

$$I_{b^-}^{\alpha}f(x) = \frac{1}{\Gamma(\alpha)}\int_x^b (t-x)^{\alpha-1}f(t)dt, \quad x < b, \tag{1.6}$$

respectively. Here, $\Gamma(\alpha)$ is the Gamma function and

$$I_{a^+}^{0}f(x) = I_{b^-}^{0}f(x) = f(x).$$

Definition 1.4 Let $f \in L_1([a,b]\times[c,d])$. The Riemann–Liouville integrals $I_{a^+,c^+}^{\alpha,\beta}$, $I_{a^+,d^-}^{\alpha,\beta}$, $I_{b^-,c^+}^{\alpha,\beta}$ and $I_{b^-,d^-}^{\alpha,\beta}$ of order $\alpha, \beta > 0$ with $a, c \geq 0$ are defined by [8]

$$I_{a^+,c^+}^{\alpha,\beta}f(x,y) = \frac{1}{\Gamma(\alpha)\Gamma(\beta)}\int_a^x\int_c^y (x-t)^{\alpha-1}(y-s)^{\beta-1}f(t,s)dsdt, \tag{1.7}$$

$$I_{a^+,d^-}^{\alpha,\beta}f(x,y) = \frac{1}{\Gamma(\alpha)\Gamma(\beta)}\int_a^x\int_y^d (x-t)^{\alpha-1}(s-y)^{\beta-1}f(t,s)dsdt, \tag{1.8}$$

$$I_{b^-,c^+}^{\alpha,\beta}f(x,y) = \frac{1}{\Gamma(\alpha)\Gamma(\beta)}\int_x^b\int_c^y (t-x)^{\alpha-1}(y-s)^{\beta-1}f(t,s)dsdt, \tag{1.9}$$

$$I_{b^-,d^-}^{\alpha,\beta}f(x,y) = \frac{1}{\Gamma(\alpha)\Gamma(\beta)}\int_x^b\int_y^d (t-x)^{\alpha-1}(s-y)^{\beta-1}f(t,s)dsdt, \tag{1.10}$$

respectively. Here, $\Gamma(\alpha)$ is the Gamma function and

$$\begin{aligned} I^{0,0}_{a^+,c^+} f(x,y) &= I^{0,0}_{a^+,d^-} f(x,y) = I^{0,0}_{b^-,c^+} f(x,y) \\ &= I^{0,0}_{b^-,d^-} f(x,y) = f(x,y). \end{aligned}$$

Also,

$$I^{1,1}_{a^+,c^+} f(x,y) = \frac{1}{\Gamma(\alpha)\Gamma(\beta)} \int\limits_a^x \int\limits_c^y f(t,s)\,ds\,dt. \tag{1.11}$$

Similar to Definitions 1.3 and 1.4, we give the following fractional integrals:

$$I^{\alpha}_{a^+} f\left(x, \frac{c+d}{2}\right) = \frac{1}{\Gamma(\alpha)} \int\limits_a^x (x-t)^{\alpha-1} f\left(t, \frac{c+d}{2}\right) dt, \tag{1.12}$$

$$I^{\alpha}_{b^-} f\left(x, \frac{c+d}{2}\right) = \frac{1}{\Gamma(\alpha)} \int\limits_x^b (t-x)^{\alpha-1} f\left(t, \frac{c+d}{2}\right) dt, \tag{1.13}$$

$$I^{\beta}_{c^+} f\left(\frac{a+b}{2}, y\right) = \frac{1}{\Gamma(\beta)} \int\limits_c^y (y-s)^{\beta-1} f\left(\frac{a+b}{2}, s\right) ds, \tag{1.14}$$

$$I^{\beta}_{d^-} f\left(\frac{a+b}{2}, y\right) = \frac{1}{\Gamma(\beta)} \int\limits_y^d (s-y)^{\beta-1} f\left(\frac{a+b}{2}, s\right) ds. \tag{1.15}$$

It is remarkable that Sarikaya et al. [13] first give the following interesting integral inequalities of Hermite–Hadamard-type involving Riemann–Liouville fractional integrals.

Theorem 1.5 *Let $f : [a,b] \to \mathbb{R}$ be positive function with $0 \le a < b$ and $f \in L_1[a,b]$. If f is a convex function on $[a,b]$ then the following inequalities hold:*

$$f\left(\frac{a+b}{2}\right) \le \frac{\Gamma(\alpha+1)}{2(b-a)^{\alpha}} \left[I^{\alpha}_{a^+} f(b) + I^{\alpha}_{b^-} f(a) \right] \le \frac{f(a)+f(b)}{2}, \tag{1.16}$$

with $\alpha > 0$.

Lemma 1.6 *Let $f : [a,b] \to \mathbb{R}$, be a differentiable mapping on (a,b) with $a < b$. If $f' \in L_1[a,b]$. Then the following equality hold* [13]*;*

$$\frac{f(a)+f(b)}{2} - \frac{\Gamma(\alpha+1)}{2(b-a)^{\alpha}} \left[J^{\alpha}_{a^+} f(b) + J^{\alpha}_{b^-} f(a) \right]$$

$$= \frac{b-a}{2}\int_0^1 \left[(1-t)^\alpha - t^\alpha\right] f'(ta+(1-t)b)dt.$$

Lemma 1.7 *Let $f : \Delta \subset \mathbb{R}^2 \to \mathbb{R}$ be a partial differentiable mapping on $\Delta := [a,b]\times[c,d]$ in $\mathbb{R}^2$ with $0 \le a < b,\ 0 \le c < d$. If $\frac{\partial^2 f}{\partial t \partial s} \in L_1(\Delta)$, then the following equality holds* [8]*:*

$$\frac{f(a,c)+f(a,d)+f(b,c)+f(b,d)}{4} + \frac{\Gamma(\alpha+1)\Gamma(\beta+1)}{4(b-a)^\alpha (d-c)^\beta}$$

$$\left[I^{\alpha,\beta}_{a^+,c^+} f(b,d) + I^{\alpha,\beta}_{a^+,d^-} f(b,c) + I^{\alpha,\beta}_{b^-,c^+} f(a,d) + I^{\alpha,\beta}_{b^-,d^-} f(a,c)\right] - A$$

$$= \frac{(b-a)(d-c)}{4}\left\{\int_0^1\int_0^1 t^\alpha s^\beta \frac{\partial^2 f}{\partial s \partial t}(ta+(1-t)b, sc+(1-s)d)dsdt\right.$$

$$-\int_0^1\int_0^1 t^\alpha (1-s)^\beta \frac{\partial^2 f}{\partial s \partial t}(ta+(1-t)b, sc+(1-s)d)dsdt$$

$$-\int_0^1\int_0^1 (1-t)^\alpha s^\beta \frac{\partial^2 f}{\partial s \partial t}(ta+(1-t)b, sc+(1-s)d)dsdt$$

$$\left.+\int_0^1\int_0^1 (1-t)^\alpha (1-s)^\beta \frac{\partial^2 f}{\partial s \partial t}(ta+(1-t)b, sc+(1-s)d)dsdt\right\},$$

where

$$A = \frac{\Gamma(\beta+1)}{4(d-c)^\beta}\left[I^\beta_{c^+} f(a,d) + I^\beta_{c^+} f(b,d) + I^\beta_{d^-} f(a,c) + I^\beta_{d^-} f(b,c)\right]$$

$$+ \frac{\Gamma(\alpha+1)}{4(b-a)^\alpha}\left[I^\alpha_{a^+} f(b,c) + I^\alpha_{a^+} f(b,d) + I^\alpha_{b^-} f(a,c) + I^\alpha_{b^-} f(a,d)\right].$$

The aim of this article is to first give a general definition of exponential co-ordinated convex functions and to establish Hermite–Hadamard type inequalities with the help of this definition. We will investigate some integral inequalities connected with the Hermite–Hadamard-type inequalities involving Riemann–Liouville fractional integrals.

2 Main Results

In this section, we introduce a new definition, which is called exponential type co-ordinated convexity. Thus, using this definition, new Hermite–Hadamard type inequalities will be obtained.

Definition 2.1 A function $f : \Delta \to R$ will be called exponential type co-ordinated convex on Δ, for all $t, s \in [0, 1]$ and $(x, y), (u, w) \in \Delta$, if the following inequality holds;

$$\begin{aligned} f(tx+(1-t)y, su+(1-s)w) \leq & \left(e^{t}-1\right)\left(e^{s}-1\right)f(x, u) \\ & + \left(e^{1-t}-1\right)\left(e^{s}-1\right)f(y, u) \\ & + \left(e^{t}-1\right)\left(e^{1-s}-1\right)f(x, w) \\ & + \left(e^{1-t}-1\right)\left(e^{1-s}-1\right)f(y, w). \end{aligned} \tag{2.1}$$

Theorem 2.2 *Let* $f \in L_1([a, b] \times [c, d])$. *Then the following inequality hold;*

$$\begin{aligned} f\left(\frac{a+b}{2}, \frac{c+d}{2}\right) \leq & \frac{\Gamma(\alpha+1)\Gamma(\beta+1)\left(\sqrt{e}-1\right)^{2}}{(b-a)^{\alpha}(d-c)^{\beta}\Omega(n, m)} \\ & \times \left[I^{\alpha,\beta}_{a^{+},c^{+}}f(b, d) + I^{\alpha,\beta}_{a^{+},d^{-}}f(b, c)\right. \\ & \left. + I^{\alpha,\beta}_{b^{-},c^{+}}f(a, d) + I^{\alpha,\beta}_{b^{-},d^{-}}f(a, c)\right] \\ & \leq f(a, c) + f(a, d) + f(b, c) + f(b, d), \end{aligned} \tag{2.2}$$

where,

$$\begin{aligned} \Omega(n, m) = & \sum_{n=0}^{\infty}\sum_{m=0}^{\infty}\left[\frac{(-1)^{n}e+1}{(n+\beta)n!}\frac{(-1)^{n}e+1}{(m+\alpha)m!}\right] \\ & - 2\sum_{n=0}^{\infty}\frac{(-1)^{n}e+1}{n!}\left[\frac{1}{\alpha(n+\beta)} + \frac{1}{\beta(n+\alpha)} - \frac{1}{2^{n}\alpha\beta}\right]. \end{aligned}$$

Proof Let

$$\begin{aligned} & f(tx+(1-t)y, su+(1-s)w) \\ & \leq \left(e^{t}-1\right)\left(e^{s}-1\right)f(x, u) + \left(e^{1-t}-1\right)\left(e^{s}-1\right)f(y, u) \\ & \quad + \left(e^{t}-1\right)\left(e^{1-s}-1\right)f(x, w) + \left(e^{1-t}-1\right)\left(e^{1-s}-1\right)f(y, w), \end{aligned}$$

if we choose $x = t_1a + (1-t_1)b$, $y = (1-t_1)a + t_1b$, $u = s_1c + (1-s_1)d$, $w = (1-s_1)c + s_1d$ and we take $t, s = \frac{1}{2}$

$$f\left(\frac{a+b}{2},\frac{c+d}{2}\right) \le (\sqrt{e}-1)^2\big[f(t_1a+(1-t_1)b, s_1c+(1-s_1)d)$$
$$+f(t_1a+(1-t_1)b,(1-s_1)c+s_1d)$$
$$+f((1-t_1)a+t_1b, s_1c+(1-s_1)d)$$
$$+f((1-t_1)a+t_1b,(1-s_1)c+s_1d)\big]. \quad (2.3)$$

Multiplying both sides of (2.3) by $t_1^{\alpha-1}s_1^{\beta-1}$, then by integrating with respect to (t_1, s_1) on $[0, 1] \times [0, 1]$, we have,

$$\int_0^1\int_0^1 t_1^{\alpha-1}s_1^{\beta-1}f\left(\frac{a+b}{2},\frac{c+d}{2}\right)ds_1dt_1$$
$$\le (\sqrt{e}-1)^2\Bigg[\int_0^1\int_0^1 t_1^{\alpha-1}s_1^{\beta-1}\big[f(t_1a+(1-t_1)b, s_1c+(1-s_1)d)$$
$$+f(t_1a+(1-t_1)b,(1-s_1)c+s_1d)\big]ds_1dt_1$$
$$+\int_0^1\int_0^1 t_1^{\alpha-1}s_1^{\beta-1}\big[f((1-t_1)a+t_1b, s_1c+(1-s_1)d)$$
$$+f((1-t_1)a+t_1b,(1-s_1)c+s_1d)\big]ds_1dt_1\Bigg].$$

Using the change of the variable, we get

$$\frac{1}{\alpha\beta}f\left(\frac{a+b}{2},\frac{c+d}{2}\right) \le \frac{1}{(b-a)^\alpha(d-c)^\beta}\Bigg\{\int_a^b\int_c^d (b-x)^{\alpha-1}$$
$$(d-y)^{\beta-1}f(x,y)dydx$$
$$+\int_a^b\int_c^d (b-x)^{\alpha-1}(y-c)^{\beta-1}f(x,y)dydx$$
$$+\int_a^b\int_c^d (x-a)^{\alpha-1}(d-y)^{\beta-1}f(x,y)dydx$$
$$+\int_a^b\int_c^d (x-a)^{\alpha-1}(y-c)^{\beta-1}f(x,y)dydx\Bigg\}$$

by which the first inequality is proved. For the proof of the second inequality in (2.1) we first note that if f is an exponential type co-ordinated convex on Δ, then, by using (1.1) with $x = a$, $y = b$, $u = c$ and $w = d$, it yields

$$
\begin{aligned}
& f(ta+(1-t)b, sc+(1-s)d) \leq \left(e^{t}-1\right)\left(e^{s}-1\right) f(a, c) \\
& \quad+\left(e^{t}-1\right)\left(e^{1-s}-1\right) f(a, d)+\left(e^{(1-t)}-1\right)\left(e^{s}-1\right) f(b, c) \\
& \quad+\left(e^{(1-t)}-1\right)\left(e^{(1-s)}-1\right) f(b, d), \\
& f(ta+(1-t)b,(1-s)c+sd) \leq \left(e^{t}-1\right)\left(e^{1-s}-1\right) f(a, c) \\
& \quad+\left(e^{1-t}-1\right)\left(e^{1-s}-1\right) f(b, c)+\left(e^{t}-1\right)\left(e^{s}-1\right) f(a, d) \\
& \quad+\left(e^{1-t}-1\right)\left(e^{s}-1\right) f(b, d), \\
& f((1-t)a+tb, sc+(1-s)d) \leq \left(e^{1-t}-1\right)\left(e^{s}-1\right) f(a, c) \\
& \quad+\left(e^{t}-1\right)\left(e^{s}-1\right) f(b, c)+\left(e^{1-t}-1\right)\left(e^{1-s}-1\right) f(a, d) \\
& \quad+\left(e^{t}-1\right)\left(e^{1-s}-1\right) f(b, d), \\
& f((1-t)a+tb,(1-s)c+sd) \leq \left(e^{1-t}-1\right)\left(e^{1-s}-1\right) f(a, c) \\
& \quad+\left(e^{t}-1\right)\left(e^{1-s}-1\right) f(b, c)+\left(e^{(1-t)}-1\right)\left(e^{s}-1\right) f(a, d) \\
& \quad+\left(e^{t}-1\right)\left(e^{s}-1\right) f(b, d).
\end{aligned}
$$

By adding these inequalities we have

$$
\begin{aligned}
& f(ta+(1-t)b, sc+(1-s)d) \\
& \quad+f(ta+(1-t)b,(1-s)c+sd) \\
& \quad+f((1-t)a+tb, sc+(1-s)d) \\
& \quad+f((1-t)a+tb,(1-s)c+sd) \\
& \leq \left[\left(e^{t}-1\right)\left(e^{s}-1\right)+\left(e^{t}-1\right)\left(e^{1-s}-1\right)\right. \\
& \quad\left.+\left(e^{1-t}-1\right)\left(e^{s}-1\right)+\left(e^{1-t}-1\right)\left(e^{1-s}-1\right)\right] \\
& \quad\times[f(a, c)+f(a, d)+f(b, c)+f(b, d)]\cdot
\end{aligned} \tag{2.4}
$$

Then, multiplying both sides of (2.4) by $t^{\alpha-1}s^{\beta-1}$ and integrating with respect to (t, s) over $[0, 1] \times [0, 1]$, we get

$$
\begin{aligned}
& \int_0^1 \int_0^1 t^{\alpha-1} s^{\beta-1}[f(ta+(1-t)b, sc+(1-s)d) \\
& \quad+f(ta+(1-t)b,(1-s)c+sd) \\
& \quad+f((1-t)a+tb, sc+(1-s)d) \\
& \quad+f((1-t)a+tb,(1-s)c+sd)]dsdt \\
& \leq \int_0^1 \int_0^1 t^{\alpha-1} s^{\beta-1}\left[\left(e^{t}-1\right)\left(e^{s}-1\right)+\left(e^{t}-1\right)\left(e^{1-s}-1\right)\right. \\
& \quad\left.+\left(e^{1-t}-1\right)\left(e^{s}-1\right)\left(e^{1-t}-1\right)\left(e^{1-s}-1\right)\right][f(a, c)
\end{aligned}
$$

$$+f(a,d)+f(b,c)+f(b,d)\big]dsdt.$$

Here, using the change of the variable we have

$$\frac{\Gamma(\alpha+1)\Gamma(\beta+1)\left(\sqrt{e}-1\right)^2}{(b-a)^{\alpha}(d-c)^{\beta}\sum_{n=0}^{\infty}\sum_{m=0}^{\infty}\left[\frac{(-1)^n e+1}{(n+\beta)n!}\times\frac{(-1)^n e+1}{(m+\alpha)m!}\right]-2\sum_{m=0}^{\infty}\frac{(-1)^n e+1}{n!}\left[\frac{1}{\alpha(n+\beta)}+\frac{1}{\beta(n+\alpha)}-\frac{1}{2^n\alpha\beta}\right]}$$

$$\times\left[I^{\alpha,\beta}_{a^+,c^+}f(b,d)+I^{\alpha,\beta}_{a^+,d^-}f(b,c)+I^{\alpha,\beta}_{b^-,c^+}f(a,d)+I^{\alpha,\beta}_{b^-,d^-}f(a,c)\right]$$
$$\leq f(a,c)+f(a,d)+f(b,c)+f(b,d).$$

The proof is completed.

Theorem 2.3 *Let $f:[a,b]\to\mathbb{R}$ be a positive function with $0\leq a<b$ and $f\in L_1$, $[a,b]$. If f is a convex function on $[a,b]$, then the following inequality hold;*

$$f\left(\frac{a+b}{2}\right)\leq\frac{\Gamma(\alpha+1)\left(\sqrt{e}-1\right)}{(b-a)^{\alpha}\Omega(n)}\left[I^{\alpha}_{a^+}f(b)+I^{\alpha}_{b^-}f(a)\right]$$
$$\leq f(a)+f(b), \tag{2.5}$$

with $\alpha>0$. Where

$$\Omega(n)=\sum_{n=0}^{\infty}\left(\frac{(-1)^n e+1}{(\alpha+n)n!}-\frac{1}{2^n\alpha}\right).$$

Proof Since f is a convex function on $[a,b]$, we have for $x,y\in[a,b]$ with $t=\frac{1}{2}$

$$f\left(\frac{x+y}{2}\right)\leq\left(\sqrt{e}-1\right)\left[f(x)+f(y)\right]$$

i.e., with $x=ta+(1-t)b$, $y=(1-t)a+tb$

$$f\left(\frac{a+b}{2}\right)\leq\left(\sqrt{e}-1\right)\left[f(ta+(1-t)b)+f((1-t)a+tb)\right]. \tag{2.6}$$

Thus, multiplying both sides of (2.6) by $t^{\alpha-1}$, then by integrating with respect to t on [0, 1], we obtain,

$$\int_0^1 t^{\alpha-1} f\left(\frac{a+b}{2}\right)dt \le (\sqrt{e}-1)\int_0^1 t^{\alpha-1}\big[f(ta+(1-t)b)+f((1-t)a+tb)\big]dt$$

$$\frac{1}{\alpha} f\left(\frac{a+b}{2}\right) \le \frac{1}{(b-a)^\alpha}\left\{\int_a^b (b-x)^{\alpha-1} f(x)dx + \int_a^b (y-a)^{\alpha-1} f(y)dy\right\}$$

$$f(ta+(1-t)b) \le \left(e^t-1\right)f(a) + \left(e^{1-t}-1\right)f(b)$$
$$f((1-t)a+tb) \le \left(e^{1-t}-1\right)f(a) + \left(e^t-1\right)f(b)$$

By adding we have

$$f(ta+(1-t)b)+f((1-t)a+tb) \le \left(e^t+e^{1-t}-2\right)\big[f(a)+f(b)\big]. \tag{2.7}$$

Then, multiplying both sides of (2.7) by $t^{\alpha-1}$ and integrating with respect to t over [0, 1], we get,

$$\int_0^1 t^{\alpha-1}\big[f(ta+(1-t)b)+f((1-t)a+tb)\big]dt$$
$$\le \int_0^1 t^{\alpha-1}\left(e^t+e^{1-t}-2\right)\big[f(a)+f(b)\big]dt$$

$$\frac{\Gamma(\alpha+1)(\sqrt{e}-1)}{(b-a)^\alpha \sum_{n=0}^{\infty}\left(\frac{(-1)^n e+1}{(\alpha+n)n!}-\frac{1}{2^n\alpha}\right)}\left[I_{a+}^\alpha f(b)+I_{b-}^\alpha f(a)\right] \le f(a)+f(b).$$

Theorem 2.4 *Let $f:\Delta\subseteq\mathbb{R}^2\to\mathbb{R}$ be an exponential type co-ordinated convex on $\Delta:=[a,b]\times[c,d]$ in $\mathbb{R}^2$ with $0\le a<b$, $0\le c<d$ and $f\in L_1(\Delta)$. Then one has the inequalities;*

$$f\left(\frac{a+b}{2},\frac{c+d}{2}\right) \le \frac{\Gamma(\alpha+1)(\sqrt{e}-1)}{(b-a)^\alpha}$$
$$\left[I_{a+}^\alpha f\left(b,\frac{c+d}{2}\right)+I_{b-}^\alpha f\left(a,\frac{c+d}{2}\right)\right]$$
$$+\frac{\Gamma(\beta+1)(\sqrt{e}-1)}{(d-c)^\beta}\left[I_{c+}^\beta f\left(\frac{a+b}{2},d\right)+I_{d-}^\beta f\left(\frac{a+b}{2},c\right)\right]$$

$$\leq \frac{\Gamma(\alpha+1)\Gamma(\beta+1)(\sqrt{e}-1)^2}{(b-a)^\alpha (d-c)^\beta \sum_{n=0}^{\infty}\left(\frac{(-1)^n e+1}{(\beta+n)n!}-\frac{1}{2^n\beta}\right)}$$
$$\left[I^{\alpha,\beta}_{a^+,c^+}f(b,d)+I^{\alpha,\beta}_{a^+,d^-}f(b,c)+I^{\alpha,\beta}_{b^-,c^+}f(a,d)+I^{\alpha,\beta}_{b^-,d^-}f(a,c)\right]$$
$$+\frac{\Gamma(\alpha+1)\Gamma(\beta+1)\left(\sqrt{e}-1\right)^2}{(b-a)^\alpha (d-c)^\beta \sum_{n=0}^{\infty}\left(\frac{(-1)^n e+1}{(\alpha+n)n!}-\frac{1}{2^n\alpha}\right)}$$
$$\left[I^{\alpha,\beta}_{a^+,c^+}f(b,d)+I^{\alpha,\beta}_{a^+,d^-}f(b,c)+I^{\alpha,\beta}_{b^-,c^+}f(a,d)+I^{\alpha,\beta}_{b^-,d^-}f(a,c)\right]$$
$$\leq \frac{\Gamma(\alpha+1)\left(\sqrt{e}-1\right)}{(b-a)^\alpha}\left[I^\alpha_{a^+}f(b,d)+I^\alpha_{a^+}f(b,c)\right.$$
$$\left.+I^\alpha_{b^-}f(a,d)+I^\alpha_{b^-}f(a,c)\right]$$
$$+\frac{\Gamma(\beta+1)\left(\sqrt{e}-1\right)}{(d-c)^\beta}\left[I^\beta_{c^+}f(a,d)+I^\beta_{c^+}f(b,d)\right.$$
$$\left.+I^\beta_{d^-}f(a,c)+I^\beta_{d^-}f(b,c)\right]$$
$$\leq f(a,c)+f(a,d)+f(b,c)+f(b,d). \tag{2.8}$$

Proof Since $f:\Delta\to\mathbb{R}$ is convex on the exponential type co-ordinated, it follows that the mapping $g_x:[c,d]\to\mathbb{R}$, $g_x(y)=f(x,y)$, is convex on $[c,d]$ for all $x\in[a,b]$. Then by using inequalities (1.16) we can write

$$g_x\left(\frac{c+d}{2}\right)\leq \frac{\Gamma(\beta+1)\left(\sqrt{e}-1\right)}{(d-c)^\beta \sum_{n=0}^{\infty}\left(\frac{(-1)^n e+1}{(\beta+n)n!}-\frac{1}{2^n\beta}\right)}\left[I^\beta_{c^+}g_x(d)+I^\beta_{d^-}g_x(c)\right]$$
$$\leq g_x(c)+g_x(d).$$

That is,

$$f\left(x,\frac{c+d}{2}\right)\leq \frac{\beta\left(\sqrt{e}-1\right)}{(d-c)^\beta \sum_{n=0}^{\infty}\left(\frac{(-1)^n e+1}{(\beta+n)n!}-\frac{1}{2^n\beta}\right)}$$
$$\left[\int_c^d (d-y)^{\beta-1}f(x,y)dy+\int_c^d (y-c)^{\beta-1}f(x,y)dy\right]$$
$$\leq f(x,c)+f(x,d), \tag{2.9}$$

for all $x\in[a,b]$. Then multiplying both sides of (2.9) by $\frac{\alpha(b-x)^{\alpha-1}\left(\sqrt{e}-1\right)}{(b-a)^\alpha}$ and $\frac{\alpha(x-a)^{\alpha-1}\left(\sqrt{e}-1\right)}{(b-a)^\alpha}$, and integrating with respect to x over $[a,b]$, respectively, we have

$$
\begin{aligned}
&\frac{\alpha\left(\sqrt{e}-1\right)}{(b-a)^{\alpha}}\int_{a}^{b}(b-x)^{\alpha-1}f\left(x,\frac{c+d}{2}\right)dx\\
&\leq\frac{\alpha\beta\left(\sqrt{e}-1\right)^{2}}{(b-a)^{\alpha}(d-c)^{\beta}\sum_{n=0}^{\infty}\left(\frac{(-1)^{n}e+1}{(\beta+n)n!}-\frac{1}{2^{n}\beta}\right)}\\
&\left[\int_{a}^{b}\int_{c}^{d}(b-x)^{\alpha-1}(d-y)^{\beta-1}f(x,y)dydx\right.\\
&\left.+\int_{a}^{b}\int_{c}^{d}(b-x)^{\alpha-1}(y-c)^{\beta-1}f(x,y)dydx\right]\\
&\leq\frac{\alpha\left(\sqrt{e}-1\right)}{(b-a)^{\alpha}}\left[\int_{a}^{b}(b-x)^{\alpha-1}f(x,c)dx+\int_{a}^{b}(b-x)^{\alpha-1}f(x,d)dx\right]
\end{aligned}
\tag{2.10}
$$

and

$$
\begin{aligned}
&\frac{\alpha(\sqrt{e}-1)}{(b-a)^{\alpha}}\int_{a}^{b}(x-a)^{\alpha-1}f\left(x,\frac{c+d}{2}\right)dx\\
&\leq\frac{\alpha\beta\left(\sqrt{e}-1\right)^{2}}{(b-a)^{\alpha}(d-c)^{\beta}\sum_{n=0}^{\infty}\left(\frac{(-1)^{n}e+1}{(\beta+n)n!}-\frac{1}{2^{n}\beta}\right)}\\
&\left[\int_{a}^{b}\int_{c}^{d}(x-a)^{\alpha-1}(d-y)^{\beta-1}f(x,y)dydx\right.\\
&\left.+\int_{a}^{b}\int_{c}^{d}(x-a)^{\alpha-1}(y-c)^{\beta-1}f(x,y)dydx\right]\\
&\leq\frac{\alpha\left(\sqrt{e}-1\right)}{(b-a)^{\alpha}}\left[\int_{a}^{b}(x-a)^{\alpha-1}f(x,c)dx\right.\\
&\left.+\int_{a}^{b}(x-a)^{\alpha-1}f(x,d)dx\right].
\end{aligned}
\tag{2.11}
$$

By similar argument applied for the mapping $g_y : [a, b] \to \mathbb{R}$, $g_y(x) = f(x, y)$, we have

$$\begin{aligned}
&\frac{\beta(\sqrt{e}-1)}{(d-c)^\beta}\int_c^d (d-y)^{\beta-1} f\left(\frac{a+b}{2}, y\right)dy \\
&\leq \frac{\alpha\beta(\sqrt{e}-1)^2}{(b-a)^\alpha (d-c)^\beta \sum_{n=0}^{\infty}\left(\frac{(-1)^n e+1}{(\alpha+n)n!}-\frac{1}{2^n\alpha}\right)} \\
&\left[\int_a^b\int_c^d (b-x)^{\alpha-1}(d-y)^{\beta-1} f(x,y)dydx\right. \\
&\left.+\int_a^b\int_c^d (x-a)^{\alpha-1}(d-y)^{\beta-1} f(x,y)dydx\right] \\
&\leq \frac{\beta(\sqrt{e}-1)}{(d-c)^\beta}\left[\int_c^d (d-y)^{\beta-1} f(a,y)dy\right. \\
&\left.+\int_c^d (d-y)^{\beta-1} f(b,y)dy\right] \qquad (2.12)
\end{aligned}$$

and

$$\begin{aligned}
&\frac{\beta(\sqrt{e}-1)}{(d-c)^\beta}\int_c^d (y-c)^{\beta-1} f\left(\frac{a+b}{2}, y\right)dy \\
&\leq \frac{\alpha\beta(\sqrt{e}-1)^2}{(b-a)^\alpha (d-c)^\beta \sum_{n=0}^{\infty}\left(\frac{(-1)^n e+1}{(\alpha+n)n!}-\frac{1}{2^n\alpha}\right)} \\
&\left[\int_a^b\int_c^d (b-x)^{\alpha-1}(y-c)^{\beta-1} f(x,y)dydx\right. \\
&\left.+\int_a^b\int_c^d (x-a)^{\alpha-1}(y-c)^{\beta-1} f(x,y)dydx\right] \\
&\leq \frac{\beta(\sqrt{e}-1)}{(d-c)^\beta}\left[\int_c^d (y-c)^{\beta-1} f(a,y)dy\right.
\end{aligned}$$

$$+\int_c^d (y-c)^{\beta-1} f(b,y)dy\Bigg]. \tag{2.13}$$

Adding the inequalities (2.10)–(2.13) we have

$$\begin{aligned}
&\frac{\Gamma(\alpha+1)\left(\sqrt{e}-1\right)}{(b-a)^{\alpha}}\left[I^{\alpha}_{a^+}f\left(b,\frac{c+d}{2}\right)+I^{\alpha}_{b^-}f\left(a,\frac{c+d}{2}\right)\right]\\
&\quad+\frac{\Gamma(\beta+1)\left(\sqrt{e}-1\right)}{(d-c)^{\beta}}\left[I^{\beta}_{c^+}f\left(\frac{a+b}{2},d\right)+I^{\beta}_{d^-}f\left(\frac{a+b}{2},c\right)\right]\\
&\leq\frac{\Gamma(\alpha+1)\Gamma(\beta+1)\left(\sqrt{e}-1\right)^2}{(b-a)^{\alpha}(d-c)^{\beta}\sum_{n=0}^{\infty}\left(\frac{(-1)^n e+1}{(\beta+n)n!}-\frac{1}{2^n\beta}\right)}\\
&\left[I^{\alpha,\beta}_{a^+,c^+}f(b,d)+I^{\alpha,\beta}_{a^+,d^-}f(b,c)+I^{\alpha,\beta}_{b^-,c^+}f(a,d)+I^{\alpha,\beta}_{b^-,d^-}f(a,c)\right]\\
&\quad+\frac{\Gamma(\alpha+1)\Gamma(\beta+1)(\sqrt{e}-1)^2}{(b-a)^{\alpha}(d-c)^{\beta}\sum_{n=0}^{\infty}\left(\frac{(-1)^n e+1}{(\alpha+n)n!}-\frac{1}{2^n\alpha}\right)}\\
&\left[I^{\alpha,\beta}_{a^+,c^+}f(b,d)+I^{\alpha,\beta}_{a^+,d^-}f(b,c)+I^{\alpha,\beta}_{b^-,c^+}f(a,d)+I^{\alpha,\beta}_{b^-,d^-}f(a,c)\right]\\
&\leq\frac{\Gamma(\alpha+1)\left(\sqrt{e}-1\right)}{(b-a)^{\alpha}}\left[I^{\alpha}_{a^+}f(b,d)+I^{\alpha}_{a^+}f(b,c)\right.\\
&\quad\left.+I^{\alpha}_{b^-}f(a,d)+I^{\alpha}_{b^-}f(a,c)\right]+\frac{\Gamma(\beta+1)\left(\sqrt{e}-1\right)}{(d-c)^{\beta}}\\
&\left[I^{\beta}_{c^+}f(a,d)+I^{\beta}_{c^+}f(b,d)+I^{\beta}_{d^-}f(a,c)+I^{\beta}_{d^-}f(b,c)\right].
\end{aligned}$$

Which give the second and the third inequalities in (2.8). Now, by using the first inequalitiy in (1.16), we also have

$$\begin{aligned}
f\left(\frac{a+b}{2},\frac{c+d}{2}\right)\leq\frac{\alpha\left(\sqrt{e}-1\right)}{(b-a)^{\alpha}}&\left[\int_a^b (b-x)^{\alpha-1}f\left(x,\frac{c+d}{2}\right)dx\right.\\
&\left.+\int_a^b (x-a)^{\alpha-1}f\left(x,\frac{c+d}{2}\right)dx\right]
\end{aligned}$$

and

$$f\left(\frac{a+b}{2},\frac{c+d}{2}\right)\leq\frac{\beta\left(\sqrt{e}-1\right)}{(d-c)^{\beta}}\left[\int_c^d (d-y)^{\beta-1}f\left(\frac{a+b}{2},y\right)dy\right.$$

$$+\int_c^d (y-c)^{\beta-1} f\left(\frac{a+b}{2}, y\right) dy\Bigg]$$

by addition,

$$f\left(\frac{a+b}{2}, \frac{c+d}{2}\right) \le \frac{\Gamma(\alpha+1)\left(\sqrt{e}-1\right)}{(b-a)^{\alpha}}\left[I_{a+}^{\alpha} f\left(b, \frac{c+d}{2}\right) + I_{b-}^{\alpha} f\left(a, \frac{c+d}{2}\right)\right]$$
$$+ \frac{\Gamma(\beta+1)\left(\sqrt{e}-1\right)}{(d-c)^{\beta}}\left[I_{c+}^{\beta} f\left(\frac{a+b}{2}, d\right) + I_{d-}^{\beta} f\left(\frac{a+b}{2}, c\right)\right]$$

which give the first inequality in (2.8). Finally, by using the second inequality in (1.16) we can also state,

$$\frac{\alpha\left(\sqrt{e}-1\right)}{(b-a)^{\alpha}}\left[\int_a^b (b-x)^{\alpha-1} f(x,c)dx + \int_a^b (x-a)^{\alpha-1} f(x,c)dx\right]$$
$$\le f(a,c) + f(b,c)$$
$$\frac{\alpha\left(\sqrt{e}-1\right)}{(b-a)^{\alpha}}\left[\int_a^b (b-x)^{\alpha-1} f(x,d)dx + \int_a^b (x-a)^{\alpha-1} f(x,d)dx\right]$$
$$\le f(a,d) + f(b,d)$$
$$\frac{\beta\left(\sqrt{e}-1\right)}{(d-c)^{\beta}}\left[\int_c^d (d-y)^{\beta-1} f(a,y)dy + \int_c^d (y-c)^{\beta-1} f(a,y)dy\right]$$
$$\le f(a,c) + f(a,d)$$
$$\frac{\beta\left(\sqrt{e}-1\right)}{(d-c)^{\beta}}\left[\int_c^d (d-y)^{\beta-1} f(b,y)dy + \int_c^d (y-c)^{\beta-1} f(b,y)dy\right]$$
$$\le f(b,c) + f(b,d)$$

which give, by addition, the last inequality in (2.8).

Theorem 2.5 *Let $f : \Delta \subseteq \mathbb{R}^2 \to \mathbb{R}$ be a partial differentiable mapping on $\Delta := [a,b] \times [c,d]$ in $\mathbb{R}^2$ with $0 \le a < b$, $0 \le c < d$. If $\left|\frac{\partial^2 f}{\partial t \partial s}\right|$ is a convex function on the exponential type co-ordinates on Δ, then one has the inequalities;*

$$
\begin{aligned}
&\left|\frac{f(a,c)+f(a,d)+f(b,c)+f(b,d)}{4}+\frac{\Gamma(\alpha+1)\Gamma(\beta+1)}{4(b-a)^{\alpha}(d-c)^{\beta}}\left[I^{\alpha,\beta}_{a^+,c^+}f(b,d)\right.\right.\\
&\left.\left.+I^{\alpha,\beta}_{a^+,d^-}f(b,c)+I^{\alpha,\beta}_{b^-,c^+}f(a,d)+I^{\alpha,\beta}_{b^-,d^-}f(a,c)\right]-A\right|\\
&\leq\frac{(b-a)(d-c)}{4}\left(\left\{\left[\sum_{n=0}^{\infty}\left(\frac{B(\alpha+1,\alpha+1)}{n!}+\frac{1}{n!(n+\alpha+1)}\right)\right]\right.\right.\\
&\times\left[\sum_{m=0}^{\infty}\left(\frac{B(\beta+1,\beta+1)}{m!}+\frac{1}{m!(m+\beta+1)}\right)\right]\\
&-\frac{2}{\beta+1}\sum_{n=0}^{\infty}\left(\frac{B(\alpha+1,\alpha+1)}{n!}+\frac{1}{n!(n+\alpha+1)}\right)\\
&-\frac{2}{\alpha+1}\sum_{m=0}^{\infty}\left(\frac{B(\beta+1,\beta+1)}{m!}+\frac{1}{m!(m+\beta+1)}\right)\\
&\left.+\frac{4}{(\alpha+1)(\beta+1)}\right\}\left|\frac{\partial^2 f}{\partial s\partial t}(a,c)\right|+\left\{e\left[\sum_{n=0}^{\infty}\left(\frac{B(\alpha+1,\alpha+1)}{n!}+\frac{1}{n!(n+\alpha+1)}\right)\right]\right.\\
&\times\left[\sum_{m=0}^{\infty}\left(\frac{(-1)^m B(\beta+1,\beta+1)}{m!}+\frac{(-1)^m}{m!(m+\beta+1)}\right)\right]\\
&-\frac{2}{\beta+1}\sum_{n=0}^{\infty}\left(\frac{B(\alpha+1,\alpha+1)}{n!}+\frac{1}{n!(n+\alpha+1)}\right)\\
&\left.-\frac{2e}{\alpha+1}\sum_{m=0}^{\infty}\left(\frac{(-1)^m B(\beta+1,\beta+1)}{m!}+\frac{(-1)^m}{m!(m+\beta+1)}\right)+\frac{4}{(\alpha+1)(\beta+1)}\right\}\\
&\left|\frac{\partial^2 f}{\partial s\partial t}(a,d)\right|+\left\{e\left[\sum_{n=0}^{\infty}\left(\frac{(-1)^n B(\alpha+1,\alpha+1)}{n!}+\frac{(-1)^n}{n!(n+\alpha+1)}\right)\right]\right.\\
&\left[\sum_{m=0}^{\infty}\left(\frac{B(\beta+1,\beta+1)}{m!}+\frac{1}{m!(m+\beta+1)}\right)\right]\\
&-\frac{2e}{\beta+1}\sum_{n=0}^{\infty}\left(\frac{(-1)^n B(\alpha+1,\alpha+1)}{n!}+\frac{(-1)^n}{n!(n+\alpha+1)}\right)\\
&-\frac{2}{\alpha+1}\sum_{m=0}^{\infty}\left(\frac{B(\beta+1,\beta+1)}{m!}+\frac{1}{m!(m+\beta+1)}\right)\\
&\left.+\frac{4}{(\alpha+1)(\beta+1)}\right\}\left|\frac{\partial^2 f}{\partial s\partial t}(b,c)\right|\\
&+\left\{e^2\left[\sum_{n=0}^{\infty}\left(\frac{(-1)^n B(\alpha+1,\alpha+1)}{n!}+\frac{(-1)^n}{n!(n+\alpha+1)}\right)\right]\right.\\
&\times\left[\sum_{m=0}^{\infty}\left(\frac{(-1)^m B(\beta+1,\beta+1)}{m!}+\frac{(-1)^m}{m!(m+\beta+1)}\right)\right]\\
&-\frac{2e}{\beta+1}\sum_{n=0}^{\infty}\left(\frac{(-1)^n B(\alpha+1,\alpha+1)}{n!}+\frac{(-1)^n}{n!(n+\alpha+1)}\right)\\
&-\frac{2e}{\alpha+1}\sum_{m=0}^{\infty}\left(\frac{(-1)^m B(\beta+1,\beta+1)}{m!}+\frac{(-1)^m}{m!(m+\beta+1)}\right)\\
&\left.\left.+\frac{4}{(\alpha+1)(\beta+1)}\right\}\left|\frac{\partial^2 f}{\partial s\partial t}(b,d)\right|\right)
\end{aligned}
\tag{2.14}
$$

where A is defined by Lemma 1.7.

Proof From Lemma 1.7, we have

$$\left|\frac{f(a,c)+f(a,d)+f(b,c)+f(b,d)}{4}+\frac{\Gamma(\alpha+1)\Gamma(\beta+1)}{4(b-a)^{\alpha}(d-c)^{\beta}}\right.$$
$$\Big[I^{\alpha,\beta}_{a^+,c^+}f(b,d)+I^{\alpha,\beta}_{a^+,d^-}f(b,c)$$
$$\left.+I^{\alpha,\beta}_{b^-,c^+}f(a,d)+I^{\alpha,\beta}_{b^-,d^-}f(a,c)\Big]-A\right|$$
$$\leq\frac{(b-a)(d-c)}{4}\left\{\int_0^1\int_0^1 t^{\alpha}s^{\beta}\left|\frac{\partial^2 f}{\partial s\partial t}(ta+(1-t)b,sc+(1-s)d)\right|dsdt\right.$$
$$-\int_0^1\int_0^1 t^{\alpha}(1-s)^{\beta}\left|\frac{\partial^2 f}{\partial s\partial t}(ta+(1-t)b,sc+(1-s)d)\right|dsdt$$
$$+\int_0^1\int_0^1 (1-t)^{\alpha}s^{\beta}\left|\frac{\partial^2 f}{\partial s\partial t}(ta+(1-t)b,sc+(1-s)d)\right|dsdt$$
$$\left.+\int_0^1\int_0^1 (1-t)^{\alpha}(1-s)^{\beta}\left|\frac{\partial^2 f}{\partial s\partial t}(ta+(1-t)b,sc+(1-s)d)\right|dsdt\right\}. \quad (2.15)$$

Since $\left|\frac{\partial^2 f}{\partial t\partial s}\right|$ is convex function on the exponential type co-ordinated convex on Δ, then one has:

$$\left|\frac{f(a,c)+f(a,d)+f(b,c)+f(b,d)}{4}+\frac{\Gamma(\alpha+1)\Gamma(\beta+1)}{4(b-a)^{\alpha}(d-c)^{\beta}}\right.$$
$$\Big[I^{\alpha,\beta}_{a^+,c^+}f(b,d)+I^{\alpha,\beta}_{a^+,d^-}f(b,c)$$
$$\left.+I^{\alpha,\beta}_{b^-,c^+}f(a,d)+I^{\alpha,\beta}_{b^-,d^-}f(a,c)\Big]-A\right|$$
$$\leq\frac{(b-a)(d-c)}{4}\int_0^1\int_0^1\left[t^{\alpha}s^{\beta}+t^{\alpha}(1-s)^{\beta}+(1-t)^{\alpha}s^{\beta}+(1-t)^{\alpha}(1-s)^{\beta}\right]$$
$$\times\left\{\left(e^t-1\right)\left(e^s-1\right)\left|\frac{\partial^2 f}{\partial s\partial t}(a,c)\right|+\left(e^t-1\right)\left(e^{1-s}-1\right)\left|\frac{\partial^2 f}{\partial s\partial t}(a,d)\right|\right.$$
$$+\left(e^{1-t}-1\right)\left(e^s-1\right)\left|\frac{\partial^2 f}{\partial s\partial t}(b,c)\right|$$
$$\left.+\left(e^{1-t}-1\right)\left(e^{1-s}-1\right)\left|\frac{\partial^2 f}{\partial s\partial t}(b,d)\right|\right\}dsdt.$$

By calculation the integral in above iequality, we have

$$
\begin{aligned}
&= \frac{(b-a)(d-c)}{4}\left(\left\{\left[\sum_{n=0}^{\infty}\left(\frac{B(\alpha+1,\alpha+1)}{n!}+\frac{1}{n!(n+\alpha+1)}\right)\right]\right.\right.\\
&\times\left[\sum_{m=0}^{\infty}\left(\frac{B(\beta+1,\beta+1)}{m!}+\frac{1}{m!(m+\beta+1)}\right)\right]\\
&-\frac{2}{\beta+1}\sum_{n=0}^{\infty}\left(\frac{B(\alpha+1,\alpha+1)}{n!}+\frac{1}{n!(n+\alpha+1)}\right)\\
&-\frac{2}{\alpha+1}\sum_{m=0}^{\infty}\left(\frac{B(\beta+1,\beta+1)}{m!}+\frac{1}{m!(m+\beta+1)}\right)\\
&\left.+\frac{4}{(\alpha+1)(\beta+1)}\right\}\times\left|\frac{\partial^2 f}{\partial s\partial t}(a,c)\right|\\
&+\left\{e\left[\sum_{n=0}^{\infty}\left(\frac{B(\alpha+1,\alpha+1)}{n!}+\frac{1}{n!(n+\alpha+1)}\right)\right]\right.\\
&\left[\sum_{m=0}^{\infty}\left(\frac{(-1)^m B(\beta+1,\beta+1)}{m!}+\frac{(-1)^m}{m!(m+\beta+1)}\right)\right]\\
&-\frac{2}{\beta+1}\sum_{n=0}^{\infty}\left(\frac{B(\alpha+1,\alpha+1)}{n!}+\frac{1}{n!(n+\alpha+1)}\right)\\
&-\frac{2e}{\alpha+1}\sum_{m=0}^{\infty}\left(\frac{(-1)^m B(\beta+1,\beta+1)}{m!}+\frac{(-1)^m}{m!(m+\beta+1)}\right)\\
&\left.+\frac{4}{(\alpha+1)(\beta+1)}\right\}\left|\frac{\partial^2 f}{\partial s\partial t}(a,d)\right|\\
&+\left\{e\left[\sum_{n=0}^{\infty}\left(\frac{(-1)^n B(\alpha+1,\alpha+1)}{n!}+\frac{(-1)^n}{n!(n+\alpha+1)}\right)\right]\right.\\
&\times\left[\sum_{m=0}^{\infty}\left(\frac{B(\beta+1,\beta+1)}{m!}+\frac{1}{m!(m+\beta+1)}\right)\right]\\
&-\frac{2e}{\beta+1}\sum_{n=0}^{\infty}\left(\frac{(-1)^n B(\alpha+1,\alpha+1)}{n!}+\frac{(-1)^n}{n!(n+\alpha+1)}\right)\\
&-\frac{2}{\alpha+1}\sum_{m=0}^{\infty}\left(\frac{B(\beta+1,\beta+1)}{m!}+\frac{1}{m!(m+\beta+1)}\right)\\
&\left.+\frac{4}{(\alpha+1)(\beta+1)}\right\}\left|\frac{\partial^2 f}{\partial s\partial t}(b,c)\right|\\
&+\left\{e^2\left[\sum_{n=0}^{\infty}\left(\frac{(-1)^n B(\alpha+1,\alpha+1)}{n!}+\frac{(-1)^n}{n!(n+\alpha+1)}\right)\right]\right.
\end{aligned}
$$

$$\times\left[\sum_{m=0}^{\infty}\left(\frac{(-1)^m B(\beta+1,\beta+1)}{m!}+\frac{(-1)^m}{m!(m+\beta+1)}\right)\right]$$
$$-\frac{2e}{\beta+1}\sum_{n=0}^{\infty}\left(\frac{(-1)^n B(\alpha+1,\alpha+1)}{n!}+\frac{(-1)^n}{n!(n+\alpha+1)}\right)$$
$$-\frac{2e}{\alpha+1}\sum_{m=0}^{\infty}\left(\frac{(-1)^m B(\beta+1,\beta+1)}{m!}+\frac{(-1)^m}{m!(m+\beta+1)}\right)$$
$$+\frac{4}{(\alpha+1)(\beta+1)}\Bigg\}\left|\frac{\partial^2 f}{\partial s\partial t}(b,d)\right|\Bigg)$$

which completes the proof.

Theorem 2.6 *Let $f:\Delta\subseteq\mathbb{R}^2\to\mathbb{R}$ be a partial differentiable mapping on $\Delta:=[a,b]\times[c,d]$ in $\mathbb{R}^2$ with $0\le a<b$, $0\le c<d$. If $\left|\frac{\partial^2 f}{\partial t\partial s}\right|^q$, $q>1$, is a convex function on the exponential type co-ordinates on Δ, then one has the equalities;*

$$\Bigg|\frac{f(a,c)+f(a,d)+f(b,c)+f(b,d)}{4}$$
$$+\frac{\Gamma(\alpha+1)\Gamma(\beta+1)}{4(b-a)^a(d-c)^\beta}\Big[I^{\alpha,\beta}_{a^+,c^+}f(b,d)+I^{\alpha,\beta}_{a^+,d^-}f(b,c)$$
$$+I^{\alpha,\beta}_{b^-,c^+}f(a,d)+I^{a,\beta}_{b^-,d^-}f(a,c)\Big]-A\Bigg|$$
$$\le\frac{(b-a)(d-c)}{\left[(\alpha p+1)(\beta p+1)\right]^{\frac{1}{p}}}\Bigg((e-1)^2\left|\frac{\partial^2 f}{\partial s\partial t}(a,c)\right|^q$$
$$+(e-1)(e-2)\left|\frac{\partial^2 f}{\partial s\partial t}(a,d)\right|^q$$
$$+(e-1)(e-2)\left|\frac{\partial^2 f}{\partial s\partial t}(b,c)\right|^q+(e-2)^2\left|\frac{\partial^2 f}{\partial s\partial t}(b,d)\right|^q\Bigg)^{\frac{1}{q}}$$

where $\frac{1}{p}+\frac{1}{q}=1$ and A is defined by Lemma 1.7.

Proof From Lemma 1.7, we have inequality (2.15). By using the well known Hölder's inequality for double integrals, we get:

$$\Bigg|\frac{f(a,c)+f(a,d)+f(b,c)+f(b,d)}{4}+\frac{\Gamma(\alpha+1)\Gamma(\beta+1)}{4(b-a)^\alpha(d-c)^\beta}$$
$$\Big[I^{a,\beta}_{a^+,c^+}f(b,d)+I^{\alpha,\beta}_{a^+,d^-}f(b,c)+I^{a,\beta}_{b^-,c^+}f(a,d)+I^{\alpha,\beta}_{b^-,d^-}f(a,c)\Big]-A\Bigg|$$
$$\le\frac{(b-a)(d-c)}{4}\Bigg\{\left(\int_0^1\int_0^1 t^{p\alpha}s^{p\beta}dsdt\right)^{\frac{1}{p}}+\left(\int_0^1\int_0^1(1-t)^{p\alpha}s^{p\beta}dsdt\right)^{\frac{1}{p}}$$

$$\times\left(\int_0^1\int_0^1 t^{p\alpha}(1-s)^{p\beta}dsdt\right)^{\frac{1}{p}}+\left(\int_0^1\int_0^1(1-t)^{p\alpha}(1-s)^{p\beta}dsdt\right)^{\frac{1}{p}}\Bigg\}$$

$$\times\left(\int_0^1\int_0^1\left|\frac{\partial^2 f}{\partial s\partial t}(ta+(1-t)b,sc+(1-s)d)\right|^q dsdt\right)^{\frac{1}{q}}.$$

Since $\left|\frac{\partial^2 f}{\partial t\partial s}\right|^q$ is a convex function on the exponential type co-ordinates on Δ, we know that for $t\in[0,1]$

$$\left|\frac{\partial^2 f}{\partial s\partial t}(ta+(1-t)b,sc+(1-s)d)\right|^q\leq\left(e^t-1\right)\left|\frac{\partial^2 f}{\partial s\partial t}(a,sc+(1-s)d)\right|^q$$
$$+\left(e^{1-t}-1\right)\left|\frac{\partial^2 f}{\partial s\partial t}(b,sc+(1-s)d)\right|^q$$

and

$$\left|\frac{\partial^2 f}{\partial s\partial t}(ta+(1-t)b,sc+(1-s)d)\right|^q$$
$$\leq\left(e^t-1\right)\left(e^s-1\right)\left|\frac{\partial^2 f}{\partial s\partial t}(a,c)\right|^q$$
$$+\left(e^t-1\right)\left(e^{1-s}-1\right)\left|\frac{\partial^2 f}{\partial s\partial t}(a,d)\right|^q$$
$$+\left(e^{1-t}-1\right)\left(e^s-1\right)\left|\frac{\partial^2 f}{\partial s\partial t}(b,c)\right|^q$$
$$+\left(e^{1-t}-1\right)\left(e^{1-s}-1\right)\left|\frac{\partial^2 f}{\partial s\partial t}(b,d)\right|^q$$

hence, it follows that

$$\Bigg|\frac{f(a,c)+f(a,d)+f(b,c)+f(b,d)}{4}+\frac{\Gamma(\alpha+1)\Gamma(\beta+1)}{4(b-a)^\alpha(d-c)^\beta}$$
$$\left[I^{\alpha,\beta}_{a^+,c^+}f(b,d)+I^{\alpha,\beta}_{a^+,d^-}f(b,c)+I^{\alpha,\beta}_{b^-,c^+}f(a,d)+I^{\alpha,\beta}_{b^-,d^-}f(a,c)\right]-A\Bigg|$$
$$\leq\frac{(b-a)(d-c)}{\left[(\alpha p+1)(\beta p+1)\right]^{\frac{1}{p}}}\left(\int_0^1\int_0^1\left\{\left(e^t-1\right)\left(e^s-1\right)\left|\frac{\partial^2 f}{\partial s\partial t}(a,c)\right|^q\right.\right.$$
$$+\left(e^t-1\right)\left(e^{1-s}-1\right)\left|\frac{\partial^2 f}{\partial s\partial t}(a,d)\right|^q+\left(e^{1-t}-1\right)\left(e^s-1\right)$$

$$\left|\frac{\partial^2 f}{\partial s \partial t}(b,c)\right|^q + \left(e^{1-t}-1\right)\left(e^{1-s}-1\right)\left|\frac{\partial^2 f}{\partial s \partial t}(b,d)\right|^q\Bigg\}dtds\Bigg)^{\frac{1}{q}}$$

$$= \frac{(b-a)(d-c)}{\left[(\alpha p+1)(\beta p+1)\right]^{\frac{1}{p}}}\Bigg((e-1)^2\left|\frac{\partial^2 f}{\partial s \partial t}(a,c)\right|^q$$

$$+ (e-1)(e-2)\left|\frac{\partial^2 f}{\partial s \partial t}(a,d)\right|^q$$

$$+(e-1)(e-2)\left|\frac{\partial^2 f}{\partial s \partial t}(b,c)\right|^q + (e-2)^2\left|\frac{\partial^2 f}{\partial s \partial t}(b,d)\right|^q\Bigg)^{\frac{1}{q}},$$

which completes the proof.

References

1. Azpeitia, A.G.: Convex functions and the Hadamard inequality. Rev. Colomb. Math. **28**, 7–12 (1994)
2. Bakula, M.K., Pecaric, J.: Note on some Hadamard-type inequalities. J. Inequal. Pure Appl. Math. **5**(3), Article 74 (2004)
3. Dragomir, S.S.: On Hadamard's inequality for convex functions on the co-ordinates in a rectangle from the plane. Taiwan. J. Math. **4**, 775–788 (2001)
4. Dragomir, S.S., Pearce, C.E.M.: Selected Topics on Hermite–Hadamard Inequalities and Applications. RGMIA Monographs, Victoria University (2000)
5. Kilbas, A.A., Srivastava, H.M., Trujillo, J.J.: Theory and Applications of Fractional Differential Equations. In: North-Holland Mathematics Studies, vol. 204. Elsevier Science B.V., Amsterdam (2006)
6. Latif, M.A., Dragomir, S.S.: On some new inequalities for differentiable co-ordinated convex functions. J. Inequal. Appl. **1**, 1–13 (2012)
7. Samko, S.G., Kilbas, A.A., Marichev, O.I.: Fractional Integrals and Derivatives, Theory and Applications. Gordon and Breach, Yverdon et alibi (1993)
8. Sarikaya, M.Z.: On the Hermite–Hadamard-type inequalities for co-ordinated convex function via fractional integrals. Integr. Transf. Spec. Funct. **25**, 134–147 (2014)
9. Sarikaya, M.Z.: Some inequalities for differentiable co-ordinated convex mappings. Asian-Eur. J. Math. **8** (2015)
10. Sarikaya, M.Z., Aktan, N.: On the generalization of some integral inequalities and their applications. Math. Comput. Model. **54**, 2175–2182 (2011)
11. Sarikaya, M.Z., Yaldiz, H.: On the Hadamard's type inequalities for L-Lipschitzian mapping. Konuralp J. Math. **1**, 33–40 (2013)
12. Sarikaya, M.Z., Set, E., Özdemir, M.E., Dragomir, S.S.: New some Hadamard's type inequalities for co-ordinated convex functions. Tamsui Oxf. J. Inf. Math. Sci. **28**, 137–152 (2012)
13. Sarikaya, M.Z., Set, E., Yaldiz, H., Basak, N.: Hermite–Hadamard's inequalities for fractional integrals and related fractional inequalities. Math. Comput. Model. **57**, 2403–2407 (2012)
14. Sarikaya, M.Z., Budak, H., Yaldiz, H.: Cebysev type inequalities for co-ordinated convex functions. Pure Appl. Math. Lett. **2**, 36–40 (2014)
15. Sarikaya, M.Z., Budak, H., Yaldiz, H.: Some new Ostrowski type inequalities for co-ordinated convex functions. Turk. J. Anal. Numb. Theory **2**, 176–182 (2014)

16. Akkurt, A., Sarıkaya, M., Budak, H., Yıldırım, H.: On the Hadamard's type inequalities for co-ordinated convex functions via fractional integrals. J. King Saud Univ. Sci. **29**(3) (2016)
17. Akkurt, A., Sarıkaya, M., Budak, H., Yıldırım, H.: On the Hermite–Hadamard type inequalities for co-ordinated convex functions. Appl. Comput. Math. Int. J. **20**(3), 408–420 (2021)

Detection of the Bifurcation of Chen's Chaotic Attractor

Gizem S. Oztepe

Abstract The investigation of bifurcations of a Chen's system is the main aim of this paper. Especially we focus on Hopf bifurcation and after obtaining the Hopf bifurcation point in terms of system parameters, the normal form of this bifurcation is obtained by using an effective method called the multiple scales method. The obtained results are also confirmed numerically.

Keywords Chen's system · Hopf bifurcation · Multiple scale method · Normal form

1 Introduction

Chaos has remained its popularity since the exploration of the famous Lorenz chaotic system, built in 1963 to model atmospheric convection [24]. Then numerous studies were devoted to this subject. Some of them were interested in discovering new chaotic models like Lü system [25], Chen system [4], Rössler system [37], Lorenz-type systems [14, 20, 23, 55], and other new chaotic systems [28, 39, 40, 43, 51] while others focused on reconsidering existing chaotic systems and exploring some previously unknown phenomena [21, 38, 44, 48, 53, 54, 56]. Chaotic systems exhibit a number of features, including a high sensitivity to the initial state, pseudo stochasticity, and unpredictability. The uncertainty in the parameters of the system and great sensitivity to even minor alterations in the initial conditions are two essential characteristics of these systems. Because of all these, this subject enables not only mathematicians but also many scientists from different branches to use chaos in their works such as biological systems [9, 16]; image encryption [29, 36]; neuroscience [11, 34]; robotics [33, 45]; secure communication [46, 50].

Among these chaotic systems, Chen's system [4] is considered deeply because of its complex dynamics. This system has been looked from different perspectives

G. S. Oztepe (✉)
Department of Mathematics, Ankara University, Ankara, Turkey
e-mail: gseyhan@ankara.edu.tr

O. Castillo et al. (eds.), *Advanced Mathematics for the Modeling and Solution of Challenging Problems in Engineering*, Studies in Systems, Decision and Control 582,
https://doi.org/10.1007/978-3-031-82342-8_15

for years. Here we want to mention the studies in the last years. In [13], the hash function, which is a very useful tool in cryptography, is built using a Chen chaotic system with a pretty high level of complexity; various properties of fractional order Chen's system are investigated in [2, 15, 18, 32, 47, 52]; in 2020, Yi et al., studied the considered system as a financial model [22] and time delay Chen system is analyzed in [12, 41, 42].

Hopf bifurcation is one of the most popular branches in nonlinear research. From this point of view, many different types of systems have been studied, and Chen's systems have also received their share from these studies [1, 8, 19, 27]. In the investigation of Hopf bifurcation many authors prefer a center manifold theory as a first step, but actually to obtain a normal form for the Hopf bifurcation, the method of multiple scales is more effective and useful. For these reason with this paper, we built the normal form of the Hopf bifurcation of the considered system via multiple scale method. The details of this method can be found in [30, 31].

The focus of this research is on the investigation of the dynamics of the following chaotic attractor called, Chen's system, which was presented in [4]

$$\begin{cases} \dot{p}(t) = a(y - p) \\ \dot{q}(t) = (c - a)p - pz + cq \\ \dot{r}(t) = pq - br \end{cases} \tag{1}$$

where a, b, c are real constants that are different from zero and "over-dot" represent a time derivative. Here, we focus on the zero and Hopf bifurcations of the system (1) and provide analytical equations for the periodic orbits resulting from this Hopf bifurcation using the multiple scales approach. Moreover, the stability of the periodic orbits is also analyzed. This kind of investigation is made for different kind of systems [3, 6, 7, 10, 17, 35, 49].

The remainder of the paper is laid out as follows: In Sect. 2, after the equilibrium points are found, linear stability analysis and conditions for zero and Hopf bifurcations at each equilibrium point are studied. The concept of multiple scales is used in Sect. 3 to obtain analytical formulas for periodic orbits originating from the Hopf bifurcation, and the resulting normal form is used to search their stability. In Sect. 4, numerical simulations are used to verify the normal form predictions. The results are briefly summarized in Sect. 5.

2 Stability Analysis

This system has three equilibrium points given by

$$E_0 = (0, 0, 0), \quad E_1 = (p_0, q_0, r_0), \quad E_2 = (-p_0, -q_0, r_0) \tag{2}$$

with

$$p_0 = q_0 = \sqrt{br_0}, \quad r_0 = 2c - a. \tag{3}$$

This section presents the linear analysis for every fixed point. At each fixed point, we look at the Jacobian matrix of the system and the associated characteristic polynomial, searching single, double, and triple-zero bifurcations. These arise when there are $\lambda = 0$ zeros at each fixed point, where λ indicates the roots of the characteristic polynomial for the Jacobian matrix, and the multiplicity varies from one to three. We also look for the Hopf bifurcation of this system which corresponds to one of the characteristic root is different from zero and the remained roots are complex conjugates with zero real part. To obtain the regular Hopf bifurcation we consider the corresponding characteristic polynomial of the Jacobian matrix for each equilibrium point in the form of

$$\lambda^3 + a_1\lambda^2 + a_2\lambda + a_3. \tag{4}$$

On the other hand, since we want to search for the Hopf bifurcation, the characteristic polynomial has roots $\lambda_0 \in R$ and $\lambda_{1,2} = \pm i\omega$ where $\omega \in R$. So, the characteristic polynomial should be written as

$$(\lambda - \lambda_0)\left(\lambda^2 + \omega^2\right) = \lambda^3 - \lambda_0\lambda^2 + \omega^2\lambda - \lambda_0\omega^2 \tag{5}$$

If Eqs. (4) and (5) are compared, we get the necessary condition

$$a_1a_2 = a_3. \tag{6}$$

The following subsections are devoted to investigate the bifurcations for each equilibrium point.

2.1 Equilibrium Point E_0

The Jacobian matrix's characteristic polynomial at E_0 is

$$\lambda^3 + (a + b - c)\lambda^2 + \left(a^2 + a(b - 2c) - bc\right)\lambda + ab(a - 2c). \tag{7}$$

When the constant term is 0, a single-zero bifurcation is produced. So this yields $a = 2c$. Inherently, double and triple-zero bifurcation can not be obtained because substituting $a = 2c$ gives

$$\lambda^3 + (b + c)\lambda^2 + bc\lambda$$

in which the coefficient of λ term is zero if and only if b or c equal to zero which is impossible.

Now, let us check the Hopf bifurcation. The roots of (7) are

$$\lambda_0 = -b, \quad \lambda_{1,2} = \frac{1}{2}\left(-a + c \pm \sqrt{-3a^2 + 6ac + c^2}\right).$$

To have pure imaginary roots, we should take $a = c$, but this causes $\lambda_{1,2} = \pm a$ which are real. So it can be concluded that there is no Hopf bifurcation.

2.2 *Equilibrium Points* $\mathbf{E}_1$ *and* $\mathbf{E}_2$

The Jacobian matrix's characteristic polynomial at E_1 is given by

$$\lambda^3 + (a + b - c)\lambda^2 + bc\lambda + 2ab(2c - a). \tag{8}$$

We get a single-zero bifurcation if we take $a = 2c$ and this choice transforms the characteristic polynomial (8) into

$$\lambda^3 + (a + b - c)\lambda^2 + bc\lambda.$$

Here the coefficient of the λ can not be zero because the constant b and c are both assumed to be different from zero. So we can not talk about double and triple zero bifurcations.

Next, we search for Hopf bifurcation: As ω is real, $bc > 0$. Using the relation (2), there should be

$$b = b_0 = \frac{-2a^2 + 3ac + c^2}{c}. \tag{9}$$

Consequently, a Hopf bifurcation at $b = b_0$ is allowed in the parameter space area fulfilling $bc > 0$.

To demonstrate that periodic solutions appear when the value of b passes over the critical values b_0, we apply the Hopf bifurcation analysis. The following well-known Hopf bifurcation theorem is used to describe the behavior of λ eigenvalues as b fluctuates.

Theorem 1 [26] *Let*

$$\frac{d\vec{x}}{dt} = \vec{F}(\vec{x}, \delta) \tag{10}$$

be an autonomous system of differential equations for each value of the parameter $\delta \in (-\delta_0, \delta)$, *where* δ_0 *is a positive number and the vector function*

$$\vec{F} \in \mathbb{C}^2(DX(-\delta_0, \delta_0))$$

where D is a domain in R^n. *Suppose that the system* (10) *has a critical point* $\overrightarrow{x_0}(\delta)$ *that is,*

$$\vec{F}\left(\overrightarrow{x_0}(\delta), \delta\right) = 0$$

Let $J(\delta)$ *be a Jacobian matrix of system* (10) *at* $\vec{x}_0(\delta)$. *Suppose that* $det(\vec{J}(\delta) - \lambda I) = 0$ *has a complex conjugate pair of solution* $\lambda(\delta)$, $\bar{\lambda}(\delta)$ *such that for* $\delta > 0$ *the* $Re\lambda(\delta) > 0$; *for* $\delta < 0$, $Re\lambda(\delta) < 0$, $\left.\frac{dRe\lambda(\delta)}{d\delta}\right|_{\delta=0} > 0$. *Assume that all other* λ*'s are distinct,* (10) *has a periodic solution in some neighborhood of* $\delta = 0$ *and* $\vec{x}$ *in some neighborhood of* $\vec{x}_0(\delta)$.

In order to apply this theorem to (1), let us define the bifurcation parameter

$$\delta = \frac{1}{b} - \frac{1}{b_0}, \quad b(\delta) = \frac{b_0}{1 + b_0\delta}$$

with $\delta = 0$ at $b = b_0$. For $\delta > 0$, $b < b_0$ and $Re\lambda_1(\delta) > 0$. For $\delta < 0$, $b > b_0$ and $Re\lambda_1(\delta) < 0$. Therefore, the theorem's first set of requirements is true, and all that's left to do is demonstrate that

$$\left.\frac{Re\lambda_1(b(\delta))}{d\delta}\right|_{\delta=0} = \left.\frac{Re\lambda_1(b(\delta))}{db}\frac{db}{d\delta}\right|_{\delta=0} > 0.$$

The specifics are omitted as the analysis is conventional. The specifics can be found in [17, 31]. Hence, we conclude that, when $b = b_0$ the system (1) undergoes to Hopf bifurcation for the equilibrium point E_1 and periodic solutions exist in a neighbourhood of the point b_0.

It should be signified that the results obtained above are also applied for the equilibrium point E_2 since the Jacobian matrix's characteristic polynomial at this point is the same as (8). So, the system (1) undergoes to Hopf bifurcation for the equilibrium point E_2 as well and periodic solutions exist in a neighbourhood of the point b_0.

3 Analytical Construction of Normal Form

The creation of analytical approximations for periodic orbits emerging from the Hopf bifurcation of the equilibrium point E_1 is the focus of this section. This construction employs the multiple scales approach with control parameter b. The limit cycle is defined, in particular, by extending about the point of equilibrium on increasingly slower time scales.

It is assumed that a solution $(p(t), q(t), r(t))$ of the system (1) has the form

$$p(t) = p_0 + \sum_{n=1}^{3} \epsilon^n p_n(s_0, s_1, s_2) + \cdots \tag{11}$$

$$q(t) = q_0 + \sum_{n=1}^{3} \epsilon^n q_n(s_0, s_1, s_2) + \cdots \tag{12}$$

$$r(t) = r_0 + \sum_{n=1}^{3} \epsilon^n r_n(s_0, s_1, s_2) + \cdots \tag{13}$$

where $s_n = \varepsilon^n t$ and ε is a small positive non-dimensional parameter. Using the chain rule, the time derivative becomes

$$\frac{d}{dt} = \Gamma_0 + \epsilon\Gamma_1 + \epsilon^2\Gamma_2 + \epsilon^3\Gamma_3 + \cdots \tag{14}$$

here $\Gamma_n = \partial/\partial s_n$. Moreover, the parameter b is considered as

$$b = b_0 + \epsilon^2 b_2 \tag{15}$$

where b_0 is given by (9). The periodic orbits resulting from an order ϵ^2 perturbation for b near the bifurcation point b = b0 are examined using this expansion [5].

Substituting (11)–(15) in (1) and equating the same powers of ε gives equations at $O(\varepsilon^i)$, $i = 1, 2, 3$ as in the following,

$$M_1(p_i, q_i, r_i) = N_{i,1} \tag{16}$$

$$M_2(p_i, q_i, r_i) = N_{i,2} \tag{17}$$

$$M_3(p_i, q_i, r_i) = N_{i,3} \tag{18}$$

with the linear operators M_i, $i = 1, 2, 3, 4$ are defined as

$$M_1(p_i, q_i, r_i) = \Gamma_0 p_i + a p_i - a q_i \tag{19}$$

$$M_2(p_i, q_i, r_i) = \Gamma_0 q_i + a p_i - c p_i - c q_i + r_0 p_i + p_0 r_i \tag{20}$$

$$M_3(p_i, q_i, r_i) = \Gamma_0 r_i - q_0 p_i - p_0 q_i + b_0 r_i \tag{21}$$

and the source terms $N_{i,j}$ for $i = 1, 2, 3$ and $j = 1, 2, 3, 4$ are given as

$$\begin{cases} N_{1,1} = N_{1,2} = N_{1,3} = 0, \\ N_{2,1} = -\Gamma_1 p_1, \\ N_{2,2} = -\Gamma_1 q_1 - p_1 r_1, \\ N_{2,3} = -\Gamma_1 r_1 - b_2 r_0 + p_1 q_1, \\ N_{3,1} = -\Gamma_2 p_1 - \Gamma_1 p_2, \\ N_{3,2} = -\Gamma_2 q_1 - \Gamma_1 q_2 - p_2 r_1 - p_1 r_2, \\ N_{3,3} = -\Gamma_2 r_1 - \Gamma_1 r_2 + p_2 q_1 + p_1 q_2 - b_2 r_1. \end{cases} \tag{22}$$

Besides, it can be easily seen that q_i can be solved in terms of p_i from Eq. (16) and r_i can be solved in terms of q_i and p_i from Eq. (17). Substituting all of these in Eq. (18), we obtain that p_i satisfies the linear ordinary differential equation in $s_0 = t$ given by

$$M_c p_i = \Theta_i \tag{23}$$

where

$$M_c = A_3 \Gamma_0^3 + A_2 \Gamma_0^2 + A_1 \Gamma_0 + A_0 \tag{24}$$

with

$$\begin{aligned} A_0 &= -\frac{ab_0 + b_0(r_0 - 2c) + p_0(p_0 + q_0)}{p_0} \\ A_1 &= -\frac{a^2 + a(b_0 - 2c + r_0) - b_0 c + p_0^2}{ap_0} \\ A_2 &= -\frac{a + b_0 - c}{ap_0} \\ A_3 &= -\frac{1}{ap_0} \end{aligned}$$

and the composite source Θ_i is

$$\Theta_i = -\frac{ab_0 N_{i2} - ap_0 N_{i3} + \left(p_0^2 - b_0 c\right) N_{i1} + (b_0 - c)\Gamma_0 N_{i1} + a\Gamma_0 N_{i2} + \Gamma_0^2 N_{i1}}{ap_0}. \tag{25}$$

In order to find and eliminate secular terms in the solutions of (16)–(18), (23) will be utilized in the future.

It is clear from (22) that for $i = 1$ the source terms $N_{1,j} = 0, j = 1, 2, 3$. So, we can consider a solution for p_1 in the form of

$$p_{1\,(s_0, s_1, s_2)} = \alpha(s_1, s_2) e^{i\omega t} + \beta(s_1, s_2) e^{-i\omega t} + \gamma(s_1, s_2) e^{\lambda_3 t} \tag{26}$$

where $\beta = \alpha$ is the complex conjugate of α and p_1 is a real valued function. It is obvious that the γ mode belongs to the stable manifold, whereas the α and β modes correspond to the center manifold where the Hopf bifurcation occurs and $\lambda_{1,2} = \pm i\omega$ are purely imaginary. As the periodic orbits are on the center manifold, we take $\gamma = 0$ in (26) in order to create and study them. Using the solution form (26) in (16)–(18) for $i = 1$, the other first order fields are found as

$$q_1(s_0, s_1, s_2) = \left(\frac{a + i\omega}{a}\right)\alpha(s_1, s_2)e^{i\omega t} + \left(\frac{a - i\omega}{a}\right)\beta(s_1, s_2)e^{-i\omega t} \tag{27}$$

and

$$\begin{aligned} r_1(s_0, s_1, s_2) = &\left[\left(2ac - a^2 + \omega^2 - ar_0\right) + i(c\omega - a\omega)\right]\alpha(s_1, s_2)e^{i\omega t} \\ &+ \left[\left(2ac - a^2 + \omega^2 - ar_0\right) - i(c\omega - a\omega)\right]\beta(s_1, s_2)e^{-i\omega t} \end{aligned} \tag{28}$$

Having the first order solutions (26), (27), (28) yields that the second-order sources $N_{2,1}$, $N_{2,2}$, $N_{2,3}$ can be calculated from (22). Setting the coefficients of the secular $e^{\pm\omega it}$ terms in these sources to zero gives

$$\frac{\partial\alpha}{\partial s_1} = 0, \quad \frac{\partial\beta}{\partial s_1} = 0. \tag{29}$$

As a next step, let us consider the second-order sources with (29). In order to achieve equilibrium between the second and zeroth harmonic terms at this level, the second-order specific solution is considered as follows:

$$p_2(s_0, s_1, s_2) = P_0(s_1, s_2) + P_1(s_1, s_2)e^{2i\omega t} + \overline{P_1}(s_1, s_2)e^{-2i\omega t} \tag{30}$$

Substituting (30) in (23) for $i = 2$, and maintaining equilibrium between the second-harmonic and time-independent terms, the coefficients of the (30) is obtained as

$$\begin{aligned} P_0(s_1, s_2) = &\left[\frac{b_2 p_0 r_0}{ab_0 + p_0(p_0 + q_0) + b_0(-2c + r_0)}\right] \\ &+ \left[\frac{2\left(a^2 b_0 - b_0\omega^2 - a\left(2b_0 c + p_0^2 - b_0 r_0\right)\right)}{ap_0(ab_0 + p_0(p_0 + q_0) + b_0(-2c + r_0))}\right]\alpha(s_1, s_2)\beta(s_1, s_2), \end{aligned} \tag{31}$$

$$P_1(s_1, s_2) = \left(\frac{r_{11} + ir_{12}}{r_{22}}\right)\alpha^2(s_1, s_2), \tag{32}$$

with

$$r_{11} = a^4\left(b_0^2 + 4\omega^2\right) + 2\omega^2\left(\left(b_0^2 + 4\omega^2\right)\left(c^2 + 2\omega^2\right) + 2\omega^2 p_0^2 - p_0^4\right)$$

$$
\begin{aligned}
&+a^3\left(2b_0^2(r_0-2c)+b_0p_0q_0+8\omega^2(r_0-2c)\right)\\
&+a\omega^2\left(b_0^2(6c-5r_0)+b_0p_0(3p_0-q_0)\right)\\
&+a\omega^2\left(2\left(c\left(12\omega^2+p_0(q_0-3p_0)\right)+r_0\left(p_0^2-10\omega^2\right)\right)\right)\\
&+a^2\left(16c^2\omega^2-12\omega^4+4\omega^2p_0^2-2\omega^2p_0q_0-p_0^4-p_0^3q_0\right)\\
&-a^2\begin{pmatrix} b_0^2\left((r_0-2c)^2-3\omega^2\right)+b_0p_0q_0(r_0-2c)\\ +16c\omega^2r_0+4\omega^2r_0^2 \end{pmatrix},
\end{aligned}
$$

$$
\begin{aligned}
r_{12}=&-\omega\left(-2\omega^2\left(b_0^2c+3b_0p_0^2+4c(\omega-p_0)(\omega+p_0)\right)\right)\\
&-\omega a^3\left(b_0^2+4\omega^2-2p_0(2p_0+q_0)\right)\\
&-\omega a^2\left(4c\left(p_0(2p_0+q_0)-3\omega^2\right)-b_0p_0q_0\right)\\
&-2\omega a^2\left(b_0^2(r_0-3c)+r_0\left(2\omega^2-p_0(2p_0+q_0)\right)\right)\\
&-\omega a\left(8c^2\omega^2+\left(2\omega^2+p_0^2\right)\left(4\omega^2+p_0(q_0-p_0)\right)\right)\\
&-\omega a\left(b_0^2\left(2\left(c^2+\omega^2\right)-cr_0\right)-4c\omega^2r_0\right)\\
&-\omega a(b_0p_0(c(q_0-3p_0)+3p_0r_0)),
\end{aligned}
$$

$$
\begin{aligned}
r_{22}=&\,a^4p_0\left(4\omega^2\left(b_0^2+4\omega^2\right)\left(c^2+4\omega^2\right)+\left(b_0^2+4\omega^2\right)\right)\\
&-8\omega^2p_0\left(p_0^2\left(b_0c+4\omega^2\right)+p_0^4\right)\\
&+8a\omega^2p_0\left(c\left(4\omega^2+p_0(q_0-p_0)\right)-b_0p_0q_0\right)\\
&+8a\omega_0^{2p}\left(b_0^2(c-r_0)+r_0\left(p_0^2-4\omega^2\right)\right)\\
&+p_0\left(2a^3\left(b_0^2(r_0-2c)+b_0p_0(p_0+q_0)+4\omega^2(r_0-2c)\right)\right)\\
&+p_0\left(a^2\left(16c^2\omega^2-16\omega^4-8\omega^2p_0q_0+p_0^4+2p_0^3q_0\right)\right)\\
&+p_0\left(a^2\left(p_0^2q_0^2-2b_0p_0(2c-r_0)(p_0+q_0)\right)\right)\\
&-\left(p_0\left(a^2\left(b_0^2\left((r_0-2c)^2-4\omega^2\right)+16c\omega^2r_0+4\omega^2r_0^2\right)\right)\right).
\end{aligned}
$$

Using (30) in (16)–(18) for $i=2$, along with $N_{2,1}$, $N_{2,2}$, $N_{2,3}$, q_2 and r_2 are obtained.

These allow us to compute the coefficients of the secular terms in the composite source Θ_3, in conjunction with the first-order results in (22) and (23). For the slow second-order time scales, the final equation for the coefficients in the linear solutions may be found by suppressing these secular components to get uniform expansions by

$$\frac{\partial\alpha}{\partial s_2}=H_1\alpha^2\beta+H_2b_2\alpha \tag{33}$$

where the very large expressions for the coefficients H_1 and H_2 are omitted for the sake of brevity. Equation (33) is the normal form, or simplified system in the center-manifold, in the vicinity of the Hopf bifurcation point(s).

Now, let us take

$$\alpha = \frac{1}{2}A\,(s_2\,)e^{i\theta\,(s_2\,)}$$

in (33). Considering $\beta = \alpha$ and separating the expression into real and imaginary parts, we get

$$\frac{\partial A}{\partial s_2} = H_{2r}b_2A + \frac{1}{4}H_{1r}A^3 \tag{34}$$

$$\frac{\partial \theta}{\partial s_2} = H_{2i}b_2A + \frac{1}{4}H_{1i}A^2 \tag{35}$$

where H_{kr} is the real part and H_{ki} is the imaginary parts of H_k, $k = 1, 2$.

The fixed points of (34) are found as

$$A_0 = 0,\;\; A_{1,2} = \pm 2\sqrt{-\frac{H_{2r}}{H_{1r}}b_2}$$

which give the amplitude of the solution $\alpha = \frac{1}{2}Ae^{i\theta}$ with $A_{1,2}$ corresponding to the bifurcating orbits. It can be easily seen that, if

$$\frac{H_{2r}}{H_{1r}}b_2 < 0 \Rightarrow sgn(H_{1r}) = -\,sgn(H_{2r})sgn(b_2), \tag{36}$$

then the fixed points $A_{1,2}$ are real valued. Hence, by $b = b_0 + \varepsilon^2 b_2$, there are periodic orbits below the Hopf bifurcation point b_0.

On the other hand, the Jacobian of (34) at the fixed point $A_{1,2}$ is evaluated as

$$J\,|_{A_{1,2}} = -\,2(H_{2r})b_2.$$

If this value is positive, then the Hopf bifurcation at $b = b_0$ subcritical, while if it is negative, then the Hopf bifurcation at $b = b_0$ becomes supercritical.

4 Numerical Results

In this section, we precisely select the values of the model's parameters and obtain some numerical results to better explain the behavior of the periodic orbits near a Hopf bifurcation. We set $p(0) = -\,10$, $q(0) = 0$, $r(0) = 37$ for all cases since they are appropriate parameters that enable us to examine the system's general behavior close to the Hopf bifurcation.

In Fig. 1, we take $a = 35$, $b = 40$, $c = 27$. Here b is chosen below the Hopf bifurcation value $b_0 = 41.2593$ (this value is calculated from (9)). Since the bifurcation is supercritical, it is seen that there exist stable limit cycles below the Hopf bifurcation value b_0 and Fig. 2 represents the periodic solutions for $r(t)$ for these parameters.

We analyze the preceding example at the Hopf parameter $b_0 = 41.2593$ in Fig. 3 and notice how the solutions to the system (1) behave in phase space when the system experiences Hopf bifurcation. It is observed that, before becoming unstable and moving away from the attractor E_1, the trajectories stay near to the periodic orbit

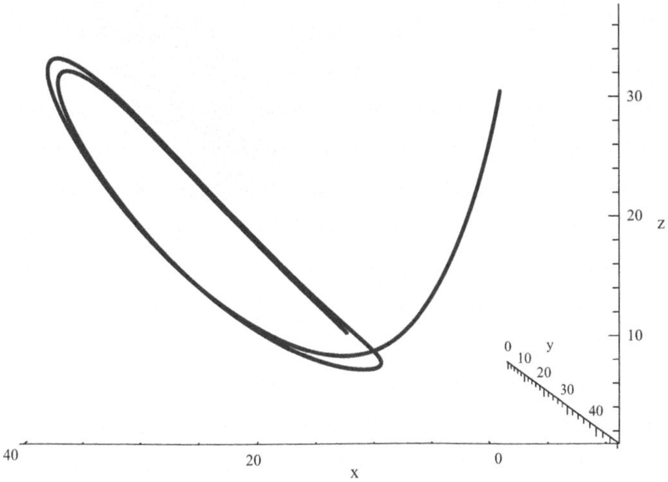

Fig. 1 Phase portrait for the system (1) for $a = 35$, $b = 40$, $c = 27$. The behaviour of the solutions for the system can be seen before a Hopf bifurcation occurs ($b < b_0$)

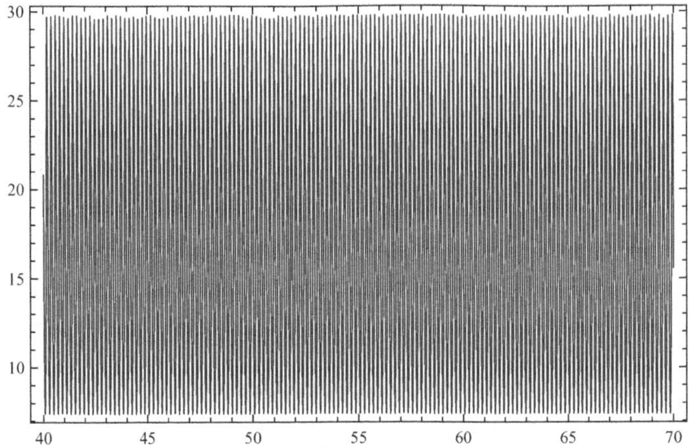

Fig. 2 Periodic solutions for $r(t)$ when $a = 35$, $b = 40$, $c = 27$, which is below the Hopf bifurcation

surrounding the attractor E_2. Also, Fig. 4 shows the oscillation for $q(t)$ at the Hopf point. There exists intermittent behaviour compared with Fig. 2 because of the small change in b.

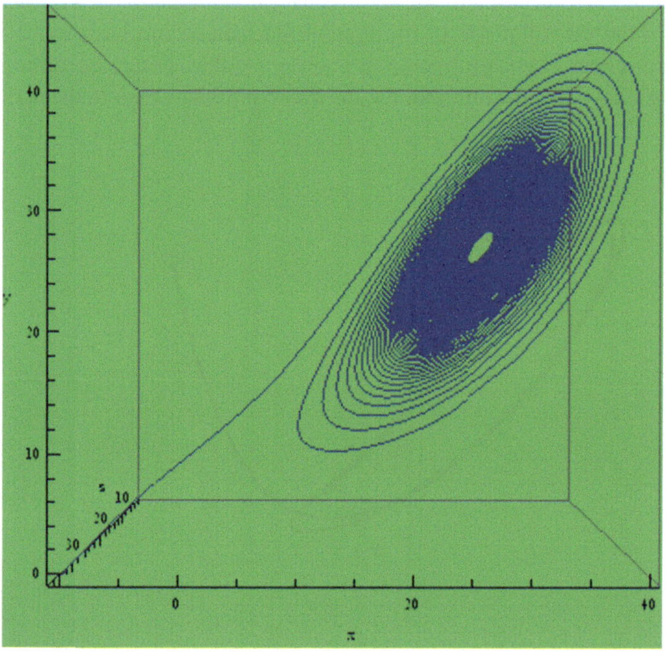

Fig. 3 The behaviour of the solutions for the system can be seen at the Hopf bifurcation point

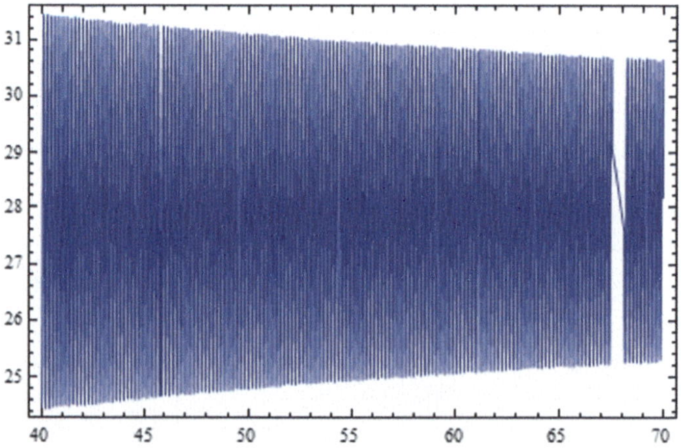

Fig. 4 Oscillations in q for $a = 35$, $b = 41.2593$, $c = 27$

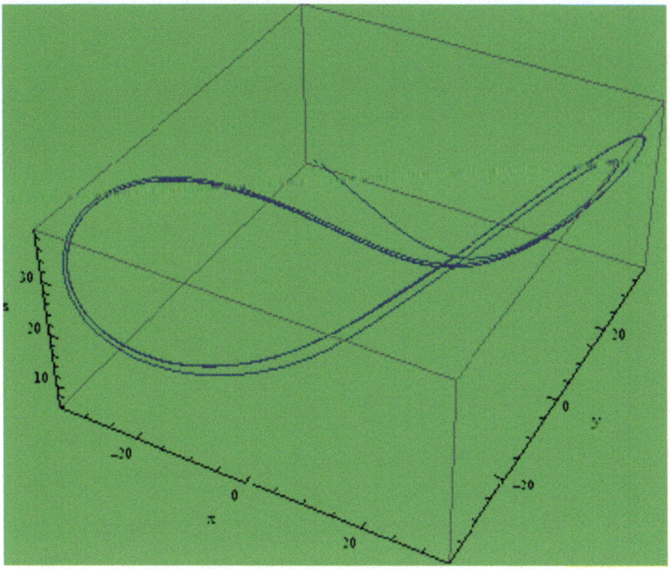

Fig. 5 A stable limit cycle which is not created because of Hopf bifurcation

Naturally, stable limit cycles may also occur in a system with many parameters, they do not have to be seen near the Hopf bifurcation point. In Fig. 5 we see an example to this case, here the parameters are taken as $a = 40$, $b = 20$, $c = 30$.

In Fig. 6, we finally examine the scenario where $a = 35$, $b = 65$, $c = 27$. As predicted, we find that the orbit is unstable.

5 Conclusions

We have completed the Hopf bifurcation investigation of a Chen's system which is a three dimensional nonlinear famous attractor in chaos theory. It is found that the Hopf bifurcation occurs at the point b_0 which is a function of the other system parameters a and c. Moreover, we were able to obtain analytical formulas for the periodic orbits arising from this Hopf bifurcation by using an effective perturbation technic called the multiple scale methods. The results found theoretically were verified also numerically. In addition to potential applications in different disciplines, the plethora of behaviors seen in such a simple model demonstrates that the study of chaotic behavior in dynamical systems is vital to understanding nonlinear systems of differential equations of many shapes and forms.

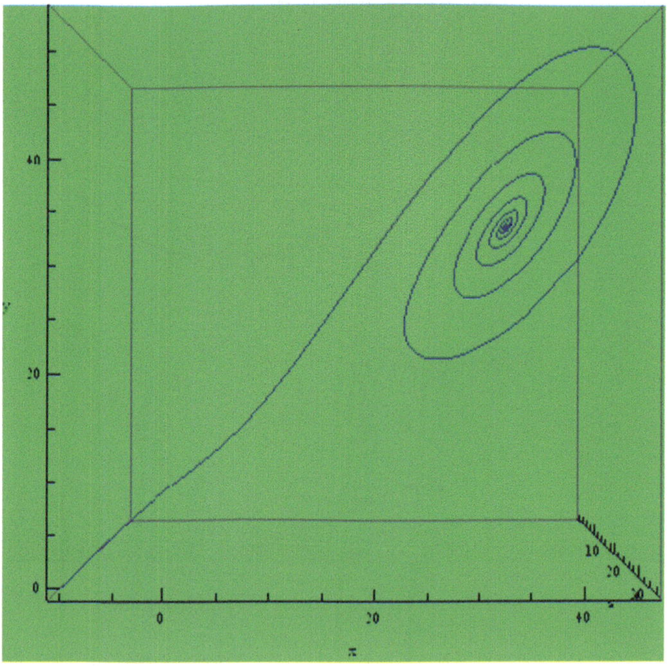

Fig. 6 Phase portrait for system (1) when $a = 35$, $b = 65$, $c = 27$

Statements and Declarations

Funding The author declares that no funds, grants, or other support were received during the preparation of this manuscript.

Conflict of Interest The authors declare that they have no conflict of interest.

References

1. Algaba, A., Domínguez-Moreno, M.C., Merino, M., Rodríguez-Luis, A.J.: Study of the Hopf bifurcation in the Lorenz, Chen and Lü systems. Nonlinear Dyn. **79**(2), 885–902 (2015)
2. Azil, S., Odibat, Z., Shawagfeh, N.: Nonlinear dynamics and chaos in Caputo-like discrete fractional Chen system. Phys. Scr. **96**(9), 095219 (2021)
3. Chen, T., Choudhury, S.R.: Bifurcations and chaos in a modified driven Chen's system. Far East J. Dyn. Syst. **181**, 53–81 (2012)
4. Chen, G., Ueta, T.: Yet another chaotic attractor. Int. J. Bifurcation Chaos **9**(07), 1465–1466 (1999)
5. Choudhury, S.R.: On bifurcations and chaos in predator-prey models with delay. Chaos Solitons Fractals **2**(4), 393–409 (1992)

6. Choudhury, S.R., Mandragona, D.: A chaotic chemical reactor with and without delay: bifurcations, competitive modes, and amplitude death. Int. J. Bifurcation Chaos Appl. Sci. Eng. **29**(2), 1950019, 15 pp. (2019)
7. Choudhury, S.R., Oztepe, G.S.: Bifurcations and dynamics in modified two population neuronal network models. J. Discontin. Nonlin. Complex **10**(02), 237–257 (2021)
8. Dong, C., Liu, H.: Analysis of periodic orbits for the Chen and the Lü systems via a variational approach. J. Korean Phys. Soc. **75**(3), 188–195 (2019)
9. Glass, L., Mackey, M.C.: From Clocks to Chaos: The Rhythms of Life, 6th edn. Princeton University Press, New Jersey (2020)
10. Van Gorder, R.A., Choudhury, S.R.: Analytical Hopf bifurcation and stability analysis of T system. Commun. Theor. Phys. **55**(4), 609–616 (2011)
11. Grosvenor, K., Jefferson, R.: The edge of chaos: quantum field theory and deep neural networks. SciPost Phys. **12**(3), 81 (2022)
12. He, H., Cui, Y., Lu, C., Sun, G.: Time delay Chen system analysis and its application. In: Tan, J. (ed.) Advances in Mechanical Design. ICMD 2019. Mechanisms and Machine Science, vol. 77, p. 77. Springer, Singapore (2020)
13. Jia, Y., Chen, H., Wang, S., Wang, Y., Yang, J., Liu, S.: Construction of hash function algorithm based on Chen chaotic system. In: IEEE 2nd International Conference on Power, Electronics and Computer Applications (ICPECA), pp. 319–323. IEEE, Shenyang, China (2022)
14. Kamdem, K.P.D., Lai, Q., Fotsin, H.: Dynamics, synchronization and electronic implementations of a new Lorenz-like chaotic system with nonhyperbolic equilibria. Int. J. Bifurcation Chaos Appl. Sci. Eng. **29**(14), 1950197, 19 pp. (2019)
15. Kayalvizhi, S., Malarvizhi, S.: A novel encrypted compressive sensing of images based on fractional order hyper chaotic Chen system and DNA operations. Multimed. Tools Appl. **79**, 3957–3974 (2020)
16. Korolj, A., Wu, H.T., Radisic, M.: A healthy dose of chaos: using fractal frameworks for engineering higher-fidelity biomedical systems. Biomaterials **219**, 119363 (2019)
17. Krise, S., Choudhury, S.R.: Bifurcations and chaos in a predator-prey model with delay and a laser-diode system with self-sustained pulsations. Chaos Solitons Fractals **16**, 59–77 (2003)
18. Li, Y.: Stability of fractional order Chen chaos based on Takagi-Sugeno fuzzy model. Oper. Res. Fuzz. **10**(2), 115–121 (2020)
19. Li, C., Chen, G.: A note on Hopf bifurcation in Chen's system. Int. J. Bifurcation Chaos **13**(6), 1609–1615 (2003)
20. Li, X., Ou, Q.: Dynamical properties and simulation of a new Lorenz-like chaotic system. Nonlinear Dyn. **65**(3), 255–270 (2011)
21. Li, X.Y., Wang, P.: Hopf bifurcation and heteroclinic orbit in a 3D autonomous chaotic system. Nonlinear Dyn. **73**, 621–632 (2013)
22. Liao, Y., Zhou, Y., Xu, F., Shu, X.B.: A study on the complexity of a new chaotic financial system. Complexity **2020**, 8821156, 5 pp. (2020)
23. Liu, Y., Yang, Q.: Dynamics of a new Lorenz-like chaotic system. Nonlinear Anal. Real World Appl. **11**(4), 2563–2572 (2010)
24. Lorenz, E.N.: Deterministic nonperiodic flow. J. Atmos. Sci. **20**(2), 130–141 (1963)
25. Lü, J., Chen, G.: A new chaotic attractor coined. Int. J. Bifurcation Chaos **12**(3), 659–661 (2002)
26. Mancas, S., Choudhury, R.S.: Periodic and chaotic traveling wave patterns in reaction-diffusion/predator-prey models with general nonlinearities. Far East J. Dyn. Syst. **11**(2), 20 (2009)
27. Marwan, M., Mehboob, M., Ahmad, S., Aqeel, M.: Hopf bifurcation of forced Chen system and its stability via adaptive control with arbitrary parameters. Soft. Comput. **24**(6), 4333–4341 (2020)
28. Matsumoto, T.: A chaotic attractor from Chua's circuit. IEEE Trans. Circuits Syst. **31**(12), 1055–1058 (1984)
29. Mostafaee, J., Mobayen, S., Vaseghi, B., Vahedi, M., Fekih, A.: Complex dynamical behaviors of a novel exponential hyper–chaotic system and its application in fast synchronization and color image encryption. Sci. Progr. **104**(1), 00368504211003388 (2021)

30. Nayfeh, A.H.: Introduction to Perturbation Techniques. Wiley, New York (1981)
31. Nayfeh, A.H., Balachandran, B.: Applied Nonlinear Dynamics. Wiley, New York (1995)
32. Nunez-Perez, J.-C., Adeyemi, V.-A., Sandoval-Ibarra, Y., Perez-Pinal, F.-J., Tlelo-Cuautle, E.: Maximizing the chaotic behavior of fractional order Chen system by evolutionary algorithms. Mathematics **9**(11), 1194 (2021)
33. Nwachioma, C., Perez-Cruz, J.H.: Analysis of a new chaotic system, electronic realization and use in navigation of differential drive mobile robot. Chaos Solitons Fractals **144**, 110684 (2021)
34. Orio, P., Gatica, M., Herzog, R., Maidana, J.P., Castro, S., Xu, K.: Chaos versus noise as drivers of multistability in neural networks. Chaos **28**(10), 106321 (2018)
35. Oztepe, G.S., Choudhury, S.R., Bhatt, A.: Multiple scales and energy analysis of coupled Rayleigh-Van der Pol oscillators with time-delayed displacement and velocity feedback: Hopf bifurcations and amplitude death. Far East J. Dyn. Syst. **26**(1), 31–59 (2015)
36. Pourasad, Y., Ranjbarzadeh, R., Mardani, A.: A new algorithm for digital image encryption based on chaos theory. Entropy **23**(3), 341, 16 pp. (2021)
37. Rössler, O.E.: An equation for continuous chaos. Phys. Lett. A **57**(5), 397–398 (1976)
38. Sparrow, C.: The Lorenz Equations: Bifurcation Chaos, and Strange Attractor. Springer-Verlag, New York (1982)
39. Sprott, J.C.: Some simple chaotic flows. Phys. Rev. E **50**(2), R647–R650 (1994)
40. Sprott, J.C.: Elegant Chaos Algebraically Simple Chaotic Flows. World Scientific Publishing, Toh Tuck Link, Singapore (2010)
41. Suqi, M.: Two-dimensional manifolds of modified Chen system with time delay. Int. J. Bifurcation Chaos **31**(09), 2150174 (2021)
42. Tian, K., Ren, H.P., Grebogi, C.: Existence of chaos in the Chen system with linear time-delay feedback. Int. J. Bifurcation Chaos **29**(09), 1950114 (2019)
43. Tigan, G., Opris, D.: Analysis of a 3D chaotic system. Chaos Solitons Fractals **36**(5), 1315–1319 (2008)
44. Uate, T., Chen, G.R.: Bifurcation analysis of Chen's equation. Int. J. Bifurcation Chaos **10**(8), 1917–1931 (2000)
45. Volos, C.K., Bardis, N.G., Kyprianidis, I.M., Stouboulos, I.N.: Implementation of mobile robot by using double-scroll chaotic attractors. In: WSEAS Recent Researches in Applications of Electrical and Computer Engineering, pp. 119–124 (2012)
46. Wang, B., Zhang, X., Dong, X.: Novel secure communication based on chaos synchronization. IEICE Trans. Fundam. Electron. Commun. Comput. Sci. **101**(7), 1132–1135 (2018)
47. Wang, B., Li, L., Wang, Y.: An efficient nonstandard finite difference scheme for chaotic fractional-order Chen system. IEEE Access **8**, 98410–98421 (2020)
48. Wang, X., Kuznetsov, N.V., Chen, G.: Chaotic Systems with Multistability and Hidden Attractors. Springer, Cham, Switzerland (2021)
49. White, C.I., Smith, T.B., Choudhury, S.R.: Generalized Hopf bifurcations and chaos in a predator-prey system. Far East J. Dyn. Syst. **20**(1), 39–62 (2012)
50. Xiu, C., Zhou, R., Liu, Y.: New chaotic memristive cellular neural network and its application in secure communication system. Chaos Solitons Fractals **141**, 110316 (2020)
51. Yang, Q., Chen, G.: A chaotic system with one saddle and two stable node-foci. Int. J. Bifurcation Chaos **18**(5), 1393–1414 (2008)
52. Yin, C.: Chaos detection of the Chen system with Caputo–Hadamard fractional derivative. Int. J. Bifurcation Chaos **31**(01), 2150016 (2021)
53. Yu, P., Han, M.A., Bai, Y.Z.: Dynamics and bifurcation study on an extended Lorenz system. J. Nonlin. Modell. Anal. **1**(1), 107–128 (2019)
54. Zhang, R., Yu, L., Jiang, D., Ding, W., Song, J., He, K., Ding, Q.: A novel plaintext-related color image encryption scheme based on cellular neural network and Chen's chaotic system. Symmetry **13**(3), 393 (2021)

55. Zhang, F., Zhou, P., Qin, J., Mu, C., Xu, F.: Dynamics of a generalized Lorenz-like chaos dynamical systems. J. Appl. Anal. Comput. **11**(3), 1577–1587 (2021)
56. Zhou, T., Tang, Y., Chen, G.: Complex dynamical behaviors of the chaotic Chen's system. Int. J. Bifurcation Chaos **13**(9), 2561–2574 (2003)

The Robust Method for Solving the Conformable Time-Fractional Klein–Gordon Equation

Halil Anaç

Abstract The present study used the conformable fractional q-Shehu homotopy analysis transform method to investigate the conformable time-fractional Klein–Gordon equations with proportional delay. The numerical answers to this problem are graphed using Maple software. The efficacy and consistency of the proposed method have been demonstrated through numerical simulations.

Keywords Klein–Gordon equation · Shehu transform · Conformable fractional derivative

1 Introduction

Fractional calculus (FC) is a developing branch of mathematics that has significant implications in various scientific and engineering disciplines. The findings were documented in multiple books or relevant review papers. Nevertheless, we are now in the early stages of Applying this highly potent tool in various study domains. Currently, fractional calculus has expanded its scope to encompass the dynamics of complicated real-world phenomena. Additionally, new ideas are being incorporated into theoretical disciplines that may not often involve practical applications. Thus, it is imperative to offer a concise overview of the triumphant implementations of FC in the fields of science and engineering. Furthermore, it is vital to acknowledge that FC is not universally applicable, but rather has specific contexts in which it is relevant. Therefore, presenting notable examples of successful FC applications can serve as a valuable reference for future studies on its implementation. Validated with actual data. Occasionally, certain patents have been issued that enhance the potential of the FC tool. Despite being established over 300 years ago and having been applied in various scientific and engineering domains, the promotion of applications remains a crucial responsibility for the fractional calculus (FC) community. When discussing FC with

H. Anaç (✉)
Gumushane University, Gumushane, Turkey
e-mail: halilanac0638@gmail.com

O. Castillo et al. (eds.), *Advanced Mathematics for the Modeling and Solution of Challenging Problems in Engineering*, Studies in Systems, Decision and Control 582,
https://doi.org/10.1007/978-3-031-82342-8_16

scientists and engineers who are not part of our community, we often encounter two common questions regarding the practical implementation and understanding of FC applications. Thus, it is imperative to offer a concise overview of the triumphant implementations of FC in the fields of science and engineering. Furthermore, it is vital to acknowledge that FC is not universally applicable, but rather has specific areas where it is relevant. Therefore, presenting notable examples of successful FC applications can serve as a valuable reference for future studies on its application [1–15].

In their recent publication, Khalil et al. proposed a novel and well-defined concept of a fractional derivative known as the conformable fractional derivative (CFD). This derivative has the property of satisfying the chain rule. The revised definition has garnered significant interest from numerous researchers. Previous studies have yielded conclusions regarding the fundamental characteristics of the CFD. Additionally, its various applications and extensions have been thoroughly examined. However, the examination of the theory of conformable fractional differential equations has only commenced in the recent past [16–20].

The nonlinear Klein–Gordon equation (KGE) discussed in this context was initially formulated by renowned physicists Klein and Gordon in 1926 to describe relativistic electrons. The KGE, on the other hand, was originally investigated by Schrodinger for quantum waves [15, 21]. The KGE finds several applications in various scientific fields. Additionally, it possesses a soliton-type solution [22–32].

There exist numerous numerical and semi-analytical techniques to solve conformable time-fractional partial differential equations (CTFPDEs) [21, 23–26, 33–35]. The aim of this study is to obtain numerical solutions of the conformable time-fractional KGE with proportional delay (CTFNKGEPD) via the conformable fractional q-Shehu homotopy analysis transform method (CFq-SHATM).

2 Main Definitions and Theorems

Remember what CFD and CFq-SHATM are and how they should be used in this work.

Definition 2.1 [18, 19, 36] Assume that $g : [0, \infty) \to \mathbb{R}$. Then, the CFD of g order α is defined as

$$D_\alpha(\rho)(x) = \lim_{\varepsilon \to 0} \frac{\rho\left(x + \varepsilon x^{1-\alpha}\right) - \rho(x)}{\varepsilon}, \tag{1}$$

for $\forall x \in (0, \infty),\ 0 < \alpha \leq 1$.

Theorem 2.1 [18, 19, 36] *Assume that* $0 < \alpha \leq 1$ *and* g, h *be* α*-differentiable at a point* $x \in (0, \infty)$. *It is acquired by*

$$i.\ D_\alpha(\rho g + \sigma h) = \rho D_\alpha(g) + \sigma D_\alpha(h), \quad \text{for } \forall \rho, \sigma \in \mathbb{R}, \tag{2}$$

$$ii.\ D_\alpha\left(x^m\right) = mx^{m-1}, \quad \text{for } \forall m \in \mathbb{R}, \tag{3}$$

$$iii.\ D_\alpha(c) = 0 \quad \text{for } \forall g(t) = c, \tag{4}$$

$$iv.\ D_\alpha(gh) = hD_\alpha(g) + gD_\alpha(h), \tag{5}$$

$$v.\ D_\alpha\left(\frac{g}{h}\right) = \frac{hD_\alpha(g) - gT_\alpha(h)}{h^2}. \tag{6}$$

Definition 2.2 [37] Assume that $\alpha \in (0, 1]$, $g : [0, \infty) \to \mathbb{R}$. The conformable fractional Shehu transform (CFST) of g is identified as

$$^{C}\mathfrak{I}_\gamma\left[g(t)\right] = V_\gamma(s; u) = \int\limits_0^\infty \exp\left(\frac{-st^\gamma}{u\gamma}\right) g(t)t^{\gamma-1}dt. \tag{7}$$

Definition 2.3 [37] Assume that $\alpha \in (0, 1]$, $g : [0, \infty) \to \mathbb{R}$. The CFST for the CFD of the function $g(t)$ is given by

$$V_\alpha\left[T_\alpha g(t)\right](v) = \frac{s}{u} V_\alpha(s; u) - g(0). \tag{8}$$

3 The Proposed Method

Consider CTFNKGEPD

$$D_t^\alpha \eta(x, t) + A\eta(\vartheta_i x, \pi_i t) + H\eta(\vartheta_i x, \pi_i t) = \omega(x, t), t > 0, n - 1 < \alpha \le n, \tag{9}$$

where A, H are linear and nonlinear operators, $\omega(x, t)$ is the nonhomogeneous term, $\vartheta_i, \pi_i \in (0, 1)$ and D_t^α is a CFD of order α.

Implementing the CFST to Eq. (9) and using the initial condition (IC), we obtain

$$\begin{aligned} &\frac{s}{u}\, ^{C}\mathfrak{I}_\alpha[\eta(x, t)] - \eta(x, 0) + {}^{C}\mathfrak{I}_\alpha[A\eta(\vartheta_i x, \pi_i t)] \\ &\quad + {}^{C}\mathfrak{I}_\alpha[H\eta(\vartheta_i x, \pi_i t)] = {}^{C}\mathfrak{I}_\alpha[\omega(x, t)]. \end{aligned} \tag{10}$$

Simplifying this equation, we get

$$^{C}\mathfrak{I}_\alpha[\eta(x, t)] - \frac{u}{s}\eta(x, 0) + \frac{u}{s}\, ^{C}\mathfrak{I}_\alpha[A\eta(\vartheta_i x, \pi_i t) + H\eta(\vartheta_i x, \pi_i t)]$$

$$-\frac{u}{s}{}^{C}\mathfrak{I}_{\alpha}[\omega(x,t)]=0. \tag{11}$$

Using the HAM, the nonlinear operator (NO) of $\Theta(x, t; q)$ is described by

$$N\big[\Theta(x,t;q)\big]={}^{C}\mathfrak{I}_{\alpha}\big[\Theta(x,t;q)\big]-\frac{u}{s}\Theta(x,t;q)\big(0^{+}\big)+\frac{u}{s}\big({}^{C}\mathfrak{I}_{\alpha}\big[A\Theta(\vartheta_{i}x,\pi_{i}t;q)\big]$$
$$+{}^{C}\mathfrak{I}_{\alpha}\big[H\Theta(\vartheta_{i}x,\pi_{i}t;q)\big]-{}^{C}\mathfrak{I}_{\alpha}[\omega(x,t)]\big), \tag{12}$$

where $0\leq q\leq\frac{1}{n}$.

A homotopy is constructed via

$$(1-nq)^{C}\mathfrak{I}_{\alpha}\big[\Theta(x,t;q)-w_{0}(x,t)\big]=hqG^{*}(x,t)H\big[\Theta(\vartheta_{i}x,\pi_{i}t;q)\big], \tag{13}$$

where, $h\neq 0$ and ${}^{C}\mathfrak{I}_{\alpha}$ represents CFST. For $q=0$ and $q=\frac{1}{n}$, the outcomes of Eq. (13) are in the following.

$$\Theta(x,t;0)=\eta_{0}(x,t),\ \Theta\left(x,t;\frac{1}{n}\right)=\eta(x,t). \tag{14}$$

Therefore, by generating q from 0 to $\frac{1}{n}$, the solution $\Theta(x, t; q)$ converges from $\eta_0(x, t)$ to the solution $\eta(x, t)$. $\Theta(x, t; q)$ is expanded via Taylor theorem, we get

$$\Theta(x,t;q)=\eta_{0}(x,t)+\sum_{i=1}^{\infty}\eta_{m}(x,t)q^{m}, \tag{15}$$

where,

$$\eta_{m}(x,t)=\frac{1}{m!}\frac{\partial^{m}\Theta(x,t;q)}{\partial q^{m}}|_{q=0}. \tag{16}$$

Equation (15) converges at $q=\frac{1}{n}$ for the convenient $\eta_0(x, t)$, n and h. Therefore, we get the solutions of Eq. (9)

$$\eta(x,t)=\eta_{0}(x,t)+\sum_{m=1}^{\infty}\eta_{m}(x,t)\left(\frac{1}{n}\right)^{m}. \tag{17}$$

The zeroth order deformation Eq. (13) m-times is differentiated with respect to q and dividing by $m!$, then for $q = 0$, we get the mth deformation equation (mDE)

$${}^{C}\mathfrak{I}_{\alpha}\big[\eta_{m}(x,t)-\mathrm{X}_{m}\eta_{m-1}(x,t)\big]=hG^{*}(x,t)\mathcal{R}_{m}(\vec{w}_{m-1}). \tag{18}$$

When Eq. (18) is applied to the inverse CFST (ICFST), we have

$$\eta_m(x,t) = \mathrm{X}_m \eta_{m-1}(x,t) + h\left({}^C\mathfrak{I}_\alpha\right)^{-1}\left[G^*(x,t)\mathcal{R}_m(\vec{\eta}_{m-1})\right], \tag{19}$$

where

$$\begin{aligned}\mathcal{R}_m(\vec{\eta}_{m-1}) = {}& {}^C\mathfrak{I}_\alpha\left[\eta_{m-1}(x,t)\right] - \left(1 - \frac{\mathrm{X}_m}{n}\right)\frac{u}{s} w_0(x,t) \\ & + \frac{u}{s}\,{}^C\mathfrak{I}_\alpha\left[A\eta_{m-1}(\vartheta_i x, \pi_i t) + H^*_{m-1}(x,t) - \omega(x,t)\right],\end{aligned} \tag{20}$$

and

$$\mathrm{X}_m = \begin{cases} 0, & m \leq 1, \\ n, & m > 1. \end{cases} \tag{21}$$

Here, G^*_m is homotopy polynomial and presented as

$$G^*_m = \frac{1}{m!}\frac{\partial^m \Theta(x,t;q)}{\partial q^m}|_{q=0}, \; \Theta(x,t;q) = \Theta_0 + q\Theta_1 + q^2\Theta_2 + \cdots. \tag{22}$$

Using Eqs. (19)–(20), one gets

$$\begin{aligned}\eta_m(x,t) = {}& (\mathrm{X}_m + h)\eta_{m-1}(x,t) - \left(1 - \frac{k_m}{n}\right)\frac{u}{s}\eta_0(x,t) \\ & + h \times \left({}^C\mathfrak{I}_\alpha\right)^{-1}\left[\left(\frac{u}{s}\,{}^C\mathfrak{I}_\alpha\left[A\eta_{m-1}(\vartheta_i x, \pi_i t) + G^*_{m-1}(x,t) - \omega(x,t)\right]\right)\right].\end{aligned} \tag{23}$$

Using CFq-SHATM, then we have

$$\eta(x,t) = \sum_{c=0}^{\infty} \eta_c(x,t). \tag{24}$$

4 Applications

4.1 Consider CTFNKGEPD [29]

$$D_t^\alpha w(x,t) - \frac{\partial^2 w(x,t)}{\partial x^2} + w^2\left(\frac{x}{2}, \frac{t}{2}\right) = 0, \; t \geq 0, \quad 0 < \alpha \leq 1, \tag{25}$$

subject to IC

$$w(x,0) = 1 + \sin(x). \tag{26}$$

Applying the CFST to Eq. (25) and using the IC, it is generated as

$$\frac{s}{u}\,{}^{C}\mathfrak{J}_{\alpha}[w(x,t)] - w(x,0) + {}^{C}\mathfrak{J}_{\alpha}\left[-\frac{\partial^2 w(x,t)}{\partial x^2} + w^2\left(\frac{x}{2},\frac{t}{2}\right)\right] = 0. \tag{27}$$

Simplifying this equation, Eq. (30) is acquired by

$${}^{C}\mathfrak{J}_{\alpha}[w(x,t)] - \frac{u}{s}w(x,0) + \frac{u}{s}\,{}^{C}\mathfrak{J}_{\alpha}\left[-\frac{\partial^2 w(x,t)}{\partial x^2} + w^2\left(\frac{x}{2},\frac{t}{2}\right)\right] = 0. \tag{28}$$

Using the HAM, the NO of $\Theta(x,t;q)$ is given as

$$\begin{aligned} N\big[\Theta(x,t;q)\big] = {}&{}^{C}\mathfrak{J}_{\alpha}\big[\Theta(x,t;q)\big] - \frac{u}{s}\Theta(x,t;q)\big(0^{+}\big) \\ &+ \frac{u}{s}\,{}^{C}\mathfrak{J}_{\alpha}\left[-\frac{\partial^2 \Theta(x,t;q)}{\partial x^2} + \Theta^2\left(\frac{x}{2},\frac{t}{2};q\right)\right], \end{aligned} \tag{29}$$

where $q \in \left[0, \frac{1}{n}\right]$.

The mDE is identified as

$${}^{C}\mathfrak{J}_{\alpha}\big[w_m(x,t) - \mathrm{X}_m w_{m-1}(x,t)\big] = h\mathcal{R}_m(\vec{w}_{m-1}), \tag{30}$$

where,

$$\begin{aligned} \mathcal{R}_m(\vec{w}_{m-1}) = {}&{}^{C}\mathfrak{J}_{\alpha}\big[w_{m-1}(x,t)\big] - \left(1 - \frac{\mathrm{X}_m}{n}\right)\frac{u}{s}w_0(x,t) \\ &+ \frac{u}{s}\,{}^{C}\mathfrak{J}_{\alpha}\left[-\frac{\partial^2 w_{m-1}(x,t)}{\partial x^2} + \sum_{j=0}^{m-1} w_j\left(\frac{x}{2},\frac{t}{2}\right)w_{1-j}\left(\frac{x}{2},\frac{t}{2}\right)\right]. \end{aligned} \tag{31}$$

Applying the ICFST to Eq. (30), we get

$$w_m(x,t) = \mathrm{X}_m w_{m-1}(x,t) + h\big({}^{C}\mathfrak{J}_{\alpha}\big)^{-1}\big[\mathcal{R}_m(\vec{w}_{m-1})\big]. \tag{32}$$

By the use of IC, then it is generated as

$$w_0(x,t) = 1 + \sin(\mathrm{x}). \tag{33}$$

and for $m = 1,\ m = 2$ in Eq. (32), Eqs. (34)–(35) are acquired by

$$w_1(x,t) = \frac{ht^{\alpha}}{\alpha}\left[-\sin(\mathrm{x}) + 1 + \sin^2\left(\frac{x}{2}\right) + 2\sin\left(\frac{x}{2}\right)\right], \tag{34}$$

$$w_2(x,t) = (n+h)\frac{ht^{\alpha}}{\alpha}\left[-\sin(\mathrm{x}) + 1 + \sin^2\left(\frac{x}{2}\right) + 2\sin\left(\frac{x}{2}\right)\right]$$

$$- \frac{1}{4\alpha^2}\Bigg(h^2 t^{2\alpha}\Bigg(1 + 2 \times \Bigg(-2 + \cos^2\Big(\frac{x}{4}\Big)\Bigg)2^{1-\alpha}$$
$$+ 8\Bigg(\Bigg(2\cos^2\Big(\frac{x}{4}\Big) + \Big(-2 - \sin\Big(\frac{x}{4}\Big)\Big)\cos\Big(\frac{x}{4}\Big)\Bigg)\sin^2\Big(\frac{x}{4}\Big)$$
$$- \sin\Big(\frac{x}{4}\Big)\Bigg)2^{-\alpha} + 2\Bigg(8\cos^3\Big(\frac{x}{4}\Big) - 5\cos\Big(\frac{x}{4}\Big)\Bigg)\sin\Big(\frac{x}{4}\Big)$$
$$+ 8\cos^4\Big(\frac{x}{4}\Big) - 8\cos^2\Big(\frac{x}{4}\Big)\Bigg)\Bigg)\Bigg). \tag{35}$$

Thus, the CFq-SHATM solution of Eq. (25) is given by

$$w(x, t) = w_0(x, t) + \sum_{m=1}^{\infty} w_m(x, t)\left(\frac{1}{n}\right)^m. \tag{36}$$

Three dimensional plots for CFq-SHATM solution by distinct α values are indicated in Fig. 1.

The two-dimensional plots for CFq-SHATM solution via different α values are indicated in Fig. 2.

4.2 Consider CTFNKGEPD [29]

$$D_t^{\alpha} w(x, t) - \frac{\partial^2 w(x, t)}{\partial x^2} + w(x, t) - w^3\left(\frac{x}{2}, \frac{t}{2}\right) = 0,\ t \geq 0, \quad 0 < \alpha \leq 1, \tag{37}$$

subject to IC

$$w(x, 0) = -\operatorname{sech} x. \tag{38}$$

Applying the CFST to this equation and using the IC, we get

$$\frac{s}{u}\, {}^{C}\mathfrak{J}_{\alpha}[w(x, t)] - w(x, 0) + {}^{C}\mathfrak{J}_{\alpha}\left[-\frac{\partial^2 w(x, t)}{\partial x^2} + w(x, t) - w^3\left(\frac{x}{2}, \frac{t}{2}\right)\right] = 0. \tag{39}$$

Simplifying this equation, Eq. (40) is acquired via

$${}^{C}\mathfrak{J}_{\alpha}[w(x, t)] - \frac{u}{s} w(x, 0) + \frac{u}{s}\, {}^{C}\mathfrak{J}_{\alpha}\left[-\frac{\partial^2 w(x, t)}{\partial x^2} + w(x, t) - w^3\left(\frac{x}{2}, \frac{t}{2}\right)\right] = 0. \tag{40}$$

Using the HAM, the NO $\psi(x, t; q)$ is given by

$$N\big[\psi(x, t; q)\big] = {}^{C}\mathfrak{J}_{\alpha}\big[\psi(x, t; q)\big] - \frac{u}{s}\psi(x, t; q)\big(0^{+}\big)$$

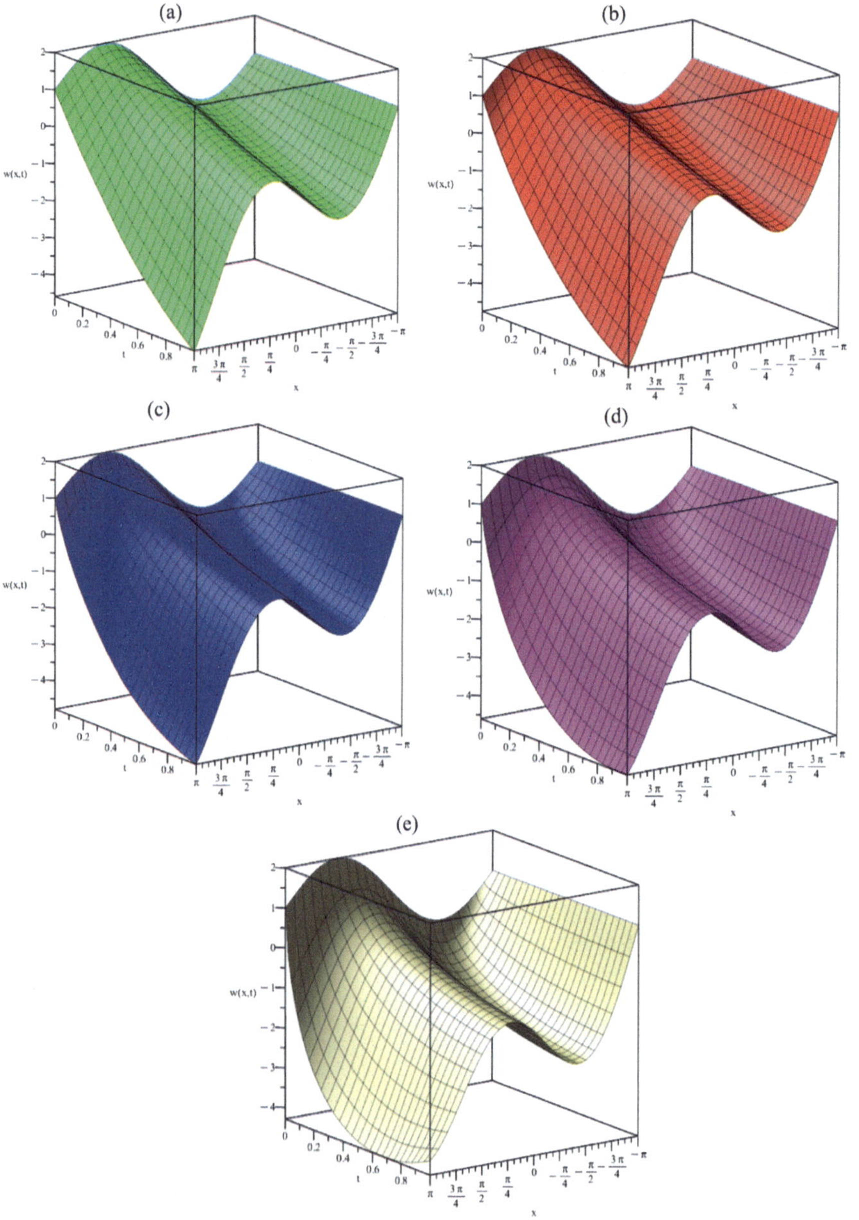

Fig. 1 **a** For $\alpha = 1$, CFq-SHATM solution, **b** for $\alpha = 0.9$, CFq-SHATM solution, **c** for $\alpha = 0.8$, CFq-SHATM solution, **d** for $\alpha = 0.7$, CFq-SHATM solution, **e** for $\alpha = 0.6$, CFq-SHATM solution for Eq. (25) at $h = -1, n = 1$

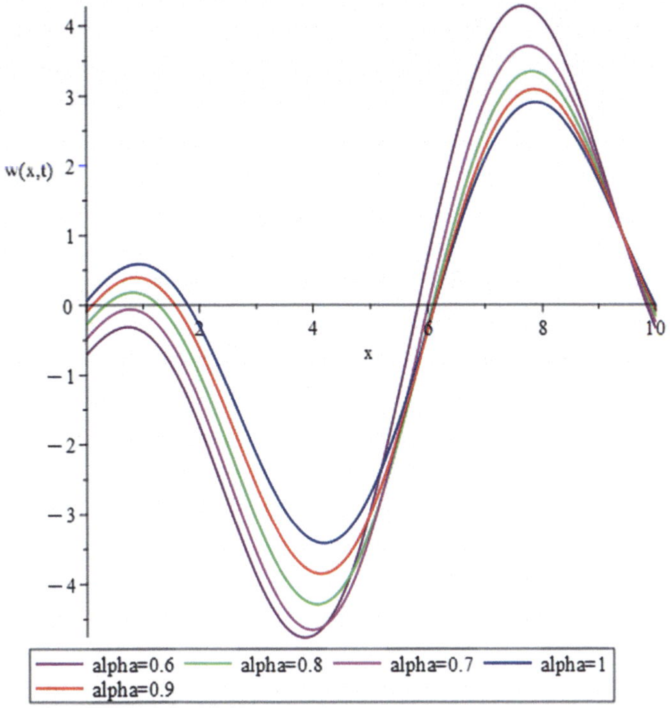

Fig. 2 The 2D graph of the CFq-SHATM solution for Eq. (26) at h = − 1, n = 1, $t = 0.5$ with distinct α

$$+\frac{u}{s}{}^{C}\mathfrak{I}_{\alpha}\left[-\frac{\partial^2\psi(x,t;q)}{\partial x^2}+\psi(x,t;q)-\psi^3\left(\frac{x}{2},\frac{t}{2};q\right)\right], \tag{41}$$

where $q\epsilon\left[0,\frac{1}{n}\right]$.

The mDE is identified by

$${}^{C}\mathfrak{I}_{\alpha}\left[w_m(x,t)-\mathrm{X}_m w_{m-1}(x,t)\right]=h\mathcal{R}_m(\vec{w}_{m-1}), \tag{42}$$

where,

$$\begin{aligned}\mathcal{R}_m(\vec{w}_{m-1}) = {}&{}^{C}\mathfrak{I}_{\alpha}\left[w_{m-1}(x,t)\right]-\left(1-\frac{\mathrm{X}_m}{n}\right)\frac{u}{s}w_0(x,t)\\&+\frac{u}{s}{}^{C}\mathfrak{I}_{\alpha}\left[-\frac{\partial^2 w_{m-1}(x,t)}{\partial x^2}+w_{m-1}(x,t)\right.\\&\left.-\sum_{r=0}^{m-1}\left(\sum_{j=0}^{r}w_r\left(\frac{x}{2},\frac{t}{2}\right)w_{r-j}\left(\frac{x}{2},\frac{t}{2}\right)\right)w_{m-1-r}\left(\frac{x}{2},\frac{t}{2}\right)\right].\end{aligned} \tag{43}$$

Applying the ICFST to Eq. (42), we have

$$w_m(x,t) = X_m w_{m-1}(x,t) + h\left({}^{C}\mathfrak{I}_{\alpha}\right)^{-1}\left[\mathcal{R}_m(\vec{w}_{m-1})\right]. \tag{44}$$

By the use of IC, then it is generated as

$$w_0(x,t) = -\operatorname{sech} x. \tag{45}$$

For $m = 1,\ m = 2$ in Eq. (44), Eqs. (46)–(47) are obtained by

$$w_1(x,t) = \frac{ht^{\alpha}}{\alpha}\left[2\operatorname{sech} x \tanh^2 x - \operatorname{sech}^3\left(\frac{x}{2}\right) - 2\operatorname{sech} x\right], \tag{46}$$

$$\begin{aligned} w_2(x,t) = {} & (n+h)\frac{ht^{\alpha}}{\alpha}\left[2\operatorname{sech} x \tanh^2 x - \operatorname{sech}^3\left(\frac{x}{2}\right) - 2\operatorname{sech} x\right] \\ & + h\Psi\left(mx, \frac{t^{n\alpha}}{\alpha^n}\right), \end{aligned} \tag{47}$$

where, $m \in \mathbb{Q}$ and $n \in \mathbb{N}$.

Three dimensional plots for CFq-SHATM solution by distinct α values are shown in Fig. 3.

The two-dimensional plots for CFq-SHATM solution via different α values are shown in Fig. 4.

5 Results and Discussion

For CTFNKGEPD, Fig. 1 shows the three dimensional plots of $w(x,t)$ solution obtained by CFq-SHATM for different values of α. The two dimensional plots of $w(x,t)$ solution obtained by CFq-SHATM for different values of α are shown in Fig. 2. Figure 3 shows the three dimensional plots of $w(x,t)$ solutions obtained by CFq-SHATM for different values of α in Eq. (37). The two dimensional plots of the $w(x,t)$ solution obtained by CFq-SHATM for different values of α at Eq. (37) are shown in Fig. 4. Table 1 indicates the $w(x,t)$ solution obtained by CFq-SHATM for different values of α for Eq. (25). Also, for Eq. (37), Table 2 exhibits the plots of the $w(x,t)$ solution obtained via CFq-SHATM for distinct values of α.

6 Conclusion

CTFNKGEPD are analysed by CFq-SHATM in this study. Also, the MAPLE software is utilizied to construct 2D–3D dimensional plots that depict the CFq-SHATM solutions of Eqs. (25)–(37) for various values of α. The Maple software generates

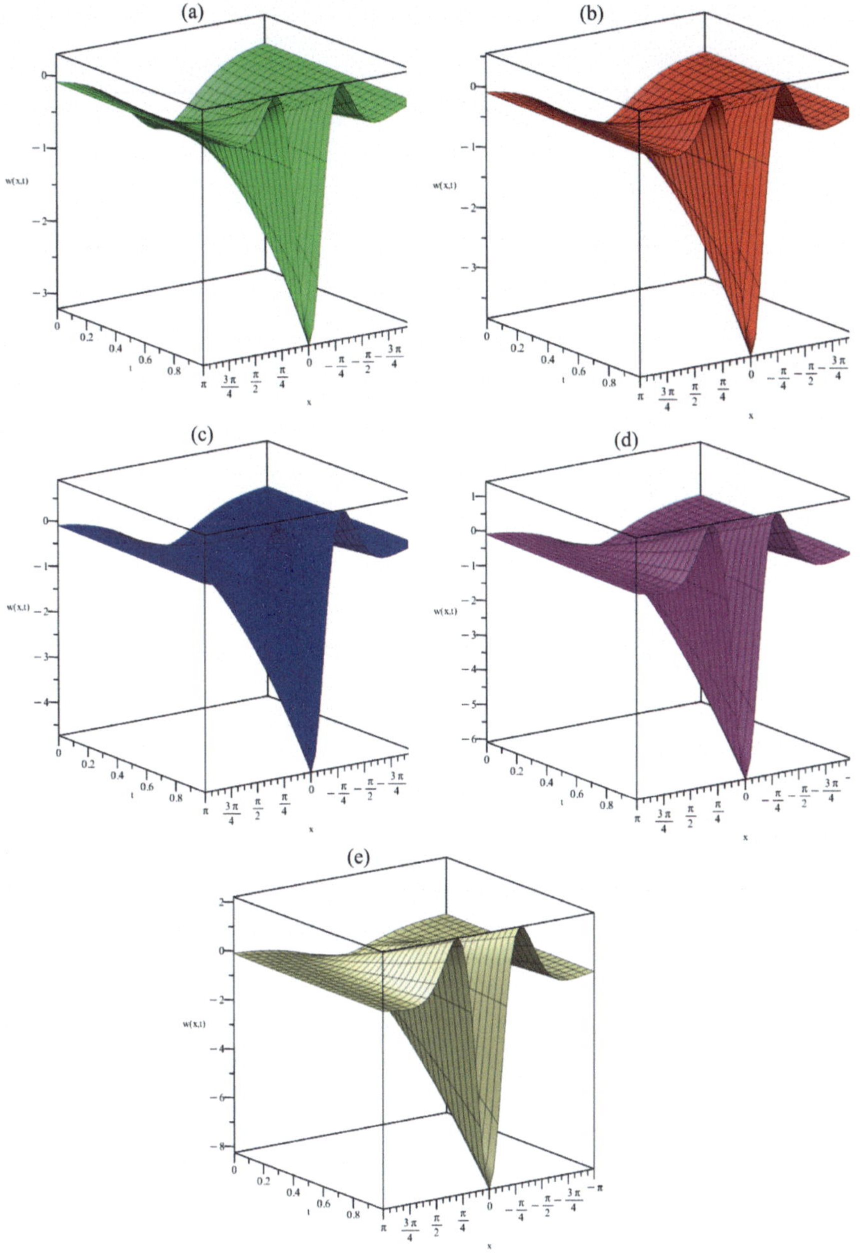

Fig. 3 **a** For $\alpha = 1$, CFq-SHATM solution, **b** for $\alpha = 0.9$, CFq-SHATM solution, **c** for $\alpha = 0.8$, CFq-SHATM solution, **d** for $\alpha = 0.7$, CFq-SHATM solution, **e** for $\alpha = 0.6$, CFq-SHATM solution for Eq. (37) at $h = -1, n = 1$

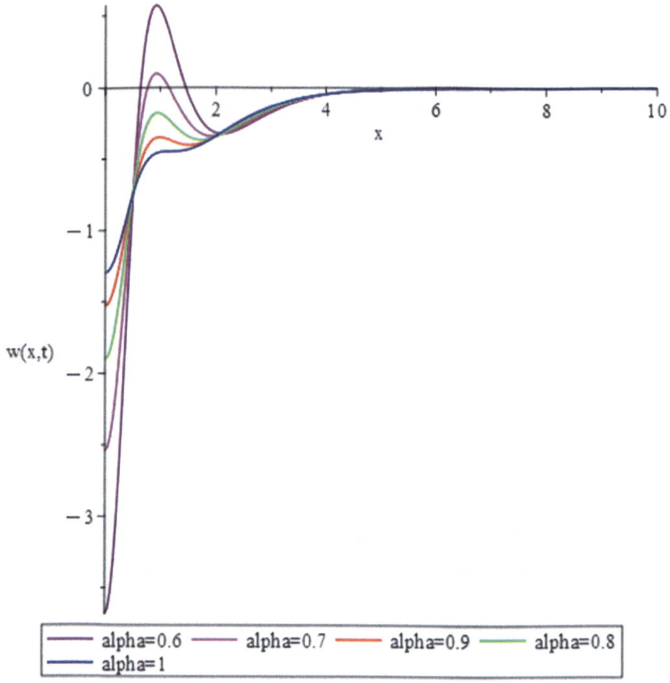

Fig. 4 The two-dimensional plot of the CFq-SHATM solution for Eq. (37) at h = − 1, n = 1, t = 0.5 with different α

surface graphs for Eqs. (25)–(37) with varying structures. The numerical outcomes of the suggested approach for various α values may be found in Tables 1 and 2. The recently discovered method for solving CTFNKGEPD can be deemed useful and effective.

Table 1 Numerical values of CTFNKGEPD via CFq-SHATM in Eq. (25) at h = − 1, n = 1 with different values of x and t for distinct α

x	t	$\alpha = 0.7$	$\alpha = 0.8$	$\alpha = 0.9$	$\alpha = 1$
0.1	0.1	0.5575589876	0.7150230937	0.8247555976	0.9017003780
	0.2	0.2486268515	0.4459072077	0.5952749113	0.7083303966
	0.3	0.0029613963	0.2153531258	0.3850350465	0.5197834727
	0.4	− 0.2045219510	0.0099892434	0.1887801311	0.3359996064
	0.5	− 0.3847126232	− 0.1762752809	0.0040066198	0.1569987978
0.2	0.1	0.6515446872	0.8104238259	0.9211538579	0.9987961496
	0.2	0.3394871850	0.5385942054	0.6893938260	0.8035622892
	0.3	0.0910948406	0.3054750208	0.4768467481	0.6129677512
	0.4	− 0.1188967655	0.0976102427	0.2782281891	0.4270125351
	0.5	− 0.3014516220	− 0.0911255355	0.0910236010	0.2456966409
0.3	0.1	0.7402610602	0.9014791963	1.0138643680	1.0926721770
	0.2	0.4233863327	0.6254039752	0.7784841242	0.8944032104
	0.3	0.1710412270	0.3884989981	0.5624723200	0.7007133093
	0.4	− 0.0423924707	0.1771284578	0.3604775900	0.5116024726
	0.5	− 0.2280305568	− 0.0149125359	0.1699552646	0.3270707005
0.4	0.1	0.8224416548	0.9870048420	1.1017622480	1.1822463660
	0.2	0.4989129928	0.7050168303	0.8613003332	0.9796863661
	0.3	0.2412909689	0.4630019323	0.6405651934	0.7817383414
	0.4	0.0234126956	0.2470398651	0.4340963886	0.5884022942
	0.5	− 0.1660742812	0.0507968765	0.2392980983	0.3996782229
0.5	0.1	0.8968897164	1.0658744120	1.1837725220	1.2664808460
	0.2	0.5647478646	0.7761904874	0.9366634640	1.0582832060
	0.3	0.3004438288	0.5276543012	0.7098600778	0.8548326195
	0.4	0.0770641616	0.3059480986	0.4977473525	0.6561290864
	0.5	− 0.1170718496	0.1045559133	0.2976555850	0.4621726071

Table 2 Numerical values of CTFNKGEPD via CFq-SHATM in Eq. (37) at h = − 1, n = 1 with different values of x and t for distinct α

x	t	$\alpha = 0.7$	$\alpha = 0.8$	$\alpha = 0.9$	$\alpha = 1$
0.1	0.1	0.5575589876	0.7150230937	0.8247555976	0.9017003780
	0.2	0.2486268515	0.4459072077	0.5952749113	0.7083503966
	0.3	0.0029613963	0.2153531258	0.3850350465	0.5197834727
	0.4	− 0.2045219510	0.0099892434	0.1887801311	0.3359996064
	0.5	− 0.3847126232	− 0.1762752809	0.0040066198	0.1569987978
0.2	0.1	0.6515446872	0.8104238259	0.9211538579	0.9987961496
	0.2	0.3394871850	0.5385942054	0.6893938260	0.8035622892
	0.3	0.0910948406	0.3054750208	0.4768467481	0.6129677512
	0.4	− 0.1188967655	0.0976102427	0.2782281891	0.4270125351
	0.5	− 0.3014516220	− 0.0911255355	0.0910236010	0.2456966409
0.3	0.1	0.7402610602	0.9014791963	1.0138643680	1.0926721770
	0.2	0.4233863327	0.6254039752	0.7784841242	0.8944032104
	0.3	0.1710412270	0.3884989981	0.5624723200	0.7007133093
	0.4	− 0.0423924707	0.1771284578	0.3604775900	0.5116024726
	0.5	− 0.2280305568	− 0.0149125359	0.1699552646	0.3270707005
0.4	0.1	0.8224416548	0.9870048420	1.1017622480	1.1822463660
	0.2	0.4989129928	0.7050168303	0.8613003332	0.9796863661
	0.3	0.2412909689	0.4630019323	0.6405651934	0.7817383414
	0.4	0.0234126956	0.2470398651	0.4340963886	0.5884022942
	0.5	− 0.1660742812	0.0507968765	0.2392980983	0.3996782229
0.5	0.1	0.8968897164	1.0658744120	1.1837725220	1.2664808460
	0.2	0.5647478646	0.7761904874	0.9366634640	1.0582832060
	0.3	0.3004438288	0.5276543012	0.7098600778	0.8548326195
	0.4	0.0770641616	0.3059480986	0.4977473525	0.6561290864
	0.5	− 0.1170718496	0.1045559133	0.2976555850	0.4621726071

References

1. Liouville, J.: Mémoire sur quelques questions de géométrie et de mécanique, et sur un nouveau genre de calcul pour résoudre ces questions. Ecole Polytech. **13**, 71–162 (1832)
2. Riemann, G.F.B.: Versuch einer allgemeinen Auffassung der Integration und Differentiation. Gesammelte Mathematische Werke, Leipzig (1896)
3. Caputo, M.: Elasticità e Dissipazione. Zanichelli, Bologna (1969)
4. Miller, K.S., Ross, B.: An Introduction to Fractional Calculus and Fractional Differential Equations. Wiley, New York (1993)
5. Podlubny, I.: Fractional Differential Equations. Academic Press, New York (1999)
6. Baleanu, D., Diethelm, K., Scalas, E., Trujillo, J.J.: Fractional Calculus: Models and Numerical Methods. World Scientific, Boston (2012)
7. Povstenko, Y.: Linear Fractional Diffusion-Wave Equation for Scientists and Engineers. Birkhäuser, New York (2015)

8. Baleanu, D., Wu, G.C., Zeng, S.D.: Chaos analysis and asymptotic stability of generalized Caputo fractional differential equations. Chaos Solitons Fractals **102**, 99–105 (2017)
9. Sweilam, N.H., Hasan, M.M.A., Baleanu, D.: New studies for general fractional financial models of awareness and trial advertising decisions. Chaos Solitons Fractals **104**, 772–784 (2017)
10. Liu, D.Y., Gibaru, O., Perruquetti, W., Laleg-Kirati, T.M.: Fractional order differentiation by integration and error analysis in noisy environment. IEEE Trans. Autom. Control **60**, 2945–2960 (2015)
11. Esen, A., Sulaiman, T.A., Bulut, H., Baskonus, H.M.: Optical solitons to the space-time fractional (1 + 1)-dimensional coupled nonlinear Schrödinger equation. Optik **167**, 150–156 (2018)
12. Veeresha, P., Prakasha, D.G., Baskonus, H.M.: New numerical surfaces to the mathematical model of cancer chemotherapy effect in Caputo fractional derivatives. Chaos **29**, 013119 (2019)
13. Caponetto, R., Dongola, G., Fortuna, L., Gallo, A.: New results on the synthesis of FO-PID controllers. Commun. Nonlinear Sci. Numer. Simul. **15**, 997–1007 (2010)
14. Prakash, A., Veeresha, P., Prakasha, D.G., Goyal, M.: A homotopy technique for fractional order multi-dimensional telegraph equation via Laplace transform. Eur. Phys. J. Plus **134**, 1–18 (2019)
15. Veeresha, P., Prakasha, D.G., Baskonus, H.M.: Novel simulations to the time-fractional Fisher's equation. Math. Sci. **13**(1), 33–42 (2019)
16. Kilbas, A.A., Srivastava, H.M., Trujillo, J.J.: Theory and Applications of Fractional Differential Equations. Elsevier, Amsterdam (2006)
17. Debnath, L.: Recent applications of fractional calculus to science and engineering. Int. J. Math. Math. Sci. **2003**(54), 3413–3442 (2003)
18. Khalil, R., Al Horani, M., Yousef, A., Sababheh, M.: A new definition of fractional derivative. J. Comput. Appl. Math. **264**, 65–70 (2014)
19. Abdeljawad, T.: On conformable fractional calculus. J. Comput. Appl. Math. **279**, 57–66 (2015)
20. Gao, F., Chi, C.: Improvement on conformable fractional derivative and its applications in fractional differential equations. J. Funct. Spaces **2020**, 5852414 (2020)
21. Keller, A.A.: Contribution of the delay differential equations to the complex economic macrodynamics. WSEAS Trans. Syst. **9**(4), 358–371 (2010)
22. Swift, J., Hohenberg, P.C.: Hydrodynamic fluctuations at the convective instability. Phys. Rev. A **15**(1), 319 (1977)
23. Vishal, K., Kumar, S., Das, S.: Application of homotopy analysis method for fractional Swift–Hohenberg equation revisited. Appl. Math. Model. **2012**(36), 3630–3637 (2012)
24. Khan, N.A., Khan, N.U., Ayaz, M., Mahmood, A.: Analytical methods for solving the time-fractional Swift–Hohenberg (S–H) equation. Comput. Math. Appl. **2011**(61), 2181–2185 (2011)
25. Prakasha, D.G., Veeresha, P., Baskonus, H.M.: Residual power series method for fractional Swift–Hohenberg equation. Fractal Fract. **3**(1), 9 (2019)
26. Vishal, K., Das, S., Ong, S.H., Ghosh, P.: On the solutions of fractional Swift–Hohenberg equation with dispersion. Appl. Math. Comput. **2013**(219), 5792–5801 (2013)
27. Pomeau, Y., Zaleski, S., Manneville, P.: Dislocation motion in cellular structures. Phys. Rev. A **27**(5), 2710 (1983)
28. Peletier, L.A., Rottschäfer, V.: Large time behaviour of solutions of the Swift–Hohenberg equation. C. R. Math. **336**(3), 225–230 (2003)
29. Golmankhaneh, A.K., Golmankhaneh, A.K., Baleanu, D.: On nonlinear fractional Klein–Gordon equation. Signal Process. **91**(3), 446–451 (2011)
30. Hohenberg, P.C., Swift, J.B.: Effects of additive noise at the onset of Rayleigh–Bénard convection. Phys. Rev. A **46**(8), 4773 (1992)
31. Cross, M.C., Hohenberg, P.C.: Pattern formation outside of equilibrium. Rev. Mod. Phys. **65**(3), 851 (1993)
32. Hoyle, R.B.: Pattern Formation. Cambridge University Press, Cambridge (2006)

33. Wu, J.: Theory and Applications of Partial Functional Differential Equations. Springer, New York (1996)
34. Chen, X., Wang, L.: The variational iteration method for solving a neutral functional-differential equation with proportional delays. Comput. Math. Appl. **59**(8), 2696–2702 (2010)
35. Singh, B.K., Kumar, P.: Fractional variational iteration method for solving fractional partial differential equations with proportional delay. Int. J. Differ. Equ. 5206380 (2017)
36. Gözütok, N.Y., Gözütok, U.: Multivariable conformable fractional calculus. arXiv preprint arXiv:1701.00616 (2017)
37. Benattia, M.E., Belghaba, K.: Shehu conformable fractional transform, theories and applications. Cankaya Univ. J. Sci. Eng. **18**(1), 24–32 (2021)

A Review on Latest Developments in Assembly and Temporary Shelters for Natural Disasters

İrem Karakaya and Alev Taşkın

Abstract Assembly and temporary shelter areas are spaces that allow people to stay in safe zones following a disaster. Therefore, studies related to assembly areas and temporary shelters, particularly within the framework of natural disasters, are of great importance. This article examines the efforts concerning assembly areas, temporary shelters, and evacuations during a natural disaster process. The purpose of this review is to identify challenges, provide a research gap, and propose potential research directions for the future. The motivation behind this study stems from the increasing number of natural disasters in recent years due to climate change and the need for ideas to ensure the well-being of disaster victims and regional authorities. For the first time, this study reviews the literature on assembly areas and temporary shelters regarding natural disasters over the past three years. Finally, following the analysis of the reviewed studies, some potential options for future research are suggested.

Keywords Assembly and temporary shelter · Natural disasters

1 Introduction

The majority of land on Earth carries various disaster risks, primarily including natural disasters. Natural disasters are uncontrollable natural events that cause harm to people, result in loss of life and property, and, most importantly, cannot be controlled. Natural disasters can be categorized as geological and meteorological. Examples of geological natural disasters include earthquakes, landslides, volcanic eruptions, and tsunamis, which originate from the Earth's crust or interior [1]. Floods,

İ. Karakaya (✉)
Department of Management and Organization, Bartin University, Bartin, Türkiye
e-mail: isahmutoglu@bartin.edu.tr

A. Taşkın
Department of Industrial Engineering, Yildiz Technical University, Istanbul, Türkiye

O. Castillo et al. (eds.), *Advanced Mathematics for the Modeling and Solution of Challenging Problems in Engineering*, Studies in Systems, Decision and Control 582,
https://doi.org/10.1007/978-3-031-82342-8_17

storms, typhoons, hurricanes, droughts, avalanches, and erosion are meteorological natural disasters caused by weather events [1]. Here, we briefly explain these disasters.

Earthquake: An earthquake is a geological disaster that occurs as a result of the energy released from faults in the Earth's crust propagating in the form of waves, causing shaking on the Earth's surface.

Flood: Floods, which are one of the meteorological natural disasters, are sudden and significant inundations that temporarily or partially submerge the surface under water.

Tsunami: Tsunami refers to the long-period, large-scale sea wave that develops as a result of geological events such as earthquakes, volcanic eruptions, associated collapses, and landslides occurring on the ocean or sea floor.

Landslide: Landslides usually occur when the underlying layer becomes loose due to excessive rainfall, causing rocks, soil, etc. to detach and move downward.

Bushfire: It involves the partial or complete burning of forests due to human or natural causes.

Volcano: Volcano is a disaster that occurs when molten rocks from the Earth's interior erupt onto the planet's surface.

Hurricane: It is a type of wind characterized by destructive force, cylindrical rotation, and often associated with cumulus clouds.

Preventing or limiting the potential damages and losses that disasters can cause in physical, social, economic, psychological, and environmental aspects is desired. Therefore, disaster risk reduction efforts, which need to be prepared and implemented before disasters occur, are of great importance. As a result, disaster assembly and temporary shelter areas are necessary in various parts of the world for various reasons [2]. The need for disaster assembly areas and temporary shelters may arise due to cultural, social, legal, or natural factors.

Assembly areas are safe locations where people can gather away from hazardous areas until temporary shelter facilities are ready after a disaster [2]. Finding a suitable location for an assembly area is of critical importance in reducing the harmful effects of disasters. Temporary shelter areas, on the other hand, are shelters that can be used to rapidly meet the housing needs of individuals and provide a safe place for disaster victims after a disaster [2]. Temporary shelters are used until a long-term solution, such as permanent housing, is found to address the housing issues of disaster victims.

Emergency assembly areas are safe locations where people need to gather before, during, and after a disaster. Emergency assembly areas are usually located closest to people's homes. These areas serve as places where disaster victims receive information, coordination of aid teams takes place, and instructions for temporary shelter are provided. The criteria taken into consideration in determining these areas include population density, accessibility, and ease of evacuation in the region. Additionally, gathering places for the elderly and disabled individuals should be as accessible as possible, away from secondary risks, and located on flat terrains [3]. Assembly areas close to residential areas are minimally affected by both structural and non-structural characteristics. Emergency assembly areas are often located near buildings that provide access to electricity, water, toilets, and other necessary supplies.

Due to the material and spiritual losses caused by disasters that significantly damage both the natural and built environment, there is an urgent need for shelter for affected individuals, and this problem needs to be resolved as soon as possible. Temporary shelters are existing structures designed to rapidly meet the housing needs of individuals and provide a safe place for disaster victims [4]. In order to improve the quality and accountability of humanitarian activities during assistance to disaster victims, the Sphere Project [5] has classified public buildings, especially schools, as temporary shelter structures. In addition to schools, sports facilities, cultural centers, hospitals, stadiums, hotels, childcare facilities, municipal buildings, religious structures, large meeting halls, military barracks, indoor playgrounds, vacant or unfinished buildings, factories, warehouses, parks, and specially constructed shelters can be used as common temporary shelters in emergencies [3].

In many countries, there have been various studies conducted at different scales and scopes for the physical planning and site selection of "post-disaster emergency assembly and temporary shelter areas" following natural disasters such as earthquakes, floods, bushfires, and hurricanes [1]. Numerous mathematical and qualitative studies have been carried out using the type of disaster, site selection criteria, and a multitude of model parameters to determine the optimal location, number, distribution, and physical characteristics of these areas.

The conceptual framework for structured systematic literature review in qualitative content analysis is depicted in Fig. 1. Firstly, the decision is made to focus on studies related to assembly areas and temporary shelters. Then, the screening process is conducted based on the filtering options provided in Fig. 2. Finally, the selection of articles to be examined is determined, and the analysis and evaluation stages are carried out.

1.1 Motivation

It is evident that after a disaster, people will require a safe shelter to meet their urgent or long-term basic needs. Therefore, scientists conduct research on topics such as post-disaster assembly areas and temporary shelter locations. This article discusses the latest developments in assembly and shelter locations involving natural disasters. The research method involves using the most recent studies containing the keywords "assembly OR temporary AND natural disasters" from numerous databases indexed by Scopus over the past three years. The mentioned keyword "natural disaster" is individually searched for all the disasters listed in Table 1. The articles from the databases listed in Table 2 and the years specified in Table 3 are examined.

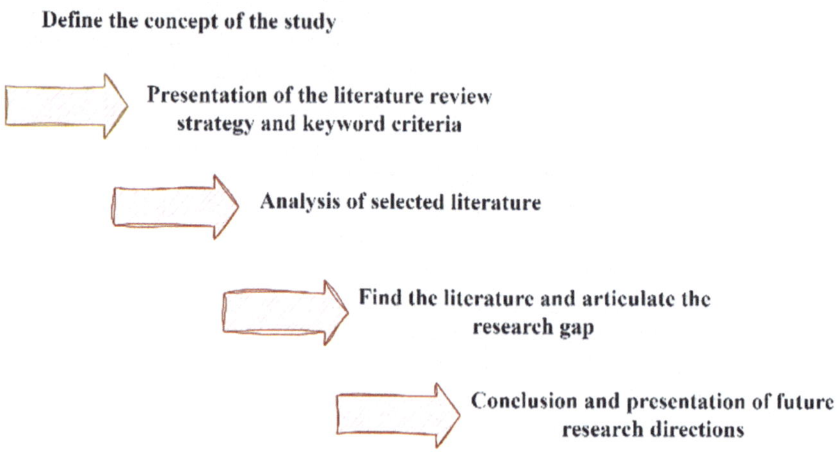

Fig. 1 Conceptual framework of the study

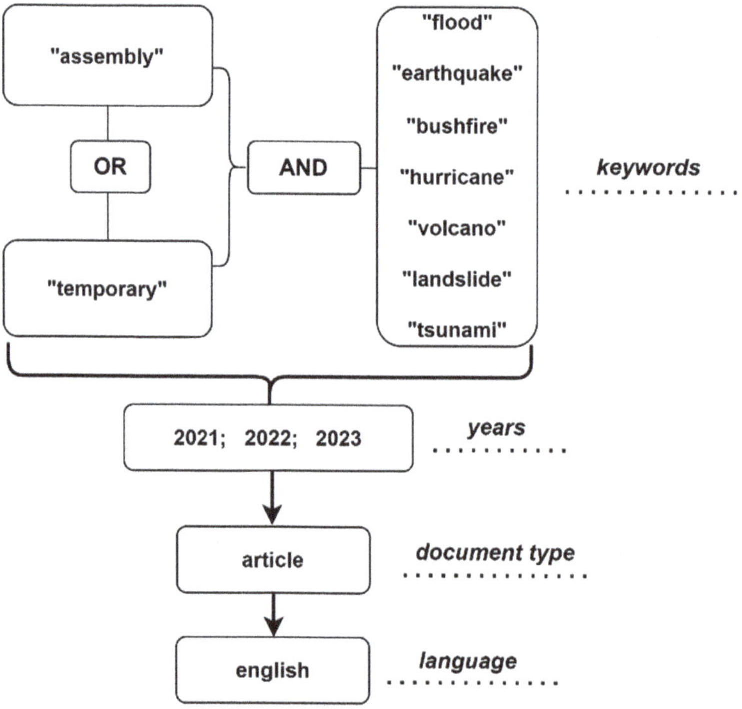

Fig. 2 Literature search filtering options

Table 1 Articles by disaster

Disaster	Papers
Earthquake	[6–17]
Flood	[17–26]
Tsunami	[17, 27–30]
Landslide	[17, 26, 31, 32]
Bushfire	[17, 33]
Volcano	[17, 34]
Hurricane	[17, 35]

Table 2 Articles by publisher

Publisher	Papers
Elsevier	[6, 11, 22, 24, 26, 33, 35]
MDPI	[15, 18–20, 30, 34]
Springer	[10, 13, 21, 27, 29, 31]
Emerald	[7, 9, 23]
SAGE Journal	[8, 12]
IEEE	[17]
Others	[14, 16, 25, 28, 32]

Table 3 Articles by year

Year	Papers
2023	[26, 27, 31, 34]
2022	[6–10, 17–24, 28–30, 32, 33, 35]
2021	[25–31]

1.2 Organization

The following sections of the article examine the latest developments and details regarding assembly and shelter locations for natural disasters in Sect. 2. Section 3 includes a discussion to provide guidelines for future research in this field. The final section presents the conclusion of the article.

2 Latest Developments in Assembly and Temporary Shelters for Natural Disasters

This section presents the latest developments in the literature regarding assembly and shelter locations for natural disasters.

2.1 Latest Developments Related to Topic

The literature on assembly and shelter locations for natural disasters in the past 3 years is being reviewed. The addressed issues, objectives, used methods, proposed models, and results are analyzed in detail.

In [6], the authors utilize the Analytic Hierarchy Process (AHP) and Geographic Information System (GIS) to determine safe assembly locations after an earthquake. Hosseini Sabzevari et al. [7] determines safe temporary shelter areas by considering temporary settlements, cultural characteristics, accessibility, terrain conditions, slope and type of land, availability, and construction materials. Wang et al. [8] introduces an agent-based simulation framework to assess the shared temporary shelter needs of contractors and displaced households. It presents a case study on housing recovery in San Francisco following hypothetical earthquakes of magnitudes M6.5, M7.2, and M7.9. Yépez and Villacreses [9] demonstrates the implementation of Temporary Safety Areas (TSAs) is demonstrated in resource-constrained cities by following a systematic decision-making process. Four pilot temporary shelter areas were constructed, and the effectiveness of the project was evaluated years later. The results indicate that the project is an effective and low-cost disaster management program for poor countries. However, if sustainable communication is not established, effective mitigation projects may not be completed on time. Ommi and Janalipour [10] utilizes seismological and environmental design methodologies to assist decision-making during earthquakes. Aftershock hazards are analyzed using the AHP and Technique for Order of Preference by Similarity to Ideal Solution (TOPSIS) methods, focusing on the recent Sarpol-e Zahab earthquake in Iran. Aghaie and Karimi [11] develops a GIS software and a mathematical model to determine temporary shelters in Tehran and to allocate injured individuals to the identified shelters. The model solves a two-stage, scenario-based optimization problem using the Non-dominated Sorting Genetic Algorithm II (NSGA-II). The output of the model demonstrates the effectiveness of the proposed methodology. Mousavi et al. [12] proposes a hybrid mathematical-simulation location-allocation model to carry out post-earthquake disaster management operations with the widest coverage. The proposed methodology first determines the best locations for temporary emergency stations and then optimally assigns injured individuals to these stations.

In [13], suitable, moderately suitable, and unsuitable areas are identified in the city center of Düzce, Turkey, based on appropriate size, accessibility, and location selection criteria for earthquake scenarios. Public green spaces are designed using GIS, and recommendations are made for earthquake preparedness techniques. Dabiri et al. [14] employs Fuzzy Delphi in the city of Sanandaj, Iran, to explore a series of significant hazards during the temporary sheltering process, both qualitatively and statistically. Subsequently, these hazards are ranked according to their relevance and importance. Kim et al. [15] conducts a case study in South Korea, utilizing interviews and analysis to establish a framework consisting of four livability concepts (security, health, sociability, and comfort) and four shelter zones (entrance, residential, service, and special needs) for temporary shelter areas. Ghaderi et al. [16] identifies primary and secondary criteria influencing the selection of temporary settlement locations in Iran through previous studies and expert opinions. Subsequently, potential losses in earthquakes were calculated using HAZUS, a GIS-based natural hazard analysis tool, and temporary site selection was made based on the results obtained from fuzzy AHP considering the damage.

In [18], an analysis of a mega flood event is conducted, and the selection of the most suitable shelters for mega flood events is determined using the AHP. Chelariu et al. [19] develops scenarios for population evacuation from river basins and integrates a multi-criteria analysis of environmental, natural, proximity, and constraining factors into a GIS environment. Zhu et al. [20] proposes an optimized evacuation route planning using an improved 3D Dijkstra algorithm for pedestrian evacuation under different flood risks. The aim is to ensure pedestrians reach evacuation shelters in the shortest possible time. Sahmutoglu et al. [21] employs the neutrosophic AHP to identify criteria affecting both floods and gathering points. Subsequently, a risk assessment of gathering points is conducted using neutrosophic CODAS. Lee et al. [22] presents a multi-criteria framework for the selection of temporary disaster waste management sites based on GIS analysis techniques. The framework contributes to the efficient determination of temporary disaster waste management sites in densely populated and disaster-prone areas of developed cities.

In [23], the article proposes techniques and solutions to improve the post-flood settlement process in Iran's Golestan province for future potential floods. These include equipping camps, providing suitable temporary shelters, monitoring material costs, and tracking the rehabilitation process. Samany et al. [24] focuses on determining safe areas or temporary shelters for emergency evacuation during and after floods by using the AHP and Weighted Linear Combination (WLC) to prepare a flood risk map. It then utilizes a modified particle swarm optimization algorithm with local search (LMPSO) to identify safe areas for flood evacuation. Comparative results demonstrate that LMPSO produces more efficient outcomes compared to modified PSO, genetic algorithm (GA), ant colony optimization (ACO), and simulated annealing (SA) algorithms. Tripathy et al. [25] aims to maximize flood evacuations and minimize infections during shelter stays in India by developing a multi-objective simulation-optimization model using the max–min approach. The SEIRS (Susceptible, Exposed, Infectious, Recovered, Susceptible) model is used to examine the effects of Covid-19 spread in the community. The probabilistic global search Laussane (PSGL) algorithm is employed to solve the multi-objective optimization model.

In [27], the article addresses three topics related to a tsunami scenario in Chile. Firstly, using an agent-based model, potential human casualties are estimated in a fully horizontal evacuation scenario. Secondly, potential vertical evacuation shelters are identified through a VR experience provided to participants. Lastly, the potential impact of vertical evacuation is assessed. Ophiyandri et al. [28] conducts an analysis of 23 potential residential buildings for pre-tsunami high-rise building studies. Evaluations based on six main criteria and 17 supporting criteria reveal that the majority of buildings are unprepared. Jitt-Aer et al. [29] calculates the coordinates of each shelter location and the minimum and maximum number of affected individuals using a multi-stage spatial interpolation. It presents a dasymetric mapping approach combining remote sensing data with a population-weighted technique. Hachiya et al. [30] designs a multi-objective model for the problem of delivering food supplies to disaster areas using drones (UAVs). The model is optimized with Q-Learning.

In [26], a Random Forest algorithm is applied for both flood and landslide susceptibility mapping, and a network-based P-median problem model is used to evaluate service coverage. Samm-A et al. [31] identifies factors that contribute to earthquake and rainfall-induced landslide occurrences and conducts scenario assessments using Monte Carlo simulation and direct estimation. Sinčić et al. [32] evaluates different scenarios by generating five scenarios under geomorphological, geological, hydrological, and anthropogenic factors that contribute to landslide occurrence. The sensitivity map with the highest success rate based on receiver operating characteristic (ROC) analysis results is selected for each scenario.

In [33], a mixed-integer nonlinear programming model is proposed to find a balance between preparedness budget and evacuation congestion for mass self-evacuation network design problems. Wigati et al. [34] suggests a model for determining the optimal location of temporary shelters using the example of Mount Merapi eruption. Potential shelters are selected from candidates that meet the criteria using GIS, fuzzy logic, AHP, and WLC. Nickdoost et al. [35] proposes an integrated simulation-based framework to determine the optimal locations of debris management areas considering both debris collection logistics planning and geographical factors. Shih et al. [17] suggests a resource-efficient shelter selection approach that takes into account the efforts of individuals in post-disaster shelters to communicate with their acquaintances, which can lead to communication and transportation congestion.

2.2 Details of Reviewed Papers

Table 4 presents details of the studies investigated regarding gathering and sheltering locations for natural disasters, including the problem addressed, the objective of the study, the methodology used, and the results.

3 Discussion

This section presents a discussion of the methods used in recent years for gathering and temporary shelter areas in natural disasters. With the increase in natural disasters worldwide in recent years, there has been a growing interest in this field. Similarly, there is a tendency for an increased interest in the studies of gathering places and temporary shelter areas before, during, and after disasters.

Table 4 Details of reviewed papers

Paper	Problem	Aim	Methodology	Results
[6]	Identifying assembly areas	Identifying safe assembly areas after an earthquake	AHP, GIS	Not all open areas are safe assembly areas
[7]	Identifying a temporary shelter	Systematically identifying safe locations and selecting suitable spaces based on basic variables	AHP, GIS	Field selection, vulnerability, and priority criteria indicate that 60% of the working area is unprotected. Additionally, open spaces such as public parks, stadiums, and vacant lots are the best locations for temporary settlements
[8]	Identifying a temporary shelter	Creating a post-disaster simulation framework for contractors and displaced households	Agent-based simulation	Preparation is necessary to address the issues of labor and temporary shelter during the recovery process
[9]	Temporary shelter areas	Applying and evaluating engineering design, implementation, and policy experiences for temporary shelter areas after an earthquake	AHP	The project demonstrates an effective and low-cost disaster management program for impoverished countries
[10]	Shelter selection	Conducting probabilistic seismic aftershock hazard assessment using remote sensing to select shelters in Iran	AHP, TOPSIS	The southern regions of Sarpol-e Zahab city are suitable locations for constructing shelters

(continued)

Table 4 (continued)

Paper	Problem	Aim	Methodology	Results
[11]	Location-allocation-routing problem for emergency shelters	Identifying temporary shelters and placing injured individuals in shelters	GIS, NSGA-II	Computational results indicate the efficiency of the proposed approach
[12]	Location and allocation of temporary emergency stations	Reducing the cost of establishing temporary emergency stations and the time required to transport the injured to the stations	Simulation-based optimization, ε-constraint method	The distance traveled by earthquake victims and associated costs decrease by 15%
[13]	Emergency assembly points and temporary shelter areas	Determining the size and location of emergency assembly points and temporary shelter areas	GIS	In the city of Düzce, it has been determined that there is insufficient space throughout the city. Mapping of emergency gathering points, temporary shelter areas, and inaccessible areas is being carried out
[14]	Crisis management of the temporary shelter process	Identifying and prioritizing major hazards associated with temporary housing procedures	Fuzzy Delphi, Lawshe method	Among the 11 key risk factors considered, climate-related issues are seen to have the greatest potential for temporary housing risk after earthquakes in the region

(continued)

Table 4 (continued)

Paper	Problem	Aim	Methodology	Results
[15]	Shelter planning	Identifying problems and unmet needs in shelter areas and providing solutions to improve shelter area designs	Interview	Temporary shelter planning should provide a large area, allow for cleanliness and indoor air quality control, and allocate spaces according to function and user type in response to victims' urgent needs
[16]	Emergency accommodation and post-disaster logistics	Determining temporary shelter locations using damage analysis results	HAZUS, fuzzy AHP, GIS	Comprehensive and full-scale damage rates in villages are very high, indicating an urgent need for suitable housing for a significant number of homeless individuals
[18]	Shelter selection	Making optimal shelter selections using an improved method to minimize losses from floods	AHP	The method is reliable for the analysis of mega-flood events and the selection of the most appropriate shelters
[19]	Shelter location and evacuation scenario	Assessing the accessibility level of evacuation areas during flood scenarios in small mountainous drainage basins	GIS-based multi criteria evaluation	91.68% of the vulnerable population reaches evacuation sites within 35 min, while 8.32% of residents reach them within 54 min

(continued)

Table 4 (continued)

Paper	Problem	Aim	Methodology	Results
[20]	Evacuation route planning	Planning optimal evacuation routes for pedestrians to reach evacuation shelters under flood risk	3D Dijkstra algorithms	The proposed method can effectively reduce pedestrian evacuation time, optimize the location of existing shelters, and provide recommendations for urban road reconstruction
[21]	Assembly area risk assessment	Identifying risk factors affecting floods and assembly points and conducting risk analyses for assembly points	Interval valued neutrosophic AHP-CODAS	The numerical results obtained with IVN-AHP and IVN-CODAS indicate that the most important sub-criterion is "proximity to flood risk areas," and the gathering point with the lowest risk score is A-9
[22]	Temporary disaster waste management area selection	Developing a multi-criteria GIS framework for selecting temporary disaster waste management areas	Multi criteria GIS	Compared to the existing temporary disaster waste management sites, the minimum, maximum, and average suitability scores of the temporary sites have increased by 27.27%, 11.54%, and 23.08%, respectively
[23]	Post-disaster shelter process	Examining the process of providing shelter during emergency, rehabilitation, and reconstruction phases	Observations, field survey activities, questionnaires and indepth interviews	Challenges include unhealthy conditions in collective camps, lack of thermal resistance in temporary shelters, and increased material costs

(continued)

Table 4 (continued)

Paper	Problem	Aim	Methodology	Results
[24]	Identification of safe areas	Identifying safe areas based on the spatial and environmental characteristics of the urban dimension (accessibility, topography, congestion, and land use)	AHP, WLC, MPSO, LMPSO	LMPSO is more efficient than other applied algorithms (MPSO, GA, ACO, SA)
[25]	Evacuation	Maximizing the number of people evacuated from flood-prone areas and minimizing the number of infections at the end of the shelter stay	Max–min optimization, SEIRS, PGSL	The proposed approach can estimate the number of residents from specific communities who need to be evacuated and prevent an uncontrolled number of new infections
[27]	Shelter layout, building criteria and evacuee decision making	Predicting potential human deaths in a horizontal evacuation scenario and identifying vertical evacuation shelters	Agent-based modelling, VR	In the horizontal scenario, deaths range from 50 to 72% of the population, while including vertical evacuation buildings in tsunami scenarios reduces losses by 6.5–13.7%
[28]	Temporary vertical evacuation shelter assessment	Identifying, analyzing, and measuring the readiness of 23 potential housing units	Observation, interview	Improvement of support facilities and infrastructure is required for seven buildings, while the remaining 16 buildings need to meet all main and supporting criteria

(continued)

Table 4 (continued)

Paper	Problem	Aim	Methodology	Results
[29]	Hazard assessment and population estimation	Determining the number of people affected by a tsunami in a specific region of Thailand to facilitate humanitarian logistics	Integrated GIS and population estimation algorithms	The results provide both numerical solutions and maps visualizing disaster intensity in the regions
[30]	Transportation of supplies to shelters	Optimizing the routing of UAVs by optimizing criteria such as speed of supplies, urgency of supplies, and equality of supply quantities	VRP, Q-learning	The proposed model for emergency aid material supply to all evacuation centers is more stable compared to other models
[26]	Shelter location and allocation	Conducting shelter site modeling under flood and landslide susceptibility mapping and scenarios	GIS, Random Forest, P-median	None of the evacuation shelters are located in safe areas regarding floods and landslides
[31]	Hazard assessment	Evaluating earthquake and rainfall-induced landslide scenarios	Monte-Carlo simulation, direct estimation	ROC curves have AUC values ranging from 85 to 93% for all scenarios evaluated at different confidence levels
[32]	Sensitivity assessment	Performing pre-landslide sensitivity analysis using 17 geofactors that cause landslide formation	The weight of evidence, bivariate statistical analysis	The study achieves a 94% success rate and a 93% prediction rate

(continued)

Table 4 (continued)

Paper	Problem	Aim	Methodology	Results
[33]	Evacuation planning	Designing a dynamic evacuation network with the smoothest traffic flow when roads become inaccessible	Bi-level programming, benders decomposition	Modifying factors such as budget and the allowed number of shelters significantly reduce the overall evacuation time and alleviate traffic congestion
[34]	Temporary shelter location	Determining the most suitable locations for temporary shelters	GIS, fuzzy logic, AHP, WLC	The identified locations for temporary shelters are highly suitable
[35]	Temporary disaster debris management site selection and debris collection logistics	Minimizing the social and economic impact of uncollected debris on people and determining optimal debris management site locations	GIS, agent-based modelling, simulation	The selection of near-optimal locations reduces the total debris collection time from 156 to 112 days
[17]	Evacuation and shelter selection	Assigning people to shelters based on their relationships and locations to prevent communication and transportation barriers	Simulation	According to the simulation results, travel costs are reduced by 23%

Current studies on assembly and temporary shelter areas mainly focus on location-allocation. However, other issues such as evacuation of people, determining evacuation routes, risk assessment, distribution of supplies to shelters, and debris collection logistics are also addressed. GIS are frequently used in solving these problems due to the geological, meteorological, and morphological aspects of natural disasters. Simulations are employed due to the rapid and devastating impacts of natural disasters. Multi-criteria decision-making techniques are particularly used in topics like site selection and risk assessment, where multiple criteria and alternative scenarios are involved. Fuzzy systems are also utilized due to the high level of uncertainty inherent in natural disaster situations.

Recent developments in assembly and temporary shelter areas focus on reducing distances, costs, and improving suitability. While earthquake studies receive the most attention, flood studies have also shown an increasing trend due to climate change. However, when examining natural disasters other than these two, such as forest fires, hurricanes, and volcanic eruptions, there is a lack of sufficient research in the literature, despite the significant material and human losses they cause. Additionally, there are gaps in the literature regarding the use of optimization and machine learning techniques in the studies of assembly and temporary shelter areas during disasters.

4 Conclusion

In today's world, issues related to post-disaster shelter, assembly and temporary shelter area suitability, evacuations, and other related topics are encountered in various fields. With the increase in natural disasters, the issue of shelter takes on a different dimension, and research is being conducted in this direction due to the inadequacy of current conditions. In this article, we present a detailed literature review of the recent developments in approaches used for assembly and temporary shelter areas for natural disasters. Recent studies focus on determining the most suitable shelter locations. In this context, geographic information systems, multi-criteria decision-making methods, and simulations are used for analysis. In recent years, deep learning-based approaches, particularly in prediction, clustering, and classification, have been increasingly used, showing promise for leveraging these approaches in the field of natural disasters and assembly and temporary shelter.

References

1. Wirasinghe, S.C., Caldera, H.J., Durage, S.W., Ruwanpura, J.Y.: Preliminary analysis and classification of natural disasters. In: 9th Annual International Conference of the International Institute for Infrastructure, Renewal and Reconstruction, Feb 2015, p. 11 (2013). https://doi.org/10.13140/RG.2.1.4283.5041
2. Çınar, A.K.: Analysing the planning criterias of emergency assembly points and temporary shelter areas: case of İzmir-Karşıyaka. J. Plan. **28**(2), 179–200 (2018). https://doi.org/10.14744/planlama.2018.07088
3. Li, Y., Liu, Y., Jiao, J.: A GIS-based suitability analysis of Xiamen's green space in park for earthquake disaster prevention and refuge. Urban Plan. Des. Res. **1**(1), 1–8 (2013). [Online]. Available: www.seipub.org/updr
4. Félix, D., Branco, J.M., Feio, A.: Temporary housing after disasters: a state of the art survey. Habitat Int. **40**, 136–141 (2013). https://doi.org/10.1016/j.habitatint.2013.03.006
5. Gostelow, L.: The sphere project: the implications of making humanitarian principles and codes work. Disasters **23**(4), 316–325 (1999). https://doi.org/10.1111/1467-7717.00121
6. Aman, D.D., Aytac, G.: Multi-criteria decision making for city-scale infrastructure of post-earthquake assembly areas: case study of Istanbul. Int. J. Disaster Risk Reduct. **67**, 102668 (2022). https://doi.org/10.1016/j.ijdrr.2021.102668
7. Hosseini Sabzevari, S.A., Mottaki, Z., Hassani, A., Zandiyeh, S., Aslani, F.: Temporary housing site selection in Soffeh Mountain, District 5 of Isfahan, Iran. Int. J. Disaster Resil. Built Environ. (2022). https://doi.org/10.1108/IJDRBE-12-2021-0162

8. Wang, C., Costa, R., Baker, J.W.: Simulating post-disaster temporary housing needs for displaced households and out-of-town contractors. Earthq. Spectra **38**(4), 2922–2940 (2022). https://doi.org/10.1177/87552930221112690
9. Yépez, F., Villacreses, J.P.: Temporary sheltering areas in case of earthquakes: engineering design, implementation and policy experiences. The case of Quito—Ecuador. Int. J. Disaster Resil. Built Environ. (2022). https://doi.org/10.1108/IJDRBE-09-2021-0129
10. Ommi, S., Janalipour, M.: Selection of shelters after earthquake using probabilistic seismic aftershock hazard analysis and remote sensing. Nat. Hazards **113**(1), 345–363 (2022). https://doi.org/10.1007/s11069-022-05303-0
11. Aghaie, S., Karimi, B.: Location-allocation-routing for emergency shelters based on geographical information system (ArcGIS) by NSGA-II (case study: earthquake occurrence in Tehran (District-1)). Socioecon. Plann. Sci. **84**, 101420 (2022). https://doi.org/10.1016/j.seps.2022.101420
12. Mousavi, S., Sajadi, S.M., AlemTabriz, A., Najafi, S.E.: Location and hierarchical allocation disaster with combined ε-constraint and simulation-based optimization approach. SIMULATION **98**(5), 407–432 (2022). https://doi.org/10.1177/00375497211063900
13. Şenik, B., Uzun, O.: An assessment on size and site selection of emergency assembly points and temporary shelter areas in Düzce. Nat. Hazards **105**(2), 1587–1602 (2021). https://doi.org/10.1007/s11069-020-04367-0
14. Dabiri, M., Oghabi, M., Sarvari, H., Sabeti, M.S., Kashefi, H., Chan, D.W.M.: Assessing the post-earthquake temporary accommodation risks in Iran using fuzzy Delphi method. Open Constr. Build. Technol. J. **15**(1), 93–105 (2021). https://doi.org/10.2174/1874836802115010093
15. Kim, M., Kim, K., Kim, E.: Problems and implications of shelter planning focusing on habitability. A case study of a temporary disaster shelter after the Pohang earthquake in South Korea (2021)
16. Ghaderi, S., Norouzi Khatiri, K., Ganjehi, S.: Emergency accommodation and post-earthquake logistics management using damage analysis results. Health Emerg. Disasters Q. **6**(2), 99–114 (2021). https://doi.org/10.32598/hdq.6.2.350.1
17. Shih, G.R., Ye, K.T., Tsai, P.H.: A resource-saving shelter selection approach for large-scale area emergency. IEEE Syst. J. **PP**, 1–11 (2022). https://doi.org/10.1109/JSYST.2022.3216041
18. Han, D., Kim, D., Kim, K., Wang, W.J., Jung, J., Kim, H.S.: Mega flood inundation analysis and the selection of optimal shelters. Water (Switzerland) **14**(7) (2022). https://doi.org/10.3390/w14071031
19. Chelariu, O.E., Iațu, C., Minea, I.: A GIS-based model for flood shelter locations and pedestrian evacuation scenarios in a rural mountain catchment in Romania. Water (Switzerland) **14**(19) (2022). https://doi.org/10.3390/w14193074
20. Zhu, Y., et al.: Optimal evacuation route planning of urban personnel at different risk levels of flood disasters based on the improved 3D Dijkstra's algorithm. Sustainability **14**(16) (2022). https://doi.org/10.3390/su141610250
21. Sahmutoglu, I., Taskin, A., Ayyildiz, E.: Assembly area risk assessment methodology for post-flood evacuation by integrated neutrosophic AHP-CODAS. Nat. Hazards **116**(1) (2022). https://doi.org/10.1007/s11069-022-05712-1
22. Lee, Y.H., Kim, Y.C., Lee, H.K.: Framework for selection of temporary disaster waste management sites for post-flood recovery in Seoul, South Korea. Int. J. Disaster Risk Reduct. **71** (2022). https://doi.org/10.1016/j.ijdrr.2022.102832
23. Alavi, M.S., Fallahi, A., Mottaki, Z., Aslani, F.: Post-disaster sheltering process after the 2019 floods, in Golestan province, Iran. Int. J. Disaster Resil. Built Environ. (2021). https://doi.org/10.1108/IJDRBE-03-2021-0023
24. Samany, N.N., Sheybani, M., Zlatanova, S.: Detection of safe areas in flood as emergency evacuation stations using modified particle swarm optimization with local search. Appl. Soft Comput. **111**, 107681 (2021)
25. Tripathy, S.S., Bhatia, U., Mohanty, M., Karmakar, S., Ghosh, S.: Flood evacuation during pandemic: a multi-objective framework to handle compound hazard. Environ. Res. Lett. **16**(3) (2021). https://doi.org/10.1088/1748-9326/abda70

26. Bera, S., Gnyawali, K., Dahal, K., Melo, R., Li-Juan, M.: Assessment of shelter location-allocation for multi-hazard emergency evacuation. Int. J. Disaster Risk Reduct. **84** (2023)
27. León, J., et al.: Increasing resilience to catastrophic near-field tsunamis: systems for capturing, modelling, and assessing vertical evacuation practices. Nat. Hazards 0123456789 (2023) https://doi.org/10.1007/s11069-022-05732-x
28. Ophiyandri, T., Istijono, B., Hidayat, B., Yunanda, R.: Readiness analysis of public buildings in Padang City for tsunami temporary evacuation shelter. Int. J. GEOMATE **22**(94), 113–120 (2022). https://doi.org/10.21660/2022.94.j2391
29. Jitt-Aer, K., Wall, G., Jones, D., Teeuw, R.: Use of GIS and dasymetric mapping for estimating tsunami-affected population to facilitate humanitarian relief logistics: a case study from Phuket, Thailand. Nat. Hazards **113**(1), 185–211 (2022). https://doi.org/10.1007/s11069-022-05295-x
30. Hachiya, D., Mas, E., Koshimura, S.: A reinforcement learning model of multiple UAVs for transporting emergency relief supplies. Appl. Sci. **12**(20) (2022). https://doi.org/10.3390/app122010427
31. Samm-A, A., Kamal, A.S.M.M., Rahman, M.Z.: Earthquake and rainfall-induced landslide hazard assessment of Kutupalong Rohingya camp using meteorological and geological information. Stoch. Environ. Res. Risk Assess. **0123456789** (2023). https://doi.org/10.1007/s00477-023-02418-z
32. Sinčić, M., Gazibara, S.B., Krkač, M., Arbanas, S.M.: Landslide susceptibility assessment of the City of Karlovac using the bivariate statistical analysis. Rud. Geol. Naft. Zb. **37**(2), 149–170 (2022). https://doi.org/10.17794/rgn.2022.2.13
33. Afkham, M., Ramezanian, R., Shahparvari, S.: Balancing traffic flow in the congested mass self-evacuation dynamic network under tight preparation budget: an Australian bushfire practice. Omega (United Kingdom) **111** (2022). https://doi.org/10.1016/j.omega.2022.102658
34. Wigati, S.S., Sopha, B.M., Asih, A.M.S., Sutanta, H.: Geographic information system based suitable temporary shelter location for mount Merapi eruption. Sustainability **15**(3) (2023). https://doi.org/10.3390/su15032073
35. Nickdoost, N., Jalloul, H., Choi, J.: An integrated framework for temporary disaster debris management sites selection and debris collection logistics planning using geographic information systems and agent-based modeling. Int. J. Disaster Risk Reduct. **80** (2022). https://doi.org/10.1016/j.ijdrr.2022.103215

Compositions Concerning Permuting n-Derivations with Commutativity Over Associative Rings

Mehsin Jabel Atteya

Abstract The principal idea of this article is to launch the compositions concerning the permuting n-derivations $\triangle$ over a ring R. This identity employs investigating facts concerning commutativity beside the centralizer permuting n-derivations to the prime and semiprime rings. The results refer to the centralizer mappings (resp. commuting mappings), and weakly semiprime ideals of a semiprime ring R have a crucial role in making the non-commutative area commutative.

Keywords n-derivation · Semiprime ring · Permuting mapping · Centralizer mapping · Weakly semiprime ideal

1 Introduction

Symmetric n-derivation were used to study commutativity of rings by many authors. In 1980, G. Maksa appeared in [1] inserted the idea of a symmetric bi-derivation over a ring R. The notion of the mappings that act as additive commuting is closely connected with the notion concerning bi-derivations. Obviously, any commuting mapping d which is additive from a ring R into a ring R gives rise to a bi-derivation over R. More details related to bi-derivations and their applications can be found in Reference [2]. Moreover, Bre˘sar [3] proved that every bi-derivation V over a non-commutative ring R works as a prime. It has the identity $V(\alpha, \beta) = \mu[\alpha, \beta]$ fulfill for all α and β belong to R. Here, μ plays the role as a fixed element from the extended centroid in R. Utilizing doubtless mapping expressions, Bre˘sar [2] expanded this outcome to covering semiprime rings too. Subsequently, various writers have discussed the symmetric 3-derivations concerning rings (read [4–6], where more references can be found). Furthermore, other authors actively studied some properties of associative rings. Like, the commutativity property of both kinds

M. J. Atteya (✉)
Department of Mathematics, College of Education, Mustansiriyah University, Al-Rusafa, Baghdad City, Iraq
e-mail: mehsinatteya88@uomustansiriyah.edu.iq

O. Castillo et al. (eds.), *Advanced Mathematics for the Modeling and Solution of Challenging Problems in Engineering*, Studies in Systems, Decision and Control 582,
https://doi.org/10.1007/978-3-031-82342-8_18

of associative rings prime and semiprime rings is based on numerous cases of the mappings. Specifically, centralizing (resp. commuting) maps over their convenient subsets (discover [4, 7, 23–25]).

Here, we should pay attention to the influence point that a permuting tri-derivation has been posted via Ozt¨urk in [8] while Foˇsner [9] delivered the idea of symmetric skew 3-derivations. It considers extending the results that belong to Jung with his colleague Park [10] for permuting which acts under the situation skew 3-derivations. She was supply the concept commutativity property in R with certain identities. Also, expanded the idea of permuting in case skew 3-derivation for the permuting skew n-derivations over associative rings with appointed multi other facts in Reference [11]. Second support was made in [12], where commutativity property idea of a ring achieving certain expressions covering the concept of the trace which related to the permuting n-derivation was proved. Later, Mohammad, et al. [13] run down the thought of the concept of symmetric when it has fashion generalized (α, β)-n-derivations with symmetric in case α-left n-centralizers over associative rings. In addition to that, they generalized the results of A. Foˇsner [11] in a distinct framework under certain appropriate torsion restrictions on the subordinate ring. Also, Xiaowei et al. [14] provide that an n-derivation ($n \geq 3$) that acts as a skew over a semiprime ring R must map into the center of a ring R.

However, Badr Nejjar et al. [7] obtained the following result: Fix a positive integer, n. Recall the mapping that acts as an n-additive Ω from R^n to R. It is named a generalized n-derivation over a ring R associated with n-derivation $\triangle$ when:

$$\begin{aligned}&\Omega\big(\alpha_1, \alpha_2, \ldots, \alpha_i'\alpha_i, \ldots, \alpha_n\big)\\&= \Omega(\alpha_1, \alpha_2, \ldots, \alpha_i, \ldots, \alpha_n)\alpha_i' + \alpha_i\Delta(\alpha_1, \alpha_2, \ldots, \alpha_i, \ldots, \alpha_n)\end{aligned}$$

where α_i', $\alpha_i \in R$, suppose that R is a prime ring equipped with a (n+1)!-torsion free property. Take ζ as a non-zero ideal in case Jordan over R. Suppose R concedes a non-zero additive mapping acts as a permuting type generalized n-derivation Ω associated with n-derivation $\triangle$, whenever the trace which related to Ω is centralizing over ζ. Consequently, R modifies to commutative status. Also, the additive subgroup in R is uttered to be a construct Jordan ideal over R when (δ o λ) in ζ, for all δ belongs to ζ while λ belongs to R. In other expression, ur + ru belongs to ζ, where u belongs to ζ while r belongs to R.

The problem was also studied in a more general setting of near rings by several authors, e. g. Ali et al in [15]. They assumed that a near ring N equipped with the property a 3!-torsion free 3-prime with ζ acts as a non-zero additive subgroup. Also, ζ is a semigroup ideal over N. When $\triangle$ acts as a permuting 3-derivation mapping with the trace δ. Moreover, ε belongs to N whenever N admits to satisfying the identity $\varepsilon\delta(y) = 0$ where y in ζ. We obtained two options. The first option is $\varepsilon = 0$ while the second is $\triangle = 0$ both option over ζ. Over and above that, Mohammad et al. [16] published the idea concerning (σ, τ)-n-derivation which acts over near-ring N. Furthermore, they scrutinized concerning somewhat other properties double up (σ, τ)-n-derivations over near-ring N under the situation a prime. They found N modifies to commutative property in near-ring. Later Eremita [17] debated the role of functional expression of degree 2 over triangular matrix rings. He found some facts using to descriptions the commuting maps with generalized inner bi-derivations in the triangular matrix rings. The attached result is belong to Yao Wang et al. [18]. Suppose $\zeta = \mathrm{Tri}(\zeta, \beta, \mu)$ forms a triangular algebra. Then there exists η lies in β such that $[\eta,[\zeta, \zeta]] = 0$.

Then the map γ_n which defined by

$$\gamma_n(\kappa_1, \kappa_2, \ldots, \kappa_n) = [\kappa_1, [\kappa_2, \ldots, [\kappa_n, \eta] \ldots]]$$

where $\kappa_1, \kappa_2, \ldots, \kappa_n$ belong to ζ acting as a permuting n-derivation over ζ.

Nevertheless, Skosyrskii [19] deemed bi-derivations for assorted reasons, called, in connection with non-commutative Jordan algebras, where Jordan algebras J is a non-associative algebra concerning a field whose multiplication achieves:

(i) $p\lambda = \lambda p$ (commutative property),
(ii) $(p\lambda)(pp) = p(\lambda(pp))$ (Jordan form),

for all p, λ in J. The product of two elements p and λ in a Jordan algebra is also denoted $p \circ \lambda$. Finally, Emine Ko¸c and Nadeem ur Rehman [20] placed emphasis on studying the permuting n-derivations over both types of rings, the prime rings or semiprime rings. Beside that a non-zero ideals. They published some results concerning a symmetric skew n-derivation Δ from R^n into R. In addition to that, Δ is related to an automorphism mapping T. They achieve single of the identities:

(i) $\eta(u) = 0$,
(ii) $[\eta(u), T(u)] = 0$ fulfill u in U, where η is the trace of Δ, then $\Delta = 0$.

During this work, R always refers to an associative ring with Z(R). Here Z(R) is the center of R. Assume the elements α, δ, ꟼ belong to R. We put the character $[\delta, \alpha]$ for the commutator yα-αy while the character $\alpha o \delta$ for anticommutator $\alpha\delta + \delta\alpha$. Beside on create use of the formulas $[\alpha\delta, ꟼ] = [\alpha, ꟼ]\,\delta + \alpha[\delta, ꟼ]$ and $[\alpha, \delta\, ꟼ] = [\alpha, \delta\, ꟼ] + \delta[\alpha, ꟼ]$. Moreover, the code $[\delta, \alpha]_{\sigma,\tau}$ indicate to the commutator $\delta\sigma(\alpha) - \tau(\alpha)\delta$ and $(\alpha o \delta)_{\sigma,\tau}$ for anti-commutator $\alpha\sigma(\delta) + \tau(\delta)\alpha$. Remember that R is prime if $aRb = 0$ leads to either $a = 0$ or $b = 0$ and is a semiprime ring whenever $aRa = (0)$ (where a in R) makes $a = 0$. Moreover, the classical relation between prime and semiprime rings refers to the prime ring as the semiprime ring. Unfortunately, the converse of the previous relation is not true always. When have an integer $n \geq 2$, a ring R is stated to be n-torsion free if $nx = 0$ (where x belong to R) reveal that $x = 0$. Whilst R acts as a commutative ring with identity. Reminiscence that U plays role as a proper ideal U (that's mean an ideal different from R) of R is named semiprime ideal when the element μ belongs to a ring R satisfies the relation μ^2 belongs to U imply μ lies in U. However, then a proper ideal U of R is named weakly semiprime if μ lies in R and $0 \neq \mu^2$ located in U. Under the action of the previous circumstances, we harvest μ belongs to U.

As an influence point on this subject, the study of additive mappings, which act as derivations, commenced in the 1950s and 1960s. Where an additive mapping d from an associative ring R to itself is represented as a derivation if $d(xy) = d(x)y + xd(y)$ whenever x, y lie in R.

The article, issued via K. H. Park [21], mentioned the principal idea concerning an n-derivation with a symmetric in case n-derivation for any positive n belong to the set of integers Z. In fact, we expanded several known outcomes once upon a time in the employing of additive mapping derivations in both kinds of associative rings. They are two types prime rings and semiprime rings. As follows: Fix n ≥ 2 and write R^n as the product of $R \times R \times \cdots \times R$. Usually, we name a map Δ from R^n into R is permuting (or sometimes symmetric) whenever the expression $\Delta(\nu_1, \nu_2, \ldots, \nu_n) = \Delta(\nu_{\pi(1)}, \nu_{\pi(2)}, \ldots, \nu_{\pi(n)})$ holds for every ν_i lies in R, as well as any permutation π of n characters. Hence, an n-additive map Δ from R^n to R (i.e., additive in every controversy) will be named an n-derivation whenever the expressions satisfy:

$$\begin{aligned}
\Delta\big(v_1v_1^\star, v_2, \cdots, v_n\big) &= \Delta(v_1, v_2, \cdots, v_n)v_1^\star + v_1\Delta\big(v_1^\star, v_2, \cdots, v_n\big),\\
\Delta\big(v_1, v_2v_2^\star, \cdots, v_n\big) &= \Delta(v_1, v_2, \cdots, v_n)v_2^\star + v_2\Delta\big(v_1, v_2^\star, \cdots, v_n\big), \ldots\\
\Delta\big(v_1, v_2, \cdots, v_nv_n^\star\big) &= \Delta(v_1, v_2, \cdots, v_n)v_n^\star + v_n\Delta\big(v_1, v_2, \cdots, v_n^\star\big)
\end{aligned}$$

are right for all v_i, $v_i^\star$ lie in R.

Definitely, the additive mapping like a derivation is a 1-derivation while a 2-derivation refers to a bi-derivation. Consequently, n = 3, in the situation, we harvest the another concept indicated by the tri-derivation. Whenever Δ acts as a symmetric, then the above qualities are $\Delta(v, v, \ldots, v)$ for all v in R. Here Δ from R^n to R plays role a symmetric map, be the trace of Δ. On the other hand, specific when Δ from R^n to R is a symmetric map which is also acting as an n-additive. Under this situation, the trace δ of Δ achieves the rapport $\delta(\gamma + \mu) = \delta(\gamma) + \delta(\mu) + \sum_{r=1}^{k-1} \binom{k}{r} \Delta(\gamma, \gamma, \ldots, \gamma, \mu, \mu \ldots, \mu)$ for all γ, μ in R, whenever μ appears r times and γ appears at k-r times.

The purpose of this article is to derive and employ a formula for compositions of n-derivations on any ring. We study some properties concerning permuting n-derivations over both types of associative rings, prime and semiprime rings.

The results which belong to this project achieve the expression aRb in Z(R), a, b in R, we mention that for more convenience when reading these results.

At the beginning, we post the following known results, which will be extensively employed to prove the outcomes in this paper:

Lemma 1.1 ([14] : Lemma 2.4). Take ꟼ as a semiprime ring. When α belongs to ꟼ, then the identity $[\alpha, [\alpha, x]] = 0$ whenever x in ꟼ if and only if α^2, 2α lie in Z(ꟼ).

Lemma 1.2 ([23]: Lemma 1). Take ꟼ as a semiprime ring. Then the center of ꟼ has no non-zero nilpotent elements.

Lemma 1.3 ([7]: Theorem). Assume ꟼ acts as a ring, ζ and δ belong to ꟼ. Then there exist integers $p = p(\zeta, \delta)$ and $q = q(\zeta, \delta)$ greater than or equal to 1 whenever

$[\zeta^p, \delta^q] = 0$. Consequently, the commutator ideal of ꟼ forms nil. When ꟼ has no nonzero nil ideals, then ꟼ modifies to commutative.

Lemma 1.4 ([25]: Proposition 8.5.3). Take ꟼ as a ring. Every intersection of prime ideals of ꟼ forms semiprime ideal. While each semiprime ideal lies in ꟼ makes an intersection of prime ideals.

2 Compositions Concerning Permuting n-Derivations △ and Commutativity Over Rings

Let us push the main definition:

Definition 2.1 For any fixed integer $n \geq 2$, take $\triangle$ as an additive mapping: $R^n \rightarrow R$ acting as a permuting n-derivation on a ring R. Then the compositions of $\triangle$ on R defined as

$$\Delta^k(\kappa_1, \kappa_2, \ldots, \kappa_i'\kappa_i, \ldots, \kappa_n) = \sum_{r=0}^{k} \binom{k-1}{r} \Delta^{k-r}(\kappa_1, \kappa_2, \ldots, \kappa_i'\kappa_i, \ldots, \kappa_n)\Delta^r(\kappa_i)$$
$$+ \Delta^r(\kappa_i')\Delta^{k-r}(\kappa_1, \kappa_2, \ldots, \kappa_i, \ldots, \kappa_n);$$

fulfill for all $\kappa_i'\kappa_i \in R$ such that $k \neq r$ and $i = 1, 2, \ldots, k$

where k and r act as a positive integers (we adopt the convention $D^0 = d^0 = id$).

For motivation and a closer look at the compositions of permuting n-derivations. We demonstrate the following evidence.

Example 1 Take R is a ring and $\triangle$ is a permuting n-derivations of R. For all x lies in R, the composition of $\triangle$ satisfies the expression:

$$\Delta^k(\kappa_1, \kappa_2, \ldots, \kappa_i'\kappa_i, \ldots, \kappa_n) = \sum_{r=0}^{k} \binom{k-1}{r} \Delta^{k-r}(\kappa_1, \kappa_2, \ldots, \kappa_i', \ldots, \kappa_n)\Delta^r(\kappa_i)$$
$$+ \Delta^r(\kappa_i')\Delta^{k-r}(\kappa_1, \kappa_2, \ldots, x_i, \ldots, x_n); \text{ such that } n \neq r \text{ and } i \geq 1$$

Assume that Leibniz's formula holds for $k = 2$, then from the right hand, we find that

$$= \binom{1}{0}(\Delta^2(\kappa_1, \kappa_2, \ldots, \kappa_i', \ldots, \kappa_n)\Delta^0(\kappa_i) + \Delta^0(\kappa_i')\Delta^2(\kappa_1, \kappa_2, \ldots, \kappa_i, \ldots, \kappa_n))$$
$$+ \binom{1}{1}(\Delta(\kappa_1, \kappa_2, \ldots, \kappa_i', \ldots, \kappa_n)$$
$$\Delta(\kappa_i) + \Delta(\kappa_i')\Delta(\kappa_1, \kappa_2, \ldots, \kappa_i, \ldots, \kappa_n))$$

$$= \Delta^2(\kappa_1, \kappa_2, \ldots, \kappa_i', \ldots, \kappa_n)\kappa_i + \kappa_i'\Delta^2(\kappa_1, \kappa_2, \ldots, \kappa_i, \ldots, \kappa_n)$$
$$+ \Delta(\kappa_1, \kappa_2, \ldots, \kappa_i', \ldots, \kappa_n)\Delta(\kappa_i) + \Delta(\kappa_i')\Delta(\kappa_1, \kappa_2, \ldots, \kappa_i, \ldots, \kappa_n).$$

The left side gave us

$$\Delta^2(\kappa_1, \kappa_2, \ldots, \kappa_i'\kappa_i, \ldots, \kappa_n) = \Delta(\Delta(\kappa_1, \kappa_2, \ldots, \kappa_i'\kappa_i, \ldots, \kappa_n))$$
$$= \Delta(\Delta(\kappa_1, \kappa_2, \ldots, \kappa_i', \ldots, \kappa_n)\kappa_i) + \Delta(\kappa_i'\Delta(\kappa_1, \kappa_2, \ldots, \kappa_i, \ldots, \kappa_n))$$

Investment of the definition of n-derivations to this expressions, we observe that

$$= \Delta^2(\kappa_1, \kappa_2, \ldots, \kappa_i', \ldots, \kappa_n)\kappa_i$$
$$+ \Delta(\kappa_1, \kappa_2, \ldots, \kappa_i', \ldots, \kappa_n)\Delta(\kappa_i)$$
$$+ \Delta(\kappa_i')\Delta(\kappa_1, \kappa_2, \ldots, \kappa_i, \ldots, \kappa_n) + \kappa_i'\Delta^2$$
$$(\kappa_1, \kappa_2, \ldots, \kappa_i, \ldots, \kappa_n).$$

From these proceedings, we see the both sides are equal.

Remark 1 The condition k is not equal to r is essential in the formula for the compositions of permuting n-derivations Δ. The following example interprets that.

Example 2 Assume

$$\Delta^k(\kappa_1, \kappa_2, \ldots, \kappa_i'\kappa_i, \ldots, \kappa_n) = \sum_{r=0}^{k} \binom{k-1}{r} \Delta^{k-r}(\kappa_1, \kappa_2, \ldots, \kappa_i', \ldots, \kappa_n)\Delta^r(\kappa_i)$$
$$+ \Delta^r(\kappa_i')\Delta^{k-r}(\kappa_1, x_2, \ldots, x_i, \ldots, x_n);$$

Taking $k = 2$, then the right side proved the terms

$$= \binom{1}{0}(\Delta^2(\kappa_1, \kappa_2, \ldots, \kappa_i', \ldots, \kappa_n)\Delta^0(\kappa_i) + \Delta^0(\kappa_i')\Delta^2(\kappa_1, \kappa_2, \ldots, \kappa_i, \ldots, \kappa_n))$$
$$+ \binom{1}{1}\Delta(x_1, x_2, \ldots, x_i', \ldots, x_n)\Delta(\kappa_i) + \Delta(\kappa_i')\Delta(\kappa_1, \kappa_2, \ldots, \kappa_i, \ldots, \kappa_n)$$
$$+ \binom{1}{2}(\Delta^0(\kappa_1, \kappa_2, \ldots, \kappa_i', \ldots, \kappa_n)\Delta^2(\kappa_i) + \Delta^2(\kappa_i')\Delta^0(\kappa_1, \kappa_2, \ldots, \kappa_i, \ldots, \kappa_n))$$
$$= \Delta^2(\kappa_1, \kappa_2, \ldots, \kappa_i', \ldots, \kappa_n)\kappa_i + \kappa_i'\Delta^2(\kappa_1, \kappa_2, \ldots, \kappa_i, \ldots, \kappa_n)$$
$$+ \Delta(\kappa_1, \kappa_2, \ldots, \kappa_i', \ldots, \kappa_n)\Delta(\kappa_i) + \Delta(\kappa_i')\Delta(\kappa_1, \kappa_2, \ldots, \kappa_i, \ldots, \kappa_n)$$
$$- \frac{1}{2}((\kappa_1, \kappa_2, \ldots, \kappa_i', \ldots, \kappa_n)\Delta^2(\kappa_i) + \Delta^2(\kappa_i')(\kappa_1, \kappa_2, \ldots, \kappa_i, \ldots, \kappa_n)),$$

where in [22], the light of the Kurba's classical case of negative integers, we have $-1! = -1$.

While from the other side, we arrive to the fact:

$$\Delta^2(\kappa_1, \kappa_2, \ldots, \kappa_i'\kappa_i, \ldots, \kappa_n) = \Delta^2(\kappa_1, \kappa_2, \ldots, \kappa_i', \ldots, \kappa_n)\kappa_i$$
$$+ \Delta(\kappa_1, \kappa_2, \ldots, \kappa_i', \ldots, \kappa_n)\Delta(\kappa_i) + \Delta(\kappa_i')\Delta(\kappa_1, \kappa_2, \ldots, \kappa_i, \ldots, \kappa_n)$$
$$+ \kappa_i'\Delta^2(\kappa_1, \kappa_2, \ldots, \kappa_i, \ldots, \kappa_n).$$

Obviously, the equality between the two sides is not satisfying.

Theorem 2.1 Take R as a semiprime ring with 2-torsion-free property and U as a maximal ideal over R. Suppose that Δ achieves the condition $\Delta^k(\kappa \circ y) \mp (\kappa \circ y) \in Z(R)$ where κ, y in R. Then

(i) Δ^k plays as centralizer mapping (resp. commuting mapping) over R.
(ii) if $\Delta = 0$ then R becomes commutative.

Proof To prove Branch (i). Firstly, let us discuss when Δ is non zero. We have the term $\Delta^k(\kappa \circ y) \mp (\kappa \circ y)$ in Z(R) where x, y belong to R. Modify the main expression as following

$$\Delta^k(\kappa_1, \kappa_2, \ldots, \kappa_i'\kappa_i, \ldots, \kappa_n \circ y_1, y_2, \ldots, y_i'y_i, \ldots, y_n)$$
$$\mp (\kappa_1, \kappa_2, \ldots, \kappa_i'\kappa_i, \ldots, \kappa_n \circ y_1, y_2, \ldots, y_i'y_i, \ldots, y_n)$$

in Z(R) for all κ_i', κ_i, y_i',y_i in R, i = 1, 2...., n. We further consider

$$\Delta^k((\kappa_1, \kappa_2, \ldots, \kappa_i'\kappa_i, \ldots, \kappa_n)(y_1, y_2, \ldots, y_i'y_i, \ldots, y_n))$$
$$+ \Delta^k((y_1, y_2, \ldots, y_i'y_i, \ldots, y_n)(\kappa_1, \kappa_2, \ldots, \kappa_i'\kappa_i, \ldots, \kappa_n)$$
$$\mp (\kappa_1, \kappa_2, \ldots, \kappa_i'\kappa_i, \ldots, \kappa_n \circ y_1, y_2, \ldots, y_i'y_i, \ldots, y_n) \text{in } Z(R)$$

where κ_i', κ_i, y_i', y_i in R, $i = 1, 2 \ldots, n$.

In the first term of this expression, we select the elements x_n and y_1 for applying the formula for the compositions of a permuting n-derivations Δ. Therefore, we write

$$= \sum_{r=0}^{k} \binom{k-1}{r} \Delta^{k-r}(\kappa_1, \kappa_2, \ldots, \kappa_i', \ldots, \kappa_n y_2, \ldots, y_i'y_i, \ldots, y_n)\Delta^r(y_1)$$
$$+ \Delta^r(\kappa n)\Delta^{k-r}(\kappa_1, \kappa_2, \ldots, \kappa_i, \ldots, \kappa_n); \ldots, \kappa_{n-1}, y_1, y_2, \ldots, y_i', \ldots, y_n)$$
$$+ \Delta^k((y_1, y_2, \ldots, y_i'y_i, \ldots, y_n)(\kappa_1, \kappa_2, \ldots, \kappa_i'\kappa_i, \ldots, \kappa_n))$$
$$\mp ((\kappa_1, \kappa_2, \ldots, \kappa_i'\kappa_i, \ldots, \kappa_n) \circ (y_1, y_2, \ldots, y_i'y_i, \ldots, y_n)) \text{in } Z(R)$$

when κ_i', κ_i, y_i', y_i in R, $i = 1, 2 \ldots, n$. Now we start the procedure concerning the first term of the previous expression, we deliver

$$\Delta^k((\kappa_1, \kappa_2, \ldots, \kappa_i'\kappa_i, \ldots, \kappa_n)(y_2, \ldots, y_i', \ldots, y_n))y_1$$
$$+ \kappa_n\Delta^k(((y_1, y_2, \ldots, y_i'y_i, \ldots, y_n)(\kappa_1, \kappa_2, \ldots, \kappa_i'\kappa_i, \ldots, \kappa_{n-1}))$$

$$+\binom{k-1}{1}(\Delta^{k-1}((\kappa_1,\kappa_2,\ldots,\kappa_i'\kappa_i,\ldots,\kappa_n)(y_2,\ldots,y_i',\ldots,y_n))\Delta(y_1)$$
$$+\Delta(x_n)\Delta^k((y_1,y_2,\ldots,y_i',\ldots,y_n)(\kappa_1,\kappa_2,\ldots,\kappa_i'\kappa_i,\ldots,\kappa_{n-1})$$
$$+\cdots+\binom{k-1}{k-1}(\Delta((\kappa_1,\kappa_2,\ldots,\kappa_i'\kappa_i,\ldots,\kappa_n)(y_2,\ldots,y_i',\ldots,y_n))\Delta^{k-1}(y_1)$$
$$+\Delta^{k-1}(\kappa_n)\Delta((y_1,y_2,\ldots,y_i'y_i,\ldots,y_n)(\kappa_1,\kappa_2,\ldots,\kappa_i'\kappa_i,\ldots,\kappa_{n-1})))$$
$$\mp((\kappa_1,\kappa_2,\ldots,\kappa_i'\kappa_i,\ldots,\kappa_n)o(y_1,y_2,\ldots,y_i'y_i\ldots,y_n))\text{in } Z(R)$$

for all κ_i', κ_i, y_i', y_i in R, $i = 1, 2\ldots, n$. Simplify this expression deduce that $[\Delta^k((\kappa_1,\kappa_2,\ldots,\kappa_i'\kappa_i,\ldots,\kappa_n)(y_2,\ldots,y_i'y_i,\ldots,y_n))y_1, r] + [\varepsilon, r] = 0$, such that $i \geq 2$ and ε represent of the other terms, i.e.

$$\varepsilon = \kappa_n\Delta^k((y_1,y_2,\ldots,y_i'x_i,\ldots,y_n)(\kappa_1,\kappa_2,\ldots,\kappa_i'\kappa_i,\ldots,\kappa_{n-1}))$$
$$+\binom{k-1}{1}(\Delta^{k-1}(\kappa_1,\kappa_2,\ldots,\kappa_i'\kappa_i,\ldots,\kappa_n)(y_2,\ldots,y_i',\ldots,y_n))\Delta(y_1)$$
$$+\Delta(x_n)\Delta^k((y_1,y_2,\ldots,y_i',\ldots,y_n)(\kappa_1,\kappa_2,\ldots,\kappa_i'\kappa_i,\ldots,\kappa_{n-1})))$$
$$+\cdots+\binom{k-1}{k-1}(\Delta((\kappa_1,\kappa_2,\ldots,\kappa_i'\kappa_i,\ldots,\kappa_n)(y_2,\ldots,y_i',\ldots,y_n))\Delta^{k-1}(y_1)$$
$$+\Delta^{k-1}(\kappa_n)\Delta((y_1,y_2,\ldots,y_i'y_i,\ldots,y_n)(\kappa_1,\kappa_2,\ldots,\kappa_i'\kappa_i,\ldots,\kappa_{n-1}))$$
$$\mp((\kappa_1,\kappa_2,\ldots,\kappa_i'\kappa_i,\ldots,\kappa_n)o(y_1,y_2,\ldots,y_i'y_i,\ldots,y_n))$$

whenever κ_i', κ_i, y_i', y_i in R. Furthermore, when κ_i', κ_i, y_i', y_i, r in R. Note that

$$\Delta^k((\kappa_1,\kappa_2,\ldots,\kappa_i'\kappa_i,\ldots,\kappa_n)(y_2,\ldots,y_i',\ldots,y_n))[y_1,r]$$
$$+[\Delta^k((\kappa_1,\kappa_2,\ldots,\kappa_i'\kappa_i,\ldots,\kappa_n)(y_2,\ldots,y_i',\ldots,y_n)),r]y_1+[\varepsilon r]=0. \tag{1}$$

We pick y_1 for r in this expression yields

$$[\Delta^k((\kappa_1,\kappa_2,\ldots,\kappa_i'\kappa_i,\ldots,\kappa_n)(y_2,\ldots,y_i',\ldots,y_n)),y_1]y_1=[y_1,\varepsilon]. \tag{2}$$

In Relation (1) putting r for ε, it followed that

$$\Delta^k((\kappa_1,\kappa_2,\ldots,\kappa_i'\kappa_i,\ldots,\kappa_n)(y_2,\ldots,y_i',\ldots,y_n))[y_1,\varepsilon]$$
$$+[\Delta^k((\kappa_1,\kappa_2,\ldots,\kappa_i'\kappa_i,\ldots,\kappa_n)(y_2,\ldots,y_i',\ldots,y_n)),\varepsilon]y_1=0. \tag{3}$$

Now substituting Relation (2) in the first term of Equation (3), we see that

$$\Delta^k((\kappa_1,\kappa_2,\ldots,\kappa_i'\kappa_i,\ldots,\kappa_n)(y_2,\ldots,y_i',\ldots,y_n))$$

$$\left[\Delta^{k}\left(\left(\kappa_1,\kappa_2,\ldots,\kappa_i'\kappa_i,\ldots,\kappa_n\right)\left(y_2,\ldots,y_i',\ldots,y_n\right)\right),y_1\right]y_1$$
$$+\left[\Delta^{k}\left(\left(\kappa_1,\kappa_2,\ldots,\kappa_i'\kappa_i,\ldots,\kappa_n\right)\left(y_2,\ldots,y_i',\ldots,y_n\right)\right),\varepsilon\right]y_1=0.$$

Putting ε instead of y_1 in Expression (3), we deliver

$$\left[\Delta^{k}\left(\left(\kappa_1,\kappa_2,\ldots,\kappa_i'\kappa_i,\ldots,\kappa_n\right)\left(y_2,\ldots,y_i',\ldots,y_n\right)\right),\varepsilon\right]=0.$$

Left-multiplying by εt and from the right side by $t\left[\Delta^{k}\left(\left(\kappa_1,\kappa_2,\ldots,\kappa_i'\kappa_i,\ldots,\kappa_n\right)\left(y_2,\ldots,y_i',\ldots,y_n\right)\right),\varepsilon\right]$, t in R with using Lemma 1.2 and the information aRb in Z(R), a, b lie in R, we harvest

$\varepsilon t\left[\Delta^{k}\left(\left(\kappa_1,\kappa_2,\ldots,\kappa_i'\kappa_i,\ldots,\kappa_n\right)\left(y_2,\ldots,y_i',\ldots,y_n\right)\right),\varepsilon\right]=0$ for all κ, y in U, t in R.

In other words,

$$\varepsilon R\left[\Delta^{k}\left(\left(\kappa_1,\kappa_2,\ldots,\kappa_i'\kappa_i,\ldots,\kappa_n\right)\left(y_2,\ldots,y_i',\ldots,y_n\right)\right),\varepsilon\right]=(0).$$

where R acts as a semiprime, we put the set $\{P_\alpha\}$ of prime ideals over R satisfy the expression $\cap P_\alpha=\{0\}$. Due to Lemma 1.4, we deduce that the intersection $\{P_\alpha\}$ of prime ideals over R forms a semiprime ideal. Basically, U acts as a maximal ideal, and $\cap P_\alpha$ is a proper semiprime ideal of R. Essentially, U plays role as a maximal ideal of R. That means there are no other ideals contained between U and R. Hence, we arrive at the result $\cap P_\alpha \subseteq U$. Whenever P is a exemplary member in $\cap P_\alpha$ and κ in U. Consequently, either ε in P or $\left[\Delta^{k}\left(\left(\kappa_1,\kappa_2,\ldots,\kappa_i'\kappa_i,\ldots,\kappa_n\right)\left(y_2,\ldots,y_i',\ldots,y_n\right)\right),\varepsilon\right]$ in P.

Arrange two additive subgroups:

$$T_1=\{\kappa\in U\backslash\varepsilon\in P\}\text{ and}$$
$$T_2=\left\{\kappa\in U\backslash\left[\Delta^{k}\left(\left(\kappa_1,\kappa_2,\ldots,\kappa_i'\kappa_i,\ldots,\kappa_n\right)\left(y_2,\ldots,y_i',\ldots,y_n\right)\right),\varepsilon\right]\in P\right\},$$

Basically, any subgroup of the additive group over a ring R forms an ideal. Hence $T_1\cup T_2=U$.

By reason of a group can not be a union of two its proper subgroups. We have two options, either $T_1 = U$ or $T_2 = U$, that is, either $\varepsilon \in P$ or $\left[\Delta^{k}\left(\left(\kappa_1,\kappa_2,\ldots,\kappa_i'\kappa_i,\ldots,\kappa_n\right)\left(y_2,\ldots,y_i',\ldots,y_n\right)\right),\varepsilon\right]\in P$. Thus, both cases yield $\varepsilon\in\cap P_\alpha$ or $\left[\Delta^{k}\left(\left(\kappa_1,\kappa_2,\ldots,\kappa_i'\kappa_i,\ldots,\kappa_n\right)\left(y_2,\ldots,y_i',\ldots,y_n\right)\right),\varepsilon\right]\in\cap P_\alpha$. In other words, $\varepsilon\in\cap P_\alpha\subseteq U$ or $\left[\Delta^{k}\left(\left(\kappa_1,\kappa_2,\ldots,\kappa_i'\kappa_i,\ldots,\kappa_n\right)\left(y_2,\ldots,y_i',\ldots,y_n\right)\right),\varepsilon\right]\in\cap P_\alpha\subseteq U$.

Consequently, either $\varepsilon \in U$ for all $\kappa \in U$ or $\left[\Delta^{k}\left(\left(\kappa_1,\kappa_2,\ldots,\kappa_i'\kappa_i,\ldots,\kappa_n\right)\left(y_2,\ldots,y_i',\ldots,y_n\right)\right),\varepsilon\right]\in U$ for all κ, $y\in U$.

Next, we partition the proof into two cases. As a matter of fact, the first case which is $\varepsilon\in U$ sufficient to continue on the proof.

Viewing the first case. Then $\varepsilon\in U$ yields $\varepsilon=0$.

Using the fact that $\varepsilon\in U$ in Relation (2). We note the previous expressions to

$$\left[\Delta^{k}\left(\left(\kappa_1, \kappa_2, \ldots, \kappa_i'\kappa_i, \ldots, \kappa_n\right)\left(y_2, \ldots, y_i', \ldots, y_n\right)\right), y_1\right]y_1 = 0.$$

Writing this relation as follows:

$$[y_1, \Omega]y_1 = 0. \tag{4}$$

Due to that Ω stand for $\Delta^{k}\left(\left(\kappa_1, \kappa_2, \ldots, \kappa_i'\kappa_i, \ldots, \kappa_n\right)\left(y_2, \ldots, y_i', \ldots, y_n\right)\right)$. Multiplying from left side by Ω and right-multiplying by $\Omega[y_1, \Omega]$, we conclude that $\Omega[y_1, \Omega]y_1\Omega^2[y_1, \Omega] = 0$.

Right-multiplying this relation by y_1 and left-multiplying by Ω yields $(\Omega^2[y_1, \Omega]y_1)^2 = 0$. In agreement with Lemma 1.1, we find that

$[\Omega^2[y_1, \Omega]y_1, [\Omega^2[y_1, \Omega]y_1, t]] =$ for all $t \in R$. Again applying Lemma 1.1, we find that $2\Omega^2[y_1, \Omega]y_1 \in Z(R)$.

Consequently, $2[\Omega^2[y_1, \Omega]y_1, r] = 0$ for all x, y in U, r in R. Basically, R is a semiprime ring has a 2-torsion free. We collect $\Omega^2[y_1, \Omega]y_1 \in Z(R)$. By Lemma 1.2, we find that

$$\Omega^2[y_1, \Omega]y_1 = 0. \tag{5}$$

In Relation (5) replacing y_1 by $y_1\Omega$, it follows that:

$$\Omega^2[y_1, \Omega]\Omega y_1 = 0. \tag{6}$$

Right-multiplying Relation (5) by Ω, we show that

$$\Omega^2[y_1, \Omega]y_1\Omega = 0. \tag{7}$$

Subtracting Relations (6) and (7), we deduce that

$$\Omega^2[y_1, \Omega]^2 = 0. \tag{8}$$

Multiplying Relation (8) from left side by $\Omega[y_1, \Omega]^2$ and right-multiplying by Ω, we see that $\Omega[y_1, \Omega]^2\Omega^2[y_1, \Omega]^2\Omega = 0$. Writing this relation as follows, $\left(\Omega[y_1, \Omega]^2\Omega\right)^2 = 0$. Applying Lemma 1.1, we obtain

$\Omega[y_1, \Omega]^2\Omega = 0$. Right-multiplying by $[y_1, \Omega]^2$ and employing Lemma 1.1, we conclude that

$$\Omega[\Omega, y_1]^2 = 0. \tag{9}$$

Now for Relation (9) right-multiplying by $R\Omega$ and left-multiplying by $[\Omega, y_1]^2R$ with employing Lemma 1.2 and the information aRb in Z(R), a, b in R, we observe that

$$[\Omega, y_1]^2 R\Omega = 0. \tag{10}$$

Replacing R by $\Omega[y_1, \Omega]^2$ and using Lemmas (1.1 and 1.2), we see that

$$[\Omega, y_1]^2 \Omega = 0. \tag{11}$$

Moreover, multiplying (11) from right side by R. Subtracting this result from Relation (10), we find that $[\Omega, y_1]^3 = 0$.

Multiplying this relation by the term $[\Omega, y_1]$ and applying Lemma 1.1, we find that $[\Omega, y_1] = 0$. Replacing y_1 by $((\kappa_1, \kappa_2, \ldots, \kappa_i'\kappa_i, \ldots, \kappa_n)(y_2, \ldots, y_i'y_i, \ldots, y_n))$, we deduce $[\Omega, ((\kappa_1, \kappa_2, \ldots, \kappa_i'\kappa_i, \ldots, \kappa_n)(y_2, \ldots, y_i'y_i, \ldots, y_n))] = 0$.

This relation implies Ω to be commuting n-derivation, i.e. $\Delta^k((\kappa_1, \kappa_2, \ldots, \kappa_i'\kappa_i, \ldots, \kappa_n)(y_2, \ldots, y_i', \ldots, y_n))$ is commuting n-derivation which yields

$$[\Delta^k((\kappa_1, \kappa_2, \ldots, \kappa_i'\kappa_i, \ldots, \kappa_n)(y_2, \ldots, y_i', \ldots, y_n)), ((\kappa_1, \kappa_2, \ldots, \kappa_i'\kappa_i, \ldots, \kappa_n)(y_2, \ldots, y_i', \ldots, y_n))] \in Z(R) \text{i.e.}$$

$\Delta^k((\kappa_1, \kappa_2, \ldots, \kappa_i'\kappa_i, \ldots, \kappa_n)(y_2, \ldots, y_i', \ldots, y_n))$ is centralizer mapping, the result as desired. If $\Delta = 0$ then we immediately obtain the relation $(\kappa \circ y) \in Z(R)$ for all κ, y lie in R. For any arbitrary element in a ring R, r lies in R, we find that $[\kappa y, r] + [y\kappa, r] = 0$ for all $\kappa, y, r \in R$.

Putting $y = \kappa$ and using the fact that R has a 2-torsion free property. We deduce $[\kappa^2, r] = 0$ for all κ, r in R. Substituting y^4 for r and employing Lemma 1.3, we find that R is commutative the result as desired. This finishes the proof.

Using the similar approach as above we can prove the corollary.

Corollary 2.2 Let R be a 2-torsion free semiprime ring. Suppose a non-zero permuting n-derivation Δ satisfies $\Delta^k([\kappa, y]) \mp [\kappa, y]$ in Z(R) for all κ, y in R. Then

(i) Δ^k acts as a commuting mapping (resp. centralizer mapping) over R.
(ii) R becomes commutative.

Theorem 2.3 Let R be a 2-torsion free semiprime ring. Suppose Δ satisfies the identity $\Delta^k([\kappa, y]) \mp (\kappa \circ y)$ in Z(R) for all κ, y in R. Then

(i) $\Delta^k(R)$ is centralizer mapping (resp. commuting mapping) over R.
(ii) if $\Delta = 0$ then R is commutative.

Proof To prove Branch (i). First, we consider the case $\Delta \neq 0$. Consequently, we have the relation $\Delta^k([\kappa, y]) \mp (\kappa \circ y) \in Z(R)$ for all $\kappa, y \in R$.

Modify the main expression as follows:

$$\Delta^k([\kappa_1, \kappa_2, \ldots, \kappa_i'\kappa_i, \ldots, \kappa_n, y_1, y_2, \ldots, y_i'y_i, \ldots, y_n]) \mp (\kappa_1, \kappa_2, \ldots, \kappa_i'\kappa_i, \ldots, \kappa_n \circ y_1, y_2, \ldots, y_i'y_i, \ldots, y_n) \in Z(R)$$

for all $\kappa_i', \kappa_i, y_i', y_i \in R$, $i = 1, 2 \ldots, k$. Moreover,

$$\begin{aligned}&\Delta^k\big(\kappa_1, \kappa_2, \ldots, \kappa_i'\kappa_i, \ldots, \kappa_n, y_1, y_2, \ldots, y_i'y_i, \ldots, y_n\big)\\&- \Delta^k\big(y_1, y_2, \ldots, y_i'y_i, \ldots, y_n, \kappa_1, \kappa_2, \ldots, \kappa_i'\kappa_i, \ldots, \kappa_n\big)\\&\mp \big(\kappa_1, \kappa_2, \ldots, \kappa_i'\kappa_i, \ldots, \kappa_n \, o \, y_1, y_2, \ldots, y_i'y_i, \ldots, y_n\big)\end{aligned}$$

in Z(R) for all $\kappa_i', \kappa_i, y_i', y_i$ in R, $i = 1, 2 \ldots, n$.

On the first term of this relation we apply the formula for the composition of permuting n-derivations.

That can be putting as

$$\begin{aligned}&= \Delta^{k-r}\big(\kappa_1, \kappa_2, \ldots, \kappa_i', \ldots, \kappa_n, y_2, \ldots, y_i', \ldots, y_n\big)\Delta^r\big(y_1\big)\\&+ \Delta^r(\kappa_n)\Delta^{k-r}\big(\kappa_1, \kappa_2, \ldots, \kappa_i, \ldots, x_{n-1}, y_1, y_2, \ldots, y_i', \ldots, y_n\big)\\&- \Delta^k\big(\big(y_1, y_2, \ldots, y_i', \ldots, y_n\big)\big(\kappa_1, \kappa_2, \ldots, \kappa_i'\kappa_i, \ldots, \kappa_n\big)\big)\\&\mp \big(\big(\kappa_1, \kappa_2, \ldots, \kappa_i'\kappa_i, \ldots, \kappa_n\big) o \big(y_1, y_2, \ldots, y_i'y_i, \ldots, y_n\big)\big)\end{aligned}$$

in Z(R) whenever $\kappa_i', \kappa_i, y_i', y_i$ in R, $i = 1, 2 \ldots, k$.

Depending on the same protocol as used in the proof of Theorem 2.1, the proof of the theorem is completed.

In the next results, we show a permuting n-derivation Δ has different powers. Here applying the same scenario, one can demonstrate the theorem.

Theorem 2.4 For any fixed integers n and q $\geq$2. Take R as a semiprime ring has a 2-torsion free property. Suppose that R admits a permuting n-derivation Δ satisfying the relation $\Delta^k([\kappa, y]) \mp \Delta^j([\kappa, y]) \mp [\kappa, y]$ in Z(R) (resp. $\Delta^k(\kappa \, o \, y) \mp \Delta^j(\kappa \, o \, y) \mp (\kappa \, o \, y)$ in Z(R)) for all κ, y in R. Then

(i) Δ^k(R) (resp. Δ^j(R)) forms a centralizer mapping (resp. commuting mapping) over R.
(ii) R acts as a commutative if $\Delta = 0$.

Theorem 2.5 For any fixed integers n and q$\geq$ 2. Take R as a semiprime ring has a 2-torsion free property. Suppose that R admits a permuting n-derivation Δ achieving $\Delta^k(\kappa \, o \, y) \mp \Delta^j([\kappa,y]) \mp ([\kappa, y]) \in Z(R)$ (resp. $\Delta^k([\kappa, y]) \mp \Delta^j(\kappa \, o \, y) \mp (\kappa \, o \, y)$ in Z(R)) for all κ, y belong to R. Then

(i) Δ^k (resp. Δ^j) forms a centralizer mapping (resp. commuting mapping) over R.
(ii) R becomes commutative ring if $\Delta = 0$.

Proof For Branch (i) when $\Delta \neq 0$, we have the expression $\Delta^k(\kappa \, o \, y) \mp \Delta^j\big(\big[\kappa, y\big]\big) \mp \big[\kappa\,,\, y\big] \in Z(R)$ when $\kappa, y \in R$. Now in the main relation replacing κ by y, we arrive to

$\Delta^k\big(y^2\big) \in Z(R)$ for all x, y in R. Moreover, for any element $r \in R$, we obtain $\big[\Delta^k\big(y^2\big), r\big] = 0$ for all $\kappa, y \in R$. Linearizing this relation and using the result $\big[\Delta^k\big(y^2\big), r\big] = 0$ yields that $\big[\Delta^k(\kappa y), r\big] + \big[\Delta^k(y\kappa), r\big] = 0$ when κ, y in R.

Writing κ by $(\kappa_1, \kappa_2, \ldots, \kappa_i'\kappa_i, \ldots, \kappa_n)$ and y by $(y_1,y_2, \ldots, y_i'y_i, \ldots, y_n)$, for all κ_i', κ_i, y_i', y_i lie in R such that $i \geq 2$, we have that

$$\begin{aligned}&\left[\Delta^k(\kappa_1, \kappa_2, \ldots, \kappa_i'\kappa_i, \ldots, \kappa_n, y_1,y_2, \ldots, y_i'y_i, \ldots, y_n), r\right]\\&+\left[\Delta^k(y_1,y_2, \ldots, y_i'y_i, \ldots, y_n, \kappa_1, \kappa_2, \ldots, \kappa_i'\kappa_i, \ldots, \kappa_n), r\right]\\&= 0 \text{ when } \kappa_i', \kappa_i, y_i', y_i \in R.\end{aligned}$$

To apply the formula for the compositions of permuting n-derivations △ on the first term, we select the elements κ_n and y_1. We note

$$\begin{aligned}&\left[\sum_{r=0}^{k}\binom{k-1}{r}\Delta^{k-r}(\kappa_1, \kappa_2, \ldots, \kappa_i'\kappa_i, \ldots, \kappa_{n-1}, y_1, y_2, \ldots, y_i'y_i, \ldots, y_n)\right.\\&\left.\Delta^r(\kappa_n)\Delta^r(y_1)\Delta^{k-r}(\kappa_1, \kappa_2, \ldots, \kappa_i'\kappa_i, \ldots, \kappa_n, y_2, \ldots, y_i'y_i, \ldots, y_n, r\right]\\&+\left[\Delta^k(y_1,y_2, \ldots, y_i'y_i, \kappa_1, \kappa_2, \ldots, \kappa_i'\kappa_i, \ldots, \kappa_n), r\right] = 0 \text{ when } \kappa_i', \kappa_i, y_i', y_i \in R.\end{aligned}$$

Clearly, employing similar arguments with the proof of Theorem 2.1. This finishes the work.

Depending on the same manner which used in the proof of Theorem 2.1, we immediately get the following corollary.

Corollary 2.6 For any fixed integers n, $q \geq 2$. Take R as a semiprime ring has a 2-torsion free property. Assume R allows a non-zero permuting n-derivation △ satisfying the relation $\left[\Delta_1^k(\kappa), \Delta_2^j(\kappa)\right] = 0$ for all κ in R. Then Δ_1^k (resp. Δ_2^j) acts as a centralizer mapping (resp. commuting mapping) over R.

We may obtain the same conclusion, when $\kappa\Delta^k(R) = 0$, for all κ belongs to R as the following corollary.

Corollary 2.7 Let R be a semiprime ring. Suppose that △ satisfies the identity $\kappa\Delta^k(R) = 0$ whenever κ in R. Then Δ^k forms a commuting mapping (resp. a centralizer mapping) over R.

Proof We begin with the hypothesis $\kappa\Delta^k(R) = 0$ for all κ belongs to R. This identity can be modified as $\Delta^k(R)R\Delta^k(R) = 0$. Depending on the fact that R acts as a semiprime ring, yielding that $\Delta^k(R)$ lies in Z(R).

We immediately arrive at the required results. The proof of corollary is completed.

Theorem 2.8 Suppose that R is a prime ring and a be a fixed element of R. If △ satisfies the identity $a\Delta^k(R) = 0$, then either $\Delta^k(R) = 0$ or $a = 0$.

Proof By our hypothesis, we find the main expression a $\Delta^k(R) = 0$.

Applying the compositions of permuting n-derivations △, we can be rewritten the main relation as follows:

$$a\left(\sum_{r=0}^{k}\binom{k-1}{r}\Delta^{k-r}(\kappa_1, \kappa_2, \ldots, \kappa_i', \ldots, \kappa_n)\right.$$

$$\left(\Delta^r(\kappa_i)\Delta^r(\kappa_i')\Delta^{k-r}(\kappa_1,\kappa_2,\ldots,\kappa_i',\ldots,\kappa_n)\right)$$
$$= 0 \text{ for all } \kappa_i', \kappa_i \in R \text{ such that } i = 1, 2...., n.$$

Proceedings this formula, we observe that

$$\begin{aligned}
&a\big(\Delta^k(\kappa_1,\kappa_2,\ldots,\kappa_i',\ldots,\kappa_n)\kappa_i + \kappa_i'\Delta^k(\kappa_1,\kappa_2,\ldots,\kappa_i,\ldots,\kappa_n)\\
&+\binom{k-1}{1}\big(\Delta^{k-1}(\kappa_1,\kappa_2,\ldots,\kappa_i',\ldots,\kappa_n)\Delta(\kappa_i)\\
&+\Delta(\kappa_i')\Delta^{k-1}(\kappa_1,\kappa_2,\ldots,\kappa_i,\ldots,\kappa_n)\big)\\
&+\cdots+\big(\Delta(\kappa_1,\kappa_2,\ldots,\kappa_i',\ldots,\kappa_n)\Delta^{k-1}(\kappa_i)\\
&+\Delta^{k-1}(\kappa_i')\Delta(\kappa_1,\kappa_2,\ldots,\kappa_i,\ldots,\kappa_n)\big)\text{for all } \kappa_i', \kappa_i \in R.
\end{aligned}$$

Employing the main relation to this relation, we arrive to

$$\begin{aligned}
&a\big(\kappa_i'\Delta^k(\kappa_1,\kappa_2,\ldots,\kappa_i,\ldots,\kappa_n)\big) + a\big(\Delta^{k-1}(\kappa_1,\kappa_2,\ldots,\kappa_i',\ldots,\kappa_n)\Delta(\kappa_i)\\
&+\Delta(\kappa_i')\Delta^{k-1}(\kappa_1,\kappa_2,\ldots,\kappa_i',\ldots,\kappa_n)\big)\\
&+\cdots+\big(\Delta(\kappa_1,\kappa_2,\ldots,\kappa_i',\ldots,\kappa_n)\Delta^{k-1}(\kappa_i)\\
&+\Delta^{k-1}(\kappa_i')\Delta(\kappa_1,\kappa_2,\ldots,\kappa_i,\ldots,\kappa_n)\big) = 0 \text{ for all } \kappa_i', \kappa_i \in R.
\end{aligned}$$

Furthermore, we conclude that
$a\big(\kappa_i'\Delta^k(\kappa_1,\kappa_2,\ldots,\kappa_i,\ldots,\kappa_n)\big) + \gamma = 0$ for all $\kappa_i', \kappa_i \in R$, where γ denoted to

$$\begin{aligned}
&a\left(\binom{k-1}{1}\big(\Delta^{k-1}(\kappa_1,\kappa_2,\ldots,\kappa_i',\ldots,\kappa_n)\Delta(\kappa_i) + \Delta(\kappa_i')\Delta^{k-1}(\kappa_1,\kappa_2,\ldots,\kappa_i,\ldots,\kappa_n)\big)\right.\\
&\left.\cdots+\big(\Delta(\kappa_1,\kappa_2,\ldots,\kappa_i',\ldots,\kappa_n)\Delta^{k-1}(\kappa_i) + \Delta^{k-1}(\kappa_i')\Delta(\kappa_1,\kappa_2,\ldots,\kappa_i,\ldots,\kappa_n)\big)\right)
\end{aligned}$$

Consequently, we find that
$[a\kappa_i'\Delta^k(\kappa_1,\kappa_2,\ldots,\kappa_i,\ldots,\kappa_n), r] = [r, \gamma]$ for all $\kappa_i', \kappa_i, r \in R$. Choosing element say r lies in R, we satisfy the identity

$$\begin{aligned}
&a\kappa_i'\big[\Delta^k(\kappa_1,\kappa_2,\ldots,\kappa_i,\ldots,\kappa_n), r\big] + a\big[\kappa_i', r\big]\Delta^k(\kappa_1,\kappa_2,\ldots,\kappa_i,\ldots,\kappa_n)\\
&+[a, r]\kappa_i'\Delta^k(\kappa_1,\kappa_2,\ldots,\kappa_i,\ldots,\kappa_n) = \big[r, \gamma\big]. \qquad (12)
\end{aligned}$$

Now from the main relation and applying Lemma 1.1 and using the information aRb in Z(R), a, b in R, we conclude that

$$\Delta^k(R)a = 0. \qquad (13)$$

In Relation (12) replacing r by $\Delta^k(\kappa_1,\kappa_2,\ldots,\kappa_i,\ldots,\kappa_n)$ without κ_i' and using the main relation with (13), we observe that

$$a\kappa_i'\Delta^k(\kappa_1, \kappa_2, \ldots, \kappa_i, \ldots, \kappa_n)^2 = -\gamma\Delta^k(\kappa_1, \kappa_2, \ldots, \kappa_i, \ldots, \kappa_n) \quad (14)$$

Applying the main relation of the right-side, we deduce that
$a\kappa_i'\Delta^k(\kappa_1, \kappa_2, \ldots, \kappa_i, \ldots, \kappa_n)^2 = 0$ for all κ_i', κ_i in R.
Hence, aR $\Delta^k(\kappa_1, \kappa_2, \ldots, \kappa_i, \ldots, \kappa_n) = (0)$.
Compatibility between the primness of R with Lemma 1.1 implies to satisfy two options; either $\Delta^k(R)^2 = 0$ implies to $\Delta^k(R) = 0$ or $a = 0$, as desired.

Using the same previous approach, we can satisfy the following:

Theorem 2.9 Suppose R is a semiprime ring with identity, U is a maximal ideal of R and a is a fixed element of R. Suppose that Δ satisfies a $\Delta^k(R) = 0$, then R contains a weakly semiprime ideal.

Proof By the same scenario which used in the proof of Theorem 2.8, we observe aR $\Delta^k(\kappa_1, \kappa_2, \ldots, \kappa_i, \ldots, \kappa_n) = (0)$.

Based on R acts as a semiprime, we have the collection $\{P_\alpha\}$ covering prime ideals of R such that $\cap P_\alpha = \{0\}$. Indicate to Lemma 1.3, we see the set of intersection $\{P_\alpha\}$ of prime ideals over R forms a semiprime ideal. Due to U plays role as a maximal ideal and $\cap P_\alpha$ is proper semiprime ideal of R. Essentially, U acts as a maximal ideal of R. Then there exist no other ideals contained between U and R. Consequently, we observe $\cap P_\alpha \subseteq U$.

When P is a standard member in $\cap P_\alpha$ and x lies in U. We have two options. Either $a \in P$ or $\Delta^k(\kappa_1, \kappa_2, \ldots, \kappa_i, \ldots, \kappa_n)^2 \in P$. Construct two additive subgroup $T_1 = \{a \in R\backslash a \in P\}$ and $T_2 = \{\kappa \in U\backslash^k(\kappa_1, \kappa_2, \ldots, \kappa_i, \ldots, \kappa_n)^2 \in P\}$. Where any ideal of a ring R forms subgroup of the additive group over R. Consequently, $T_1 \cup T_2 = U$. Because of a group can not be a union of two its proper subgroups. That's mean, we have two options, either $T_1 = U$ or $T_2 = U$. That is, either $a \in P$ or $\Delta^k(\kappa_1, \kappa_2, \ldots, \kappa_i, \ldots, \kappa_n)^2 \in P$. Thus, both cases yield $a \in \cap P_\alpha$ or $\Delta^k(\kappa_1, \kappa_2, \ldots, \kappa_i, \ldots, \kappa_n)^2 \in \cap P_\alpha$. In other words, $a \in \cap P_\alpha \subseteq U$ or $\Delta^k(\kappa_1, \kappa_2, \ldots, \kappa_i, \ldots, \kappa_n)^2 \in \cap P_\alpha \subseteq U$. Hence, either $a \in U$ for all κ, $w \in U$ or $\Delta^k(\kappa_1, \kappa_2, \ldots, \kappa_i, \ldots, \kappa_n)^2 \in U$ for all κ lies in U.

We separate the proof into two parts:

Case 1: If $a \in U$ then $a = 0$. For element $t \in U$ such that $ta = (0)$, where U acts as a maximal ideal. Then U is a prime ideal. Basically, (0) is a prime ideal implies to either $a = 0$ or $t = 0$. Second option yields a contradiction, so we find that $t \neq 0$.

By reason of U plays his role as a semiprime ideal over R, we write $t^2 \in U$, for all $t \in R$.

Employing the information of the hypothesis, we find that U is an ideal in R has identity which means $U = R$. Due to that $a \in U = R \subseteq Z(R)$. Consequently, a semiprime ring R modifies to commutative.

Thus, U forms a weakly semiprime ideal of R. By this step, we harvest the required result.

Case 2: When $\Delta^k(\kappa_1, \kappa_2, \ldots, \kappa_i, \ldots, \kappa_n)^2$ belongs to U, directly $\Delta^k(\kappa_1, \kappa_2, \ldots, \kappa_i, \ldots, \kappa_n)^2 = 0$. Applying the similar approach of the proof of Case (1) and using Lemmas (1.2 and 1.1). It is sufficient to satisfy the required result . This completes the proof of the theorem.

In the next results, a permuting n-derivation Δ has different powers.

Theorem 2.10 Suppose R is a semiprime ring with $1 \neq 0$ and U acts as a maximal ideal of R. Assume that R admits a non-zero permuting n-derivation Δ satisfying $\left[\Delta^k([x, y]), \Delta^j(x \circ y)\right] \in Z(R)$ for all $x, y \in R$. Then R has a weakly semiprime ideal.

Proof We take place with the assumption $\left[\Delta^k\left[(x, y)\right], \Delta^j(x \circ y)\right] \in Z(R)$ when x, y lie in R. Simple calculations this formula yield

$$\begin{aligned}
&\Delta^k(xy)\Delta^k(xy) + \Delta^k(xy)\Delta^j(yx) - \Delta^j(xy)\Delta^k(xy)\\
&- \Delta^j(yx)\Delta^k(xy) - \Delta^k(yx)\Delta^j(xy) - \Delta^k(yx)\Delta^j(yx)\\
&+ \Delta^j(xy)\Delta^k(yx) + \Delta^j(yx)\Delta^k(yx) \in Z(R) \text{ for all } x, y \text{ in } R.
\end{aligned}$$

This expression modifies to

$\Delta^k(xy)\Delta^j(xy) + \gamma \in Z(R)$ for all $x, y \in R$.

Where γ stand for $(\Delta^k(xy)\Delta^j(yx) - \Delta^j(xy)\Delta^k(xy) - \Delta^k(yx)\Delta^j(xy) - \Delta^k(yx)\Delta^j(yx + \Delta^j(xy)\Delta^k(yx) + \Delta^k(yx)\Delta^k(yx)$.

We employ the compositions of permuting n-derivations Δ on the first term which is $\Delta^k(xy)$ of this relation and putting $x = \left(x_1, x_2, \ldots, x_i'x_i, \ldots, x_n\right)$, also $y = \left(y_1, y_2, \ldots, y_i'y_i, \ldots, y_n\right)$, we note

$$\begin{aligned}
&\left(\sum_{r=0}^{k}\binom{k-1}{r}\Delta^{k-r}\left(x_1, x_2, \ldots, x_i'x_i, \ldots, x_n, y_2, \ldots, y_i'y_i, \ldots, y_n\right)\right.\\
&\Delta^r\left(y_1\right) + \Delta^r(x_n)\Delta^{k-r}\left(x_1, x_2, \ldots, x_i'x_i, \ldots, x_{n-1}, y_1, y_2, \ldots, y_i'y_i, \ldots, y_n)\right)\\
&\Delta^j(xy) + \gamma \in Z(R)
\end{aligned}$$

for all $x_i', x_i, y_i', y_i \in R$ such that $i = 1, 2, \ldots, n$.

Simplifying this expression, it follows that R achieves the expression

$$\begin{aligned}
&\left(\Delta^k\left(x_1, x_2, \ldots, x_i'x_i, \ldots, x_n, y_2, \ldots, y_i'y_i, \ldots, y_n\right)y_1\right.\\
&+ x_n\Delta^k\left(x_1, x_2, \ldots, x_i'x_i, \ldots, x_{n-1}, y_1, y_2, \ldots, y_i'y_i, \ldots, y_n\right)\\
&+ \binom{k-1}{r}\Delta^{k-1}\left(x_1, x_2, \ldots, x_i'x_i, \ldots, x_n, y_2, \ldots, y_i'y_i, \ldots, y_n\right)\Delta\left(y_1\right)\\
&+ \Delta(x_n)\Delta^{k-1}\left(x_1, x_2, \ldots, x_i'x_i, \ldots, x_{n-1}, y_1, y_2, \ldots, y_i'y_i, \ldots, y_n\right)\\
&+ \cdots + \Delta\left(x_1, x_2, \ldots, x_i'x_i, \ldots, x_n, y_2, \ldots, y_i'y_i, \ldots, y_n\right)\Delta^{k-1}\left(y_1\right)\\
&+ \Delta^{k-1}(x_n)\Delta(x_1, x_2, \ldots, x_i'x_i, \ldots, x_{n-1}, y_1, y_2, \ldots, y_i'y_i, \ldots, y_n)))\Delta^j(xy) + \gamma
\end{aligned}$$

in Z(R) for all x'_i, x_i, y'_i, y_i in R.

Additionally, we show that

$$\Delta^k(x_1, x_2, \ldots, x'_i x_i, \ldots, x_n, y_2, \ldots, y'_i y_i, \ldots, y_n) y_1 \Delta^j(xy) + \vartheta \in Z(R). \quad (15)$$

We write ϑ instead of

$$\begin{aligned}&\big(x_n \Delta^k(x_1, x_2, \ldots, x'_i x_i, \ldots, x_{n-1}, y_1, y_2, \ldots, y'_i y_i, \ldots, y_n)\\&\Delta^{k-1}(x_1, x_2, \ldots, x'_i x_i, \ldots, x_n, y_2, \ldots, y'_i y_i, \ldots, y_n)\\&\Delta(y_1) + \binom{k-1}{1}\Delta(x_n)\Delta^{k-1}(x_1, x_2, \ldots, x'_i x_i, \ldots, x_{n-1}, y_1, y_2, \ldots, y'_i y_i, \ldots, y_n)\Big)\\&+ \cdots + \Delta(x_1, x_2, \ldots, x'_i x_i, \ldots, x_n, y_2, \ldots, y'_i y_i, \ldots, y_n)\Delta^{k-1}(y_1)\\&+ \Delta^{k-1}(x_n)\Delta(x_1, x_2, \ldots, x'_i x_i, \ldots, x_{n-1}, y_1, y_2, \ldots, y'_i y_i, \ldots, y_n)))\Delta^j(xy) + \gamma\end{aligned}$$

of the this relation. By virtue of (15) for any element $r \in R$. We observe $[\Delta^k(x_1, x_2, \ldots, x'_i x_i, \ldots, x_n, y_2, \ldots, y'_i y_i, \ldots, y_n) y_1 \Delta^j(xy), r] + [\vartheta, r] = 0$ for all x'_i, x_i, y'_i, y_i, r in R. Analogously, in the last expression taking ϑ instead of r. Deliver $[\Delta^k(x_1, x_2, \ldots, x'_i x_i, \ldots, x_n, y_2, \ldots, y'_i y_i, \ldots, y_n) y_1 \Delta^j(xy), \vartheta] = 0$ when x'_i, x_i, y'_i, y_i in R.

It is a small modification of the proof of Theorem 2.9, right multiplying the above relation by $\Delta^k(t_1, t_2, \ldots, t'_i x_i, \ldots, t_n, s_2, \ldots, s'_i s_i, \ldots, s_n)^2$ for all t'_i, t_i, s'_i, s_i in R whenever $i = 1, 2, \ldots, n$.

Concluding that R has a weakly semiprime ideal as desired. The proof of the theorem is completed.

The following corollary is an immediate consequence of Theorem 2.10.

Corollary 2.11 Take R as a semiprime ring with $1 \neq 0$ and U as a maximal ideal of R. Suppose that R admits a non-zero permuting n-derivation Δ satisfying $[\Delta^k([x, y]), \Delta^j([x, y])]$ in Z(R) (resp., $[\Delta^k(x \circ y), \Delta^j(x \circ y)] \in Z(R)$) for all x, y in R. Then R contains a weakly semiprime ideal.

Theorem 2.12 Take R as a semiprime ring with $1 \neq 0$ and U as a maximal ideal of R. Suppose that R admits a permuting n-derivations Δ satisfying $\Delta^k(x)R\Delta^j(x) = 0$ for all x lies in R. Then R has a weakly semiprime ideal.

Proof From our hypothesis, we find that $\Delta^k(x)R\Delta^j(x) = 0$. Suppose $x = (x_1, x_2, \ldots, x'_i x_i, \ldots, x_n)$. Hence, the main relation imply to $\Delta^k(x_1, x_2, \ldots, x'_i x_i, \ldots, x_n) r \Delta^j(x_1, x_2, \ldots, x'_i x_i, \ldots, x_n) = 0$ for all $x'_i, x_i, y'_i, y_i \in R$, such that $i \geq 1$.

Applying the composition of permuting n-derivations Δ on the first term Δ of this relation and simple calculations, yields

$$\Delta^k(x_1, x_2, \ldots, x'_i, \ldots, x_n) x_i r \Delta^j(x_1, x_2, \ldots, x'_i x_i, \ldots, x_n) + \vartheta = 0. \quad (16)$$

We write ϑ instead of

$$\left(x_i'\Delta^k(x_1, x_2, \ldots, x_i, \ldots, x_n) + \binom{k-1}{1}\Delta^{k-1}\left(x_1, x_2, \ldots, x_i', \ldots, x_n\right)\right.$$
$$\Delta(x_i) + \Delta\left(x_i'\right)\Delta^{k-1}(x_1, x_2, \ldots, x_i, \ldots, x_n) + \cdots + \Delta\left(x_1, x_2, \ldots, x_i', \ldots, x_n\right)$$
$$\left.\Delta^{k-1}(x_i) + \Delta^{k-1}\left(x_i'\right)\Delta(x_1, x_2, \ldots, x_i, \ldots, x_n)\right)r\Delta^j\left(x_1, x_2, \ldots, x_i', \ldots, x_n\right)$$

In the main expression putting $x = \left(x_1, x_2, \ldots, x_i', \ldots, x_n\right)$ with out the element x_i, it implies to

For any element t in R. Multiplying from left side by $\Delta^j\left(x_1, x_2, \ldots, x_i'x_i, \ldots, x_n\right)t$ and multiplying from right side via $t\Delta^k\left(x_1, x_2, \ldots, x_i^{\cdot}x_i, \ldots, x_n\right)$ with using the information that R is semiprime ring, we arrive to

$$\Delta^j\left(x_1, x_2, \ldots, x_i', \ldots, x_n\right)t\Delta^k\left(x_1, x_2, \ldots, x_i', \ldots, x_n\right) = 0. \tag{17}$$

Left-multiplying (16) by $\Delta^j\left(x_1, x_2, \ldots, x_i', \ldots, x_n\right)t$ and applying (17), we deduce that

$\Delta^j\left(x_1, x_2, \ldots, x_i', \ldots, x_n\right)t\vartheta = 0$ for all $x_i', x_i, t \in R$.

Depending on the identical approach utilized in the proof of Theorem 2.9, we satisfy the outcome as desired. The proof is complete.

The following corollary is an immediate continuation of the proof of Theorem 2.1.

Corollary 2.13 Take R as a semiprime ring has a 2-torsion free property with $1 \neq 0$ and U as a maximal ideal over R. Suppose that Δ satisfies the expression $\Delta^k(x \circ y) \mp (x \circ y)$ in Z(R) (resp., $\Delta^k[x, y] \mp [x, y]$ in Z(R)) for all $x, y \in R$. Then R has a weakly semiprime ideal.

We now write the definition of the inner a permuting n-derivations of a ring R.

Definition 2.2 Suppose R acts as a ring and Δ is an additive mapping $\Delta : R^n \to R^2$. Then, for a fixed a∈R the inner a permuting n-derivation over a ring R defined as

$$\Delta\left(\kappa_1, \kappa_2, \ldots, \kappa_i'\kappa_i, \ldots, \kappa_n\right) = \left[\kappa_1, \kappa_2, \ldots, \kappa_i'\kappa_i, \ldots, \kappa_n, a\right]$$
$$= \prod_{i=1}^{n} [\kappa_i, a] \text{ for all } \kappa_i', \kappa_i \in R, i = 1, 2, \ldots, n.$$

The following example offers that in previous results the event that U is an ideal over R with $1 \neq 0$ be necessary. We can not exclude it from the hypothesis. Hence, we exams it for Theorem 2.10 and for other cases one can exam that by a similar technique .

Example 3 Take a ring of matrices R as $= \left\{\begin{pmatrix} x_1 & x_2 & 0 & \ldots\ldots & 0 \\ 0 & 0 & 0 & \ldots\ldots & 0 \\ \vdots & \vdots & \vdots & \ldots\ldots & \vdots \\ 0 & 0 & 0 & \ldots\ldots & 0 \end{pmatrix}\right.$ / for all $x_1, x_2 \in$ F} over a field F.

Suppose that $\triangle$ is an additive mapping induced by a fixed element $a \in R$, that is,

$$\Delta(x_1, x_2, \ldots, x_i'x_i, \ldots, x_n) = [x_1, x_2, \ldots, x_i'x_i, \ldots, x_n, a]$$
$$= \prod_{i=1}^{n} [x_i, a] \text{ for all } x_i', x_i \in R,$$

whenever $i = 1, 2, \ldots, n$,

$$= (x_1, x_2, \ldots, x_i'x_i, \ldots, x_n)$$
$$\begin{pmatrix} 1\,0\,0\ldots0 \\ 0\,0\,0\ldots0 \\ \vdots\;\vdots\;\vdots\ldots\vdots \\ 0\,0\,\vdots\ldots0 \end{pmatrix} - \begin{pmatrix} 1\,0\,0\ldots0 \\ 0\,0\,0\ldots0 \\ \vdots\;\vdots\;\vdots\ldots\vdots \\ 0\,0\,\vdots\ldots0 \end{pmatrix} (x_1, x_2, \ldots, x_i'x_i, \ldots, x_n),$$

where $a = \begin{pmatrix} 1\,0\,0\ldots0 \\ 0\,0\,0\ldots0 \\ \vdots\;\vdots\;\vdots\ldots\vdots \\ 0\,0\,\vdots\ldots0 \end{pmatrix}$

3 The Action of the Trace of Permuting n-Derivations

In this part, we examine certain results concerning the trace δ of $\triangle$ of semiprime rings satisfying certain conditions, where R has a torsion free property.

Theorem 3.1 Take R as a 2-torsion free semiprime ring and U as a maximal ideal of R. Suppose that R admits a non-zero permuting n-derivations $\triangle$ with trace $\delta : R \to R$ satisfying the identity $[\Delta^k(x), \delta(y)]$ in Z(R) for all x, y in R. Then either Δ^k acts as a centralizer mapping (resp. commuting mapping) of R or $[[x_i, \Delta^k(x)], \delta(y)] = [[x_i, \delta(y)], \Delta^k(x)]$ such that $i \geq 1$.

Proof In agreement with the postulate, we see that $[\Delta^k(x), \delta(y)]$ in Z(R) for all x, y in R. Applying the formula of $\triangle$ yields

$$\left[\sum_{r=0}^{k} \binom{k-1}{r} \Delta^{k-r}(x_1, x_2, \ldots, x_i'x_i, \ldots, x_n)\Delta^r(x_i) \right.$$
$$\left. + \Delta^r(x_i')\Delta^{k-r}(x_1, x_2, \ldots, x_i, \ldots, x_n), \delta(y)\right]$$

belongs to Z(R) for all $x_i'x_i, y \in R$.

Now, pick an element $r \in R$, this relation can be written as

$$\left[\left[\sum_{r=0}^{k}\binom{k-1}{r}\Delta^{k-r}(x_1, x_2, \ldots, x_i'x_i, \ldots, x_n)\Delta^{r}(x_i) + \Delta^{r}(x_i')\Delta^{k-r}(x_1, x_2, \ldots, x_i, \ldots, x_n), \delta(y)\right], r\right] = 0$$

for all $x_i', x_i, y, r \in R$ such that $i \geq 1$.

The consequence of the processing of $\Delta^k(x_1, x_2, \ldots, x_i^{`}x_i, \ldots, x_n)$, we find

$$\begin{aligned}&[(\Delta^k(x_1, x_2, \ldots, x_i^{`}, \ldots, x_n)x_i + x_i^{`}\Delta^k(x_1, x_2, \ldots, x_i, \ldots, x_n))\\ &+ \cdots + (\Delta(x_1, x_2, \ldots, x_i^{`}, \ldots, x_n)\\ &\Delta^{k-1}(x_i) + \Delta^{k-1}(x_i^{`})\Delta(x_1, x_2, \ldots, x_i, \ldots, x_n)), \delta(y)], r] = 0\end{aligned}$$

for all $x_i'x_i, y, r \in R$. In this expression, we put $\Delta^k(x_1, x_2, \ldots, x_i', \ldots, x_n)$ for r, we notice that

$$\begin{aligned}&\Delta^k(x_1, x_2, \ldots, x_i', \ldots, x_n)[[x_i, \delta(y)], \Delta^k(x_1, x_2, \ldots, x_i', \ldots, x_n)]\\ &+ [\zeta, \Delta^k(x_1, x_2, \ldots, x_i'x_i, \ldots, n)] = 0.\end{aligned} \tag{18}$$

Clearly, ζ denotes to following the bracket

$$\begin{aligned}&[x_i'\Delta^k(x_1, x_2, \ldots, x_i, \ldots, x_n)\\ &+ \cdots + (\Delta(x_1, x_2, \ldots, x_i', \ldots, x_n)\Delta^{k-1}(x_i))\\ &+ \Delta^{k-1}(x_i')\Delta(x_1, x_2, \ldots, x_i, \ldots, x_n)), \delta(y)].\end{aligned}$$

From the first bracket of (18), we establish

$$\begin{aligned}&\Delta^k(x_1, x_2, \ldots, x_i', \ldots, x_n)x_i[\delta(y), \Delta^k(x_1, x_2, \ldots, x_i'x_i, \ldots, x_n)]\\ &+ \Delta^k(x_1, x_2, \ldots, x_i', \ldots, x_n)[x_i, \Delta^k(x_1, x_2, \ldots, x_i'x_i, \ldots,\\ &x_n)]\delta(y) - \Delta^k(x_1, x_2, \ldots, x_i', \ldots, x_n)\delta(y)[x_i, \Delta^k(x_1, x_2, \ldots, x_i'x_i, \ldots, x_n)]\\ &- \Delta^k(x_1, x_2, \ldots, x_i', \ldots, x_n)[\delta(y), \Delta^k(x_1), (x_2, \ldots, x_i'x_i, \ldots, x_n)]x_i\\ &+ [\zeta, \Delta^k(x_1, x_2, \ldots, x_i'x_i, \ldots, x_n)] = 0.\end{aligned}$$

Using the main expression $[\Delta^k(x)\delta(y)] \in Z(R)$ after writing $x = (x_1, x_2, \ldots, x_i'x_i, \ldots, x_n)$ with simple proceeding, this equation reduces to

$$\begin{aligned}&\Delta^k(x_1, x_2, \ldots, x_i', \ldots, x_n)[[x_i, \Delta^k(x_1, x_2, \ldots, x_i'x_i, \ldots, x_n)], \delta(y)]\\ &+ [\zeta, \Delta^k(x_1, x_2, \ldots, x_i'x_i, \ldots, n)] = 0.\end{aligned} \tag{19}$$

Particularly, one can easily from the latter identity obtaining

$$\Delta^k(x_1, x_2, \ldots, x_i', \ldots, x_n)(x_i\Delta^k(x_1, x_2, \ldots, x_i'x_i, \ldots, x_n)\delta(y)$$

$$
\begin{aligned}
&- \Delta^k(x_1, x_2, \ldots, x_i'x_i, \ldots, x_n) \\
&x_i\delta(y) - \delta(y)x_i\Delta^k(x_1, x_2, \ldots, x_i'x_i, \ldots, x_n) \\
&+\delta(y)\Delta^k(x_1, x_2, \ldots, x_i'x_i, \ldots, x_n)x_i) \\
&+ [\zeta, \Delta^k(x_1, x_2, \ldots, x_i'x_i, \ldots, x_n)] = 0.
\end{aligned}
$$

Again, depending on the identity $[\Delta^k(x_1, x_2, \ldots, x_i', \ldots, x_n), \delta(y)] \in Z(R)$. Observation

$$
\begin{aligned}
&\Delta^k(x_1, x_2, \ldots, x_i', \ldots, x_n)[[x_i, \delta(y)], \Delta^k(x_1, x_2, \ldots, x_i'x_i, \ldots, x_n)] \\
&+ [\zeta, \Delta^k(x_1, x_2, \ldots, x_i'x_i, \ldots, x_n)] = 0.
\end{aligned} \tag{20}
$$

Combining the Relations (19) and (20) yields the following relation

$$
\begin{aligned}
&\Delta^k(x_1, x_2, \ldots, x_i', \ldots, x_n)([[x_i, \delta(y)], \Delta^k(x_1, x_2, \ldots, x_i'x_i, \ldots, x_n)] \\
&- [[x_i, \Delta^k(x_1, x_2 \ , \ldots, x_i'x_i, \ldots, x_n)], \delta(y)]) = 0.
\end{aligned}
$$

We arrange this relation as below

$$
\begin{aligned}
&\Delta^k(x_1, x_2, \ldots, x_i', \ldots, x_n)([[x_i, \Delta^k(x_1, x_2, \ldots, x_i'x_i, \ldots, x_n)], \delta(y)] \\
&+[\Delta^k(x_1, x_2, \ldots, x_i'x_i, \ldots, x_n), [x_i, \delta(y)]]) = 0.
\end{aligned}
$$

Moreover, multiplying from left side by

$$
\begin{aligned}
&([[x_i, \Delta^k(x_1, x_2, \ldots, x_i'x_i, \ldots, x_n), \delta(y)] \\
&+ [\Delta^k(x_1, x_2, \ldots, x_i'x_i, \ldots, x_n), \ [x_i, \delta(y)]])R
\end{aligned}
$$

and multiplying from right side by $R\Delta^k(x_1, x_2, \ldots, x_i', \ldots, x_n)$. Applying Lemma 1.2 and aRb in Z(R), a, b belong to R, one can rewrite this expression as

$$
\begin{aligned}
&([[x_i, \Delta^k(x_1, x_2, \ldots, x_i'x_i, \ \ldots, x_n)], \delta(y)] \\
&+ ([\Delta^k(x_1, x_2, \ldots, x_i'x_i, \ldots, x_n), [x_i, \delta(y)]])R\Delta^k(x_1, x_2, \ldots, x_i', \ldots, x_n) = (0).
\end{aligned}
$$

Basically, R is a ring that acts as a semiprime, we consider the collection $\{P_\alpha\}$ of prime ideals over R such that $\cap P_\alpha = \{0\}$.

Depend on Lemma 1.3, we observe the intersection $\{P_\alpha\}$ of prime ideals of R forms a semiprime ideal. Due to U plays role as a maximal ideal and $\cap P_\alpha$ is proper semiprime ideal in R. Essentially, U has a role as a maximal ideal in R. Then there are no other ideals contained between U and R. Consequently, applying the same scenario which used in Theorem 2.1, we obtain

either $\Delta^k(x_1, x_2, \ldots, x_i', \ldots, x_n) \in U$ or

$([[x_i, \Delta^k(x_1, x_2, \ldots, x_i'x_i, \ldots, x_n)], \delta(y)] + [\Delta^k(x_1, x_2, \ldots, x_i'x_i, \ldots, x_n), [x_i, \delta(y)]) \in U$. Without a doubt, the second case proved that $[[x_i, \Delta^k(x_1, x_2, \ldots, x_i'x_i, \ldots, x_n)], \delta(y)]\,[[x_i, \delta(y)], \Delta^k(x_1, x_2, \ldots, x_i'x_i, \ldots, x_n)]$.

The first case gives us $\Delta^k(x_1, x_2, \ldots, x_i', \ldots, x_n) \in U$ yields that $[\Delta^k(x_1, x_2, \ldots, x_i', \ldots, x_n), x_i] = 0$ for all $x_i', x_i \in R$. Replacing x_i' by x_i', x_i, we deduce that,

$[\Delta^k(x_1, x_2, \ldots, x_i'x_i, \ldots, x_n), x_i] = 0$ for all $x_i', x_i \in R$.

Obviously, Δ^k is centralizer mapping (resp. commuting mapping) of R, as desired. This completes the proof.

In the following by using a similar approach than from the first case of the proof of Theorems (3.1 and 2.9), we can prove to attach corollary.

Corollary 3.2 Take R as a semiprime ring and U as a maximal ideal of R has $1 \neq 0$. Suppose that R admits a non-zero permuting n-derivations Δ. When the trace $\delta : R \to R$ satisfies the expression $[\Delta^n(x), \delta(y)] \in Z(R)$ for all x, y lie in R. Then R has a weakly semiprime ideal.

Theorem 3.3 For any fixed integer $q \geq 1$. Suppose R is a semiprime ring has a q-torsion free property and U is a maximal ideal of R. When R admits a non-zero permuting n-derivations Δ and the trace $\delta : R \to R$ satisfying the identity $\Delta^n(\delta(R)) = 0$. Then Δ^{n+1} is centralizer mapping (resp. commuting mapping) of R.

Proof We are given that $\Delta^k(\delta(x)) = 0$ for all $x \in R$. Of course, the linearization of the element x lead to the trace δ of the permuting n-derivations Δ satisfies the identity

$$\delta(x + y) = \delta(x) + \delta(y) + \sum_{r=1}^{k-1} \binom{k}{r} \Delta(x, x, \ldots, x, y, y \ldots, y)$$

where x, y in R. Here y appears r times and x appears at k-r times. Substituting this identity in the main relation $\Delta(\delta(R)) = 0$ and reusing the main relation. Hence, it straightforward to see that

$$\sum_{r=1}^{k-1} \binom{k}{r} \Delta^k(\Delta(x_1, x_2, \ldots, x_{n-r-1}, y_{r-1}, y_{r-2} \ldots, y_r)x_{n-r}$$
$$+y_{r-1}\Delta(x_1, x_2, \ldots, x_{n-r}, y_{r-2}, y_{r-3} \ldots, y_r)) = 0$$

Rewriting this relation as follows:

$$\sum_{r=0}^{k-1} \binom{k}{r} \Delta^k(\Delta(x_1, x_2, \ldots, x_{n-r-1}, y_{r-1}, y_{r-2} \ldots, y_r)x_{n-r})$$
$$+ \Delta^k(y_{r-1}\, \Delta\, (x_1, x_2, \ldots, x_{n-r}, y_{r-2}, y_{r-3} \ldots, y_r)) = 0.$$

Employing the formula for the compositions of permuting n-derivations Δ on the first term in this relation, after simplification, we find that

$$\left(\sum_{r=1}^{k-1}\binom{k}{r}\right)\left(\sum_{r=0}^{k}\binom{k-1}{r}\right)\left(\Delta^{k+1}(x_1, x_2, \ldots, x_{n-r-1}, y_{r-1}, y_{r-2} \ldots, y_r)x_{n-r} + \mu\right)$$
$$= 0,$$

where μ denote for the terms

$$\Delta(x, x, \ldots, x_{n-r-1}, y_{r-1}, y_{r-2} \ldots, y)\Delta^k(x_{n-r})$$
$$+ \Delta^k(y_{r-1}\Delta(x_1, x_2, \ldots, x_{n-r-1}, x_{n-r}, y_{r-2}, y_{r-3}, \ldots, y_r))).$$

Basically, R has a q-torsion free property.

Thus, this expression modifies to $\Delta^{k+1}(x_1, x_2, \ldots, x_{n-r-1}, y_{r-1}, y_{r-2} \ldots, y_r)x_{n-r}+$ $\mu = 0$. Suppose the element r lies in R, we deliver

$$\Delta^{k+1}(x_1, x_2, \ldots, x_{n-1}, y_{r-1}, y_{r-1} \ldots, y_r)[x_{n-r}, r]$$
$$+ [\Delta^{k+1}(x_1, x_2, \ldots, x_{n-r-1}, y_{r-1}, y_{r-2} \ldots, y_r), r]x_{n-r} + [\mu, r] = 0. \quad (21)$$

Since Relation (21) is similar to (1) in Theorem 2.1. Hence, by the similar argument of the proof of Theorem 2.1, we acquire the required result.

One can prove the following corollary using the same technique as above.

Corollary 3.4 For any fixed integer $q \geq 1$. Suppose R is a semiprime ring has a q-torsion free property and U is a maximal ideal over R. When R allows a non-zero permuting n-derivations Δ and the trace $\delta : R \to R$ achieving the expression $\Delta^k(\delta(R)) \in Z(R)$. Then Δ^{k+1} is centralizer mapping (resp. commuting mapping) of R.

Theorem 3.5 For any fixed integer $q \geq 1$. Suppose R is a semiprime ring has a q-torsion free property and U is a maximal ideal of R. When R allows a trace $\delta : R \to R$ satisfying $[\delta(x), \delta(y)] = 0$ for all x, y in R. Then Δ acts as a centralizer mapping (resp. commuting mapping) of R.

Proof The linearization of the element x of our hypothesis $[\delta(x), \delta(y)] = 0$. Using the information δ of Δ has the formula

$$\delta(x + y) = \delta(x) + \delta(y) + \sum_{r=1}^{k-1}\binom{k}{r}\Delta(x, x, \ldots, x, y, y \ldots, y)$$

where $x, y \in R$. Moreover, y appears r times and x appears at k-r times. Again, the main condition yields that

$$\left[\sum_{r=1}^{k-1}\binom{k}{r}\Delta(x_1, x_2, \ldots, x_{n-r}, t_{r-1}, t_{r-2}\ldots, t_r), \delta(y)\right] = 0.$$

Based on the hypothesis that Δ is permuting n-derivations. Then, the last expression yields that

$$\left[\sum_{r=1}^{k-1}\binom{k}{r}(\Delta(x_1, x_2, \ldots, x_{n-r-1}, t_{r-1}, t_{r-2}\ldots, t_r)x_{n-r} \right.$$
$$\left. + t_{r-1}\Delta(x_1, x_2, \ldots, x_{n-r}, t_{r-2}\ldots, t_r)), \delta(y)\right] = 0.$$

Basically, R has a q-torsion free and using simple proceeding this identity becomes

$$\left[\Delta(x_1, x_2, \ldots, x_{n-r-1}, t_{r-1}, t_{r-2}\ldots, t_r)x_{n-r}, \delta(y)\right]$$
$$+ \left[t_{r-1}\Delta(x_1, x_2, \ldots, x_{n-r}, t_{r-2}\ldots, t_r)), \delta(y)\right] = 0 \text{ for all } x, t \in R.$$

Subsequently, we see that

$$\Delta(x_1, x_2, \ldots, x_{n-r-1}, t_1, t_2\ldots, t_r)\left[x_{n-r}, \delta(y)\right] + \left[\Delta(x_1, x_2, \ldots, x_{n-r-1}, t_{r-1}, t_{r-2}\ldots, t_r),\right.$$

$\delta(y)]x_{n-r} + \kappa = 0$ for all $x, t \in R$, and κ denotes to the term $\left[t_{r-1}\Delta(x_1, x_2, \ldots, x_{n-r}, t_{r-2}\ldots, t_t)), \delta(y)\right]$.

Replacing x_{n-r} by $\delta(y)$ in this expression, whence it follows $\left[\Delta(x_1, x_2, \ldots, x_{n-r-1}, t_{r-1}, t_{r-2}\ldots, t_r), \delta(y)\right]\delta(y) + \kappa = 0$.

For any element r belongs to R. We find the result

$$\left[\Delta(x_1, x_2, \ldots, x_{n-r-1}, t_1, t_2\ldots, t_r), \delta(y)\right][\delta(y), r]$$
$$+ \left[\left[\Delta(x_1, x_2, \ldots, x_{n-r-1}, t_{r-1}, t_{r-2}\ldots, t_r), \delta(y)\right], r\right]\delta(y) + [\kappa, r] = 0.$$

Obviously, this result like Relation (1) in Theorem 2.1. Applying a similar approach as used in the proof of Theorem 2.1, we obtain the required result.

Using a similar approach as above we can prove the following.

Theorem 3.6 For any fixed integer $q \geq 1$. Take R as a q-torsion free semiprime ring and U as a maximal ideal of R. Suppose that R admits a trace $\delta : R \rightarrow R$ satisfying the identity $[\delta_1(x), \delta_2(x)] = 0$ for all x lies in R. Then Δ acts as a centralizer mapping (resp. commuting mapping) of R.

The following example shows that the condition of R has q-torsion free of our results of this section not superfluous. However, the following example focuses on the main condition of Theorem 3.6 and by the same technique, one can satisfy that for other results.

Example 4 Firstly, we consider Z the set of integers. Take R as a ring of matrices such that $R = \begin{pmatrix} \frac{Z}{2Z} & \frac{Z}{2Z} & \cdots & \frac{Z}{2Z} \\ \frac{Z}{2Z} & \frac{Z}{2Z} & \cdots & \frac{Z}{2Z} \\ \vdots & \vdots & \vdots & \vdots \\ \frac{Z}{2Z} & \frac{Z}{2Z} & \cdots & \frac{Z}{2Z} \end{pmatrix}$ over a field F and R has characteristic (for short char. R) $= n$. Suppose Δ is the additive mapping of R, defined via:

$$\Delta_1(x_1, x_2, \ldots, x_i' x, \ldots, x_n) = \Delta_1([(x_1, x_2, \ldots, x_i' x_i, \ldots, x_n), a])$$
$$= (x_1, x_2, \ldots, x_i' x_i, \ldots, x_n) \begin{pmatrix} 1\ 0\ 0 \ldots 0 \\ 0\ 0\ 0 \ldots 0 \\ \vdots\ \vdots\ \vdots \ldots \vdots \\ 0\ 0\ \vdots \ldots 0 \end{pmatrix} - \begin{pmatrix} 1\ 0\ 0 \ldots 0 \\ 0\ 0\ 0 \ldots 0 \\ \vdots\ \vdots\ \vdots \ldots \vdots \\ 0\ 0\ \vdots \ldots 0 \end{pmatrix}$$
$$(x_1, x_2, \ldots, x_i' x_i, \ldots, x_n) = \begin{pmatrix} 0 & \frac{-Z^2}{4Z^2} & \cdots & \frac{-Z^2}{4Z^2} \\ \frac{Z^2}{4Z^2} & 0 & \cdots & 0 \\ \vdots & \vdots & \vdots & \vdots \\ \frac{Z^2}{4Z^2} & 0 & \cdots & 0 \end{pmatrix}, i = 1, 2, \ldots, n,$$

for all $x_i', x_i \in R$. Additionally,

$$\Delta_2(x_1, x_2, \ldots, x_i' x_i, \ldots, x_n) = \Delta_2([(x_1, x_2, \ldots, x_i' x_i, \ldots, x_n), a])$$
$$= (x_1, x_2, \ldots, x_i' x_i, \ldots, x_n) \begin{pmatrix} 1\ 0\ 0 \ldots 0 \\ 0\ 0\ 0 \ldots 0 \\ \vdots\ \vdots\ \vdots \ldots \vdots \\ 0\ 0\ \vdots \ldots 0 \end{pmatrix} - \begin{pmatrix} 1\ 0\ 0 \ldots 0 \\ 0\ 0\ 0 \ldots 0 \\ \vdots\ \vdots\ \vdots \ldots \vdots \\ 0\ 0\ \vdots \ldots 0 \end{pmatrix}$$
$$(x_1, x_2, \ldots, x_i' x_i, \ldots, x_n) = \begin{pmatrix} \frac{-Z^2}{4Z^2} & 0 & \frac{-Z^2}{4Z^2} & \cdots & \frac{-Z^2}{4Z^2} \\ 0 & \frac{Z^2}{4Z^2} & 0 \ldots\ldots & & 0 \\ \vdots & \vdots & \vdots & & \vdots \\ 0 & \frac{Z^2}{4Z^2} & 0 \ldots\ldots & & 0 \end{pmatrix},$$

$i = 1, 2, \ldots, n$, for all $x_i', x_i \in R$.

We set $x = x_1 = x_2 = \ldots = x_i' = x_i = \ldots = x_n$. Also, $y = y_1 = y_2 = \ldots = y_i' = y_i = \ldots = y_n$. Here, for any element in R and using this expression of Δ, we find the value of

$\delta_1(x) = \Delta_1(x_1, x_2, \ldots, x_i' x_i, \ldots, x_n)$ and $\delta_2(x) = \Delta_2(x_1, x_2, \ldots, x_i' x_i, \ldots, x_n)$.

Applying the previous values of $\delta_1(x)$ and $\delta_2(x)$ and using the fact that R has char.R $= n$. It follows $[\delta_1(x), \delta_2(x)] = 0$ for all x in R. Based on the information that char.R $= 2$. Then Δ acts as a centralizer mapping (resp. commuting mapping) over R.

Acknowledgment The author is grateful to Department of Mathematics, College of Education, Mustansiriyah University, Baghdad, Republic of Iraq and beholden to the reviewer(s) for his/ their accuracy with professionally reading the article.

References

1. Maksa, G.: A remark on symmetric biadditive functions having non-negative diagonaliza-tion. Glasnik Mat. **15**(35), 279–282 (1980)
2. Breˇsar, M.: Commuting maps: a survey. Taiwanese J. Math. **8**, 361–397 (2004)
3. Breˇsar, M., Martindale, W.S. III, Miers, C.R.: Centralizing maps in prime rings with involution. J. Algebra **161**, 342–357 (1993)
4. Ashraf, M.: On symmetric bi-derivations in rings. Rend. Istit. Math. Univ. Trieste. **31**(1–2), 25–36 (1999)
5. Hasret, Y.: Permuting triderivations of prime and semiprime rings. Miskolc Math. Notes **18**(1):489-497 (2017). https://doi.org/10.18514/MMN.2017.1647
6. Atteya, M.J.: Permuting 3-derivations of semiprime rings. In: Proceedings of the 7th Annual Canadian Young Researchers Conference in Mathematics and Statistics (2010). http://www.math.ualberta.ca/game/CYRC10/talks/MehsinAtteya.pdf
7. Herstein, I.N.: A commutativity theorem. J. Algebra **38**(1), 112–118 (1976)
8. Ozt¨urk, M.A.: Permuting tri-derivations in prime and semiprime rings. East Asian. Math. J. **15**(2), 177–190 ((1999))
9. Foˇsner, A.: Prime and semiprime rings with symmetric skew 3-derivations. Aequationes Math. **87**(1-2), 191–200 (2014) . https://doi.org/10.1007/s00010-013-0208-8
10. Jung, Y.S., Park, K.H.: On prime and semi-prime rings with permuting 3-derivations. Bull. Korean Math. Soc. **44**, 789–794 (2007)
11. Foˇsner, A.: Prime and semiprime rings with symmetric skew n-derivations. Colloq. Math. **134**(2), 245–253 (2014)
12. Ashraf, M., Parveen, N., Jamal, M.R.: Traces of permuting n-derivations and commutativity of rings. SE Asian Bull Math **38**, 321–332 (2014)
13. Ashraf, M., Parveen, N.: On generalized (α, β)-n-derivations in rings. SE Asian Bull Math **40**, 783–796 (2016)
14. Xu, X., Liu, Y., Zhang, W.: Skew N-derivationson semiprime rings. Bull. Korean Math. Soc. **50**(6), 1863–1871 (2013).https://doi.org/10.4134/BKMS.2013.50.6.1863
15. Ali, A., Atteya, M.J., Miyan, P., Ali F.: Semigroup ideals and per-muting 3-generalized derivations in prime near rings. Ital. J. Pure Appl. Math. **35**, 207–226 (2015)
16. Ashraf M, Siddeeque MA (2013) On (σ, τ)-n-derivations in near-rings. Asian-Eur. J. Math. **06**(04):1350051.https://doi.org/10.1142/S1793557113500514(14pages)
17. Eremita, D.: Functional identities of degree 2 in triangular rings. Linear Algebra Appl. **438**, 584–597 (2013)
18. Yao, W., Wang, Y., Du. Y.: n-derivations of triangular algebras. Linear Algebra Appl. **439**, 463–471 (2013)
19. Skosyrskii, V.G.: Strongly prime noncommutative Jordan algebras. Trudy Inst Mat (Novosibirsk) **16**, 131–164 (1989) (in Russian)
20. Emine Koç, E., ur-Rehman, N.: Notes on symmetric skew n-derivation in rings. Commun. Korean Math. Soc. **33**(4):1113–1121 (2018). https://doi.org/10.4134/CKMS.c170454
21. Park, K.H.: On prime and semiprime rings with symmetric n-derivations. J Chungcheong Math Soc **22**, 451–458 (2009)
22. Ibrahim, A.M.: Extension of factorial concept to negative numbers. Notes Number Theory Discrete Math. **19**(2), 30–42 (2013)
23. Foˇsner, A., Atteya, M.J.: Study of (σ, τ)-generalized derivations with their composition of semiprime rings. Kragujevac J. Math. **43**(4), 535–558 (2019)

24. Bell, H.E., Matindale, W.S., III.: Centralizing mappings of semiprime rings. Canad. Math. Bull. **30**(1), 42–60 (1987)
25. Cohn, P.M.: Further Algebra and Applications, 1st edn. Springer, London Berlin Heidelberg (2003). SBN 978-1-4471-1120-7

Comparison for the Dispersion Matrices of BLUPs in an Over-Parameterized Linear Model and Its Related Models

Nesrin Güler and Melek Eriş Büyükkaya

Abstract In this paper, we delineate the results to present a comprehensive methodology for an over-parameterized linear model. This model is derived by augmenting new regressors to a linear model. The analysis includes an over-parameterized linear model, its appropriately reduced model, and a base linear model of the over-parameterized model. The predictors and estimators known as the best linear unbiased predictors (BLUPs) and the best linear unbiased estimators (BLUEs) are discussed in relation to their analytical expressions and qualities within the context of these three models. We analyze the comparison problems regarding the dispersion matrices of the BLUPs/BLUEs inside specified models utilizing matrix algebra. Specifically, we utilize formulas for matrix rank and inertia, along with simple block matrix operations. Specifically, we establish many equations and inequalities to compare the dispersion matrices of the BLUPs/BLUEs of a general vector that contains joint unknown vectors under the given models.

Keywords BLUP · Dispersion matrix · Over-parameterized model

1 Introduction

This work focuses on comparing the dispersion matrices of predictors in an over-parameterized linear model and its related models. The following notations are introduced to concisely represent mathematical expressions in the context of matrix algebra. $R^{t,s}$ represents the collection of all $t \times s$ real matrices. The symbol $(.)'$ represents the transpose operation on a matrix. Additionally, the symbol $r(.)$ is used to describe the rank of a matrix and $\mathcal{C}(.)$ is used to represent the column space of a

N. Güler
Sakarya University, Sakarya, Turkey
e-mail: nesring@sakarya.edu.tr

M. Eriş Büyükkaya (✉)
Karadeniz Technical University, Trabzon, Turkey
e-mail: melekeris@ktu.edu.tr

O. Castillo et al. (eds.), *Advanced Mathematics for the Modeling and Solution of Challenging Problems in Engineering*, Studies in Systems, Decision and Control 582,
https://doi.org/10.1007/978-3-031-82342-8_19

matrix. The symbol $\Upsilon^{\perp} = I_t - \Upsilon\Upsilon^{+}$ denotes the orthogonal projectors associated with $\Upsilon \in R^{t,s}$, where Υ^{+} represents the Moore–Penrose generalized inverse of Υ and $I_t \in R^{t,t}$ denotes the identity matrices. The symbols $i_{+}(.)$ and $i_{-}(.)$ express the positive and negative inertias of a symmetric matrix, respectively. On the other hand, $i_{\pm}(.)$ and $i_{\mp}(.)$ are used to collectively represent both positive and negative inertias. The inequality $\Upsilon_1 - \Upsilon_2 \prec (\preccurlyeq, \succ, \succcurlyeq) 0$ or $\Upsilon_1 \prec (\preccurlyeq, \succ, \succcurlyeq) \Upsilon_2$ indicates that the difference $\Upsilon_1 - \Upsilon_2$ is a negative definite (negative semi-definite, positive definite, positive semi-definite) matrix in the Löwner partial ordering (LPO) for the symmetric matrices Υ_1 and Υ_2 of the same size. The property of LPO, which was first introduced in [27], is highly advantageous and valuable when applied to symmetric matrices. To obtain further information regarding the inertias and ranks of symmetric matrices, as well as the relationships between inertias and the LPO of symmetric matrices, refer to the sources [32, 47, 38, 45]. In addition, the notation $E(\cdot)$ is employed to represent the expectation of a random vector $cov(\cdot,\cdot)$ and $D(\cdot)$ are used to describe the covariance and dispersion matrices of random vectors, respectively.

Linear models stand out as widely recognized and extensively utilized statistical tools, appreciated from both theoretical and practical perspectives. The formulation commonly adopted for linear models takes a typical form, characterized by:

$$\mathcal{M} : y = X\alpha + \varepsilon \text{ with } E(\varepsilon) = 0 \text{ and } D(\varepsilon) = \sigma^2\Sigma, \tag{1}$$

where $y \in R^{n,1}$ is a vector of observable response variables, $X \in R^{n,k}$ is a known matrix of any rank, $\alpha \in R^{k,1}$ is a vector of fixed but unknown parameters, $\varepsilon \in R^{n,1}$ is an unobservable vector of random errors, σ^2 is a positive unknown parameter, and $\Sigma \in R^{n,n}$ is a known positive semi-definite matrix of any rank. The assumptions stated in Eq. (1) result in the expected value of y being equal to $X\alpha$ and the dispersion matrices of y being equal to $\sigma^2\Sigma$.

Researchers may find it necessary to explore similarities or distinctions among competing linear statistical models, particularly those involving jointly unknown vectors, to make informed choices during data analysis. Comparisons between models can be approached in various ways, and one method involves a systematic comparison by introducing or removing regressors from the models while considering the same observed vector. This approach is particularly relevant because numerous situations arise where multiple alternative models vie for fitting the same observed data. Considering $\mathcal{M}$ as a base or real model, introducing a new component with additional regressors on the right side transforms $\mathcal{M}$ into an over-parameterized model, expressed as:

$$\mathcal{N} : y = X\alpha + Z\gamma + \varepsilon = \begin{bmatrix} X & Z \end{bmatrix} \begin{bmatrix} \alpha \\ \gamma \end{bmatrix} + \varepsilon. \tag{2}$$

Here, $Z \in R^{n,p}$ is a known matrix of arbitrary rank and $\gamma \in R^{p,1}$ is a vector of fixed but unknown parameters. Under model $\mathcal{N}$, $E(y) = X\alpha + Z\gamma$ and $D(y) = \sigma^2\Sigma$. This expansion allows for the incorporation of additional features through the matrix Z, contributing to the model's over-parameterization.

The relation between $\mathcal{M}$ and $\mathcal{N}$ can be explained by considering one of the models is correct and the other is mis-specified; see, e.g., [25]. In other words, if we take $\mathcal{M}$ as a correct model then $\mathcal{N}$ becomes a mis-specified model of $\mathcal{M}$. On the contrary, if we take $\mathcal{N}$ as a correct model then $\mathcal{M}$ becomes a mis-specified model of $\mathcal{N}$ by considering the model matrix in $\mathcal{M}$ as $\begin{bmatrix} X & 0 \end{bmatrix}$. For relations between correct and mis-specified linear models; see, e.g., [30]. If we want to describe the relation between models $\mathcal{M}$ and $\mathcal{N}$ from a different perspective, we can say that the model $\mathcal{M}$ is a small model of the partitioned linear model $\mathcal{N}$. For this kind of work, we may refer [3, 4], among others. We can also use various transformations on $\mathcal{N}$ to characterize the results of prediction/estimation of unknown vectors including partial parameters. One of these transformations is the one that encompasses only one of the partial parameters. In view of the fact that $Z^{\perp}Z = 0$, pre-multiplying both sides of $\mathcal{N}$ by $Z^{\perp}$ yielding the following reduced model of $\mathcal{N}$:

$$\mathcal{R} : Z^{\perp}y = Z^{\perp}X\alpha + Z^{\perp}\varepsilon, \tag{3}$$

also known as the correctly-reduced model of $\mathcal{N}$; see, e.g., [4, 5, 12]. The model $\mathcal{R}$ is the other alternative model for estimating the partial unknown parameter vector α as well as predicting the random vector ε in $\mathcal{N}$ independently. It is further seen that the model $\mathcal{R}$ is a transformed model of $\mathcal{M}$. For transformation approach to linear models, we may refer [6, 13, 14, 43, 48], among others.

The three models $\mathcal{M}$, $\mathcal{N}$, and $\mathcal{R}$ are applicable for making statistical inferences on the jointly and separately unknown vectors α and ε. In order to obtain accurate predictions estimations for certain functions that encompass all joint unknown vectors in three models, the vector

$$r = F\alpha + H\varepsilon = \begin{bmatrix} F & 0 \end{bmatrix}\begin{bmatrix} \alpha \\ \gamma \end{bmatrix} + H\varepsilon \tag{4}$$

can be formulated for $F \in R^{s,k}$ and $H \in R^{s,n}$, which are given matrices. This vector allows for a comprehensive approach to prediction and estimation outcomes across the models. The vector r encompasses joint unknown vectors in $\mathcal{M}$, $\mathcal{N}$, and $\mathcal{R}$ as its special cases by making specific choices for the matrices F and H. For instance, when setting $F = X$ and $H = I_n$, r transforms into the observed response vector in $\mathcal{M}$ and also in $\mathcal{N}$. This choice essentially captures the response variable y in both models $\mathcal{M}$ and $\mathcal{N}$. The flexibility in defining matrices F and H allows for the adaptation of r to various scenarios within the context of these models. From the assumptions in (1),

$$E(r) = F\alpha = \hat{F}\begin{bmatrix} \alpha \\ \gamma \end{bmatrix}, \; D(r) = \sigma^2 H\Sigma H', \tag{5}$$

$$cov(r, y) = \sigma^2 H\Sigma, \; cov\big(r, Z^{\perp}y\big) = \sigma^2 H\Sigma Z^{\perp}, \tag{6}$$

where $\hat{F} = \big[\, F \; 0 \,\big]$.

In this study, the linear model $\mathcal{M}$ and the related over-parameterized linear model $\mathcal{N}$ with its correctly-reduced model $\mathcal{R}$ are considered. Obviously, different inference results are obtained from the models $\mathcal{M}$, $\mathcal{N}$, and $\mathcal{R}$. Models $\mathcal{M}$ and $\mathcal{N}$ produce the same observable random vector, however statistical inferences drawn for unknown parameters using these competing models might not always agree. Furthermore, it can be possible to predict or estimate all or some part of unknown vectors in $\mathcal{N}$ using the linear transformation $Z^{\perp}y$ of the observable random vector y in $\mathcal{R}$, even though the observable random vector in models $\mathcal{N}$ and $\mathcal{R}$ is different. Therefore, it is only reasonable to compare the inference outcomes from these three models and look for possible correlations. It would be wise to go more into the models to learn more about the connections and consequences of the statistical conclusions drawn from each one, given the apparent parallels and variations in the models' underlying structures and observable vectors.

One of the essential areas of inquiry in linear models is describing the relationships between predictors and estimators after a linear model has been formulated and unknown vectors inside the model have been predicted or estimated. In linear statistical models, unknown vectors can be estimated using a number of different predictors/estimators, each with their own unique set of characteristics. Although unbiasedness is an important quality in estimators and predictors, unbiased predictors and estimators do not have to be unique. Among all unbiased choices, finding the one with the least dispersion matrix is a well-established way to find the best predictor or estimator for unknown vectors. Both the best linear unbiased predictors (BLUPs) and the best linear unbiased estimators (BLUEs) are well-known in their fields; they are known for both unbiasedness and lowest dispersion matrix structure. For statistical applications and investigations involving linear models, their simplicity and optimum features make them a popular choice. Dispersion matrices of the BLUPs and the BLUEs serve as criteria for comparing optimality with other types of unbiased predictors and estimators. The comparison of dispersion matrices of predictors and estimators using the method of matrix ranks and inertias is a well-established approach, with references including [6, 15, 41, 42, 44], among others.

Considering the models $\mathcal{M}$, $\mathcal{N}$, and $\mathcal{R}$ together is eloquent for obtaining results both separately and simultaneously, facilitating the dual objectives of estimating α and predicting ε. In the present study, we derive some further results in the context of over-parameterized linear models by considering the following inference problems under the assumptions in (1)–(3). We establish computational formulas for the BLUPs of r under $\mathcal{M}$, $\mathcal{N}$, and $\mathcal{R}$, respectively, using rank/inertia method and give some of their properties. Especially, the fundamental equations of the BLUPs and the BLUEs are derived by using a constrained quadratic matrix-valued function optimization problem. The primary objective of our study is to explore the relationships among the dispersion matrices of the BLUPs of r under three competing models $\mathcal{M}$, $\mathcal{N}$, and $\mathcal{R}$, given as follows:

$$D[r - BLUP_{\mathcal{N}}(r)] = (\preccurlyeq, \succ, \prec, \succcurlyeq)D[r - BLUP_{\mathcal{M}}(r)],$$
$$D[r - BLUP_{\mathcal{M}}(r)] = (\preccurlyeq, \succ, \prec, \succcurlyeq)D[r - BLUP_{\mathcal{R}}(r)],$$
$$D[r - BLUP_{\mathcal{R}}(r)] = (\preccurlyeq, \succ, \prec, \succcurlyeq)D[r - BLUP_{\mathcal{N}}(r)],$$

by using rank/inertia method together with elementary block matrix operations. The use of the rank/inertia method, coupled with elementary block matrix operations, provides a powerful approach to simplify complex matrix expressions involving the Moore–Penrose generalized inverses of matrices. These operations help streamline and manipulate matrices effectively, contributing to a more manageable and computationally efficient analysis. The results are also derived for special cases of r.

There are many related studies to the topic considered in this paper in the statistical literature. In various sources, an over-parameterized model is alternatively referred to as a richly-parameterized model, an over-fitted model, or a mis-specified model. The statistical inference of this model has garnered considerable attention, and the paper contributes to this body of work. Recent studies have delved into comparison problems concerning BLUEs under two competing statistical models. These investigations incorporate certain restrictions on unknown parameters, assuming one model is true, and the other is mis-specified with the inclusion of superfluous variables, see [24]. Some notable contributions include necessary and sufficient conditions for the equality of BLUEs in competing constrained linear models, with the addition or deletion of unknown parameters, see [26]. Researchers have also explored conditions under which the BLUPs of joint unknown vectors are equivalent under a real model and its over-parameterized counterpart, see [10]. These conditions provide valuable insights into the relationships between predictors and estimators in different model specifications. Additionally, there are studies considering the equality of BLUEs under an original linear model and its related mis-specified model, particularly in partitioned linear models, see [16, 17, 25]. The present study stands out by offering a theoretically comprehensive exploration of the comparison of dispersion matrices of BLUPs for unknown vectors in the context of an over-parameterized linear model and its related models. It builds upon and extends prior work, particularly studies involving the addition of new regressors in linear models. Theoretical contributions in this area can be found in both previous and recent literature, such as [2, 3, 8, 9, 18–23, 29, 33, 34, 36, 37, 46, 49]. These works collectively contribute to advancing our understanding of statistical inferences in linear models with the inclusion of new regressors.

2 Preliminaries

The utilization of known rank and inertia formulas is a valuable strategy for simplifying matrix expressions, including those involving the Moore–Penrose generalized inverses of matrices, when establishing results. Matrix algebra formulas play a crucial role in simplifying complex matrix equations that arise during the derivation of statistical inferences for predictions and estimations.

The assortment of rank and inertia formulas for block matrices, as delineated in the three lemmas, referenced in [47] and employed in the demonstrations of the principal findings in the study, serves as a testament to the importance of these tools. These formulas, which are frequently derived from known principles in matrix algebra, offer a methodical and effective way of dealing with complex mathematical expressions that arise in statistical research.

Lemma 2.1 *Let* $\Upsilon,\ \Phi \in R^{m,n}$*, or, let* $\Upsilon = \Upsilon',\ \Phi = \Phi' \in R^{m,m}$*. Then,*

(a) $\Upsilon = \Phi \Leftrightarrow r(\Upsilon - \Phi) = 0.$
(b) $\Upsilon \succ \Phi \Leftrightarrow i_{+}(Y - \Phi) = m$ and $\Upsilon \prec \Phi \Leftrightarrow i_{-}(\Upsilon - \Phi) = m.$
(c) $\Upsilon \succcurlyeq \Phi \Leftrightarrow i_{-}(\Upsilon - \Phi) = 0$ and $\Upsilon \preccurlyeq \Phi \Leftrightarrow i_{+}(\Upsilon - \Phi) = 0.$

Lemma 2.2 *Let* $\Upsilon = \Upsilon' \in R^{m,m},\ \Phi = \Phi' \in R^{n,n},\ \Psi \in R^{m,n}$*, and* $v \in R$*. Then,*

(a) $r(\Upsilon) = i_{+}(\Upsilon) + i_{-}(\Upsilon)$.
(b) $i_{\pm}(v\Upsilon) = \begin{cases} i_{\pm}(\Upsilon) \ \textit{if} \ v > 0 \\ i_{\mp}(\Upsilon) \ \textit{if} \ v < 0 \end{cases}.$
(c) $i_{\pm}\begin{bmatrix} \Upsilon & \Psi \\ \Psi' & \Phi \end{bmatrix} = i_{\pm}\begin{bmatrix} \Upsilon & -\Psi \\ -\Psi' & \Phi \end{bmatrix} = i_{\mp}\begin{bmatrix} -\Upsilon & \Psi \\ \Psi' & -\Phi \end{bmatrix}.$
(d) $i_{\pm}\begin{bmatrix} \Upsilon & 0 \\ 0 & \Phi \end{bmatrix} = i_{\pm}(\Upsilon) + i_{\pm}(\Phi),\ i_{+}\begin{bmatrix} 0 & \Psi \\ \Psi' & 0 \end{bmatrix} = i_{-}\begin{bmatrix} 0 & \Psi \\ \Psi' & 0 \end{bmatrix} = r(\Psi)$.

Lemma 2.3 *Let* $\Upsilon = \Upsilon' \in R^{m,m},\ \Lambda = \Lambda' \in R^{n,n}$*, and* $\Phi \in R^{m,n}$*. Then,*

$$i_{\pm}\begin{bmatrix} \Upsilon & \Phi \\ \Phi' & 0 \end{bmatrix} = r(\Phi) + i_{\pm}\big(\Phi^{\perp}\Upsilon\Phi^{\perp}\big). \tag{7}$$

$$i_{+}\begin{bmatrix} \Upsilon & \Phi \\ \Phi' & 0 \end{bmatrix} = r\begin{bmatrix} \Upsilon & \Phi \end{bmatrix} \text{ and } i_{-}\begin{bmatrix} \Upsilon & \Phi \\ \Phi' & 0 \end{bmatrix} = r(\Phi) \text{ if } \Upsilon \succcurlyeq 0. \tag{8}$$

$$i_{\pm}\begin{bmatrix} \Upsilon & \Phi \\ \Phi' & \Lambda \end{bmatrix} = i_{\pm}(\Upsilon) + i_{\pm}\big(\Lambda - \Phi'\Upsilon^{+}\Phi\big), \text{ if } \mathcal{C}(\Phi) \subseteq \mathcal{C}(\Upsilon). \tag{9}$$

The subsequent renowned outcome was provided by [31].

Lemma 2.4 *Let* $\Upsilon X = \Lambda$ *be a system of linear equations. This system is consistent* $\Leftrightarrow r\begin{bmatrix} \Upsilon & \Lambda \end{bmatrix} = r(\Upsilon)$*, or equivalently,* $\Upsilon\Upsilon^{+}\Lambda = \Lambda$*.* $X = \Upsilon^{+}\Lambda + \big(I - \Upsilon^{+}\Upsilon\big)V$ *is the representation for the general solution of the system* $\Upsilon X = \Lambda$*, where* V *is any arbitrary matrix.*

3 BLUPs' Computations Under $\mathcal{M}, \mathcal{N}$, and $\mathcal{R}$

This section provides an overview of the ideas and definitions related to the consistency of models, estimability, predictability, and the BLUPs of the general vector of unknown vectors in models $\mathcal{M}$, $\mathcal{N}$, and $\mathcal{R}$. In addition, we obtain basic equations for the BLUPs through solving constrained optimization problems involving quadratic matrix-valued functions. Furthermore, we provide an overview of some of the attributes associated with the BLUPs.

We make statistical inferences of the models $\mathcal{M}, \mathcal{N}$, and $\mathcal{R}$ under the assumptions that these three models are consistent. We outline the consistency issues with the models under consideration as follows; see, e.g., [35]. The model $\mathcal{M}$ is said to be consistent $\Leftrightarrow$

$$y \in \mathcal{C}\begin{bmatrix} X & \Sigma \end{bmatrix} \text{ holds with probability (wp) 1.}$$

Let $G = \begin{bmatrix} X & Z \end{bmatrix}$. In this instance, $y \in \mathcal{C}\begin{bmatrix} G & \Sigma \end{bmatrix}$ holds wp 1, i.e., $\mathcal{N}$ is consistent under the assumption of consistency of $\mathcal{M}$. Further, we point out that $\mathcal{R}$ is consistent under the assumption of consistency of $\mathcal{N}$, i.e., $Z^{\perp}y \in \mathcal{C}\begin{bmatrix} Z^{\perp}X & Z^{\perp}\Sigma Z^{\perp} \end{bmatrix}$ holds wp 1.

Let r be as given in (4). r is said to be predictable under $\mathcal{N}$ if there exists a matrix $L \in R^{s,n}$ such that $E(Ly - r) = 0$. The predictability requirements for all models are compiled in the following definition, see, e.g., [1, 10, 28].

Definition 3.1 The predictability and estimability requirements of r and its special cases in $\mathcal{M}, \mathcal{N}$, and $\mathcal{R}$ are given as follows. Let $G = \begin{bmatrix} X & Z \end{bmatrix}$ and $\hat{F} = \begin{bmatrix} F & 0 \end{bmatrix}$.

(a) r is said to be predictable in $\mathcal{M} \Leftrightarrow \mathcal{C}(F') \subseteq \mathcal{C}(X') \Leftrightarrow r\begin{bmatrix} X \\ F \end{bmatrix} = r(X) \Leftrightarrow F\alpha$ is also estimable in $\mathcal{M}$.

(b) $X\alpha + \varepsilon$ is always predictable in $\mathcal{M}$ and $X\alpha$ is always estimable in $\mathcal{M}$.

(c) α is said to be estimable in $\mathcal{M} \Leftrightarrow r(X) = k$.

(d) r is said to be predictable in $\mathcal{N} \Leftrightarrow \mathcal{C}(\hat{F}') \subseteq \mathcal{C}(G') \Leftrightarrow r\begin{bmatrix} G \\ \hat{F} \end{bmatrix} = r(G) \Leftrightarrow \mathcal{C}(F') \subseteq \mathcal{C}\left((Z^{\perp}X)'\right) = \mathcal{C}(X'Z^{\perp}) \Leftrightarrow F\alpha$ is also estimable in $\mathcal{N}$.

(e) $X\alpha+\varepsilon$ is said to be predictable in $\mathcal{N} \Leftrightarrow \mathrm{r}(G) = r(X)+r(Z) \Leftrightarrow \mathcal{C}(X) \cap \mathcal{C}(Z) = \{0\} \Leftrightarrow X\alpha$ is also estimable in $\mathcal{N}$.

(f) α is said to be estimable in $\mathcal{N} \Leftrightarrow r(Z^{\perp}X) = k$.

(g) r is said to be predictable in $\mathcal{R} \Leftrightarrow \mathcal{C}(F') \subseteq \mathcal{C}(X'Z^{\perp}) \Leftrightarrow r\begin{bmatrix} Z^{\perp}X \\ F \end{bmatrix} = r(Z^{\perp}X) \Leftrightarrow F\alpha$ is also estimable in $\mathcal{R}$.

(h) $X\alpha + \varepsilon$ is said to be predictable in $\mathcal{R} \Leftrightarrow \mathcal{C}(X') \subseteq \mathcal{C}(X'Z^{\perp}) \Leftrightarrow r(G) = r(X) + r(Z) \Leftrightarrow r(X) = r(Z^{\perp}X) \Leftrightarrow X\alpha$ is also estimable in $\mathcal{R}$.

(i) α is said to be estimable in $\mathcal{R} \Leftrightarrow r(Z^{\perp}\mathrm{X}) = k$.

It should be added that if r is predictable under $\mathcal{R}$ then it is also predictable under both $\mathcal{N}$ and $\mathcal{M}$. Further, if r is predictable under $\mathcal{N}$ then it is predictable under $\mathcal{M}$.

Suppose that r is predictable in $\mathcal{N}$. In this instance, there exists L, which may not be unique, that satisfies the equation $E(Ly - r) = 0$. Therefore, it is logical to employ an optimal criterion to determine the most suitable option from the set of $Ly - r$ that satisfies $E(Ly - r) = 0$, in other words, to select an element from the set of all unbiased predictors of r according to an ideal criterion. A commonly used approach in this instance is to minimize the dispersion matrix of $Ly - r$ in the LPO. The case in question is explicated in the subsequent definition; refer to [11, 32] for more information.

Definition 3.2 Suppose that r is predictable in $\mathcal{N}$. If there exists Ly such that

$$D(Ly - r) = min \text{ s.t. } E(Ly - r) = 0 \tag{10}$$

holds in the LPO, Ly is defined to be the BLUP of r under $\mathcal{N}$ and is denoted by

$$Ly = BLUP_{\mathcal{N}}(r) = BLUP_{\mathcal{N}}(F\alpha + H\varepsilon).$$

Moreover, Ly corresponds to the BLUE of $F\alpha$, denoted by $BLUE_{\mathcal{N}}(F\alpha)$, by writing $H = 0$ in r. Further, if $F = 0$ and $H = I_n$ in r, Ly corresponds to the BLUP of ε, i.e., $BLUP_{\mathcal{N}}(\varepsilon)$, under $\mathcal{N}$.

For the models $\mathcal{M}$ and $\mathcal{R}$, this definition can also be stated in a similar manner. In light of this definition, the statement in (10) is expressed as follows for models $\mathcal{M}$ and $\mathcal{R}$, respectively. There exists a matrix T such that

$$Ty = BLUP_{\mathcal{M}}(r) \Leftrightarrow D(Ty - r) = min \text{ s.t. } E(Ty - r) = 0$$

and there exists a matrix K such that

$$KZ^{\perp}y = BLUP_{\mathcal{R}}(r) \Leftrightarrow D\big(KZ^{\perp}y - r\big) = min \text{ s.t. } E\big(KZ^{\perp}y - r\big) = 0.$$

The following theorem provides deductions on the existences and expressions of the BLUP under $\mathcal{N}$, see, [14], which was also proved in some other ways, e.g., [7, 32, 35, 40].

Theorem 3.1 *Let r be predictable under $\mathcal{N}$. Suppose that Ly and Sy are two unbiased linear predictors for r under $\mathcal{N}$. Then*

$$\max_{E(Sy-r)=0} i_{+}(D(Ly - r) - D(Sy - r)) = r\begin{pmatrix} \begin{bmatrix} L & -I_s \end{bmatrix}\begin{bmatrix} I_n \\ H \end{bmatrix}\Sigma\begin{bmatrix} I_n \\ H \end{bmatrix}' \\ \begin{bmatrix} G \\ \hat{F} \end{bmatrix}' \end{pmatrix} - r\begin{bmatrix} G \\ \hat{F} \end{bmatrix}$$

$$= r\left(\begin{bmatrix} L & -I_s \end{bmatrix}\begin{bmatrix} I_n \\ H \end{bmatrix}\Sigma\begin{bmatrix} I_n \\ H \end{bmatrix}'\begin{bmatrix} G \\ \hat{F} \end{bmatrix}^{\perp}\right) \tag{11}$$

is the maximal positive inertia of $D(Ly - r) - D(Sy - r)$ s.t. $SG = \hat{F}$, where $G = \begin{bmatrix} X & Z \end{bmatrix}$ and $\hat{F} = \begin{bmatrix} F & 0 \end{bmatrix}$. Hence,

$$D(Ly - r) = min \Leftrightarrow Ly = BLUP_{\mathcal{N}}(r) \Leftrightarrow L\begin{bmatrix} G & \Sigma G^{\perp} \end{bmatrix} = \begin{bmatrix} \hat{F} & H\Sigma G^{\perp} \end{bmatrix}. \quad (12)$$

The equation in (12) is consistent and

$$BLUP_{\mathcal{N}}(r) = Ly = (\begin{bmatrix} \hat{F} & H\Sigma G^{\perp} \end{bmatrix} W_n^{+} + U_n W_n^{\perp})y, \quad (13)$$

is the expression corresponding to the general solution of the equation expressed in parametric form, where $U_n \in R^{s,n}$ is an arbitrary matrix, and $W_n = \begin{bmatrix} G\,\Sigma G^{\perp} \end{bmatrix}$. Especially,

(i) *L is unique $\Leftrightarrow r(W_n) = n$.*
(ii) *$BLUP_{\mathcal{N}}(r)$ is unique $\Leftrightarrow \mathcal{N}$ is consistent.*
(iii) *W_n satisfies*

$$r\begin{bmatrix} G & \Sigma G^{\perp} \end{bmatrix} = r\begin{bmatrix} G & \Sigma \end{bmatrix} = r\begin{bmatrix} G & G^{\perp}\Sigma \end{bmatrix},$$

$$\mathcal{C}\begin{bmatrix} G & \Sigma G^{\perp} \end{bmatrix} = \mathcal{C}\begin{bmatrix} G & \Sigma \end{bmatrix} = \mathcal{C}\begin{bmatrix} G & G^{\perp}\Sigma \end{bmatrix}.$$

(iv) *The following dispersion and covariance matrices equalities hold.*

$$D[BLUP_{\mathcal{N}}(r)] = \sigma^2\begin{bmatrix} \hat{F} & H\Sigma G^{\perp} \end{bmatrix} W_n^{+}\Sigma(\begin{bmatrix} \hat{F} & H\Sigma G^{\perp} \end{bmatrix} W_n^{+})', \quad (14)$$

$$cov\{BLUP_{\mathcal{N}}(r), r\} = \sigma^2\begin{bmatrix} \hat{F} & H\Sigma G^{\perp} \end{bmatrix} W_n^{+}\Sigma H', \quad (15)$$

$$D[r - BLUP_{\mathcal{N}}(r)] = \sigma^2(\begin{bmatrix} \hat{F} & H\Sigma G^{\perp} \end{bmatrix} W_n^{+} - H)\Sigma(\begin{bmatrix} \hat{F} & H\Sigma G^{\perp} \end{bmatrix} W_n^{+} - H)'. \quad (16)$$

Proof Suppose that Ly and Sy are two unbiased linear predictors for r in $\mathcal{N}$. Then,

$$E(Ly - r) = 0 \Leftrightarrow LG = \hat{F}, \text{ namely, } \begin{bmatrix} L & -I_s \end{bmatrix}\begin{bmatrix} G \\ \hat{F} \end{bmatrix} = 0, \quad (17)$$

$$D(Ly - r) = \sigma^2(L - H)\Sigma(L - H)' = \sigma^2\begin{bmatrix} L & -I_s \end{bmatrix}\begin{bmatrix} I_n \\ H \end{bmatrix}\Sigma\begin{bmatrix} I_n \\ H \end{bmatrix}'\begin{bmatrix} L & -I_s \end{bmatrix}'. \quad (18)$$

By setting S in place of L, the expressions that are similar to those in (17) and (18) can be written for Sy. Definition 3.2 describes the matrix minimization problem for determining the BLUP under $\mathcal{N}$. In light of this, this problem can be stated as finding solution L of the linear matrix equation $LG = \hat{F}$ that is consistent such that

$$\begin{bmatrix} L & -I_s \end{bmatrix}\begin{bmatrix} I_n \\ H \end{bmatrix}\Sigma\begin{bmatrix} I_n \\ H \end{bmatrix}'\begin{bmatrix} L & -I_s \end{bmatrix}' \preccurlyeq \begin{bmatrix} S & -I_s \end{bmatrix}\begin{bmatrix} I_n \\ H \end{bmatrix}\Sigma\begin{bmatrix} I_n \\ H \end{bmatrix}'\begin{bmatrix} S & -I_s \end{bmatrix}' \quad \text{s.t. } SG = \hat{F}. \tag{19}$$

(19) is standard constrained quadratic matrix-valued function optimization problems in the LPO; for details see, [14, Lemma 3.1, 25, Lemmas 2.1–2.2, 38]. From (17)–(19), we obtain (11). With (11), L satisfies both

$$\begin{bmatrix} L & -I_s \end{bmatrix}\begin{bmatrix} G \\ \hat{F} \end{bmatrix} = 0 \text{ and } \begin{bmatrix} L & -I_s \end{bmatrix}\begin{bmatrix} I_n \\ H \end{bmatrix}\Sigma\begin{bmatrix} I_n \\ H \end{bmatrix}'\begin{bmatrix} G \\ \hat{F} \end{bmatrix}^{\perp} = 0. \tag{20}$$

Combining two equations in (20) gives us (12). It is seen easily from Lemma 2.4 that the general solution of the consistent matrix equation in (12) is given as in (13). (i) and (ii) follow from (13). For (iii), see [39], Lemma 2.1 (a)]. (14) is seen from (13). (15) is seen from (5) and (13). (16) is obtained from (14) and (15). □

According to Theorem 3.1, we obtain the following well-known conclusion easily by setting $Z = 0$; see, e.g., [46].

Corollary 3.1 *Suppose that r is predictable under $\mathcal{M}$. Then*

$$D(Ty - r) = min \Leftrightarrow Ty = BLUP_{\mathcal{M}}(r) \Leftrightarrow T\begin{bmatrix} X & \Sigma X^{\perp} \end{bmatrix} = \begin{bmatrix} F & H\Sigma X^{\perp} \end{bmatrix}.$$

This equation is consistent and

$$BLUP_{\mathcal{M}}(r) = Ty = \left(\begin{bmatrix} F & H\Sigma X^{\perp} \end{bmatrix} W^{+} + UW^{\perp}\right)y, \tag{21}$$

is the expression corresponding to the general solution of the equation expressed in parametric form, where $U \in R^{s,n}$ is an arbitrary matrix and $W = \begin{bmatrix} X & \Sigma X^{\perp} \end{bmatrix}$. Especially,

(i) *T is unique $\Leftrightarrow$ $r(W) = n$.*
(ii) *$BLUP_{\mathcal{M}}(r)$ is unique $\Leftrightarrow$ $\mathcal{M}$ is consistent.*
(iii) *The matrix W satisfies*

$$r\begin{bmatrix} X & \Sigma X^{\perp} \end{bmatrix} = r\begin{bmatrix} X & \Sigma \end{bmatrix} = r\begin{bmatrix} X & X^{\perp}\Sigma \end{bmatrix},$$

$$\mathcal{C}\begin{bmatrix} X & \Sigma X^{\perp} \end{bmatrix} = \mathcal{C}\begin{bmatrix} X & \Sigma \end{bmatrix} = \mathcal{C}\begin{bmatrix} X & X^{\perp}\Sigma \end{bmatrix}.$$

(iv) *The following dispersion and covariance matrices equalities hold.*

$$D[BLUP_{\mathcal{M}}(r)] = \sigma^2\begin{bmatrix} F & H\Sigma X^{\perp} \end{bmatrix} W^{+}\Sigma\left(\begin{bmatrix} F & H\Sigma X^{\perp} \end{bmatrix} W^{+}\right)', \tag{22}$$

$$cov\{BLUP_{\mathcal{M}}(r), r\} = \sigma^2\begin{bmatrix} F & H\Sigma X^{\perp} \end{bmatrix} W^{+}\Sigma H', \tag{23}$$

$$D[r - BLUP_{\mathcal{M}}(r)] = \sigma^2([F\ H\Sigma X^{\perp}]W^{+} - H)\Sigma([F\ H\Sigma X^{\perp}]W^{+} - H)'. \tag{24}$$

The following conclusion can be derived from the above results as can be obtained from Lemma 3.1 in [14] *by taking* $Z^{\perp}$ *as transformation matrix and by setting appropriate modifications in this lemma.*

Corollary 3.2 *Suppose that r is predictable under* $\mathcal{R}$. *Then*

$$D(KZ^{\perp}y - r) = min \Leftrightarrow KZ^{\perp}y = BLUP_{\mathcal{R}}(r)$$
$$\Leftrightarrow K\Big[Z^{\perp}X\ Z^{\perp}\Sigma Z^{\perp}(Z^{\perp}X)^{\perp}\Big] = \Big[F\ H\Sigma Z^{\perp}(Z^{\perp}X)^{\perp}\Big].$$

This equation is consistent and

$$BLUP_{\mathcal{R}}(r) = KZ^{\perp}y = \Big(\Big[F\ H\Sigma Z^{\perp}(Z^{\perp}X)^{\perp}\Big]W_r^{+} + U_r W_r^{\perp}\Big)Z^{\perp}y, \tag{25}$$

is the expression corresponding to the general solution of the equation expressed in parametric form, where $U_r \in R^{s,n}$ *is an arbitrary matrix and* $W_r = \Big[Z^{\perp}X\ Z^{\perp}\Sigma Z^{\perp}(Z^{\perp}X)^{\perp}\Big]$. *In particular,*

(i) *K is unique* $\Leftrightarrow r(W_r) = n$.
(ii) $BLUP_{\mathcal{R}}(r)$ *is unique* $\Leftrightarrow$ $\mathcal{R}$ *is consistent.*
(iii) *The matrix* W_r *satisfies*

$$r\Big[Z^{\perp}X\ Z^{\perp}\Sigma Z^{\perp}(Z^{\perp}X)^{\perp}\Big] = r[Z^{\perp}X\ Z^{\perp}\Sigma Z^{\perp}] = r\Big[Z^{\perp}X\ (Z^{\perp}X)^{\perp}Z^{\perp}\Sigma Z^{\perp}\Big],$$

$$\mathcal{C}\Big[Z^{\perp}X\ Z^{\perp}\Sigma Z^{\perp}(Z^{\perp}X)^{\perp}\Big] = \mathcal{C}[Z^{\perp}X\ Z^{\perp}\Sigma Z^{\perp}] = \mathcal{C}\Big[Z^{\perp}X\ (Z^{\perp}X)^{\perp}Z^{\perp}\Sigma Z^{\perp}\Big].$$

(iv) *The following dispersion and covariance matrices equalities hold.*

$$D[BLUP_{\mathcal{R}}(r)] = \sigma^2\Big[F\ H\Sigma Z^{\perp}(Z^{\perp}X)^{\perp}\Big]W_r^{+}Z^{\perp}\Sigma Z^{\perp}$$
$$\Big(\Big[F\ H\Sigma Z^{\perp}(Z^{\perp}X)^{\perp}\Big]W_r^{+}\Big)', \tag{26}$$

$$cov\{BLUP_{\mathcal{R}}(r), r\} = \sigma^2\Big[F\ H\Sigma Z^{\perp}(Z^{\perp}X)^{\perp}\Big]W_r^{+}Z^{\perp}\Sigma H', \tag{27}$$

$$D[r - BLUP_{\mathcal{R}}(r)] = \sigma^2\Big(\Big[F\ H\Sigma Z^{\perp}(Z^{\perp}X)^{\perp}\Big]W_r^{+}Z^{\perp} - H\Big)\Sigma$$
$$\times \Big(\Big[F\ H\Sigma Z^{\perp}(Z^{\perp}X)^{\perp}\Big]W_r^{+}Z^{\perp} - H\Big)'. \tag{28}$$

4 Dispersion Matrices of BLUPs: Equalities and Inequalities

This section displays the results of comparing the predictors' dispersion matrices under models $\mathcal{M}, \mathcal{N}$, and $\mathcal{R}$. In certain circumstances, the findings are demonstrated by the use of block matrix rank and inertia equations.

Theorem 4.1 *Suppose that r is predictable under $\mathcal{N}$ (also predictable under $\mathcal{M}$). Denote*

$$P = \begin{bmatrix} \Sigma & \Sigma & \Sigma H' & 0 & 0 & X \\ \Sigma & 0 & 0 & X & Z & 0 \\ H\Sigma & 0 & 0 & F & 0 & 0 \\ 0 & X' & F' & 0 & 0 & 0 \\ 0 & Z' & 0 & 0 & 0 & 0 \\ X' & 0 & 0 & 0 & 0 & 0 \end{bmatrix}.$$

Then,

$$i_+(D[r - BLUP_{\mathcal{M}}(r)] - D[r - BLUP_{\mathcal{N}}(r)]) = i_+(P) - r\begin{bmatrix} X & \Sigma \end{bmatrix} - r(G), \quad (29)$$

$$i_-(D[r - BLUP_{\mathcal{M}}(r)] - D[r - BLUP_{\mathcal{N}}(r)]) = i_-(P) - r\begin{bmatrix} G & \Sigma \end{bmatrix} - r(X), \quad (30)$$

$$r(D[r - BLUP_{\mathcal{M}}(r)] - D[r - BLUP_{\mathcal{N}}(r)]) = r(P) - r\begin{bmatrix} X & \Sigma \end{bmatrix} - r(G) - r\begin{bmatrix} G & \Sigma \end{bmatrix} - r(X). \quad (31)$$

Here $G = \begin{bmatrix} X & Z \end{bmatrix}$. Then,

(a) $D[r - BLUP_{\mathcal{N}}(r)] \succ D[r - BLUP_{\mathcal{M}}(r)] \Leftrightarrow i_-(P) = r\begin{bmatrix} G & \Sigma \end{bmatrix} + r(X) + s.$
(b) $D[r - BLUP_{\mathcal{N}}(r)] \prec D[r - BLUP_{\mathcal{M}}(r)] \Leftrightarrow i_+(P) = r\begin{bmatrix} X & \Sigma \end{bmatrix} + r(G) + s.$
(c) $D[r - BLUP_{\mathcal{N}}(r)] \succcurlyeq D[r - BLUP_{\mathcal{M}}(r)] \Leftrightarrow i_+(P) = r\begin{bmatrix} X & \Sigma \end{bmatrix} + r(G)\,.$
(d) $D[r - BLUP_{\mathcal{N}}(r)] \preccurlyeq D[r - BLUP_{\mathcal{M}}(r)] \Leftrightarrow i_-(P) = r\begin{bmatrix} G & \Sigma \end{bmatrix} + r(X)\,.$
(e) $D[r - BLUP_{\mathcal{N}}(r)] = D[r - BLUP_{\mathcal{M}}(r)] \Leftrightarrow r(P) = r\begin{bmatrix} X & \Sigma \end{bmatrix} + r(G) + \begin{bmatrix} G & \Sigma \end{bmatrix} r + r(X).$

Proof Let $\Lambda = D[r - BLUP_{\mathcal{M}}(r)]$. By using (9) and (17),

$$\begin{aligned} & i_{\pm}(D[r - BLUP_{\mathcal{M}}(r)] - D[r - BLUP_{\mathcal{N}}(r)]) \\ &= \mathrm{i}_{\pm}\Big(\Lambda - ([\hat{F}\ H\Sigma G^{\perp}]W_n^{+} - H)\Sigma\Sigma^{+}\Sigma([\hat{F}\ H\Sigma G^{\perp}]W_n^{+} - H)'\Big) \\ &= \mathrm{i}_{\pm}\begin{bmatrix} \Sigma & \Sigma([\hat{F}\ H\Sigma G^{\perp}]W_n^{+} - H)' \\ ([\hat{F}\ H\Sigma G^{\perp}]W_n^{+} - H)\Sigma & \Lambda \end{bmatrix} - \mathrm{i}_{\pm}(\Sigma) \end{aligned}$$

$$= \mathrm{i}_{\pm}\left(\begin{bmatrix} \Sigma & -\Sigma H' \\ -H\Sigma & \Lambda \end{bmatrix} + \begin{bmatrix} \Sigma & 0 \\ 0 & [\hat{F}\ H\Sigma G^{\perp}] \end{bmatrix}\begin{bmatrix} 0 & W_n \\ W_n' & 0 \end{bmatrix}^{+}\begin{bmatrix} \Sigma & 0 \\ 0 & [\hat{F}\ H\Sigma G^{\perp}]' \end{bmatrix}\right)$$
$$- \mathrm{i}_{\pm}(\Sigma), \tag{32}$$

is obtained, where $\hat{F} = [\,F\ 0\,]$ and $W_n = [\,G\ \Sigma G^{\perp}\,]$. We can apply (9) to (32) since

$$\mathcal{C}(\Sigma) \subseteq \mathcal{C}(W_n) \text{ and } \mathcal{C}([\,\hat{F}\ H\Sigma G^{\perp}\,]') \subseteq \mathcal{C}(W_n')$$

hold. Then (32) is equivalently expressed as follows:

$$\mathrm{i}_{\pm}\begin{bmatrix} 0 & -G & -\Sigma G^{\perp} & \Sigma & 0 \\ -G' & 0 & 0 & 0 & \hat{F}' \\ -G^{\perp}\Sigma & 0 & 0 & 0 & G^{\perp}\Sigma H' \\ \Sigma & 0 & 0 & \Sigma & -\Sigma H' \\ 0 & \hat{F} & H\Sigma G^{\perp} & -H\Sigma & \Lambda \end{bmatrix} - r[\,G\ \Sigma G^{\perp}\,] - \mathrm{i}_{\pm}(\Sigma)$$
$$= i_{\pm}\begin{bmatrix} -\Sigma & -G & -\Sigma G^{\perp} & \Sigma H' \\ -G' & 0 & 0 & \hat{F}' \\ -G^{\perp}\Sigma & 0 & 0 & G^{\perp}\Sigma H' \\ H\Sigma & \hat{F} & H\Sigma G^{\perp} & \Lambda - H\Sigma H' \end{bmatrix} - r[\,G\ \Sigma\,]$$
$$= \mathrm{i}_{\mp}\begin{bmatrix} \Sigma & G & \Sigma H' \\ G' & 0 & \hat{F}' \\ H\Sigma & \hat{F} & H\Sigma H' - \Lambda \end{bmatrix} + \mathrm{i}_{\pm}(G^{\perp}\Sigma G^{\perp}) - r[\,G\ \Sigma\,]$$
$$= i_{\mp}\left(\begin{bmatrix} \Sigma & \Sigma H' & G \\ H\Sigma & H\Sigma H' & \hat{F} \\ G' & \hat{F}' & 0 \end{bmatrix} - \begin{bmatrix} 0 \\ I_s \\ 0 \end{bmatrix}\Lambda[\,0\ I_s\ 0\,]\right) - r[\,\mathrm{G}\ \Sigma\,]$$
$$+ \mathrm{i}_{\pm}(G^{\perp}\Sigma G^{\perp}). \tag{33}$$

We can reapply (9) to (33) after setting $\Lambda = D[r - BLUP_{\mathcal{M}}(r)]$ in (24). Then in a similar way to obtaining (32), (33) is equivalently expressed as follows

$$\mathrm{i}_{\mp}\left(\begin{bmatrix} \Sigma & 0 & -\Sigma H' & 0 \\ 0 & \Sigma & \Sigma H' & G \\ -H\Sigma & H\Sigma & H\Sigma H' & \hat{F} \\ 0 & G' & \hat{F}' & 0 \end{bmatrix} + \begin{bmatrix} \Sigma & 0 \\ 0 & 0 \\ 0 & [F\ H\Sigma X^{\perp}] \\ 0 & 0 \end{bmatrix}\begin{bmatrix} 0 & W \\ W' & 0 \end{bmatrix}^{+}\right.$$
$$\left.\times\begin{bmatrix} \Sigma & 0 & 0 & 0 \\ 0 & 0 & [F\ H\Sigma X^{\perp}]' & 0 \end{bmatrix}\right) + \mathrm{i}_{\pm}(G^{\perp}\Sigma G^{\perp}) - r[\,G\ \Sigma\,] - \mathrm{i}_{\mp}(\Sigma). \tag{34}$$

(9) can be applied to (34) since

$$\mathcal{C}(\Sigma) \subseteq \mathcal{C}(W) \text{ and } \mathcal{C}\Big(\big[\, F \;\; H\Sigma X^{\perp} \,\big]'\Big) \subseteq \mathcal{C}(W'),$$

where $W = \big[\, X \;\; \Sigma X^{\perp} \,\big]$. From Lemmas 2.2 and 2.3, and some congruence operations, (34) is equivalently written as

$$\begin{aligned}
&\mathrm{i}_{\mp}\begin{bmatrix} 0 & -X & -\Sigma X^{\perp} & \Sigma & 0 & 0 & 0 \\ -X' & 0 & 0 & 0 & 0 & F' & 0 \\ -X^{\perp}\Sigma & 0 & 0 & 0 & 0 & X^{\perp}\Sigma H' & 0 \\ \Sigma & 0 & 0 & \Sigma & 0 & -\Sigma H' & 0 \\ 0 & 0 & 0 & 0 & \Sigma & \Sigma H' & G \\ 0 & F & H\Sigma X^{\perp} & -H\Sigma & H\Sigma & H\Sigma H' & \hat{F} \\ 0 & 0 & 0 & 0 & G' & \hat{F}' & 0 \end{bmatrix} + \mathrm{i}_{\pm}(G^{\perp}\Sigma G^{\perp}) - r\big[\, G \;\; \Sigma \,\big] \\
&\quad - \mathrm{i}_{\mp}(\Sigma) - r\big[\, X \;\; \Sigma X^{\perp} \,\big] \\
&= \mathrm{i}_{\mp}\begin{bmatrix} -\Sigma & -X & -\Sigma X^{\perp} & 0 & \Sigma H' & 0 \\ -X' & 0 & 0 & 0 & F' & 0 \\ -X^{\perp}\Sigma & 0 & 0 & 0 & X^{\perp}\Sigma H' & 0 \\ 0 & 0 & 0 & \Sigma & \Sigma H' & G \\ H\Sigma & F & H\Sigma X^{\perp} & H\Sigma & 0 & \hat{F} \\ 0 & 0 & 0 & G' & \hat{F}' & 0 \end{bmatrix} + \mathrm{i}_{\pm}(G^{\perp}\Sigma G^{\perp}) - r\big[\, G \;\; \Sigma \,\big] \\
&\quad - r\big[\, X \;\; \Sigma \,\big] \\
&= \mathrm{i}_{\mp}\begin{bmatrix} -\Sigma & -X & 0 & \Sigma H' & 0 \\ -X' & 0 & 0 & F' & 0 \\ 0 & 0 & \Sigma & \Sigma H' & G \\ H\Sigma & F & H\Sigma & 0 & \hat{F} \\ 0 & 0 & G' & \hat{F}' & 0 \end{bmatrix} + \mathrm{i}_{\pm}(G^{\perp}\Sigma G^{\perp}) - r\big[\, G \;\; \Sigma \,\big] - r\big[\, X \;\; \Sigma \,\big] \\
&\quad + \mathrm{i}_{\mp}\big(X^{\perp}\Sigma X^{\perp}\big) \\
&= \mathrm{i}_{\pm}\begin{bmatrix} \Sigma & \Sigma & \Sigma H' & 0 & X \\ \Sigma & 0 & 0 & G & 0 \\ H\Sigma & 0 & 0 & \hat{F} & 0 \\ 0 & G' & \hat{F}' & 0 & 0 \\ X' & 0 & 0 & 0 & 0 \end{bmatrix} + \mathrm{i}_{\pm}\big(G^{\perp}\Sigma G^{\perp}\big) - r\big[\, G \;\; \Sigma \,\big] - r\big[\, X \;\; \Sigma \,\big] \\
&\quad + \mathrm{i}_{\mp}\big(X^{\perp}\Sigma X^{\perp}\big). \qquad (35)
\end{aligned}$$

In consequence, by using (7) and (8), we obtain (29) and (30). According to Lemma 2.2 (a), adding the positive and negative inertia equalities in (29) and (30) yields (31). When Lemma 2.1 is applied to (29)–(31), (a)–(e) are obtained. □

The proofs of the following theorems are obtained by using similar way to proof of Theorem 4.1, and are omitted.

Theorem 4.2 *Suppose that r is predictable under $\mathcal{R}$ (also predictable under $\mathcal{M}$). Denote*

$$Q = \begin{bmatrix} \Sigma & \Sigma Z^{\perp} & \Sigma H' & 0 & X \\ Z^{\perp}\Sigma & 0 & 0 & Z^{\perp}X & 0 \\ H\Sigma & 0 & 0 & F & 0 \\ 0 & X'Z^{\perp} & F' & 0 & 0 \\ X' & 0 & 0 & 0 & 0 \end{bmatrix}. \tag{36}$$

Then,

$$i_{+}(D[r - BLUP_{\mathcal{M}}(r)] - D[r - BLUP_{\mathcal{R}}(r)]) = i_{+}(Q) - r\begin{bmatrix} X & \Sigma \end{bmatrix} - r\left(Z^{\perp}X\right),$$
$$i_{-}(D[r - BLUP_{\mathcal{M}}(r)] - D[r - BLUP_{\mathcal{R}}(r)]) = i_{-}(Q) - r\begin{bmatrix} Z^{\perp}X & Z^{\perp}\Sigma Z^{\perp} \end{bmatrix} - r(X),$$
$$r(D[r - BLUP_{\mathcal{M}}(r)] - D[r - BLUP_{\mathcal{R}}(r)]) = r(Q) - r\begin{bmatrix} X & \Sigma \end{bmatrix} - r\left(Z^{\perp}X\right) - r\begin{bmatrix} Z^{\perp}X & Z^{\perp}\Sigma Z^{\perp} \end{bmatrix} - r(X).$$

Further,

(a) $D[r - BLUP_{\mathcal{R}}(r)] \succ D[r - BLUP_{\mathcal{M}}(r)] \Leftrightarrow i_{-}(Q) = r\begin{bmatrix} Z^{\perp}X & Z^{\perp}\Sigma Z^{\perp} \end{bmatrix} + r(X) + s.$

(b) $D[r - BLUP_{\mathcal{R}}(r)] \prec D[r - BLUP_{\mathcal{M}}(r)] \Leftrightarrow i_{+}(Q) = r\begin{bmatrix} X & \Sigma \end{bmatrix} + r\left(Z^{\perp}X\right) + s.$

(c) $D[r - BLUP_{\mathcal{R}}(r)] \succcurlyeq D[r - BLUP_{\mathcal{M}}(r)] \Leftrightarrow i_{+}(Q) = r\begin{bmatrix} X & \Sigma \end{bmatrix} + r\left(Z^{\perp}X\right).$

(d) $D[r - BLUP_{\mathcal{R}}(r)] \preccurlyeq D[r - BLUP_{\mathcal{M}}(r)] \Leftrightarrow i_{-}(Q) = r\begin{bmatrix} Z^{\perp}X & Z^{\perp}\Sigma Z^{\perp} \end{bmatrix} + r(X).$

(e) $D[r - BLUP_{\mathcal{R}}(r)] = D[r - BLUP_{\mathcal{M}}(r)] \Leftrightarrow r(Q) = r\begin{bmatrix} X & \Sigma \end{bmatrix} + r\left(Z^{\perp}X\right) + r\begin{bmatrix} Z^{\perp}X & Z^{\perp}\Sigma Z^{\perp} \end{bmatrix} + r(X).$

Theorem 4.3 *Suppose that r is predictable under $\mathcal{R}$ (also predictable under $\mathcal{N}$). Consider the matrix Q in* (36) *and let $\hat{Z}' = \begin{bmatrix} Z' & 0 & 0 & 0 & 0 & 0 \end{bmatrix}$. Denote*

$$R = \begin{bmatrix} Q & \hat{Z} \\ \hat{Z}' & 0 \end{bmatrix}.$$

Then

$$i_{+}(D[r - BLUP_{\mathcal{N}}(r)] - D[r - BLUP_{\mathcal{R}}(r)]) = i_{+}(R) - r\begin{bmatrix} G & \Sigma \end{bmatrix} - r\left(Z^{\perp}X\right),$$

$$i_{-}(D[r - BLUP_{\mathcal{N}}(r)] - D[r - BLUP_{\mathcal{R}}(r)]) = i_{-}(R) - r\begin{bmatrix} Z^{\perp}X & Z^{\perp}\Sigma Z^{\perp} \end{bmatrix} - r(G),$$

$$r(D[r - BLUP_{\mathcal{N}}(r)] - D[r - BLUP_{\mathcal{R}}(r)]) = r(R) - r\begin{bmatrix} G & \Sigma \end{bmatrix} - r\left(Z^{\perp}X\right) - r\begin{bmatrix} Z^{\perp}X & Z^{\perp}\Sigma Z^{\perp} \end{bmatrix} - r(G).$$

Here $G = \begin{bmatrix} X & Z \end{bmatrix}$. *Further,*

(a) $D[r - BLUP_{\mathcal{R}}(r)] \succ D[r - BLUP_{\mathcal{N}}(r)] \Leftrightarrow i_-(R) = r\begin{bmatrix} Z^\perp X & Z^\perp \Sigma Z^\perp \end{bmatrix} + r(G) + s.$
(b) $D[r - BLUP_{\mathcal{R}}(r)] \prec D[r - BLUP_{\mathcal{N}}(r)] \Leftrightarrow i_+(R) = r\begin{bmatrix} G & \Sigma \end{bmatrix} + r(Z^\perp X) + s.$
(c) $D[r - BLUP_{\mathcal{R}}(r)] \succcurlyeq D[r - BLUP_{\mathcal{N}}(r)] \Leftrightarrow i_+(R) = r\begin{bmatrix} G & \Sigma \end{bmatrix} + r(Z^\perp X).$
(d) $D[r - BLUP_{\mathcal{R}}(r)] \preccurlyeq D[r - BLUP_{\mathcal{N}}(r)] \Leftrightarrow i_-(R) = r\begin{bmatrix} Z^\perp X & Z^\perp \Sigma Z^\perp \end{bmatrix} + r(G).$
(e) $D[r - BLUP_{\mathcal{R}}(r)] = D[r - BLUP_{\mathcal{N}}(r)] \Leftrightarrow r(R) = r\begin{bmatrix} Z^\perp X & Z^\perp \Sigma Z^\perp \end{bmatrix} + r(G) + r\begin{bmatrix} G & \Sigma \end{bmatrix} + r(Z^\perp X).$

Within the framework of the transformation approach to linear regression models, [13] *examined the relationships between the BLUPs of unknown vectors under two transformed linear random-effects models, and* [14] *established a number of equality and inequality conditions for comparing the covariance matrices of BLUPs of unknown vectors under an original linear mixed model that incorporates both fixed and random effects, and its transformed linear mixed model. Hence, we observe that particular forms of* [13, Theorem 4.2, 14, Theorem 4.1] *correspond to Theorems* 4.2 *and* 4.3, *respectively. As a direct result of varying the matrices F and H in r, as stated in Theorems* 4.1–4.3, *we get the following outcomes.*

Corollary 4.1 *Suppose that* $F\alpha$ *is estimable under* $\mathcal{R}$ *(also estimable under both* $\mathcal{M}$ *and* $\mathcal{N}$*). Let* $\hat{Z}' = \begin{bmatrix} Z' & 0 & 0 & 0 & 0 & 0 \end{bmatrix}$ *and* $G = \begin{bmatrix} X & Z \end{bmatrix}$. *Denote*

$$P = \begin{bmatrix} \Sigma & \Sigma & 0 & 0 & 0 & X \\ \Sigma & 0 & 0 & X & Z & 0 \\ 0 & 0 & 0 & F & 0 & 0 \\ 0 & X' & F' & 0 & 0 & 0 \\ 0 & Z' & 0 & 0 & 0 & 0 \\ X' & 0 & 0 & 0 & 0 & 0 \end{bmatrix}, Q = \begin{bmatrix} \Sigma & \Sigma Z^\perp & 0 & 0 & X \\ Z^\perp \Sigma & 0 & 0 & Z^\perp X & 0 \\ 0 & 0 & 0 & F & 0 \\ 0 & X' Z^\perp & F' & 0 & 0 \\ X' & 0 & 0 & 0 & 0 \end{bmatrix},$$
$$R = \begin{bmatrix} Q & \hat{Z} \\ \hat{Z}' & 0 \end{bmatrix}.$$

Then, the following results hold

(a) $D[BLUE_{\mathcal{N}}(F\alpha)] \succ D[BLUE_{\mathcal{M}}(F\alpha)] \Leftrightarrow i_-(P) = r\begin{bmatrix} G & \Sigma \end{bmatrix} + r(X) + s.$
(b) $D[BLUE_{\mathcal{N}}(F\alpha)] \prec D[BLUE_{\mathcal{M}}(F\alpha)] \Leftrightarrow i_+(P) = r\begin{bmatrix} X & \Sigma \end{bmatrix} + r(G) + s.$
(c) $D[BLUE_{\mathcal{N}}(F\alpha)] \succcurlyeq D[BLUE_{\mathcal{M}}(F\alpha)] \Leftrightarrow i_+(P) = r\begin{bmatrix} X & \Sigma \end{bmatrix} + r(G).$
(d) $D[BLUE_{\mathcal{N}}(F\alpha)] \preccurlyeq D[BLUE_{\mathcal{M}}(F\alpha)] \Leftrightarrow i_-(P) = r\begin{bmatrix} G & \Sigma \end{bmatrix} + r(X).$
(e) $D[BLUE_{\mathcal{N}}(F\alpha)] = D[BLUE_{\mathcal{M}}(F\alpha)] \Leftrightarrow r(P) = r\begin{bmatrix} X & \Sigma \end{bmatrix} + r(G) + r(X) + r\begin{bmatrix} G & \Sigma \end{bmatrix}.$
(f) $D[BLUE_{\mathcal{R}}(F\alpha)] \succ D[BLUE_{\mathcal{M}}(F\alpha)] \Leftrightarrow i_-(Q) = r\begin{bmatrix} Z^\perp X & Z^\perp \Sigma Z^\perp \end{bmatrix} + r(X) + s.$
(g) $D[BLUE_{\mathcal{R}}(F\alpha)] \prec D[BLUE_{\mathcal{M}}(F\alpha)] \Leftrightarrow i_+(Q) = r\begin{bmatrix} X & \Sigma \end{bmatrix} + r(Z^\perp X) + s.$

(h) $D[BLUE_{\mathcal{R}}(F\alpha)] \succcurlyeq D[BLUE_{\mathcal{M}}(F\alpha)] \Leftrightarrow i_+(Q) = r\begin{bmatrix} X & \Sigma \end{bmatrix} + r(Z^{\perp}X)$.

(i) $D[BLUE_{\mathcal{R}}(F\alpha)] \preccurlyeq D[BLUE_{\mathcal{M}}(F\alpha)] \Leftrightarrow i_-(Q) = r\begin{bmatrix} Z^{\perp}X & Z^{\perp}\Sigma Z^{\perp} \end{bmatrix} + r(X)$.

(j) $D[BLUE_{\mathcal{R}}(F\alpha)] = D[BLUE_{\mathcal{M}}(F\alpha)] \Leftrightarrow r(Q) = r\begin{bmatrix} X & \Sigma \end{bmatrix} + r(Z^{\perp}X) + r(X) + r\begin{bmatrix} Z^{\perp}X & Z^{\perp}\Sigma Z^{\perp} \end{bmatrix}$.

(k) $D[BLUE_{\mathcal{R}}(F\alpha)] \succ D[BLUE_{\mathcal{N}}(F\alpha)] \Leftrightarrow i_-(R) = r\begin{bmatrix} Z^{\perp}X & Z^{\perp}\Sigma Z^{\perp} \end{bmatrix} + r(G) + s$.

(l) $D[BLUE_{\mathcal{R}}(F\alpha)] \prec D[BLUE_{\mathcal{N}}(F\alpha)] \Leftrightarrow i_+(R) = r\begin{bmatrix} G & \Sigma \end{bmatrix} + r(Z^{\perp}X) + s$.

(m) $D[BLUE_{\mathcal{R}}(F\alpha)] \succcurlyeq D[BLUE_{\mathcal{N}}(F\alpha)] \Leftrightarrow i_+(R) = r\begin{bmatrix} G & \Sigma \end{bmatrix} + r(Z^{\perp}X)$.

(n) $D[BLUE_{\mathcal{R}}(F\alpha)] \preccurlyeq D[BLUE_{\mathcal{N}}(F\alpha)] \Leftrightarrow i_-(R) = r\begin{bmatrix} Z^{\perp}X & Z^{\perp}\Sigma Z^{\perp} \end{bmatrix} + r(G)$.

(o) $D[BLUE_{\mathcal{R}}(F\alpha)] = D[BLUE_{\mathcal{N}}(F\alpha)] \Leftrightarrow r(R) = r\begin{bmatrix} Z^{\perp}X & Z^{\perp}\Sigma Z^{\perp} \end{bmatrix} + r(G) + r\begin{bmatrix} G & \Sigma \end{bmatrix} + r(Z^{\perp}X)$.

Corollary 4.2 *Let* $\hat{Z}' = \begin{bmatrix} Z' & 0 & 0 & 0 & 0 & 0 \end{bmatrix}$ *and* $G = \begin{bmatrix} X & Z \end{bmatrix}$. *Denote*

$$P = \begin{bmatrix} \Sigma & \Sigma & 0 & X \\ \Sigma & 0 & Z & 0 \\ 0 & Z' & 0 & 0 \\ X' & 0 & 0 & 0 \end{bmatrix}, \quad Q = \begin{bmatrix} \Sigma & \Sigma Z^{\perp} & X \\ Z^{\perp}\Sigma & 0 & 0 \\ X' & 0 & 0 \end{bmatrix}, \quad R = \begin{bmatrix} Q & \hat{Z} \\ \hat{Z}' & 0 \end{bmatrix}.$$

(i) *Suppose that* $X\alpha$ *is estimable under* $\mathcal{R}$ *(also estimable under both* $\mathcal{M}$ *and* $\mathcal{N}$*), in this case* $r(G) = r(X) + r(Z)$*, i.e.,* $r(X) = r(Z^{\perp}X)$*. Then,*

  (a) $D[BLUE_{\mathcal{N}}(X\alpha)] \succ D[BLUE_{\mathcal{M}}(X\alpha)] \Leftrightarrow i_-(P) = r\begin{bmatrix} G & \Sigma \end{bmatrix} + n$.

  (b) $D[BLUE_{\mathcal{N}}(X\alpha)] \prec D[BLUE_{\mathcal{M}}(X\alpha)] \Leftrightarrow i_+(P) = r\begin{bmatrix} X & \Sigma \end{bmatrix} + r(Z) + n$.

  (c) $D[BLUE_{\mathcal{N}}(X\alpha)] \succcurlyeq D[BLUE_{\mathcal{M}}(X\alpha)] \Leftrightarrow i_+(P) = r\begin{bmatrix} X & \Sigma \end{bmatrix} + r(Z)$.

  (d) $D[BLUE_{\mathcal{N}}(X\alpha)] \preccurlyeq D[BLUE_{\mathcal{M}}(X\alpha)] \Leftrightarrow i_-(P) = r\begin{bmatrix} G & \Sigma \end{bmatrix}$.

  (e) $D[BLUE_{\mathcal{N}}(X\alpha)] = D[BLUE_{\mathcal{M}}(X\alpha)] \Leftrightarrow r(P) = r\begin{bmatrix} X & \Sigma \end{bmatrix} + r\begin{bmatrix} G & \Sigma \end{bmatrix} + r(Z)$.

  (f) $D[BLUE_{\mathcal{R}}(X\alpha)] \succ D[BLUE_{\mathcal{M}}(X\alpha)] \Leftrightarrow i_-(Q) = r\begin{bmatrix} Z^{\perp}X & Z^{\perp}\Sigma Z^{\perp} \end{bmatrix} + n$.

  (g) $D[BLUE_{\mathcal{R}}(X\alpha)] \prec D[BLUE_{\mathcal{M}}(X\alpha)] \Leftrightarrow i_+(Q) = r\begin{bmatrix} X & \Sigma \end{bmatrix} + n$.

  (h) $D[BLUE_{\mathcal{R}}(X\alpha)] \succcurlyeq D[BLUE_{\mathcal{M}}(X\alpha)] \Leftrightarrow i_+(Q) = r\begin{bmatrix} X & \Sigma \end{bmatrix}$.

  (i) $D[BLUE_{\mathcal{R}}(X\alpha)] \preccurlyeq D[BLUE_{\mathcal{M}}(X\alpha)] \Leftrightarrow i_-(Q) = r\begin{bmatrix} Z^{\perp}X & Z^{\perp}\Sigma Z^{\perp} \end{bmatrix}$.

  (j) $D[BLUE_{\mathcal{R}}(X\alpha)] = D[BLUE_{\mathcal{M}}(X\alpha)] \Leftrightarrow r(Q) = r\begin{bmatrix} X & \Sigma \end{bmatrix} + r\begin{bmatrix} Z^{\perp}X & Z^{\perp}\Sigma Z^{\perp} \end{bmatrix}$.

  (k) $D[BLUE_{\mathcal{R}}(X\alpha)] \succ D[BLUE_{\mathcal{N}}(X\alpha)] \Leftrightarrow i_-(R) = r\begin{bmatrix} Z^{\perp}X & Z^{\perp}\Sigma Z^{\perp} \end{bmatrix} + r(Z) + n$.

  (l) $D[BLUE_{\mathcal{R}}(X\alpha)] \prec D[BLUE_{\mathcal{N}}(X\alpha)] \Leftrightarrow i_+(R) = r\begin{bmatrix} G & \Sigma \end{bmatrix} + n$.

  (m) $D[BLUE_{\mathcal{R}}(X\alpha)] \succcurlyeq D[BLUE_{\mathcal{N}}(X\alpha)] \Leftrightarrow i_+(R) = r\begin{bmatrix} G & \Sigma \end{bmatrix}$.

  (n) $D[BLUE_{\mathcal{R}}(X\alpha)] \preccurlyeq D[BLUE_{\mathcal{N}}(X\alpha)] \Leftrightarrow i_-(R) = r\begin{bmatrix} Z^{\perp}X & Z^{\perp}\Sigma Z^{\perp} \end{bmatrix} + r(Z)$.

(o) $D[BLUE_{\mathcal{R}}(X\alpha)] = D[BLUE_{\mathcal{N}}(X\alpha)] \Leftrightarrow r(R) = r\big[Z^{\perp}X\ Z^{\perp}\Sigma Z^{\perp}\big] + r(Z) + r\big[G\ \Sigma\big]$.

(ii) *Suppose that* α *is estimable under* $\mathcal{R}$ *(also estimable under both* $\mathcal{M}$ *and* $\mathcal{N}$*), in this case* $r\big(Z^{\perp}X\big) = k$. *Then,*

(a) $D[BLUE_{\mathcal{N}}(\alpha)] \succ D[BLUE_{\mathcal{M}}(\alpha)] \Leftrightarrow i_{-}(P) = r\big[G\ \Sigma\big] + k$.
(b) $D[BLUE_{\mathcal{N}}(\alpha)] \prec D[BLUE_{\mathcal{M}}(\alpha)] \Leftrightarrow i_{+}(P) = r\big[X\ \Sigma\big] + r(Z) + k$.
(c) $D[BLUE_{\mathcal{N}}(\alpha)] \succcurlyeq D[BLUE_{\mathcal{M}}(\alpha)] \Leftrightarrow i_{+}(P) = r\big[X\ \Sigma\big] + r(Z)$.
(d) $D[BLUE_{\mathcal{N}}(\alpha)] \preccurlyeq D[BLUE_{\mathcal{M}}(\alpha)] \Leftrightarrow i_{-}(P) = r\big[G\ \Sigma\big]$.
(e) $D[BLUE_{\mathcal{N}}(\alpha)] = D[BLUE_{\mathcal{M}}(\alpha)] \Leftrightarrow r(P) = r\big[X\ \Sigma\big] + r\big[G\ \Sigma\big] + r(Z)$.
(f) $D[BLUE_{\mathcal{R}}(\alpha)] \succ D[BLUE_{\mathcal{M}}(\alpha)] \Leftrightarrow i_{-}(Q) = r\big[Z^{\perp}X\ Z^{\perp}\Sigma Z^{\perp}\big] + k$.
(g) $D[BLUE_{\mathcal{R}}(\alpha)] \prec D[BLUE_{\mathcal{M}}(\alpha)] \Leftrightarrow i_{+}(Q) = r\big[X\ \Sigma\big] + k$.
(h) $D[BLUE_{\mathcal{R}}(\alpha)] \succcurlyeq D[BLUE_{\mathcal{M}}(\alpha)] \Leftrightarrow i_{+}(Q) = r\big[X\ \Sigma\big]$.
(i) $D[BLUE_{\mathcal{R}}(\alpha)] \preccurlyeq D[BLUE_{\mathcal{M}}(\alpha)] \Leftrightarrow i_{-}(Q) = r\big[Z^{\perp}X\ Z^{\perp}\Sigma Z^{\perp}\big]$.
(j) $D[BLUE_{\mathcal{R}}(\alpha)] = D[BLUE_{\mathcal{M}}(\alpha)] \Leftrightarrow r(Q) = r\big[X\ \Sigma\big] + r\big[Z^{\perp}X\ Z^{\perp}\Sigma Z^{\perp}\big]$.
(k) $D[BLUE_{\mathcal{R}}(\alpha)] \succ D[BLUE_{\mathcal{N}}(\alpha)] \Leftrightarrow i_{-}(R) = r\big[Z^{\perp}X\ Z^{\perp}\Sigma Z^{\perp}\big] + r(Z) + k$.
(l) $D[BLUE_{\mathcal{R}}(\alpha)] \prec D[BLUE_{\mathcal{N}}(\alpha)] \Leftrightarrow i_{+}(R) = r\big[G\ \Sigma\big] + k$.
(m) $D[BLUE_{\mathcal{R}}(\alpha)] \succcurlyeq D[BLUE_{\mathcal{N}}(\alpha)] \Leftrightarrow i_{+}(R) = r\big[G\ \Sigma\big]$.
(n) $D[BLUE_{\mathcal{R}}(\alpha)] \preccurlyeq D[BLUE_{\mathcal{N}}(\alpha)] \Leftrightarrow i_{-}(R) = r\big[Z^{\perp}X\ Z^{\perp}\Sigma Z^{\perp}\big] + r(Z)$.
(o) $D[BLUE_{\mathcal{R}}(\alpha)] = D[BLUE_{\mathcal{N}}(\alpha)] \Leftrightarrow r(R) = r\big[Z^{\perp}X\ Z^{\perp}\Sigma Z^{\perp}\big] + r(Z) + r\big[G\ \Sigma\big]$.

5 Summary Comments

Analyzing parallels or equivalences across many linear models may be necessary for drawing statistical conclusions from linear regression models. There are two possible outcomes here: either the models are well-suited to match the same observable random vector, or they are correlated via specific linear transformations, making them rival models. Although there are relationships between the models, the inference outcomes that each model produces are not identical. In such instances, comparing the inference outcomes produced by these competing models would be a worthwhile endeavor.

Within the scope of this research, we have examined related models as well as a linear model that has been over-parameterized by including additional regressors. The over-parameterized linear model is related to both the actual linear model and an appropriately reduced version of it. In the preceding sections, we have looked into a few issues with comparing the three competing models' statistical inference. The dispersion matrices of BLUPs are utilized as the comparison operations. In addition, we simplify complex matrix operations by using established rank and inertia procedures to derive the findings. In order to compare dispersion matrices of BLUPs and BLUEs of unknown vectors under the models, we have gathered a number of equalities and inequalities. In the framework of over-parameterized linear models, we have provided a thorough theoretical examination into comparing dispersion matrices of BLUPs of unknown vectors. In our opinion, this type of theoretical element offers a substantial contribution and a generic framework for comparing the models' predictors, as it allows us to derive links between three competing models.

References

1. Alalouf, I.S., Styan, G.P.H.: Characterizations of estimability in the general linear model. Ann. Statist. **7**(1), 194–200 (1979)
2. Baksalary, J.K.: A study of the equivalence between a Gauss–Markoff model and its augmentation by nuisance parameters. Ser. Statist. **15**(1), 3–35 (1984)
3. Bhimasankaram, P., Jammalamadaka, S.R.: Updates of statistics in a general linear model: a statistical interpretation and applications. Commun. Stat. Simul. Comput. **23**(3), 789–801 (1994)
4. Bhimasankaram, P., SahaRay, R.: On a partitioned linear model and some associated reduced models. Linear Algebra Appl. **264**, 329–339 (1997)
5. Chu, K.L., Isotalo, J., Puntanen, S., Styan, G.P.H.: On decomposing the Watson efficiency of ordinary least squares in a partitioned weakly singular linear model. Sankhya Indian J. Stat. **66**(4), 634–651 (2004)
6. Dong, B., Guo, W., Tian, Y.: On relations between BLUEs under two transformed linear models. J. Multivar. Anal. **131**, 279–292 (2014)
7. Drygas, H.: The Coordinate-Free Approach to Gauss–Markov Estimation. Springer, Heidelberg (1970)
8. Dube, M.: Mixed regression estimator under inclusion of some superfluous variables. Test **8**(2), 411–417 (1999)
9. Dube, M., Srivastava, V.K., Toutenburg, H., Wijekoon, P.: Stein-rule estimator under inclusion of superfluous variables in linear regression models. Commun. Statist. Theory Meth. **20**, 2009–2022 (1991)
10. Gan, S., Sun, Y., Tian, Y.: Equivalence of predictors under real and over-parameterized linear models. Commun. Statist. Theory Meth. **46**(11), 5368–5383 (2017)
11. Goldberger, A.S.: Best linear unbiased prediction in the generalized linear regression model. J. Am. Stat. Assoc. **57**(298), 369–375 (1962)
12. Groß, J., Puntanen, S.: Estimation under a general partitioned linear model. Linear Algebra Appl. **321**(1), 131–144 (2000)
13. Güler, N.: On relations between BLUPs under two transformed linear random-effects models. Commun. Stat. Simul. Comput. **51**(19), 5099–5125 (2022)
14. Güler, N., Büyükkaya, M.E.: Inertia and rank approach in transformed linear mixed models for comparison of BLUPs. Commun. Statist. Theory Meth. **52**(9), 3108–3123 (2023)
15. Güler, N., Büyükkaya, M.E., Yiğit, M.: Comparison of covariance matrices of predictors in seemingly unrelated regression models. Indian J. Pure Appl. Math. **53**, 801–809 (2022)
16. Güler, N., Puntanen, S., Özdemir, H.: On the BLUEs in two linear models via C. R. Rao's pandora's box. Commun. Statist. Theory Meth. **43**(5), 921–931 (2014)
17. Haslett, S.J., Puntanen, S.: Effect of adding regressors on the equality of the BLUEs under two linear models. J. Stat. Plan. Inference **140**(1), 104–110 (2010)
18. Hodges, J.S.: Richly Parameterized Linear Models: Additive, Time Series, and Spatial Models Using Random Effects. CRC Press, Boca Raton, FL (2014)

19. Isotalo, J., Puntanen, S., Styan, G.P.H.: Effect of adding regressors on the equality of the OLSE and BLUE. Int. J. Stat. Sci. **6**, 193–201 (2007)
20. Jammalamadaka, S.R., Sengupta, D.: Changes in the general linear model: a unified approach. Linear Algebra Appl. **289**(1), 225–242 (1999)
21. Jammalamadaka, S.R., Sengupta, D.: Inclusion and exclusion of data or parameters in the general linear model. Stat. Probab. Lett. **77**(12), 1235–1247 (2007)
22. Jun, S.J., Pinkse, J.: Adding regressors to obtain efficiency. Econom. Theory **25**(1), 298–301 (2009)
23. Kadiyala, K.: Mixed regression estimator under misspecification. Econ. Lett. **21**(1), 27–30 (1986)
24. Li, W., Tian, Y., Yuan, R.: Statistical analysis of a linear regression model with restrictions and superfluous variables. J. Ind. Manag. Optim. **19**(5), 3107–3127 (2023)
25. Lu, C., Gan, S., Tian, Y.: Some remarks on general linear model with new regressors. Stat. Probab. Lett. **97**, 16–24 (2015)
26. Lu, C., Sun, Y., Tian, Y.: A comparison between two competing fixed parameter constrained general linear models with new regressors. Statistics **52**(4), 769–781 (2018)
27. Löwner, K.: Über monotone Matrixfunktionen. Math. Z. **38**, 177–216 (1934)
28. Ma, R., Tian, Y.: A matrix approach to a general partitioned linear model with partial parameter restrictions. Linear Multilinear Algebra **70**(13), 2513–2532 (2022)
29. Magnus, J.R., Durbin, J.: Estimation of regression coefficients of interest when other regression coefficients are of no interest. Econometrica **67**(3), 639–643 (1999)
30. Mitra, S.K., Moore, B.J.: Gauss–Markov estimation with an incorrect dispersion matrix. Sankhya Indian J. Stat. Ser. A **35**(2), 139–152 (1973)
31. Penrose, R.: A generalized inverse for matrices. Math. Proc. Camb. Philos. Soc. **51**(3), 406–413 (1955)
32. Puntanen, S., Styan, G.P.H., Isotalo, J.: Matrix Tricks for Linear Statistical Models: Our Personal Top Twenty. Springer, Heidelberg (2011)
33. Qian, H., Tian, Y.: Partially superfluous observations. Econ. Theory **22**, 529–536 (2006)
34. Rao, C.R.: Representations of best linear unbiased estimators in the Gauss–Markoff model with a singular dispersion matrix. J. Multivar. Anal. **3**(3), 276–292 (1973)
35. Srivastava, V.K., Dube, M., Singh, V.: Ordinary least squares and Stein-rule predictions in regression models under inclusion of some superfluous variables. Stat. Pap. **37**, 253–265 (1996)
36. Tian, Y.: On equalities for BLUEs under misspecified Gauss–Markov models. Acta Math. Sin. Engl. Ser. **25**(11), 1907–1920 (2009)
37. Tian, Y.: Equalities and inequalities for inertias of Hermitian matrices with applications. Linear Algebra Appl. **433**(1), 263–296 (2010)
38. Tian, Y.: On properties of BLUEs under general linear regression models. J. Stat. Plan. Inference **143**(4), 771–782 (2013)
39. Tian, Y.: A new derivation of BLUPs under random-effects model. Metrika **78**, 905–918 (2015)
40. Tian, Y.: Some equalities and inequalities for covariance matrices of estimators under linear model. Stat. Pap. **58**, 467–484 (2017)
41. Tian, Y.: Matrix rank and inertia formulas in the analysis of general linear models. Open Math. **15**(1), 126–150 (2017)
42. Tian, Y.: Transformation approaches of linear random-effects models. Stat. Methods Appl. **26**(4), 583–608 (2017)
43. Tian, Y., Guo, W.: On comparison of dispersion matrices of estimators under a constrained linear model. Stat. Methods Appl. **25**(4), 623–649 (2016)
44. Tian, Y., Jiang, B.: Matrix rank/inertia formulas for least-squares solutions with statistical applications. Spec. Matrices **4**, 130–140 (2016)
45. Tian, Y., Jiang, B.: A new analysis of the relationships between a general linear model and its mis-specified forms. J. Korean Stat. Soc. **46**(2), 182–193 (2017)
46. Tian, Y., Puntanen, S.: On the equivalence of estimations under a general linear model and its transformed models. Linear Algebra Appl. **430**(10), 2622–2641 (2009)

47. Tian, Y.: Solving optimization problems on ranks and inertias of some constrained nonlinear matrix functions via an algebraic linearization method. Nonlinear Anal. Theory Methods Appl. **75**(2), 717–734 (2012)
48. Trenkler, G., Wijekoon, P.: Mean squared error matrix superiority of the mixed regression estimator under misspecification. Statistica (Bologna) **49**(1), 65–71 (1989)
49. Wijekoon, P., Trenkler, G.: Mean square error matrix superiority of estimators under linear restrictions and misspecification. Econ. Lett. **30**(2), 141–149 (1989)

Statistical Inference of a Stochastically Restricted Linear Model with Superfluous Variables

Melek Eriş Büyükkaya and Nesrin Güler

Abstract In this study, we have investigated some comparison problems of the best linear unbiased predictor (BLUP) for a unified form of all unknown parameters in the statistical inference of three competing models according to the mean squared error matrix (MSEM) criteria. Out of the three competing models that we consider: the first model is a linear model with stochastic restriction, the second model is a stochastically restricted linear model with the inclusion of superfluous variables which is obtained from new additions to the first model, and the last model is a stochastically restricted reduced linear model which is a reduced form of the second model. As the comparison operations, we have used the methodology of block matrix inertias and ranks. Furthermore, comparison results for special cases have been demonstrated. We have also illustrated our findings with numerical examples.

Keywords BLUP · Inertia · Linear model · MSEM · Rank · Stochastic restriction · Superfluous variables

1 Introduction

One of the most commonly used fundamental statistical techniques, both in theory and in practice, is the use of linear models. In inference problems involving linear models, there may be additional information along with the sample information. This additional information is often available in advance for the unknown parameters in addition to those contained in the data. Additional information is generally added to the model assumptions, and it can be expressed as linear exact or linear

M. Eriş Büyükkaya (✉)
Department of Statistics and Computer Sciences, Karadeniz Technical University, Trabzon, Turkey
e-mail: melekeris@ktu.edu.tr

N. Güler
Department of Econometrics, Sakarya University, Sakarya, Turkey
e-mail: nesring@sakarya.edu.tr

O. Castillo et al. (eds.), *Advanced Mathematics for the Modeling and Solution of Challenging Problems in Engineering*, Studies in Systems, Decision and Control 582,
https://doi.org/10.1007/978-3-031-82342-8_20

stochastic restrictions on unknown parameters. In some circumstances, such as the linear hypothesis testing on parameters, the exact knowledge of a single component of an unknown parameter vector, and the known knowledge of the rations between specific coefficients in the models, exact restrictions on parameters can occur; for more information, see [1]. If additional information comes from previous investigations and long associations with the relevant studies, linear stochastic restrictions can occur; see, e.g., [1, 2]. There are many studies on these kinds of models in the statistical literature. We may refer to the studies [1–10], among others, in which a linear model with a linear stochastic restriction on unknown parameters has been considered from different perspectives. In the statistical literature, linear models with exact restrictions on known parameters have been extensively studied from a variety of approaches. We may refer to recent studies, e.g., [11–16], among others, and the references therein.

In addition to the situations mentioned above, many different models that are candidates for fitting the same vector of observable random variables can emerge during the statistical inference of regression models. These models often arise by adding or deleting some regressors to models. Adding new parts, including new regressors, to a given linear model results in an overparameterized model of the model. In other words, a linear model with inclusion of superfluous variables is obtained with the addition of new regressors to a given linear model. There are many studies on these kinds of models in the statistical literature; see, e.g., [3, 17–29].

Establishing relationships between two or more different linear models is one of the classic research topics in linear regression analysis. One of these problems is to investigate the relationship between two linear models that are structured for the same vector of the observable random variables while another is to investigate the relationship between two linear models having the same partial parameters. In accordance with what has been stated above for the investigation of the relationships between different types of models, in the present study, we consider the following three models with general assumptions: a linear model with stochastic restriction, a stochastically restricted linear model with the inclusion of superfluous variables which is obtained from new additions to the first model, and a stochastically restricted reduced linear model which is a reduced form of the second model. Although these three models have some common parts, the inference results of them are not necessarily the same, but there may be relations between the prediction and estimation results under these models. Therefore, the problems derived from these models on connections of predictors (or, also estimators) of unknown vectors need to be characterized and it is noteworthy to compare the various inference results of these models.

The present paper describes how to establish exact formulas for the comparisons of predictors for the same general vector of all unknown vectors under considered models. After converting these models to the implicitly stochastically restricted form, we establish various equalities and inequalities for the comparisons of any predictor with the best linear unbiased predictor (BLUP) for the same unified form of all unknown parameters. Within the framework of these issues, we derive various formulas for the comparisons of any predictor and the BLUP by considering the mean squared error matrix (MSEM) criterion by utilizing the methodology of block matrix inertias and ranks. Afterward, we establish various comparisons between MSEMs of BLUPs through inertia and rank formulas under considered models. We also derive comparison results for special cases of a unified form of all unknown parameters under the models. The operations we encounter within the framework of the comparisons include complex matrix expressions. Therefore, the block matrix inertia and rank methodology are utilized to simplify these expressions. The comparisons discussed in the study provide us to determine the optimal unbiased predictor or estimator under a linear model since the definition of BLUP of the best linear unbiased estimator (BLUE) is based on the requirement of a minimum dispersion matrix of unbiased predictors or estimators of unknown parameters. In a series of studies given by [11, 30–36], and further, in the studies [16, 37–39], among others, the methodology of inertia and rank of block matrices has been widely used to characterize the comparison of dispersion matrices for predictors or estimators under linear models.

The remainder of this article is organized as follows. In Sect. 2, we give a variety of basic formulas, results, and facts in matrix theory, and we also briefly review the concepts of some principles of predictors/estimators. In Sect. 3, firstly, we introduce the considered models. Then, we discuss some of the basic characteristics of predictors/estimators to motivate the approaches taken in the paper. In Sect. 4, we derive comparison results of any predictor and the BLUP for all unknown parameters in a unified form under the considered models in the sense of the MSEM criterion. In Sect. 5, we use the comparison results obtained in the previous section for the comparison of BLUPs and BLUEs under the models. Furthermore, a series of results for special cases are given. In addition, numerical examples are given in Sect. 6. Finally, some remarks and related issues are addressed in Sect. 7.

2 Notation and Preliminary Results

We demonstrate some notations before starting the next sections. We will write $\Omega \in \mathbb{R}^{m\times n}$ if Ω is an $m \times n$ real matrix, $\Omega \in \mathbb{R}_s^{m\times m}$ if $\Omega \in \mathbb{R}^{m\times m}$ and is symmetric, $\Omega \in \mathbb{R}_{\succcurlyeq}^{m\times m}$ if $\Omega \in \mathbb{R}_s^{m\times m}$ and is positive semi-definite. We will use the symbols Ω', $r(\Omega)$, $\Re(\Omega)$, and Ω^+ to present the transpose, the rank, the column space, and the Moore–Penrose generalized inverse of $\Omega \in \mathbb{R}^{m\times n}$, respectively. $\Omega^\perp = I_m - \Omega\Omega^+$ stands for the orthogonal projector, where $I_m \in \mathbb{R}_s^{m\times m}$ is the identity matrix. $i_+(\Omega)$ and $i_-(\Omega)$ denote the positive and the negative inertias of $\Omega \in \mathbb{R}_s^{m\times m}$, respectively. Additionally, the symbols $i_\pm(\Omega)$ and $i_\mp(\Omega)$ are used to collectively denote both the positive and negative inertias for the matrix $\Omega \in \mathbb{R}_s^{m\times m}$. These notations help to express the quantities related to the inertias of the matrix and provide a concise way to discuss both positive and negative inertias in the context of matrix algebra. The inequality $\Omega_1 - \Omega_2 \succcurlyeq 0$ or $\Omega_1 \succcurlyeq \Omega_2$ means $\Omega_1 - \Omega_2 \in \mathbb{R}_{\succcurlyeq}^{m\times m}$ in the Löwner sense for $\Omega_1, \Omega_2 \in \mathbb{R}_s^{m\times m}$. This notation and inequality provide a way to express the order and relations between matrices in terms of their positive semi-definiteness. $E(\omega)$ and $cov(\omega, \omega)$ denote the expectation and the covariance matrix of a random vector $\omega \in \mathbb{R}^{k\times 1}$, respectively, and further, $D(\omega) = cov(\omega, \omega)$ denotes the dispersion matrix of ω.

We give the following two lemmas, which are required in the sections that follow.

Lemma 1 ([40]) Let $\Omega_1, \Omega_2 \in \mathbb{R}^{m\times n}$, or, let $\Omega_1, \Omega_2 \in \mathbb{R}_s^{m\times m}$. Then, $i_-(\Omega_1 - \Omega_2) = 0 \Leftrightarrow \Omega_1 \succcurlyeq \Omega_2$ and $i_+(\Omega_1 - \Omega_2) = 0 \Leftrightarrow \Omega_1 \preccurlyeq \Omega_2$.

Lemma 2 ([40]) Let $\Omega_1 \in \mathbb{R}_s^{m\times m}$, $\Omega_2 \in \mathbb{R}_s^{n\times n}$, $\Phi \in \mathbb{R}^{m\times n}$, and $a \in \mathbb{R}$. Then,

$$r(\Omega_1) = i_+(\Omega_1) + i_-(\Omega_1). \tag{1}$$

$$i_\pm(a\Omega_1) = i_\pm(\Omega_1) \text{ if } a > 0 \text{ and } i_\pm(a\Omega_1) = i_\mp(\Omega_1) \text{ if } a < 0 \tag{2}$$

$$i_\pm\begin{bmatrix} \Omega_1 & \Phi \\ \Phi' & \Omega_2 \end{bmatrix} = i_\pm\begin{bmatrix} \Omega_1 & -\Phi \\ -\Phi' & \Omega_2 \end{bmatrix} = i_\mp\begin{bmatrix} -\Omega_1 & \Phi \\ \Phi' & -\Omega_2 \end{bmatrix}. \tag{3}$$

$$i_\pm\begin{bmatrix} \Omega_1 & 0 \\ 0 & \Omega_2 \end{bmatrix} = i_\pm(\Omega_1) + i_\pm(\Omega_2),\ i_+\begin{bmatrix} 0 & \Phi \\ \Phi' & 0 \end{bmatrix} = i_-\begin{bmatrix} 0 & \Phi \\ \Phi' & 0 \end{bmatrix} = r(\Phi) \tag{4}$$

$$i_\pm\begin{bmatrix} \Omega_1 & \Phi \\ \Phi' & 0 \end{bmatrix} = r(\Phi) + i_\pm\left(\Phi^\perp \Omega_1 \Phi^\perp\right). \tag{5}$$

$$i_+\begin{bmatrix} \Omega_1 & \Phi \\ \Phi' & 0 \end{bmatrix} = r\left[\Omega_1, \Phi\right] \text{ and } i_+\begin{bmatrix} \Omega_1 & \Phi \\ \Phi' & 0 \end{bmatrix} = r(\Phi) \text{ if } \Omega_1 \succcurlyeq 0. \tag{6}$$

$$i_\pm\begin{bmatrix} \Omega_1 & \Phi \\ \Phi' & \Omega_2 \end{bmatrix} = i_\pm(A_1) + i_\pm\left(\Omega_2 - \Phi'\Omega_1^+\Phi\right) \text{ if } \Re(\Phi) \subseteq \Re(\Omega_1). \tag{7}$$

We introduce the following definitions of the notions of MSEM, consistency of models, predictability, BLUP, and BLUE.

Definition 1 ([41, 42]) Suppose that $\tilde{\omega}$ is any predictor of ω and $\hat{\delta}$ is any estimator of δ in a linear model. The MSEM of predictor $\tilde{\omega}$ is defined as matrix

$$\begin{aligned} MSEM(\tilde{\omega}) &= E(\tilde{\omega} - \omega)(\tilde{\omega} - \omega)' \\ &= D(\tilde{\omega}) + D(\omega) - cov(\tilde{\omega}, \omega) - cov(\tilde{\omega}, \omega) + [bias(\tilde{\omega})bias(\tilde{\omega})]', \end{aligned} \tag{8}$$

where $bias(\tilde{\omega}) = E(\tilde{\omega} - \omega)$. The MSEM of estimator $\hat{\delta}$ is defined as matrix

$$MSEM(\hat{\delta}) = E(\hat{\delta} - \delta)(\hat{\delta} - \delta)' = D(\hat{\delta}) + \left[bias(\hat{\delta})bias(\hat{\delta})\right]'$$

where $bias(\hat{\delta}) = E(\hat{\delta}) - \delta$.

Definition 2 ([41, 42]) Let consider the difference

$$\mathrm{MSEM}(\omega_1) - \mathrm{MSEM}(\omega_2) := \Delta(\omega_1, \omega_2) \tag{9}$$

for two given predictors or estimators ω_1 and ω_2 in a linear model. The ω_2 is said to be superior to ω_1 according to the MSEM criterion iff

$$\Delta(\omega_1, \omega_2) \succcurlyeq 0, \text{ namely, } \mathrm{MSEM}(\omega_1) \succcurlyeq \mathrm{MSEM}(\omega_2)$$

Definition 3 ([43]) Let consider a linear model $\mathcal{M} : \mathrm{y} = \mathrm{X}\alpha + \varepsilon$ with $\mathrm{E}(\varepsilon) = 0$ and $\mathrm{D}(\varepsilon) = \Omega$. $\mathcal{M}$ is said to be consistent if $\mathrm{y} \in \Re\left[\mathrm{X}, \Omega\right]$ holds with probability 1.

Definition 4 ([44, 45]) Let consider a linear model $\mathcal{M} : \mathrm{y} = \mathrm{X}\alpha + \varepsilon$ with $\mathrm{E}(\varepsilon) = 0$ and $\mathrm{D}(\varepsilon) = \Omega$. Suppose that ω is any vector in the form of $\omega = \mathrm{K}\alpha + \mathrm{L}\varepsilon$. If a matrix F exists with $\mathrm{E}(\mathrm{Fy} - \omega) = 0$, then ω is defined to be predictable under $\mathcal{M}$, namely, ω is predictable under $\mathcal{M} \Leftrightarrow \Re\left(\mathrm{K}'\right) \subseteq \Re\left(\mathrm{X}'\right) \Leftrightarrow \mathrm{K}\alpha$ is estimable under $\mathcal{M}$. Further, ε is always predictable and $\mathrm{X}\alpha$ is always estimable under $\mathcal{M}$.

Definition 5 ([46, 47]) Let consider a linear model $\mathcal{M} :\ \mathrm{y} = \mathrm{X}\alpha + \varepsilon$ with $\mathrm{E}(\varepsilon) = 0$ and $\mathrm{D}(\varepsilon) = \Omega$. Suppose that $\omega = \mathrm{K}\alpha + \mathrm{L}\varepsilon$ is predictable under $\mathcal{M}$. Fy is said to be the BLUP of ω and is represented by $\mathrm{Fy} = \tilde{\omega}_{\mathrm{BLUP}_{\mathcal{M}}}$ if there exists Fy such that $\mathrm{D}(\mathrm{Fy} - \omega) = \min$ s.t. $\mathrm{E}(\mathrm{Fy} - \omega) = 0$ holds in the Löwner sense. If $\mathrm{L} = 0$ in ω, Fy is the well-known BLUE of $K\alpha$, represented by $\widehat{K\alpha}_{\mathrm{BLUE}_{\mathcal{M}}}$.

3 Mathematical Framework on the Models and Predictors

3.1 Linear Models with Stochastic Restrictions

Let consider the following stochastically restricted linear models that are structured for the same vector of an observable random variable:

$$\mathrm{y} = \mathrm{X}\alpha + \varepsilon \text{ with stochastic restriction } \mathrm{R}\,\alpha + \mathrm{e} = \mathrm{r} \tag{10}$$

and

$$\mathrm{y} = \mathrm{X}\alpha + \mathrm{Z}\gamma + \varepsilon = \left[\mathrm{X}, \mathrm{Z}\right]\begin{bmatrix}\alpha \\ \gamma\end{bmatrix} + \varepsilon \text{ with stochastic restriction } \mathrm{R}\,\alpha + \mathrm{e} = \mathrm{r}, \tag{11}$$

where

$$\mathrm{E}(\varepsilon) = 0,\ \mathrm{D}(\varepsilon) = \sigma^2\Gamma,\ \mathrm{E}(\mathrm{e}) = 0,\ \mathrm{D}(\mathrm{e}) = \sigma^2\Delta,\ \mathrm{cov}(\varepsilon, \mathrm{e}) = \sigma^2\mathrm{V}, \tag{12}$$

and $y \in \mathbb{R}^{n\times 1}$ is a vector of responses, $X \in \mathbb{R}^{n\times k}$, $Z \in \mathbb{R}^{n\times p}$, $R \in \mathbb{R}^{m\times k}$ are known matrices of arbitrary ranks and $r \in \mathbb{R}^{m\times 1}$ is a known vector, $\alpha \in \mathbb{R}^{k\times 1}$ and $\gamma \in R^{p\times 1}$ are parameter vectors of fixed effects, $\varepsilon \in \mathbb{R}^{n\times 1}$ and $e \in \mathbb{R}^{m\times 1}$ are vectors of random errors, $\Gamma \in \mathbb{R}^{n\times n}_{\succcurlyeq}$ and $\Lambda \in \mathbb{R}^{m\times m}_{\succcurlyeq}$ of arbitrary ranks, $V \in \mathbb{R}^{n\times m}$, σ^2 is a positive unknown scalar.

By taking the linear model with stochastic restriction given in (10) as a real model and adding a new part including new regressors on the right-hand side of this model, the model in (10) becomes the over-parameterized model with the identical stochastic restriction on parameters as the model given in (11). The model in (11) is also known as a stochastically restricted linear model with the inclusion of superfluous variables. From a different perspective, the model in (11) can also be considered as a partitioned linear model. Partitioned linear regression models can be considered when our interest focuses on estimation and prediction problems on general parametric functions of partial parameters. For this reason, these models can be considered with their reduced models. To predict the general vector containing only the partial unknown parameter vector α in (11) independently, we pre-multiply both sides of the model in (11) with $Z^{\perp}$ and note that $Z^{\perp}Z = 0$ to obtain the following sub-model associated with the models in (10) and (11):

$$\begin{aligned} Z^{\perp}y &= Z^{\perp}X\alpha + Z^{\perp}\varepsilon \text{ with stochastic restriction } R\,\alpha + e = r, \\ E(Z^{\perp}\varepsilon) &= 0, D(Z^{\perp}\varepsilon) = \sigma^2 Z^{\perp}\Gamma Z^{\perp} \end{aligned} \tag{13}$$

The model in (13) is a reduced model of the model in (11). This model is also known as a correctly-reduced model of the model in (11). The model in (13) contains one of the partial unknown parameters in the partitioned linear model in (11). The model in (13) is also a transformed version of the models in (10) and (11). For transformation approach to linear models, we refer to the studies [31, 35, 48–50].

The models in (10), (11), and (13) can merge with stochastic restriction $R\,\alpha + e = r$, into the following combined forms:

$$\mathcal{R} : \begin{bmatrix} y \\ r \end{bmatrix} = \begin{bmatrix} X \\ R \end{bmatrix}\alpha + \begin{bmatrix} \varepsilon \\ e \end{bmatrix}, \text{ or, } y_c = X_c\alpha + \varepsilon_c, \tag{14}$$

$$\mathcal{S} : \begin{bmatrix} y \\ r \end{bmatrix} = \begin{bmatrix} X \\ R \end{bmatrix}\alpha + \begin{bmatrix} Z \\ 0 \end{bmatrix}\gamma + \begin{bmatrix} \varepsilon \\ e \end{bmatrix}, \text{ or, } y_c = X_c\alpha + Z_c\gamma + \varepsilon_c, \tag{15}$$

$$\mathcal{C} : \begin{bmatrix} Z^{\perp}y \\ r \end{bmatrix} = \begin{bmatrix} Z^{\perp}X \\ R \end{bmatrix}\alpha + \begin{bmatrix} Z^{\perp}\varepsilon \\ e \end{bmatrix}, \text{ or, } \tilde{Z}y_c = \tilde{Z}X_c\alpha + \tilde{Z}\varepsilon_c, \tag{16}$$

where

$$y_c = \begin{bmatrix} y \\ r \end{bmatrix}, X_c = \begin{bmatrix} X \\ R \end{bmatrix}, Z_c = \begin{bmatrix} Z \\ 0 \end{bmatrix}, \tilde{Z} = \begin{bmatrix} Z^{\perp} & 0 \\ 0 & I_m \end{bmatrix}, \varepsilon_c = \begin{bmatrix} \varepsilon \\ e \end{bmatrix},$$

and, further, according to assumptions in (12)

$$E(\varepsilon_c) = E\begin{bmatrix} \varepsilon \\ e \end{bmatrix} = 0, D(\varepsilon_c) = D\begin{bmatrix} \varepsilon \\ e \end{bmatrix} = \sigma^2 \begin{bmatrix} \Gamma & V \\ V' & \Lambda \end{bmatrix} := \sigma^2 \Delta, D(\varepsilon_c) = \sigma^2 \tilde{Z} \Delta \tilde{Z}, \tag{17}$$

where $\Delta \in \mathbb{R}_{\succcurlyeq}^{(n+m)\times(n+m)}$ of arbitrary rank with known entries. In this way, explicitly stochastically restricted linear models given in (10), (11), and (13) are transformed into the implicitly stochastically restricted linear models $\mathcal{R}$, $\mathcal{S}$, and $\mathcal{C}$ given in (14)–(16), respectively. In what follows, the considered models in this study are assumed to be consistent. According to Definition 3, we note that if $\mathcal{R}$ is consistent, then the models $\mathcal{S}$ and $\mathcal{C}$ are consistent as well. Furthermore, the consistency of $\mathcal{S}$ implies the consistency of $\mathcal{C}$.

In order to produce conclusions on predictors and estimators of all joint unknown vectors under $\mathcal{R}$, $\mathcal{S}$, and $\mathcal{C}$, we take into consideration the following general vector:

$$\lambda = K\alpha + L\varepsilon_c = \begin{bmatrix} K, 0 \end{bmatrix} \begin{bmatrix} \alpha \\ \gamma \end{bmatrix} + \begin{bmatrix} L_1, L_2 \end{bmatrix} \begin{bmatrix} \varepsilon \\ e \end{bmatrix}, \tag{18}$$

where $K \in \mathbb{R}^{t\times k}$, $L_1 \in \mathbb{R}^{t\times n}$, and $L_2 \in \mathbb{R}^{t\times m}$ with $L = \begin{bmatrix} L_1, L_2 \end{bmatrix}$, are given matrices of arbitrary ranks. Then, from (17),

$E(\lambda) = K\alpha$, $D(\lambda) = \sigma^2 L\Delta L'$, $\mathrm{cov}(\lambda, y_c) = \sigma^2 L\Delta$, and $\mathrm{cov}(\lambda, \tilde{Z}y_c) = \sigma^2 L\Delta\tilde{Z}$.

The predictability conditions of λ and its special cases under $\mathcal{R}$, $\mathcal{S}$, and $\mathcal{C}$ are obtained easily from Definition 4. We note that if λ is predictable under $\mathcal{C}$ then it is predictable under both the models $\mathcal{R}$ and $\mathcal{S}$. Also, it can be said that if λ is predictable under $\mathcal{S}$ then it is predictable under $\mathcal{R}$.

3.2 BLUPs in Linear Models with Stochastic Restriction

The fundamental equations of BLUP and BLUE, expressions of them under the models, and related properties are summarized in this section. The BLUP and the BLUE under linear models have been widely studied from different approaches in the statistical literature; see, e.g., [35, 43, 47].

Lemma 3 Suppose that λ in (18) is predictable under $\mathcal{S}$ in (15).

There exists a matrix F such that $D(Fy_c - \lambda) = \min$ s.t. $E(Fy_c - \lambda) = 0$ iff

$$F\begin{bmatrix} X_c, Z_c, \Delta\begin{bmatrix} X_c, Z_c \end{bmatrix}^{\perp} \end{bmatrix} = \begin{bmatrix} K, 0, L\Delta\begin{bmatrix} X_c, Z_c \end{bmatrix}^{\perp} \end{bmatrix}. \tag{19}$$

(19) is consistent and the BLUP of λ under $\mathcal{S}$ can be written as follows by considering the general solution of this equation:

$$\tilde{\lambda}_{\mathrm{BLUP}_{\mathcal{S}}} = \mathrm{Fy_c} = \left(\left[\mathrm{K},\, 0,\, \mathrm{L}\Delta\left[\mathrm{X_c},\, \mathrm{Z_c}\right]^{\perp}\right]\mathrm{W}_{\mathcal{S}}^{+} + \mathrm{U_s}\mathrm{W}_{\mathcal{S}}^{\perp}\right)\mathrm{y_c}, \tag{20}$$

where $\mathrm{W_s} = \left[\mathrm{X_c},\, \mathrm{Z_c},\, \Delta\left[\mathrm{X_c},\, \mathrm{Z_c}\right]^{\perp}\right]$ and $\mathrm{U_s} \in \mathbb{R}^{t\times(n+m)}$ is an arbitrary matrix. In particular, $\mathfrak{R}\left[\mathrm{X_c},\, \mathrm{Z_c},\, \Delta\left[\mathrm{X_c},\, \mathrm{Z_c}\right]^{\perp}\right] = \mathbb{R}^{(n+m)\times 1} \Leftrightarrow$ F is unique and $\mathcal{S}$ is consistent $\Leftrightarrow \tilde{\lambda}_{\mathrm{BLUP}_{\mathcal{S}}}$ is unique. Further, from (20),

$$\begin{aligned} \mathrm{D}\left(\lambda - \tilde{\lambda}_{\mathrm{BLUP}_{\mathcal{S}}}\right) =&\sigma^2\left(\left[\mathrm{K},\, 0,\, \mathrm{L}\Delta\left[\mathrm{X_c},\, \mathrm{Z_c}\right]^{\perp}\right]\mathrm{W_s^{+}} - \mathrm{L}\right)\Delta \\ &\times\left(\left[\mathrm{K},\, 0,\, \mathrm{L}\Delta\left[\mathrm{X_c},\, \mathrm{Z_c}\right]^{\perp}\right]\mathrm{W_s^{+}} - \mathrm{L}\right)'. \end{aligned} \tag{21}$$

From (8) and (21), the MSEM of $\tilde{\lambda}_{\mathrm{BLUP}_{\mathcal{S}}}$ is written as.

$$\mathrm{MSEM}\left(\tilde{\lambda}_{\mathrm{BLUP}_{\mathcal{S}}}\right) = \mathrm{D}\left(\lambda - \tilde{\lambda}_{\mathrm{BLUP}_{\mathcal{S}}}\right). \tag{22}$$

There exists a matrix H such that $\mathrm{D}(\mathrm{Hy_c}) = \min$ t. $\mathrm{E}(\mathrm{Hy_c}) = \mathrm{K}\alpha$ iff

$$\mathrm{H}\left[\mathrm{X_c},\, \mathrm{Z_c},\, \Delta\left[\mathrm{X_c},\, \mathrm{Z_c}\right]^{\perp}\right] = \left[\mathrm{K},\, 0,\, 0\right]. \tag{23}$$

Then, the general solution of (23) is written as

$$\widehat{\mathrm{K}\alpha}_{\mathrm{BLUE}_{\mathcal{S}}} = \mathrm{Hy_c} = (\left[\mathrm{K},\, 0,\, 0\right]\mathrm{W_s^{+}} + \mathrm{U_s}\mathrm{W_s^{\perp}})\mathrm{y_c},$$

where $\mathrm{W_s} = \left[\mathrm{X_c},\, \mathrm{Z_c},\, \Delta\left[\mathrm{X_c},\, \mathrm{Z_c}\right]^{\perp}\right]$ and $\mathrm{U_s} \in \mathbb{R}^{t\times(n+m)}$ is an arbitrary matrix. Further,

$$\mathrm{D}\left(\widehat{\mathrm{K}\alpha}_{\mathrm{BLUE}_{\mathcal{S}}}\right) = \sigma^2(\left[\mathrm{K},\, 0,\, 0\right]\mathrm{W}_{\mathcal{S}}^{+})\Delta(\left[\mathrm{K},\, 0,\, 0\right]\mathrm{W_s^{+}})'$$

The MSEM of $\widehat{\mathrm{K}\alpha}_{\mathrm{BLUE_S}}$ is

$$\mathrm{MSEM}\left(\widehat{\mathrm{K}\alpha}_{\mathrm{BLUE_S}}\right) = \mathrm{D}\left(\widehat{\mathrm{K}\alpha}_{\mathrm{BLUE_S}}\right)$$

We note that various related results under $\mathcal{S}$ can be derived from Lemma 3 from special choices of the matrices K and L in λ.

The fundamental equation of BLUP of λ and related results under $\mathcal{R}$ can be derived from Lemma 3 by setting $\mathrm{Z_c} = 0$. Now, assume that λ is predictable under $\mathcal{R}$. Then

$$\begin{aligned} &\mathrm{D}(\mathrm{Fy_c} - \lambda) = \min \text{ s.t. } \mathrm{E}(\mathrm{Fy_c} - \lambda) = 0 \Leftrightarrow \mathrm{F}\left[\mathrm{X_c},\, \Delta X_c^{\perp}\right] = \left[\mathrm{K},\, L\Delta X_c^{\perp}\right], \\ &\tilde{\lambda}_{\mathrm{BLUP}_{\mathcal{R}}} = \mathrm{Fy_c} = \left(\left[\mathrm{K},\, L\Delta X_c^{\perp}\right]\left[\mathrm{X_c},\, \Delta X_c^{\perp}\right]^{+} + \mathrm{U_r}\left[\mathrm{X_c},\, \Delta X_c^{\perp}\right]^{\perp}\right)\mathrm{y_c}, \\ &\mathrm{D}\left(\lambda - \tilde{\lambda}_{\mathrm{BLUP}_{\mathcal{R}}}\right) = \sigma^2(\left[\mathrm{K},\, L\Delta X_c^{\perp}\right]\mathrm{W_r^{+}} - \mathrm{L})\Delta(\left[\mathrm{K},\, L\Delta X_c^{\perp}\right]\mathrm{W_r^{+}} - \mathrm{L})', \end{aligned} \tag{24}$$

where $U_r \in \mathbb{R}^{t\times(n+m)}$ is an arbitrary matrix and $W_r = \left[X_c, \Delta X_c^{\perp}\right]$.

The following conclusions are obtained from Lemma 3.1 in [35] by taking $\tilde{Z}$ as the transformation matrix and by setting appropriate modifications in this lemma. Now, assume that λ is predictable under $\mathcal{C}$. Then,

$$D(F\tilde{Z}y_c - \lambda) = \min \text{ s.t. } E(F\tilde{Z}y_c - \lambda) = 0$$

$$\Leftrightarrow F\left[\tilde{Z}X_c, \tilde{Z}\Delta\tilde{Z}(\tilde{Z}X_c)^{\perp}\right] = \left[K, L\Delta\tilde{Z}(\tilde{Z}X_c)^{\perp}\right]$$

$$\tilde{\lambda}_{BLUP_C} = F\tilde{Z}y_c = \left(\left[K, L\Delta\tilde{Z}(\tilde{Z}X_c)^{\perp}\right]W_c^{+} + U_cW_c^{\perp}\right)\tilde{Z}y_c$$

$$D(\lambda - \tilde{\lambda}_{BLUP_C}) = \sigma^2\left(\left[K, L\Delta\tilde{Z}(\tilde{Z}X_c)^{\perp}\right]W_c^{+}\tilde{Z} - L\right)\Delta\left(\left[K, L\Delta\tilde{Z}(\tilde{Z}X_c)^{\perp}\right]W_c^{+}\tilde{Z} - L\right)'$$

where $W_c = \left[\tilde{Z}X_c, \tilde{Z}\Delta\tilde{Z}(\tilde{Z}X_c)^{\perp}\right]$ and $U_c \in \mathbb{R}^{t\times(n+m)}$ is an arbitrary matrix.

4 Comparison Under the Models

The results of the relationships between any predictor and the BLUP for the same unknown parameters under the models by considering the MSEM sense are examined in the theorems given below.

Theorem 1 Suppose that λ in (18) is predictable under $\mathcal{S}$ in (15). Let $\tilde{\lambda}$ be any predictor for λ under $\mathcal{S}$. Denote

$$\Delta(\tilde{\lambda}, \tilde{\lambda}_{BLUP_{\mathcal{S}}}) = MSEM(\tilde{\lambda}) - MSEM(\tilde{\lambda}_{BLUP_{\mathcal{S}}})$$

and

$$E_1 = \begin{bmatrix} \Delta & \Delta L' & [X_c, Z_c] \\ L\Delta & L\Delta L' - \sigma^{-2}MSEM(\tilde{\lambda}) & [K, 0] \\ [X_c, Z_c] & [K, 0]' & 0 \end{bmatrix}$$
$$= \begin{bmatrix} D(y_c) & cov(y_c, \lambda) & [X_c, Z_c] \\ cov(\lambda, y_c) & \sigma^{-2}(D(\lambda) - MSEM(\tilde{\lambda})) & [K, 0] \\ [X_c, Z_c]' & [K, 0]' & 0 \end{bmatrix}.$$

Then,

1. $\tilde{\lambda}_{BLUP_{\mathcal{S}}}$ is superior to $\tilde{\lambda}$ according to the MSEM criterion, i.e., $\Delta(\tilde{\lambda}, \tilde{\lambda}_{BLUP_{\mathcal{S}}}) \succcurlyeq 0$, namely, $MSEM(\tilde{\lambda}) \succcurlyeq MSEM(\tilde{\lambda}_{BLUP_{\mathcal{S}}}) \Leftrightarrow i_{+}(E_1) = r[X_c, Z_c, \Delta]$.
2. $\tilde{\lambda}$ is superior to $\tilde{\lambda}_{BLUP_{\mathcal{S}}}$ according to the MSEM criterion, i.e., $\Delta(\tilde{\lambda}_{BLUP_{\mathcal{S}}}, \tilde{\lambda}) \succcurlyeq 0$, namely, $MSEM(\tilde{\lambda}_{BLUP_{\mathcal{S}}}) \succcurlyeq MSEM(\tilde{\lambda}) \Leftrightarrow i_{-}(E_1) = r[X_c, Z_c]$.

3. $\Delta(\tilde{\lambda}, \tilde{\lambda}_{\mathrm{BLUP}_S}) = 0$, namely, $\mathrm{MSEM}(\tilde{\lambda}) = \mathrm{MSEM}(\tilde{\lambda}_{\mathrm{BLUP}_S}) \Leftrightarrow \mathrm{r}(\mathrm{E}_1) = \mathrm{r}[\mathrm{X}_c, \mathrm{Z}_c, \Delta] + \mathrm{r}[\mathrm{X}_c, \mathrm{Z}_c]$.

Proof. Let say $\mathrm{G} = [\mathrm{X}_c, \mathrm{Z}_c]$, $\tilde{\mathrm{K}} = [\mathrm{K}, 0]$, and $\mathrm{C} = \mathrm{MSEM}(\tilde{\lambda})$. Then, using (9), (21), and (22),

$$\begin{aligned}
\mathrm{i}_{\pm}(\Delta(\tilde{\lambda}, \tilde{\lambda}_{\mathrm{BLUP}_S})) &= \mathrm{i}_{\pm}(\mathrm{MSEM}(\tilde{\lambda}) - \mathrm{MSEM}(\tilde{\lambda}_{\mathrm{BLUP}_S})) \\
&= \mathrm{i}_{\pm}(\mathrm{C} - \mathrm{D}(\lambda - \tilde{\lambda}_{\mathrm{BLUP}_S})) \\
&= \mathrm{i}_{\pm}\left(\mathrm{C} - \sigma^2([\tilde{\mathrm{K}}, \mathrm{L}\Delta\mathrm{G}^{\perp}]\mathrm{W}_s^{+} - \mathrm{L})\Delta([\tilde{\mathrm{K}}, \mathrm{L}\Delta\mathrm{G}^{\perp}]\mathrm{W}_s^{+} - \mathrm{L})'\right)
\end{aligned} \tag{25}$$

$$\begin{aligned}
&= \mathrm{i}_{\pm}\begin{bmatrix} \Delta & \Delta(\mathrm{W}_s^{+})'[\tilde{\mathrm{K}}, \mathrm{L}\Delta\mathrm{G}^{\perp}]' - \Delta\mathrm{L}' \\ [\tilde{\mathrm{K}}, \mathrm{L}\Delta\mathrm{G}^{\perp}]\mathrm{W}_s^{+}\Delta - \mathrm{E}\Delta & \sigma^{-2}\mathrm{C} \end{bmatrix} - \mathrm{i}_{\pm}(\Delta) \\
&= \mathrm{i}_{\pm}\left(\begin{bmatrix} \Delta & -\Delta\mathrm{L}' \\ -\mathrm{L}\Delta & \sigma^{-2}\mathrm{C} \end{bmatrix} + \begin{bmatrix} \Delta & 0 \\ 0 & [\tilde{\mathrm{K}}, \mathrm{L}\Delta\mathrm{G}^{\perp}] \end{bmatrix}\begin{bmatrix} 0 & \mathrm{W}_s \\ \mathrm{W}_s' & 0 \end{bmatrix}^{+}\begin{bmatrix} \Delta & 0 \\ 0 & [\tilde{\mathrm{K}}, \mathrm{L}\Delta\mathrm{G}^{\perp}]' \end{bmatrix}\right)
\end{aligned} \tag{26}$$

$$\begin{aligned}
&= \mathrm{i}_{\pm}\begin{bmatrix} 0 & -\mathrm{W}_s & \Delta & 0 \\ -\mathrm{W}_s' & 0 & 0 & [\tilde{\mathrm{K}}, \mathrm{L}\Delta\mathrm{G}^{\perp}] \\ \Delta & 0 & \Delta & -\Delta\mathrm{L}' \\ 0 & [\tilde{\mathrm{K}}, \mathrm{L}\Delta\mathrm{G}^{\perp}] & -\mathrm{L}\Delta & \sigma^{-2}\mathrm{C} \end{bmatrix} - i_{\mp}\begin{bmatrix} 0 & \mathrm{W}_s \\ \mathrm{W}_s' & 0 \end{bmatrix} - \mathrm{i}_{\pm}(\Delta) \\
&= \mathrm{i}_{\pm}\begin{bmatrix} -\Delta & -\mathrm{W}_s & \Delta\mathrm{L}' \\ -\mathrm{W}_s' & 0 & [\tilde{\mathrm{K}}, \mathrm{L}\Delta\mathrm{G}^{\perp}]' \\ \mathrm{L}\Delta & [\tilde{\mathrm{K}}, \mathrm{L}\Delta\mathrm{G}^{\perp}] & \sigma^{-2}\mathrm{C} - \mathrm{L}\Delta\mathrm{L}' \end{bmatrix} - \mathrm{r}(\mathrm{W}_s),
\end{aligned} \tag{27}$$

where $\mathrm{W}_s = [\mathrm{G}, \Delta G^{\perp}]$. Here (7) is applied to both expressions in (25) and (26) by using the fact $\Delta = \Delta\Delta^{+}\Delta$ and the column space inclusion $\Re\left([\tilde{\mathrm{K}}, \mathrm{L}\Delta\mathrm{G}^{\perp}]'\right) \subseteq \Re(\mathrm{W}_s')$, respectively, and further, (2) and (4) are used with the elementary block matrix operations while the expression in (27) is obtained.

By setting $\mathrm{W}_s = [\mathrm{G}, \Delta G^{\perp}]$ into (27) and using (3)–(5) with the elementary block matrix operations and also from the fact $\mathrm{r}[\mathrm{G}, \Delta G^{\perp}] = \mathrm{r}[\mathrm{G}, \Delta]$, the expression in (27) is equivalently written as follows:

$$
\mathrm{i}_{\pm}\begin{bmatrix} -\Delta & -\mathrm{G} & -\Delta \mathrm{G}^{\perp} & \Delta \mathrm{L}' \\ -\mathrm{G}' & 0 & 0 & \tilde{\mathrm{K}}' \\ -\mathrm{G}^{\perp}\Delta & 0 & 0 & \mathrm{G}^{\perp}\Delta \mathrm{L}' \\ \mathrm{L}\Delta & \tilde{\mathrm{K}} & \mathrm{L}\Delta G^{\perp} & \sigma^{-2}\mathrm{C} - \mathrm{L}\Delta \mathrm{L}' \end{bmatrix} - r[\mathrm{G}, \Delta]
$$

$$
= \mathrm{i}_{\pm}\begin{bmatrix} -\Lambda & -\mathrm{G} & \Delta \mathrm{L}' \\ -\mathrm{G}' & 0 & \tilde{\mathrm{K}}' \\ \mathrm{L}\Delta & \tilde{\mathrm{K}} & \sigma^{-2}\mathrm{C} - \mathrm{L}\Delta \mathrm{L} \end{bmatrix} + i_{\pm}(\mathrm{G}^{\perp}\Delta \mathrm{G}^{\perp}) - r[\mathrm{G}, \Delta] \tag{28}
$$

$$
= \mathrm{i}_{\mp}\begin{bmatrix} \Delta & \Delta L' & \mathrm{G} \\ \mathrm{L}\Delta & \mathrm{L}\Delta \mathrm{L}' - \sigma^{-2}\mathrm{C} & \tilde{\mathrm{K}} \\ \mathrm{G}' & \tilde{\mathrm{K}}' & 0 \end{bmatrix} + i_{\pm}\begin{bmatrix} \Delta & \mathrm{G} \\ \mathrm{G}' & 0 \end{bmatrix} - r(\mathrm{G}) - r[\mathrm{G}, \Delta].
$$

According to (6),

$$
\mathrm{i}_{+}\begin{bmatrix} \Delta & \mathrm{G} \\ \mathrm{G}' & 0 \end{bmatrix} = \mathrm{r}[\mathrm{G}, \Delta] \text{ and } \mathrm{i}_{-}\begin{bmatrix} \Delta & \mathrm{G} \\ \mathrm{G}' & 0 \end{bmatrix} = \mathrm{r}(\mathrm{G}) \tag{29}
$$

and then, combining (28) with (29) and writing E_1 instead of the first matrix in (28),

$$
\mathrm{i}_{+}\left(\mathrm{MSEM}(\tilde{\lambda}) - \mathrm{MSEM}(\tilde{\lambda}_{\mathrm{BLUP}_{\mathcal{S}}})\right) = \mathrm{i}_{-}(\mathrm{E}_1) - \mathrm{r}\left[\mathrm{X}_c, \mathrm{Z}_c\right], \tag{30}
$$

$$
\mathrm{i}_{-}\left(\mathrm{MSEM}(\tilde{\lambda}) - \mathrm{MSEM}(\tilde{\lambda}_{\mathrm{BLUP}_{\mathcal{S}}})\right) = \mathrm{i}_{+}(\mathrm{E}_1) - \mathrm{r}\left[\mathrm{X}_c, \mathrm{Z}_c, \Delta\right] \tag{31}
$$

are obtained. According to (1), adding the equalities in (30) and (31) yields

$$
\mathrm{r}\left(\mathrm{MSEM}(\tilde{\lambda}) - \mathrm{MSEM}(\tilde{\lambda}_{\mathrm{BLUP}_{\mathcal{S}}})\right) = \mathrm{r}(\mathrm{E}_1) - \mathrm{r}\left[\mathrm{X}_c, \mathrm{Z}_c\right] - \mathrm{r}\left[\mathrm{X}_c, \mathrm{Z}_c, \Delta\right]. \tag{32}
$$

Applying Lemma 1 to (30)–(32) and setting $\mathrm{C} = \mathrm{MSEM}(\tilde{\lambda})$ and $\mathrm{G} = \left[\mathrm{X}_c, \mathrm{Z}_c\right]$ in the first matrix in (28) yields 1–3. □

The proofs of the following theorems are obtained by using a similar way to the proof of Theorem 1 and are omitted.

Theorem 2 Suppose that λ in (18) is predictable under $\mathcal{R}$ in (14). Let $\tilde{\lambda}$ be any predictor for λ under $\mathcal{R}$. Denote

$$
\Delta(\tilde{\lambda}, \tilde{\lambda}_{\mathrm{BLUP}_{\mathcal{R}}}) = \mathrm{MSEM}(\tilde{\lambda}) - \mathrm{MSEM}(\tilde{\lambda}_{\mathrm{BLUP}_{\mathcal{R}}})
$$

and

$$
\mathrm{E}_2 = \begin{bmatrix} \Delta & \Delta \mathrm{L}' & \mathrm{X}_c \\ \mathrm{L}\Delta & \mathrm{L}\Delta \mathrm{L}' - \sigma^{-2}\mathrm{MSEM}(\tilde{\lambda}) & \mathrm{K} \\ \mathrm{X}_c' & \mathrm{K}' & 0 \end{bmatrix}
$$

$$= \begin{bmatrix} D(y_c) & cov(y_c, \lambda) & X_c \\ cov(\lambda, y_c) & \sigma^{-2}(D(\lambda) - MSEM(\lambda)) & K \\ X_c' & \tilde{K} & 0 \end{bmatrix}$$

Then,

1. $\tilde{\lambda}_{BLUP_{\mathcal{R}}}$ is superior to $\tilde{\lambda}$ according to the MSEM criterion, i.e., $\Delta(\tilde{\lambda}, \tilde{\lambda}_{BLUP_{\mathcal{R}}}) \succcurlyeq 0$, namely, $MSEM(\tilde{\lambda}) \succcurlyeq MSEM(\tilde{\lambda}_{BLUP_{\mathcal{R}}}) \Leftrightarrow i_+(E_2) = r[X_c, \Delta]$.
2. $\tilde{\lambda}$ is superior to $\tilde{\lambda}_{BLUP_{\mathcal{R}}}$ according to the MSEM criterion, i.e., $\Delta(\tilde{\lambda}_{BLUP_{\mathcal{R}}}, \tilde{\lambda}) \succcurlyeq 0$, namely, $MSEM(\tilde{\lambda}_{BLUP_{\mathcal{R}}}) \succcurlyeq MSEM(\tilde{\lambda}) \Leftrightarrow i_-(E_2) = r(X_c)$.
3. $\Delta(\tilde{\lambda}, \tilde{\lambda}_{BLUP_{\mathcal{R}}}) = 0$, namely, $MSEM(\tilde{\lambda}) = MSEM(\tilde{\lambda}_{BLUP_{\mathcal{R}}}) \Leftrightarrow r(E_2) = r[X_c, \Delta] + r(X_c)$.

Theorem 3 Suppose that λ in (18) is predictable under $\mathcal{C}$ in (16). Let $\tilde{\lambda}$ be any predictor for λ under $\mathcal{C}$. Denote

$$\Delta(\tilde{\lambda}, \tilde{\lambda}_{BLUP_{\mathcal{C}}}) = MSEM(\tilde{\lambda}) - MSEM(\tilde{\lambda}_{BLUP_{\mathcal{C}}})$$

and

$$E_3 = \begin{bmatrix} \tilde{Z}\Delta\tilde{Z} & \tilde{Z}\Delta L' & \tilde{Z}X_c \\ L\Delta\tilde{Z} & L\Delta L' - \sigma^{-2}MSEM(\tilde{\lambda}) & K \\ (\tilde{Z}X_c)' & K' & 0 \end{bmatrix}$$

$$= \begin{bmatrix} D(y_c) & cov(y_c, \lambda) & \tilde{Z}X_c \\ cov(\lambda, y_c) & \sigma^{-2}(D(\lambda) - MSEM(\tilde{\lambda})) & K \\ (\tilde{Z}X_c)' & K' & 0 \end{bmatrix}$$

Then,

1. $\tilde{\lambda}_{BLUP_{\mathcal{C}}}$ is superior to $\tilde{\lambda}$ according to the MSEM criterion, i.e., $\Delta(\tilde{\lambda}, \tilde{\lambda}_{BLUP_{\mathcal{C}}}) \succcurlyeq 0$, namely, $MSEM(\tilde{\lambda}) \succcurlyeq MSEM(\tilde{\lambda}_{BLUP_{\mathcal{C}}}) \Leftrightarrow i_+(E_3) = r[\tilde{Z}X_c, \tilde{Z}\Delta\tilde{Z}]$.
2. $\tilde{\lambda}$ is superior to $\tilde{\lambda}_{BLUP_{\mathcal{C}}}$ according to the MSEM criterion, i.e., $\Delta(\tilde{\lambda}_{BLUP_{\mathcal{C}}}, \tilde{\lambda}) \succcurlyeq 0$, namely, $MSEM(\tilde{\lambda}_{BLUP_{\mathcal{C}}}) \succcurlyeq MSEM(\tilde{\lambda}) \Leftrightarrow i_-(E_3) = r(\tilde{Z}X_c)$.
3. $\Delta(\tilde{\lambda}, \tilde{\lambda}_{BLUP_{\mathcal{C}}}) = 0$, namely, $MSEM(\tilde{\lambda}) = MSEM(\tilde{\lambda}_{BLUP_{\mathcal{C}}}) \Leftrightarrow r(E_3) = r[\tilde{Z}X_c, \tilde{Z}\Delta\tilde{Z}] + r(\tilde{Z}X_c)$.

The following results are easily derived from the above theorems. However, it would be useful to review some of predictability and estimability conditions beforehand. The conditions, obtained from Definition 4, are listed below.

1. $X_c\alpha$ is always estimable under $\mathcal{R}$.
2. $X_c\alpha$ is estimable under $\mathcal{S} \Leftrightarrow r[X_c, Z_c] = r(X_c) + r(Z_c)$.
3. $X_c\alpha$ is estimable under $\mathcal{C} \Leftrightarrow r(X_c) = k$.

Corollary 1 Let $\mathcal{R}$, $\mathcal{S}$, and $\mathcal{C}$ be as given in (14)–(16), respectively. Assume that

$\mu = X_c\alpha$ is estimable under $\mathcal{C}$ (also estimable under both $\mathcal{R}$ and $\mathcal{S}$). Let $\widehat{\mu_{\mathcal{S}}}$, $\widehat{\mu_{\mathcal{R}}}$ and $\widehat{\mu_{\mathcal{C}}}$ be any predictors for μ under $\mathcal{C}$, $\mathcal{R}$, and $\mathcal{S}$, respectively. Then, the following results hold. Denote

$$E_4 = \begin{bmatrix} \Delta & 0 & X_c & Z_c \\ 0 & \sigma^{-2}\mathrm{MSEM}\left(\widehat{\mu_{\mathcal{S}}}\right) & X_c & 0 \\ X_c' & X_c' & 0 & 0 \\ Z_c' & 0 & 0 & 0 \end{bmatrix},$$

$$E_5 = \begin{bmatrix} \Delta & 0 & X_c \\ 0 & -\sigma^{-2}\mathrm{MSEM}\left(\widehat{\mu_{\mathcal{R}}}\right) & X_c \\ X_c' & X_c' & 0 \end{bmatrix}, E_6 = \begin{bmatrix} \widetilde{Z}\Delta\widetilde{Z} & 0 & \widetilde{Z}X_c \\ 0 & -\sigma^{-2}\mathrm{MSEM}\left(\widehat{\mu_{\mathcal{C}}}\right) & X_c \\ (\widetilde{Z}X_c)' & X_c' & 0 \end{bmatrix}.$$

Then,

1.

$$\mathrm{MSEM}\left(\widehat{\mu_{\mathcal{S}}}\right) \succcurlyeq \mathrm{MSEM}\left(\hat{\mu}_{BLUE_{\mathcal{S}}}\right) \Leftrightarrow i_+(E_4) = r\left[X_c,\ Z_c,\ \Delta\right].$$

2.

$$\mathrm{MSEM}\left(\hat{\mu}_{BLUE_{\mathcal{S}}}\right) \succcurlyeq \mathrm{MSEM}\left(\widehat{\mu_{\mathcal{S}}}\right) \Leftrightarrow i_-(E_4) = r\left[X_c,\ Z_c\right] = r(X_c) + r(Z_c),$$

3.

$$\begin{aligned} \mathrm{MSEM}\left(\widehat{\mu_{\mathcal{S}}}\right) = \mathrm{MSEM}\left(\hat{\mu}_{\mathrm{BLUE_S}}\right) \Leftrightarrow r(E_4) &= r\left[X_c,\ Z_c,\ \Delta\right] + r\left[X_c,\ Z_c\right] \\ &= r\left[X_c,\ Z_c,\ \Delta\right] + r(X_c) + r(Z_c) \end{aligned}$$

4.

$$MSEM\ \left(\widehat{\mu_{\mathcal{R}}}\right) \succcurlyeq MSEM\ \left(\hat{\mu}_{BLUE_{\mathcal{R}}}\right) \Leftrightarrow i_+(E_5) = r\left[X_c,\ \Delta\right].$$

5.

$$\mathrm{MSEM}\left(\hat{\mu}_{\mathrm{BLUE}_{\mathcal{R}}}\right) \succcurlyeq \mathrm{MSEM}\left(\widehat{\mu_{\mathcal{R}}}\right) \Leftrightarrow i_-(E_5) = r(X_c).$$

6.

$$\mathrm{MSEM}\left(\widehat{\mu_{\mathcal{R}}}\right) = \mathrm{MSEM}\left(\hat{\mu}_{\mathrm{BLUE}_{\mathcal{R}}}\right) \Leftrightarrow r(E_5) = r\left[X_c,\ \Delta\right] + r(X_c).$$

7.

$$\mathrm{MSEM}\left(\widehat{\mu_{\mathcal{C}}}\right) \succcurlyeq \mathrm{MSEM}\left(\hat{\mu}_{\mathrm{BLUE_C}}\right) \Leftrightarrow i_+(E_6) = r\left[\widetilde{Z}X_c,\ \widetilde{Z}\Delta\widetilde{Z}\right].$$

8.

$$\mathrm{MSEM}(\hat{\mu}_{\mathrm{BLUE}_C}) \succcurlyeq \mathrm{MSEM}(\widehat{\mu_C}) \Leftrightarrow \mathrm{i}_-(\mathrm{E}_6) = \mathrm{r}(\widetilde{\mathrm{Z}}\mathrm{X}_c) = \mathrm{r}(\mathrm{X}_c).$$

9.

$$\begin{aligned}\mathrm{MSEM}(\widehat{\mu_C}) = \mathrm{MSEM}(\hat{\mu}_{\mathrm{BLUE}_C}) &\Leftrightarrow \mathrm{r}(\mathrm{E}_6)\\ &= \mathrm{r}\big[\,\tilde{\mathrm{Z}}\mathrm{X}_c,\ \tilde{\mathrm{Z}}\Delta\tilde{\mathrm{Z}}\,\big] + \mathrm{r}(\tilde{\mathrm{Z}}\mathrm{X}_c) = \mathrm{r}\big[\,\tilde{\mathrm{Z}}\mathrm{X}_c,\ \tilde{\mathrm{Z}}\Delta\tilde{\mathrm{Z}}\,\big] + \mathrm{r}(\mathrm{X}_c).\end{aligned}$$

5 Dispersion Matrices' Comparison of BLUPs

The results of the relationships between the BLUPs for the same unknown parameters under the models by considering the MSEM sense are examined in the theorems given below.

Theorem 4 Suppose that λ in (18) is predictable under $\mathcal{R}$ and $\mathcal{S}$ in (14) and (15), respectively. Denote

$$\Delta\big(\tilde{\lambda}_{\mathrm{BLUP}_\mathcal{S}}, \tilde{\lambda}_{\mathrm{BLUP}_\mathcal{R}}\big) = \mathrm{MSEM}\big(\tilde{\lambda}_{\mathrm{BLUP}_\mathcal{S}}\big) - \mathrm{MSEM}\big(\tilde{\lambda}_{\mathrm{BLUP}_\mathcal{R}}\big)$$

and

$$\mathrm{E}_7 = \begin{bmatrix} \Delta & \Delta & \Delta\mathrm{L}' & 0 & 0 & \mathrm{X}_c \\ \Delta & 0 & 0 & \mathrm{X}_c & \mathrm{Z}_c & 0 \\ \mathrm{L}\Delta & 0 & 0 & \mathrm{K} & 0 & 0 \\ 0 & \mathrm{X}_c' & \mathrm{K}' & 0 & 0 & 0 \\ 0 & \mathrm{Z}_c' & 0 & 0 & 0 & 0 \\ \mathrm{X}_c' & 0 & 0 & 0 & 0 & 0 \end{bmatrix}.$$

Then,

1. $\tilde{\lambda}_{\mathrm{BLUP}_\mathcal{R}}$ is superior to $\tilde{\lambda}_{\mathrm{BLUP}_\mathcal{S}}$ according to the MSEM criterion, i.e., $\Delta\big(\tilde{\lambda}_{\mathrm{BLUP}_\mathcal{S}}, \tilde{\lambda}_{\mathrm{BLUP}_\mathcal{R}}\big) \succ 0$, namely, $\mathrm{MSEM}\big(\tilde{\lambda}_{\mathrm{BLUP}_\mathcal{S}}\big) \succ \mathrm{MSEM}\big(\tilde{\lambda}_{\mathrm{BLUP}_\mathcal{R}}\big) \Leftrightarrow \mathrm{i}_+(\mathrm{E}_7) = \mathrm{r}\big[\,\mathrm{X}_c,\ \mathrm{Z}_c\,\big] + \mathrm{r}\big[\,\mathrm{X}_c,\ \Delta\,\big]$.
2. $\tilde{\lambda}_{\mathrm{BLUP}_\mathcal{S}}$ is superior to $\tilde{\lambda}_{\mathrm{BLUP}_\mathcal{R}}$ according to the MSEM criterion, i.e., $\Delta\big(\tilde{\lambda}_{\mathrm{BLUP}_\mathcal{R}}, \tilde{\lambda}_{\mathrm{BLUP}_\mathcal{S}}\big) \succ 0$, namely, $\mathrm{MSEM}\big(\tilde{\lambda}_{\mathrm{BLUP}_\mathcal{R}}\big) \succ \mathrm{MSEM}\big(\tilde{\lambda}_{\mathrm{BLUP}_\mathcal{S}}\big) \Leftrightarrow \mathrm{i}_-(\mathrm{E}_7) = \mathrm{r}\big[\,\mathrm{X}_c,\ \mathrm{Z}_c,\ \Delta\,\big] + \mathrm{r}(\mathrm{X}_c)$.
3. $\Delta\big(\tilde{\lambda}_{\mathrm{BLUP}_\mathcal{R}}, \tilde{\lambda}_{\mathrm{BLUP}_\mathcal{S}}\big) = 0$, namely, $\mathrm{MSEM}\big(\tilde{\lambda}_{\mathrm{BLUP}_\mathcal{R}}\big) = \mathrm{MSEM}\big(\tilde{\lambda}_{\mathrm{BLUP}_\mathcal{S}}\big) \Leftrightarrow \mathrm{r}(E_7) = \mathrm{r}\big[\,\mathrm{X}_c,\ \mathrm{Z}_c,\ \Delta\,\big] + \mathrm{r}(\mathrm{X}_c) + \mathrm{r}\big[\,\mathrm{X}_c,\ \mathrm{Z}_c\,\big] + \mathrm{r}\big[\,\mathrm{X}_c,\ \Delta\,\big]$.

Proof Let $\mathrm{G} = \big[\,\mathrm{X}_c,\ \mathrm{Z}_c\,\big]$ and $\tilde{\mathrm{K}} = \big[\,\mathrm{K},\ 0\,\big]$. By taking $\mathrm{C} = \mathrm{MSEM}\big(\tilde{\lambda}_{\mathrm{BLUP}_\mathcal{R}}\big) = \mathrm{D}\big(\lambda - \tilde{\lambda}_{\mathrm{BLUP}_\mathcal{R}}\big)$ in (28), we obtain

$$
\begin{aligned}
& i_{\pm}\left(\Delta\left(\tilde{\lambda}_{\mathrm{BLUP}_{\mathcal{R}}}, \tilde{\lambda}_{\mathrm{BLUP}_{\mathcal{S}}}\right)\right) \\
& = i_{\mp}\left(\begin{bmatrix} \Delta & \Delta L' & G \\ L\Delta & L\Delta L' & \tilde{K} \\ G' & \tilde{K}' & 0 \end{bmatrix} - \begin{bmatrix} 0 \\ I_t \\ 0 \end{bmatrix} D\left(\lambda - \tilde{\lambda}_{\mathrm{BLUP}_{\mathcal{R}}}\right)\left[0,\ I_t,\ 0\right]\right)
\end{aligned}
\tag{33}
$$

We can reapply (7)–(33) after setting $D\left(\lambda - \tilde{\lambda}_{\mathrm{BLUP}_{\mathcal{R}}}\right)$ in (24). Then in a similar way to obtaining (28), (33) is equivalently written as

$$
\begin{aligned}
& i_{\mp}\left(\begin{bmatrix} \Delta & 0 & -\Delta L' & 0 \\ 0 & \Delta & \Delta L' & G \\ -L\Delta & L\Delta & L\Delta L' & \tilde{K} \\ 0 & G' & \tilde{K}' & 0 \end{bmatrix} + \begin{bmatrix} \Delta & 0 \\ 0 & 0 \\ 0 & \left[K,\ L\Delta X_c^{\perp}\right] \\ 0 & 0 \end{bmatrix} \begin{bmatrix} 0 & W_r \\ W_r' & 0 \end{bmatrix}^{+}\right. \\
& \left.\times \begin{bmatrix} \Delta\ 0 & 0\ 0 \\ 0\ 0\ [K, & L\Delta X_c^{\perp}]\ 0 \end{bmatrix}\right) + i_{\pm}\begin{bmatrix} \Delta & G \\ G' & 0 \end{bmatrix} - r(G) - r\left[G\ \Delta\right] - i_{\mp}(\Delta).
\end{aligned}
\tag{34}
$$

We can apply (7) to the first inertia expression of (34) since.

$$
\Re(\Delta) \subseteq \Re(W_r) \text{ and } \Re\left(\left[K,\ L\Delta X_c^{\perp}\right]'\right) \subseteq \Re\left(W_r'\right),
$$

where $W_r = \left[X_c,\ \Delta X_c^{\perp}\right]$. From Lemma 2, and some as

$$
\begin{aligned}
& i_{\mp}\begin{bmatrix} 0 & -X_c & -\Delta X_c^{\perp} & \Delta & 0 & 0 & 0 \\ -X_c' & 0 & 0 & 0 & 0 & K' & 0 \\ -X_c^{\perp}\Delta & 0 & 0 & 0 & 0 & X_c^{\perp}\Delta L' & 0 \\ \Delta & 0 & 0 & \Delta & 0 & -\Delta L' & 0 \\ 0 & 0 & 0 & 0 & \Delta & \Delta L' & G \\ 0 & K & L\Delta X_c^{\perp} & -L\Delta & L\Delta & L\Delta L' & \tilde{K} \\ 0 & 0 & 0 & 0 & G' & \tilde{K}' & 0 \end{bmatrix} \\
& + i_{\pm}\begin{bmatrix} \Delta & G \\ G' & 0 \end{bmatrix} - r(G) - r\left[G,\ \Delta\right] - i_{\mp}(\Delta) - r\left[X_c,\ \Delta X_c^{\perp}\right] \\
& = i_{\mp}\begin{bmatrix} -\Delta & -X_c & -\Delta X_c^{\perp} & 0 & \Delta L' & 0 \\ -X_c' & 0 & 0 & 0 & K' & 0 \\ -X_c^{\perp}\Delta & 0 & 0 & 0 & X_c^{\perp}\Delta L' & 0 \\ 0 & 0 & 0 & \Delta & \Delta L' & G \\ L\Delta & K & L\Delta X_c^{\perp} & L\Delta & 0 & \tilde{K} \\ 0 & 0 & 0 & G' & \tilde{K}' & 0 \end{bmatrix} + i_{\pm}\begin{bmatrix} \Delta & G \\ G' & 0 \end{bmatrix} \\
& - r(G) - r\left[G,\ \Delta\right] - r\left[X_c,\ \Delta\right]
\end{aligned}
$$

$$= \mathrm{i}_{\mp}\begin{bmatrix} -\Delta & -X_c & 0 & \Delta L' & 0 \\ -X_c' & 0 & 0 & K' & 0 \\ 0 & 0 & \Delta & \Delta L' & G \\ L\Delta & K & L\Delta & 0 & \tilde{K} \\ 0 & 0 & G' & \tilde{K}' & 0 \end{bmatrix} + \mathrm{i}_{\pm}\begin{bmatrix} \Delta & G \\ G' & 0 \end{bmatrix} - r(G)$$
$$- r[G, \Delta] - r[X_c, \Delta] + \mathrm{i}_{\mp}(X_c^{\perp}\Delta X_c^{\perp})$$

$$= \mathrm{i}_{\pm}\begin{bmatrix} \Delta & \Delta & \Delta L' & 0 & X_c \\ \Delta & 0 & 0 & G & 0 \\ L\Delta & 0 & 0 & \tilde{K} & 0 \\ 0 & G' & \tilde{K}' & 0 & 0 \\ X_c' & 0 & 0 & 0 & 0 \end{bmatrix} + \mathrm{i}_{\pm}\begin{bmatrix} \Delta & G \\ G' & 0 \end{bmatrix} - r(G)$$
$$- r[G, \Delta] - r[X_c, \Delta] + \mathrm{i}_{\mp}(X_c^{\perp}\Delta X_c^{\perp})$$

$$= \mathrm{i}_{\pm}\begin{bmatrix} \Delta & \Delta & \Delta L' & 0 & 0 & X_c \\ \Delta & 0 & 0 & X_c & Z_c & 0 \\ L\Delta & 0 & 0 & K & 0 & 0 \\ 0 & X_c' & K' & 0 & 0 & 0 \\ 0 & Z_c' & 0 & 0 & 0 & 0 \\ X_c' & 0 & 0 & 0 & 0 & 0 \end{bmatrix} + \mathrm{i}_{\pm}\begin{bmatrix} \Delta & X_c & Z_c \\ X_c' & 0 & 0 \\ Z_c' & 0 & 0 \end{bmatrix} - r[X_c, Z_c]$$
$$-r[X_c, Z_c, \Delta] - r[X_c, \Delta] + \mathrm{i}_{\pm}\begin{bmatrix} \Delta & X_c \\ X_c' & 0 \end{bmatrix} - r(X_c) \tag{35}$$

According to (5) and (6), writing E_7 instead of the first matrix in (35),

$$\mathrm{i}_{+}\big(\mathrm{MSEM}(\tilde{\lambda}_{\mathrm{BLUP}_{\mathcal{R}}}) - \mathrm{MSEM}(\tilde{\lambda}_{\mathrm{BLUP}_{\mathcal{S}}})\big) = \mathrm{i}_{+}(E_7) - r[X_c, Z_c] - r[X_c, \Delta] \tag{36}$$

$$\mathrm{i}_{-}\big(\mathrm{MSEM}(\tilde{\lambda}_{\mathrm{BLUP}_{\mathcal{R}}}) - \mathrm{MSEM}(\tilde{\lambda}_{\mathrm{BLUP}_{\mathcal{S}}})\big) = \mathrm{i}_{-}(E_7) - r\Big[\, X_c, Z_c, \Delta \,\Big] - r(X_c) \tag{37}$$

are obtained. According to (1), adding the equalities in (36) and (37) yields

$$\begin{aligned} &r\big(\mathrm{MSEM}(\tilde{\lambda}_{\mathrm{BLUP}_{\mathcal{R}}}) - \mathrm{MSEM}(\tilde{\lambda}_{\mathrm{BLUP}_{\mathcal{S}}})\big) = r(E_7) - r[X_c, Z_c] - r(X_c) \\ &-r[X_c, \Delta] - r[X_c, Z_c\ \Delta] \end{aligned} \tag{38}$$

Applying Lemma 1 to (36)–(38) yields 1–3. □

Theorem 5 Suppose that λ in (18) is predictable under $\mathcal{R}$ and $\mathcal{C}$ in (14) and (16), respectively. Denote

$$\Delta\left(\tilde{\lambda}_{\mathrm{BLUP}_{\mathcal{R}}}, \tilde{\lambda}_{\mathrm{BLUP}_{\mathcal{C}}}\right) = \mathrm{MSEM}\left(\tilde{\lambda}_{\mathrm{BLUP}_{\mathcal{R}}}\right) - \mathrm{MSEM}\left(\tilde{\lambda}_{\mathrm{BLUP}_{\mathcal{C}}}\right)$$

and

$$\mathrm{E}_8 = \begin{bmatrix} \Delta & \Delta\widetilde{\mathrm{Z}} & \Delta\mathrm{L}' & 0 & \mathrm{X_c} \\ \widetilde{\mathrm{Z}}\Delta & 0 & 0 & \widetilde{\mathrm{Z}}\mathrm{X_c} & 0 \\ \mathrm{L}\Delta & 0 & 0 & \mathrm{K} & 0 \\ 0 & \mathrm{X_c'}\widetilde{\mathrm{Z}} & \mathrm{K}' & 0 & 0 \\ X_c' & 0 & 0 & 0 & 0 \end{bmatrix}.$$

Then,

1. $\tilde{\lambda}_{\mathrm{BLUP}_{\mathcal{R}}}$ is superior to $\tilde{\lambda}_{\mathrm{BLUP}_{\mathcal{C}}}$ according to the MSEM criterion, i.e., $\Delta\left(\tilde{\lambda}_{\mathrm{BLUP}_{\mathcal{C}}}, \tilde{\lambda}_{\mathrm{BLUP}_{\mathcal{R}}}\right)\succcurlyeq 0$, namely, $\mathrm{MSEM}\left(\tilde{\lambda}_{\mathrm{BLUP_C}}\right)\succcurlyeq\mathrm{MSEM}\left(\tilde{\lambda}_{\mathrm{BLUP}_{\mathcal{R}}}\right) \Leftrightarrow \mathrm{i}_{+}(\mathrm{E}_8) = \mathrm{r}\left[\mathrm{X_c},\ \Delta\right] + \mathrm{r}\left(\widetilde{\mathrm{Z}}\mathrm{X_c}\right)$.
2. $\tilde{\lambda}_{\mathrm{BLUP}_{\mathcal{C}}}$ is superior to $\tilde{\lambda}_{\mathrm{BLUP}_{\mathcal{R}}}$ according to the MSEM criterion, i.e., $\Delta\left(\tilde{\lambda}_{\mathrm{BLUP}_{\mathcal{R}}}, \tilde{\lambda}_{\mathrm{BLUP}_{\mathcal{C}}}\right)\succcurlyeq 0$, namely, $\mathrm{MSEM}\left(\tilde{\lambda}_{\mathrm{BLUP}_{\mathcal{R}}}\right)\succcurlyeq\mathrm{MSEM}\left(\tilde{\lambda}_{\mathrm{BLUP_C}}\right) \Leftrightarrow \mathrm{i}_{-}(\mathrm{E}_8) = \mathrm{r}(\mathrm{X_c}) + \mathrm{r}\left[\widetilde{\mathrm{Z}}\mathrm{X_c},\ \widetilde{\mathrm{Z}}\Delta\widetilde{\mathrm{Z}}\right]$.
3. $\Delta\left(\tilde{\lambda}_{\mathrm{BLUP}_{\mathcal{R}}}, \tilde{\lambda}_{\mathrm{BLUP}_{\mathcal{C}}}\right) = 0$, namely, $\mathrm{MSEM}\left(\tilde{\lambda}_{\mathrm{BLUP}_{\mathcal{R}}}\right) = \mathrm{MSEM}\left(\tilde{\lambda}_{\mathrm{BLUP_C}}\right) \Leftrightarrow \mathrm{r}(\mathrm{E}_8) = \mathrm{r}\left[\mathrm{X_c},\ \Delta\right] + \mathrm{r}\left(\tilde{\mathrm{Z}}\mathrm{X_c}\right) + \mathrm{r}(\mathrm{X_c}) + \mathrm{r}\left[\tilde{\mathrm{Z}}\mathrm{X_c},\ \tilde{\mathrm{Z}}\Delta\tilde{\mathrm{Z}}\right]$.

Theorem 6 Suppose that λ in (18) is predictable under $\mathcal{S}$ and $\mathcal{C}$ in (15) and (16), respectively. Denote

$$\Delta\left(\tilde{\lambda}_{\mathrm{BLUP}_{\mathcal{S}}}, \tilde{\lambda}_{\mathrm{BLUP}_{\mathcal{C}}}\right) = \mathrm{MSEM}\left(\tilde{\lambda}_{\mathrm{BLUP}_{\mathcal{S}}}\right) - \mathrm{MSEM}\left(\tilde{\lambda}_{\mathrm{BLUP}_{\mathcal{C}}}\right)$$

and

$$\mathrm{E}_9 = \begin{bmatrix} \Delta & \Delta & \Delta\mathrm{L}' & 0 & 0 & \mathrm{X_c} \\ \Delta & 0 & 0 & \mathrm{X_c} & \mathrm{Z_c} & 0 \\ \mathrm{L}\Delta & 0 & 0 & \mathrm{K} & 0 & 0 \\ 0 & \mathrm{X_c'} & \mathrm{K}' & 0 & 0 & 0 \\ 0 & \mathrm{Z_c'} & 0 & 0 & 0 & 0 \\ \mathrm{X_c'} & 0 & 0 & 0 & 0 & 0 \end{bmatrix}.$$

Then,

1. $\tilde{\lambda}_{\mathrm{BLUP}_{\mathcal{S}}}$ is superior to $\tilde{\lambda}_{\mathrm{BLUP}_{\mathcal{C}}}$ according to the MSEM criterion, i.e., $\Delta\left(\tilde{\lambda}_{\mathrm{BLUP}_{\mathcal{C}}}, \tilde{\lambda}_{\mathrm{BLUP}_{\mathcal{S}}}\right)\succcurlyeq 0$, namely, $\mathrm{MSEM}\left(\tilde{\lambda}_{\mathrm{BLUP_C}}\right)\succcurlyeq\mathrm{MSEM}\left(\tilde{\lambda}_{\mathrm{BLUP_S}}\right) \Leftrightarrow \mathrm{i}_{+}(\mathrm{E}_9) = \mathrm{r}\left[\mathrm{X_c},\ \mathrm{Z_c},\ \Delta\right] + r\left(\widetilde{\mathrm{Z}}\mathrm{X_c}\right)$.
2. $\tilde{\lambda}_{\mathrm{BLUP}_{\mathcal{C}}}$ is superior to $\tilde{\lambda}_{\mathrm{BLUP}_{\mathcal{S}}}$ according to the MSEM criterion, i.e., $\Delta\left(\tilde{\lambda}_{\mathrm{BLUP}_{\mathcal{S}}}, \tilde{\lambda}_{\mathrm{BLUP}_{\mathcal{C}}}\right)\succcurlyeq 0$, namely, $\mathrm{MSEM}\left(\tilde{\lambda}_{\mathrm{BLUP_S}}\right)\succcurlyeq\mathrm{MSEM}\left(\tilde{\lambda}_{\mathrm{BLUP_C}}\right) \Leftrightarrow \mathrm{i}_{-}(\mathrm{E}_9) = \mathrm{r}\left[\mathrm{X_c},\ \mathrm{Z_c}\right] + \mathrm{r}\left[\widetilde{\mathrm{Z}}\mathrm{X_c},\ \widetilde{\mathrm{Z}}\Delta\widetilde{\mathrm{Z}}\right]$.
3. $\Delta\left(\tilde{\lambda}_{\mathrm{BLUP}_{\mathcal{C}}}, \tilde{\lambda}_{\mathrm{BLUP}_{\mathcal{S}}}\right) = 0$, namely, $\mathrm{MSEM}\left(\tilde{\lambda}_{\mathrm{BLUP_C}}\right) = \mathrm{MSEM}\left(\tilde{\lambda}_{\mathrm{BLUP_S}}\right) \Leftrightarrow \mathrm{r}(\mathrm{E}_9) = \mathrm{r}\left[\mathrm{X_c},\ \mathrm{Z_c},\ \Delta\right] + \mathrm{r}\left(\tilde{\mathrm{Z}}\mathrm{X_c}\right) + \mathrm{r}\left[\mathrm{X_c},\ \mathrm{Z_c}\right] + \mathrm{r}\left[\tilde{\mathrm{Z}}\mathrm{X_c},\ \tilde{\mathrm{Z}}\Delta\tilde{\mathrm{Z}}\right]$.

Corollary 2 Let $\mathcal{R}$, $\mathcal{S}$, and $\mathcal{C}$ be as given in (14)–(16), respectively. Assume that $\mu = X_c\alpha$ is estimable under $\mathcal{R}$, $\mathcal{S}$, and $\mathcal{C}$. Then, the following results hold. Denote

$$E_{10} = \begin{bmatrix} \Delta & \Delta & 0 & X_c \\ \Delta & 0 & Z_c & 0 \\ 0 & Z_c' & 0 & 0 \\ X_c' & 0 & 0 & 0 \end{bmatrix}, E_{11} = \begin{bmatrix} \Delta & \Delta\tilde{Z} & X_c \\ \tilde{Z}\Delta & 0 & 0 \\ X_c' & 0 & 0 \end{bmatrix}$$

$$E_{12} = \begin{bmatrix} \Delta & \Delta\tilde{Z} & X_c & Z_c \\ \tilde{Z}\Delta & 0 & 0 & 0 \\ X_c' & 0 & 0 & 0 \\ Z_c' & 0 & 0 & 0 \end{bmatrix}.$$

Then,

$$\text{MSEM}(\hat{\mu}_{\text{BLUE}_\mathcal{S}}) \succcurlyeq \text{MSEM}(\hat{\mu}_{\text{BLUE}_\mathcal{R}}) \Leftrightarrow i_+(E_{10}) = r(Z_c) + r[\, X_c,\ \Delta \,]$$

$$\text{MSEM}(\hat{\mu}_{\text{BLUE}_\mathcal{R}}) \succcurlyeq \text{MSEM}(\hat{\mu}_{\text{BLUE}_\mathcal{S}}) \Leftrightarrow i_-(E_{10}) = r[\, X_c,\ Z_c,\ \Delta \,]$$

$$\text{MSEM}(\hat{\mu}_{\text{BLUE}_\mathcal{S}}) = \text{MSEM}(\hat{\mu}_{\text{BLUE}_\mathcal{R}}) \Leftrightarrow r(E_{10}) = r(Z_c) + r\big[\, X_c,\ \Delta \,\big] + r\big[\, X_c,\ Z_c,\ \Delta \,\big].$$

$$\text{MSEM}(\hat{\mu}_{\text{BLUE}_\mathcal{C}}) \succcurlyeq \text{MSEM}(\hat{\mu}_{\text{BLUE}_\mathcal{R}}) \Leftrightarrow i_+(E_{11}) = r[\, X_c,\ \Delta \,].$$

$$\text{MSEM}(\hat{\mu}_{\text{BLUE}_\mathcal{R}}) \succcurlyeq \text{MSEM}(\hat{\mu}_{\text{BLUE}_\mathcal{C}}) \Leftrightarrow i_-(E_{11}) = r[\, \tilde{Z}X_c,\ \tilde{Z}\Delta\tilde{Z} \,].$$

$$\text{MSEM}(\hat{\mu}_{\text{BLUE}_\mathcal{R}}) = \text{MSEM}(\hat{\mu}_{\text{BLUE}_\mathcal{C}}) \Leftrightarrow r(E_{11}) = r[\, \tilde{Z}X_c,\ \tilde{Z}\Delta\tilde{Z} \,] + r[\, X_c,\ \Delta \,].$$

$$\text{MSEM}(\hat{\mu}_{\text{BLUE}_\mathcal{C}}) \succcurlyeq \text{MSEM}(\hat{\mu}_{\text{BLUE}_\mathcal{S}}) \Leftrightarrow i_+(E_{12}) = r[\, X_c,\ Z_c,\ \Delta \,].$$

$$\text{MSEM}(\hat{\mu}_{\text{BLUE}_\mathcal{S}}) \succcurlyeq \text{MSEM}(\hat{\mu}_{\text{BLUE}_\mathcal{C}}) \Leftrightarrow i_-(E_{12}) = r[\, \tilde{Z}X_c,\ \tilde{Z}\Delta\tilde{Z} \,] + r(Z_c).$$

$$\text{MSEM}(\hat{\mu}_{\text{BLUE}_\mathcal{C}}) = \text{MSEM}(\hat{\mu}_{\text{BLUE}_\mathcal{S}}) \Leftrightarrow r(E_{12}) = r\big[\, \tilde{Z}X_c,\ \tilde{Z}\Delta\tilde{Z} \,\big] + r(Z_c) + r\big[\, X_c,\ Z_c,\ \Delta \,\big].$$

Remark 1 When we consider the linear model in (10) with a linear exact restriction, i.e.,

$$y = X\alpha + \varepsilon \text{ with, } R\alpha = r \tag{39}$$

where $E(\varepsilon) = 0$ and $D(\varepsilon) = \sigma^2\Gamma$, and thereby, the combined form of (39) is expressed as

$$y_c = X_c\alpha + \varepsilon_c,\ y_c = \begin{bmatrix} y \\ r \end{bmatrix},\ X_c = \begin{bmatrix} X \\ R \end{bmatrix},\ \varepsilon_c = \begin{bmatrix} \varepsilon \\ 0 \end{bmatrix},$$

where $D(\varepsilon_c) = \sigma^2 \begin{bmatrix} \Gamma & 0 \\ 0 & 0 \end{bmatrix} := \sigma^2 \tilde{\Delta}$. Then, Theorem 2 reduces to [16, Theorem 3.1] and the items 4–6 of Corollary 1 reduce to [39, Theorem 3.3] by setting $\tilde{\Delta}$ instead of Δ in our results.

Remark 2 When we consider the linear model in (10) without any restriction, i.e.,

$$y = X\alpha + \varepsilon,$$

where $D(\varepsilon) = \sigma^2\Gamma$, and thereby, the combined form of (39) is expressed as

$$y_c = X_c\alpha + \varepsilon_c,\ y_c = \begin{bmatrix} y \\ r \end{bmatrix},\ X_c = \begin{bmatrix} X \\ R \end{bmatrix},\ \varepsilon_c = \begin{bmatrix} \varepsilon \\ 0 \end{bmatrix},$$

where $D(\varepsilon) = \sigma^2\Gamma$, Theorem 2 reduces to [38, Theorem 4.1] and the items 4–6 of Corollary 1 reduce to [37, Theorem 1] by setting $\tilde{\Delta} = \begin{bmatrix} \Gamma & 0 \\ 0 & 0 \end{bmatrix}$ instead of Δ, $\begin{bmatrix} X \\ 0 \end{bmatrix}$ instead of X_c, and $\begin{bmatrix} y \\ 0 \end{bmatrix}$ instead of y_c in our results.

6 Numerical Examples

We now use Example 4.10.2 from Sect. 4.10 of [51] to illustrate our theoretical findings. Consider the models $y = X\alpha + \varepsilon$ and $y = X\alpha + Z\gamma + \varepsilon$ with

$$X = \begin{bmatrix} \mathbf{1} & \mathbf{0} \\ \mathbf{1} & \mathbf{0} \\ \mathbf{0} & \mathbf{1} \\ \mathbf{0} & \mathbf{1} \end{bmatrix},\ Z = \begin{bmatrix} \mathbf{1} & \mathbf{1} & \mathbf{0} \\ \mathbf{1} & \mathbf{0} & \mathbf{1} \\ \mathbf{1} & \mathbf{1} & \mathbf{0} \\ \mathbf{1} & \mathbf{0} & \mathbf{1} \end{bmatrix},\ \alpha = \begin{bmatrix} \tau_1 \\ \tau_2 \end{bmatrix},\ \text{and } \gamma = \begin{bmatrix} \mu \\ \beta_1 \\ \beta_2 \end{bmatrix}$$

where $\mathbf{1} \in \mathbb{R}^{10\times 1}$ and $\mathbf{0} \in \mathbb{R}^{10\times 1}$, and consider the stochastic restriction $R\alpha + e = r$ with $R = \begin{bmatrix} 1 & -2 \\ 1 & 1 \end{bmatrix}$ and $D(e) = \Lambda = I_2$. This restriction may be a known fact from the theory or experiment view. Let consider

$$\mathrm{D}(\varepsilon) = \Gamma = \begin{bmatrix} 3\mathbf{11}' & 2\mathbf{11}' & 2\mathbf{11}' & \mathbf{11}' \\ 2\mathbf{11}' & 3\mathbf{11}' & \mathbf{11}' & 2\mathbf{11}' \\ 2\mathbf{11}' & \mathbf{11}' & 3\mathbf{11}' & 2\mathbf{11}' \\ \mathbf{11}' & 2\mathbf{11}' & 2\mathbf{11}' & 3\mathbf{11}' \end{bmatrix}$$

with $\sigma^2 = 1$ as in [1, 52]. This designed set-up is typical in agricultural experiments where several treatments are applied to various blocks of land. The experiment is often conducted to assess the differential impact of the treatments.

As is well-known, two types of commonly used estimators under general linear models are the BLUEs and the ordinary least squares estimators (OLSEs). Both of these estimators have a variety of simple and remarkable properties. Since these estimators are defined by different optimality criteria and thereby their expressions and properties are not necessarily the same, it is natural to seek possible connections between them. We use the above numerical example to illustrate the comparison results obtained in Sect. 4 by considering the BLUE and the OLSE as two estimators of $\mathrm{X}_c\alpha$ under $\mathcal{S}$. Note that $\mathrm{X}_c\alpha$ is estimable under $\mathcal{S}$. The relationship between the BLUE and the OLSE of $\mathrm{X}_c\alpha$, denoted as $\widehat{\mathrm{X}_c\alpha}_{\mathrm{OLSE}_{\mathcal{S}}}$, under the model $\mathcal{S}$ are presented as follows. Direct calculations show that

$$\mathrm{MSEM}\left(\widehat{X_c\alpha}_{\mathrm{OLSE}_{\mathcal{S}}}\right) = \begin{bmatrix} 1.13\mathbf{11}' & 1.13\mathbf{11}' & -0.03\mathbf{11}' & -0.03\mathbf{11}' & 1.19\mathbf{1} & 1.10\mathbf{1} \\ 1.13\mathbf{11}' & 1.13\mathbf{11}' & -0.03\mathbf{11}' & -0.03\mathbf{11}' & 1.19\mathbf{1} & 1.10\mathbf{1} \\ -0.03\mathbf{11}' & -0.03\mathbf{11}' & 0.26\mathbf{11}' & 0.26\mathbf{11}' & -0.55\mathbf{1} & 0.23\mathbf{1} \\ -0.03\mathbf{11}' & -0.03\mathbf{11}' & 0.26\mathbf{11}' & 0.26\mathbf{11}' & -0.55\mathbf{1} & 0.23\mathbf{1} \\ 1.19\mathbf{1}' & 1.19\mathbf{1}' & -0.55\mathbf{1}' & -0.55\mathbf{1}' & 2.29 & 0.64 \\ 1.10\mathbf{1}' & 1.10\mathbf{1}' & 0.23\mathbf{1}' & 0.23\mathbf{1}' & 0.64 & 1.32 \end{bmatrix}$$

and

$$\mathrm{MSEM}\left(\widehat{X_c\alpha}_{\mathrm{BLUE}_{\mathcal{S}}}\right) = \begin{bmatrix} 0.48\mathbf{11}' & 0.48\mathbf{11}' & 0.13\mathbf{11}' & 0.13\mathbf{11}' & 0.22\mathbf{1} & 0.61\mathbf{1} \\ 0.48\mathbf{11}' & 0.48\mathbf{11}' & 0.13\mathbf{11}' & 0.13\mathbf{11}' & 0.22\mathbf{1} & 0.61\mathbf{1} \\ 0.13\mathbf{11}' & 0.13\mathbf{11}' & 0.22\mathbf{11}' & 0.22\mathbf{11}' & -0.30\mathbf{1} & 0.35\mathbf{1} \\ 0.13\mathbf{11}' & 0.13\mathbf{11}' & 0.22\mathbf{11}' & 0.22\mathbf{11}' & -0.30\mathbf{1} & 0.35\mathbf{1} \\ 0.22\mathbf{1}' & 0.22\mathbf{1}' & -0.30\mathbf{1}' & -0.30\mathbf{1}' & 0.83 & -0.09 \\ 0.61\mathbf{1}' & 0.61\mathbf{1}' & 0.35\mathbf{1}' & 0.35\mathbf{1}' & -0.09 & 0.96 \end{bmatrix}.$$

Using the NumPy library in Python and setting the above considerations and findings in the matrix E_4 in Corollary 1, we find the eigenvalues of the matrix E_4 as

$$-25.81, -9.33, -1.09, -0.95, -0.77, 2.05, 3.65, 20.95, 21, 80.99.$$

Thus, we can see that $\mathrm{i}_+(\mathrm{E}_4) = 5$, $\mathrm{i}_-(\mathrm{E}_4) = 5$ and thereby $\mathrm{r}(\mathrm{E}_4) = 10$. We also obtain $\mathrm{r}\left[\mathrm{X}_c, \mathrm{Z}_c, \Delta\right] = 5$ and $\mathrm{r}\left[\mathrm{X}_c, \mathrm{Z}_c\right] = 4$. Therefore, $\mathrm{i}_+(\mathrm{E}_4) = \mathrm{r}\left[\mathrm{X}_c, \mathrm{Z}_c, \Delta\right] \Leftrightarrow \mathrm{MSEM}\left(\widehat{X_c\alpha}_{\mathrm{OLSE}_{\mathcal{S}}}\right) \succcurlyeq \mathrm{MSEM}\left(\widehat{X_c\alpha}_{\mathrm{BLUE}_{\mathcal{S}}}\right)$ holds. We note that this inequality always holds. This is a well-known result in statistical theory. Since $\mathrm{r}\left[\mathrm{X}_c,\ \mathrm{Z}_c\right] = 4$, i.e., $\mathrm{i}_-(\mathrm{E}_4) \neq \mathrm{r}\left[\mathrm{X}_c,\ \mathrm{Z}_c\right]$, $\mathrm{MSEM}\left(\widehat{X_c\alpha}_{\mathrm{BLUE}_\mathcal{S}}\right) \succcurlyeq \mathrm{MSEM}\left(\widehat{X_c\alpha}_{\mathrm{OLSE}_\mathcal{S}}\right)$ is not provided. It is also noted that since it is easily seen from the calculations that the difference $\mathrm{MSEM}\left(\widehat{\mathrm{X}_c\alpha}_{\mathrm{OLSE}_\mathcal{S}}\right) - \mathrm{MSEM}\left(\widehat{\mathrm{X}_c\alpha}_{\mathrm{BLUE}_\mathcal{S}}\right)$ is positive semi-definite, we observe that the inequality $\mathrm{MSEM}\left(\widehat{X_c\alpha}_{\mathrm{OLSE}_\mathcal{S}}\right) \succcurlyeq \mathrm{MSEM}\left(\widehat{X_c\alpha}_{\mathrm{BLUE}_\mathcal{S}}\right)$ holds.

Now, we use again the above numerical example to illustrate the comparison results obtained in Sect. 5 by considering the BLUEs of $\mathrm{X}_c\alpha$ under the models $\mathcal{R}$ and $\mathcal{S}$. Direct calculations show that

$$\mathrm{MSEM}\left(\widehat{\mathrm{X}_c\alpha}_{\mathrm{BLUE}_\mathcal{R}}\right) = \begin{bmatrix} 0.44\mathbf{1}\mathbf{1}' & 0.44\mathbf{1}\mathbf{1}' & 0.11\mathbf{1}\mathbf{1}' & 0.11\mathbf{1}\mathbf{1}' & 0.22\mathbf{1} & 0.54\mathbf{1} \\ 0.44\mathbf{1}\mathbf{1}' & 0.44\mathbf{1}\mathbf{1}' & 0.11\mathbf{1}\mathbf{1}' & 0.11\mathbf{1}\mathbf{1}' & 0.22\mathbf{1} & 0.54\mathbf{1} \\ 0.11\mathbf{1}\mathbf{1}' & 0.11\mathbf{1}\mathbf{1}' & 0.2\mathbf{1}\mathbf{1}' & 0.2\mathbf{1}\mathbf{1}' & -0.30\mathbf{1} & 0.31\mathbf{1} \\ 0.11\mathbf{1}\mathbf{1}' & 0.11\mathbf{1}\mathbf{1}' & 0.2\mathbf{1}\mathbf{1}' & 0.2\mathbf{1}\mathbf{1}' & -0.30\mathbf{1} & 0.31\mathbf{1} \\ 0.22\mathbf{1}' & 0.22\mathbf{1}' & -0.30\mathbf{1}' & -0.30\mathbf{1}' & 0.83 & -0.08 \\ 0.54\mathbf{1}' & 0.54\mathbf{1}' & 0.31\mathbf{1}' & 0.31\mathbf{1}' & -0.08 & 0.85 \end{bmatrix}.$$

Using the NumPy library in Python and setting the above considerations and findings in the matrix E_{10} in Corollary 2, we find the eigenvalues of the matrix E_{10} as

$$-\,50.41, -13.45, -12.82, -2.02, -1.24, -0.07,$$
$$0.81, 0.91, 2.23, 3.03, 32.53, 32.81, 129.68.$$

Thus, we can see that $\mathrm{i}_+(\mathrm{E}_{10}) = 7$, $\mathrm{i}_-(\mathrm{E}_{10}) = 6$ and thereby $\mathrm{r}(\mathrm{E}_{10}) = 13$. We also obtain $\mathrm{r}\left[\mathrm{X}_c,\ \mathrm{Z}_c,\ \Delta\right] = 5$, $\mathrm{r}\left[\mathrm{X}_c,\ \Delta\right] = 5$, and $\mathrm{r}(\mathrm{Z}_c) = 2$. Therefore, $\mathrm{i}_+(\mathrm{E}_{10}) = 7 = \mathrm{r}(\mathrm{Z}_c) + \mathrm{r}\left[\mathrm{X}_c,\ \Delta\right]$, i.e., $\mathrm{MSEM}\left(\widehat{X_c\alpha}_{\mathrm{BLUE}_\mathcal{S}}\right) \succcurlyeq \mathrm{MSEM}\left(\widehat{X_c\alpha}_{\mathrm{BLUE}_\mathcal{R}}\right)$ holds. Furthermore, $\mathrm{MSEM}\left(\widehat{X_c\alpha}_{\mathrm{BLUE}_\mathcal{R}}\right) \succcurlyeq \mathrm{MSEM}\left(\widehat{X_c\alpha}_{\mathrm{BLUE}_\mathcal{S}}\right)$ is not provided since $\mathrm{i}_-(\mathrm{E}_{10}) \neq \mathrm{r}\left[\mathrm{X}_c,\ \mathrm{Z}_c,\ \Delta\right]$. We also mention that since it is easily seen from the calculations that the difference $\mathrm{MSEM}\left(\widehat{\mathrm{X}_c\alpha}_{\mathrm{BLUE}_\mathcal{S}}\right) - \mathrm{MSEM}\left(\widehat{\mathrm{X}_c\alpha}_{\mathrm{BLUE}_\mathcal{R}}\right)$ is positive semi-definite, we observe that the inequality $\mathrm{MSEM}\left(\widehat{X_c\alpha}_{\mathrm{BLUE}_\mathcal{S}}\right) \succcurlyeq \mathrm{MSEM}\left(\widehat{X_c\alpha}_{\mathrm{BLUE}_\mathcal{R}}\right)$ already holds.

7 Conclusions

The results of this study provide a broad view of stochastically restricted linear models with superfluous variables in the context of comparability issues of predictors/estimators in accordance with the MSEM criterion using the block matrix inertias and ranks technique. In order to compare any predictor with the BLUP for the same unified form of all unknown parameters under these models, in which explicitly stochastically restricted linear models are transformed into implicitly stochastically restricted linear models, the MSEM criterion is taken into consideration. We then compare the MSEMs of two BLUPs under the models. Similar comparisons are also derived for using the BLUEs under the models in the MSEM sense. Furthermore, we illustrate our findings with numerical examples. We also note that, as stated in Remarks 1 and 2, we carry forward some of the outcomes of [16, 37–39].

The effectiveness of a new biased or unbiased predictor/estimator according to the dispersion matrix of BLUP/BLUE can be tested utilizing a variety of inequalities and equalities derived among MSEMs based on inertias and ranks. Since the conclusions in this study are all presented through certain explicit expressions and equivalences, we believe that the study can serve as a group of theoretical references in the statistical analysis of various subsequent problems regarding three competing models.

References

1. Ren, X.: Corrigendum to "On the equivalence of the BLUEs under a general linear model and its restricted and stochastically restricted models" [Statist. Probab. Lett. 90 (2014) 1–10]. Stat. Probab. Lett. 104, 181–185 (2015)
2. Haupt, H., Oberhofer, W.: Stochastic response restrictions. J. Multivar. Anal. **95**(1), 66–75 (2005)
3. Baksalary, J.K.: Comparing stochastically restricted estimators in a linear regression model. Biom. J. **26**(5), 555–557 (1984)
4. Haslett, S.J., Puntanen, S.: Equality of BLUEs or BLUPs under two linear models using stochastic restrictions. Stat. Pap. **51**, 465–475 (2010)
5. Liski, E.P.: Comparing stochastically restricted linear estimators in a regression model. Biom. J. **31**(3), 313–316 (1989)
6. Ren, X.: The equalities of estimations under a general partitioned linear model and its stochastically restricted model. Commun. Stat. – Theory Methods 45(22), 6495–6509 (2016)
7. Theil, H.: On the use of incomplete prior information in regression analysis. J. Am. Stat. Assoc. **58**(302), 401–414 (1963)
8. Theil, H., Goldberger, A.S.: On pure and mixed statistical estimation in economics. Int. Econ. Rev. **2**(1), 65–78 (1961)
9. Trenkler, G.: A note on comparing stochastically restricted linear estimators in a regression model. Biom. J. **35**(1), 125–128 (1993)
10. Xu, J., Yang, H.: Estimation in singular linear models with stochastic linear restrictions. Commun. Stat. - Theory Methods **36**(10), 1945–1951 (2007)
11. Güler, N., Büyükkaya, M.E.: Further remarks on constrained over-parameterized linear models. Stat. Pap., 65(2), 975–988 (2024)
12. Haupt, H., Oberhofer, W.: Fully restricted linear regression: a pedagogical note. Econ. Bull. **3**(1), 1–7 (2002)
13. Jiang, B., Tian, Y.: On best linear unbiased estimation and prediction under a constrained linear random-effects model. J. Ind. Manag. Optim. **19**(2), 852–867 (2023)
14. Li, W., Tian, Y., Yuan, R.: Statistical analysis of a linear regression model with restrictions and superfluous variables. J. Ind. Manag. Optim. **19**(5), 3107–3127 (2023)
15. Ma, R., Tian, Y.: A matrix approach to a general partitioned linear model with partial parameter restrictions. Linear Multilinear Algebra **70**(13), 2513–2532 (2022)

16. Tian, Y., Wang, J.: Some remarks on fundamental formulas and facts in the statistical analysis of a constrained general linear model. Commun. Stat. - Theory Methods **49**(5), 1201–1216 (2020)
17. Bhimasankaram, P., Jammalamadaka, S.R.: Updates of statistics in a general linear model: A statistical interpretation and applications. Commun. Stat. - Simul. Comput. **23**(3), 789–801 (1994)
18. Dube, M.: Mixed regression estimator under inclusion of some superfluous variables. TEST **8**(2), 411–417 (1999)
19. Güler, N., Puntanen, S., Özdemir, H.: On the BLUEs in two linear models via C. R. Rao's Pandora's Box. Commun. Statist. Theory Meth. 43(5), 921–931 (2014)
20. Hodges, J.S.: Richly Parameterized Linear Models: Additive, Time Series, and Spatial Models Using Random Effects. CRC Press, Boca Raton (2014)
21. Isotalo, J., Puntanen, S., Styan, G.P.H.: Effect of adding regressors on the equality of the OLSE and BLUE. Int. J. Stat. Sci. **6**, 193–201 (2007)
22. Jammalamadaka, S.R., Sengupta, D.: Inclusion and exclusion of data or parameters in the general linear model. Stat. Probab. Lett. **77**(12), 1235–1247 (2007)
23. Jun, S.J., Pinkse, J.: Adding regressors to obtain efficiency. Econom. Theory **25**(1), 298–301 (2009)
24. Kadiyala, K.: Mixed regression estimator under misspecification. Econ. Lett. **21**(1), 27–30 (1986)
25. Mitra, S.K., Moore, B.J.: Gauss–Markov estimation with an incorrect dispersion matrix. Sankhya: Indian J. Stat. 35(2), 139–152 (1973)
26. Qian, H., Tian, Y.: Partially superfluous observations. Econom. Theory **22**(3), 529–536 (2006)
27. Tian, Y.: On equalities for BLUEs under misspecified Gauss-Markov models. Acta Math. Sin. Engl. Ser. **25**(11), 1907–1920 (2009)
28. Tian, Y., Jiang, B.: A new analysis of the relationships between a general linear model and its mis-specified forms. J. Korean Stat. Soc. **46**(2), 182–193 (2017)
29. Wijekoon, P., Trenkler, G.: Mean square error matrix superiority of estimators under linear restrictions and misspecification. Econ. Lett. **30**(2), 141–149 (1989)
30. Büyükkaya, M.E.: Characterizing relationships between BLUPs under linear mixed model and some associated reduced models. Commun. Stat. -Simul. Comput. **52**(8), 3438–3451 (2023)
31. Güler, N.: On relations between BLUPs under two transformed linear random-effects models. Commun. Stat. Simul. Comput. **51**(9), 5099–5125 (2022)
32. Güler, N., Büyükkaya, M.E.: Notes on comparison of covariance matrices of BLUPs under linear random-effects model with its two sub-sample models. Iranian J. Sci. Tech. Transactions A **43**(6), 2993–3002 (2019)
33. Güler, N., Büyükkaya, M.E.: Rank and inertia formulas for covariance matrices of BLUPs in general linear mixed models. Commun. Stat. - Theory Methods **50**(21), 4997–5012 (2021)
34. Güler, N., Büyükkaya, M.E.: Some remarks on comparison of predictors in seemingly unrelated linear mixed models. Appl. Math. **67**, 525–542 (2022)
35. Güler, N., Büyükkaya, M.E.: Inertia and rank approach in transformed linear mixed models for comparison of BLUPs. Commun. Statist. Theory Methods **52**, 3108–3123 (2023)
36. Güler, N., Büyükkaya, M.E., Yiğit, M.: Comparison of covariance matrices of predictors in seemingly unrelated regression models. Indian J. Pure Appl. Math. **53**, 801–809 (2022)
37. Tian, Y.: Some equalities and inequalities for covariance matrices of estimators under linear model. Stat. Pap. **58**, 467–484 (2017)
38. Tian, Y.: Matrix rank and inertia formulas in the analysis of general linear models. Open Math. **15**(1), 126–150 (2017)
39. Tian, Y., Guo, W.: On comparison of dispersion matrices of estimators under a constrained linear model. Stat. Methods Appl. **25**(4), 623–649 (2016)
40. Tian, Y.: Equalities and inequalities for inertias of Hermitian matrices with applications. Linear Algebra Appl. **433**(1), 263–296 (2010)
41. Rao, C.R., Shalabh, Toutenburg, H., C., H.: Linear Models and Generalizations: Least Squares and Alternatives. Springer, (2008)

42. Yang, H., Ye, H., Xue, K.: A further study of predictions in linear mixed models. Commun. Stat. - Theory Methods **43**(20), 4241–4252 (2014)
43. Rao, C.R.: Representations of best linear unbiased estimators in the Gauss-Markoff model with a singular dispersion matrix. J. Multivar. Anal. **3**(3), 276–292 (1973)
44. Alalouf, I.S., Styan, G.P.H.: Characterizations of estimability in the general linear model. Ann. Stat. **7**(1), 194–200 (1979)
45. Lu, C., Gan, S., Tian, Y.: Some remarks on general linear model with new regressors. Stat. Probab. Lett. **97**, 16–24 (2015)
46. Goldberger, A.S.: Best linear unbiased prediction in the generalized linear regression model. J. Am. Stat. Assoc. **57**(298), 369–375 (1962)
47. Puntanen, S., Styan, G.P.H., Isotalo, J.: Matrix Tricks for Linear Statistical Models: Our Personal Top Twenty. Springer, Heidelberg (2011)
48. Dong, B., Guo, W., Tian, Y.: On relations between BLUEs under two transformed linear models. J. Multivar. Anal. **131**, 279–292 (2014)
49. Tian, Y.: Transformation approaches of linear random-effects models. Stat. Methods Appl. **26**, 583–608 (2017)
50. Tian, Y., Puntanen, S.: On the equivalence of estimations under a general linear model and its transformed models. Linear Algebra Appl. **430**(10), 2622–2641 (2009)
51. Sengupta, D., Jammalamadaka, S.R.: Linear Models: An Integrated Approach. World Scientific Press, Singapore (2003)
52. Güler, N., Büyükkaya, M.E.: Statistical inference of a stochastically restricted linear mixed model. AIMS Mathematics **8**(10), 24401–24417 (2023)

Approximation by Bivariate Complex Stancu–Schurer Polynomials in Compact Disks

Nesibe Manav Mutlu

Abstract This paper focuses on investigating the ability of the tensor product type bivariate complex Schurer–Stancu operators to approximate analytic functions on compact polydisks. The study derives the simultaneous approximation order and Voronovskaja-type outcomes for these polynomials using a qualitative estimate. The results provide insights into the accuracy of these polynomials in approximating analytic functions and their associated compact polydisks.

Keywords Stancu–Schurer operators · Bivariate complex operators · Rate of convergence · Exact order

1 Introduction

The problem of approximation to the analytic function $f(z)$ of the Bernstein sequence of polynomials was studied by Wright [8], Kantorovich [5] and then by Bernstein [2]. Lorentz was the first to investigate the complex Bernstein polynomials' approximation characteristics [6]. Gal just completed a noteworthy piece of work in this area and collected the key works in his book [4].

We had some inspiration points on the way to our main goal. The first of these was the study [7], in which Stancu defined the operators of Bernstein type, known by his own name using a probabilistic method as follows. These operators depend on a non-negative integer r:

$$L_{n,r}(f,x) = \sum_{k=0}^{n} w_{n,k,r}(x) f\left(\frac{k}{n}\right), \quad x \in [0,1] \tag{1}$$

where

N. Manav Mutlu (✉)
İstanbul Nişantaşı University, İstanbul, Türkiye
e-mail: nesibe.manavm@gmail.com

O. Castillo et al. (eds.), *Advanced Mathematics for the Modeling and Solution of Challenging Problems in Engineering*, Studies in Systems, Decision and Control 582,
https://doi.org/10.1007/978-3-031-82342-8_21

$$w_{n,k,r}(x) = \begin{cases} (1-x)p_{n-r,k}(x); & 0 \le k < r, \\ (1-x)p_{n-r,k}(x)(x) + xp_{n-r,k-r}(x); & r \le k \le n-r, \\ xp_{n-r,k-r}(x) & n-r < k < n, \end{cases}$$

with $p_{n-r,k}(x) = \binom{n-r}{k} x^k (1-x)^{n-r-k}$, f is a continuous function on the interval [0, 1], x is a value in the same interval and n must be a natural number bigger than $2r$. It should be noted that they decrease to the Bernstein polynomials for the specific cases $r = 0$ and $r = 1$.

Recently, in [3], the Schurer form of a complex generalization of $L_{n,r}$ operators were given with the following equation,

$$L^r_{n,p}(f,z) = \sum_{k=0}^{n+p-r} p_{n+p-r,k}(z)\left[(1-z)f\left(\frac{k}{n}\right) + zf\left(\frac{k+r}{n}\right)\right] \tag{2}$$

where r is a non-negative integer, n is a natural number such that $n + p > 2r$, $p \in N \cup 0$, $z \in C$ and f is a complex-valued analytic function in an open disk centered at the origin with radius $R > 1$. The author obtained quantitative upper bounds for the generalized complex Stancu–Schurer operators, which are attached to analytic functions and simultaneous approximation to determine the exact order of approximation. Additionally, the author established certain geometrical properties of these operators too.

Investigating the approximation characteristics of bivariate complex Stancu–Schurer operators of the tensor product kind is the purpose of this research. The research will provide people the chance to practice from a more comprehensive perspective and to operate in a greater area in this way. To examine the approximation characteristics, we extend to the bivariate situation the approximation findings from the univariate case for the complex operators studied in [3].

Initially, we introduce a few bivariate notions that are expansions of the common univariate concepts. These terms will be used throughout the entire work as defined here. Let $D_{R_i} := \{z_i \in C : |z_i| < R_i,\ i = 1, 2\}$ and $D_R = D_{R_1} \times D_{R_2}$ represents a polydisk with a center of 0 and a radius of R where $R = (R_1, R_2)$ and $|z_1| \le r_1$, $|z_2| \le r_2$, $r_1 < R_1$ with $r_2 < R_2$. Define $\overline{D_R} := \overline{D(0,R)} = \left\{(z_1, z_2) \in C^2 : |z_i| \le R_i,\ i = 1, 2\right\}$ as the closed polydisk as well.

The following is our definition of the bivariate complex Schurer–Stancu operators:

$$L^{r,s}_{n,m,p,q}(f)(z_1, z_2) = \sum_{k=0}^{n+p-r} \sum_{l=0}^{m+q-s} w_{n,k,r}(z_1) w_{m,l,s}(z_2) f\left(\frac{k}{n}, \frac{l}{m}\right) \tag{3}$$

where, in the polydisk $\overline{D_R}$, f is an analytical function of two complex variables (z_1, z_2), $w_{n,k,r}(z_1)$, $w_{m,l,s}(z_2)$ are similar to the previous Eq. (1), r, s are non-negative integers, $n, m \in N$ such that $n + p > 2r$, $m + q > 2s$, and $p, q \in N \cup \{0\}$.

In this study, we try to discover the exact order of approximation for the operators provided by (3) on compact polydiscs. Considering this, we provide the Voronovskaja type theorems and the approximation order with qualitative estimates for the operators $L^{r,s}_{n,m,p,q}(f)$. These results allow us to use the operators $L^{r,s}_{n,m,p,q}(f)$ to estimate the precise order.

2 Approximation by Bivariate Complex Schurer–Stancu Operators

Theorem 2.1 *Let r, s are a non-negative integers, $n, m \in N$ such that $n + p > 2r$, $m + q > 2s$ and $f : D_R \to C$ is analytic in D_R with $f(z_1, z_2) = \sum_{k=0}^{\infty}\sum_{l=0}^{\infty} c_{k,l} z_1^k z_2^l$, and $R_1 > p + 1$, $R_2 > q + 1$, for fixed $p, q \in N \cup \{0\}$. We have,*

(i) *Let $1 \le \rho_1 \le \rho_2(p + 1)$, $1 \le \rho_2 \le \rho_2(q + 1)$ and $\rho_1(p + 1) < R_1$, $\rho_2(q + 1) < R_2$ be arbitrary fixed. For all $|z_1| \le \rho_1$, $|z_2| \le \rho_2$ we have*

$$\left|L^{r,s}_{n,m,p,q}(f)(z_1, z_2) - f(z_1, z_2)\right| \le M^{r,s}_{\rho_1,\rho_2,n,m,p,q}(f)$$

where

$$\begin{aligned}
M^{r,s}_{\rho_1,\rho_2,n,m,p,q}(f) =& \frac{2\rho_1(p+1)^2}{n+p}\sum_{k=2}^{\infty}|c_k|k(k-1)\big[\rho_1(p+1)\big]^{k-2} \\
&+ \frac{4r\rho_1(p+1)}{n+p}\sum_{k=1}^{\infty}|c_k|k\big[\rho_1(p+1)\big]^{k-1} \\
&+ \frac{1}{n}\sum_{k=0}^{\infty}|c_k|\big[\rho_1(p+1)\big]^{k} - \frac{1}{n}\sum_{k=0}^{\infty}|c_k|\rho_1^k \\
&+ \frac{4\big[\rho_1(p+1)\big]^2}{n+p}\sum_{k=2}^{\infty}|c_k|k(k-1)\big[\rho_1(p+1)\big]^{k-2} \\
&+ \frac{2\rho_2(q+1)^2}{m+q}\sum_{l=2}^{\infty}|c_l|l(l-1)\big[\rho_2(q+1)\big]^{l-2} \\
&+ \frac{4s\rho_2(q+1)}{m+q}\sum_{l=1}^{\infty}|c_l|l\big[\rho_2(q+1)\big]^{l-1} \\
&+ \frac{1}{m}\sum_{l=0}^{\infty}|c_l|\big[\rho_2(q+1)\big]^{l} \\
&- \frac{1}{m}\sum_{l=0}^{\infty}|c_l|\rho_2^l + \frac{4\big[\rho_2(q+1)\big]^2}{m+q}
\end{aligned}$$

$$+\sum_{l=2}^{\infty}|c_l|l(l-1)\big[\rho_2(q+1)\big]^{l-2}<\infty.$$

(ii) *Let* $1\le\rho_1<\rho_1^*\le\rho_1(p+1)$, $1\le\rho_2<\rho_2^*\le\rho_2(q+1)$ *and* $\rho_1(q+1)<R_1$, $\rho_2(q+1)<R_2$ *then for all* $|z_1|\le\rho_1$, $|z_2|\le\rho_2$ *and* $n,m,j_1,j_2\in N\cup\{0\}$ *be with,* $j_1+j_2\ge 1$ *we have*

$$\left|\frac{\partial^{j_1+j_2}\mathrm{L}^{r,s}_{n,m,p,q}(f)}{\partial z_1^{j_1}\partial z_2^{j_2}}(z_1,z_2)-\frac{\partial^{j_1+j_2}f}{\partial z_1^{j_1}\partial z_2^{j_2}}(z_1,z_2)\right|$$
$$\le M^{r,s}_{\rho_1^*,\rho_2^*,n,m,p,q}(f)\frac{j_1!\rho_1}{\left(\rho_1^*-\rho_1\right)^{j_1+1}}\frac{j_2!\rho_1}{\left(\rho_2^*-\rho_2\right)^{j_2+1}},$$

where $M^{r,s}_{\rho_1^*,\rho_2^*,n,m,p,q}(f)$ *is given as the above.*

Proof

(i) By denoting $e_{k,l}(z_1,z_2)=e_k(z_1)\cdot e_l(z_2)$ where $e_k(z)=z^k$. From definition of (3) we get

$$\left|L^{r,s}_{n,m,p,q}(f)(z_1,z_2)-f(z_1,z_2)\right|$$
$$\le\sum_{k=0}^{\infty}\sum_{l=0}^{\infty}|c_{k,l}|\left|L^{r,s}_{n,m,p,q}\big(e_{k,l}\big)(z_1,z_2)-e_{k,l}(z_1,z_2)\right|.$$

We can write using a simple computation,

$$\left|L^{r,s}_{n,m,p,q}\big(e_{k,l}\big)(z_1,z_2)-e_{k,l}(z_1,z_2)\right|=\left|L^{r}_{n,p}(e_k,z_1)\cdot L^{s}_{m,q}(e_l,z_2)-z_1^kz_2^l\right|.$$

Considering the estimates for two cases $0\le k\le n+p-r$ and $k>n+p-r$ from [3] (proof of Theorem 2.1) we can write

$$\left|L^{r}_{n,p}(e_k,z_1)-e_k(z_1)\right|\le 4\rho_1^k(p+1)^k\frac{k(k-1)+2k(r)}{2(n+p)}+\frac{\rho_1^k}{n}\big[(p+1)^k-1\big],$$

$$\left|L^{r}_{n,p}(e_k,z_1)-e_k(z_1)\right|\le 4\frac{k(k-1)}{n+p-r}\big[\rho_1(1+p)\big]^k,$$

$$\left|L^{r}_{n,p}(e_k)(z_1)\right|\le 3\rho_1^{k+1}(p+1)^k,$$

and

$$\left|L^{s}_{m,q}(e_l,z_2)-e_l(z_2)\right|\le 4\rho_2^l(q+1)^l\frac{l(l-1)+2l(s)}{2(m+q)}+\frac{\rho_2^l}{m}\big[(q+1)^l-1\big],$$

$$\left|L^{s}_{m,q}(e_l,z_2)-e_l(z_2)\right|\leq 4\frac{l(l-1)}{m+q-s}\left[\rho_2(1+q)\right]^l$$

for $k\in N$, $|z|\leq\rho_1$, $\rho_1\geq 1$ and for all $|z_1|\leq\rho_1$ and $|z_2|\leq\rho_2$ we obtain,

$$\begin{aligned}
&\left|L^{r}_{n,p}(e_k,z_1)\cdot L^{s}_{m,q}(e_l,z_2)-z_1^k z_2^l\right|\\
&\quad\leq\left|L^{r}_{n,p}(e_k,z_1)\cdot L^{s}_{m,q}(e_l,z_2)-L^{r}_{n,p}(e_k,z_1)\cdot z_2^l\right|\\
&\qquad+\left|L^{r}_{n,p}(e_k,z_1)\cdot z_2^l-z_1^k z_2^l\right|\\
&\quad\leq\left|L^{r}_{n,p}(e_k,z_1)\right|\cdot\left|L^{s}_{m,q}(e_l,z_2)-z_2^l\right|+\left|z_2^l\right|\cdot\left|L^{r}_{n,p}(e_k,z_1)-z_1^k\right|\\
&\quad\leq 3\rho_1^{k+1}(p+1)^k\left[+\frac{2\rho_2^{l-1}(q+1)^l}{m+q}l(l-1)+\frac{4s\left[\rho_2(q+1)\right]^l}{m+q}l\right.\\
&\qquad+\frac{1}{m}\left[\rho_2(q+1)\right]^l-\frac{1}{m}\rho_2^l+\frac{4\left[\rho_2(q+1)\right]^l}{m+q}l(l-1)\\
&\qquad+\rho_2^l\left[\frac{2\rho_1^{k-1}(p+1)^k}{n+p}k(k-1)+\frac{4r\left[\rho_1(p+1)\right]^k}{n+p}k\right.\\
&\qquad\left.+\frac{1}{n}\left[\rho_1(p+1)\right]^k-\frac{1}{n}\rho_1^k+\frac{4\left[\rho_1(p+1)\right]^k}{n+p}k(k-1)\right],
\end{aligned}$$

which the coefficients' criteria imply according to $c_{k,l}$

$$\begin{aligned}
&\left|L^{r,s}_{n,m,p,q}(f)(z_1,z_2)-f(z_1,z_2)\right|\\
&\quad\leq\sum_{k=0}^{\infty}\sum_{l=0}^{\infty}|c_{k,l}|\left|L^{r,s}_{n,m,p,q}(e_{k,l})(z_1,z_2)-e_{k,l}(z_1,z_2)\right)\\
&\quad\leq\sum_{k=0}^{\infty}\sum_{l=0}^{\infty}|c_{k,l}|\left\{3\rho_1^{k+1}(p+1)^k\left[\frac{2\rho_2^{l-1}(q+1)^l}{m+q}l(l-1)\right.\right.\\
&\qquad+\frac{4s\left[\rho_2(q+1)\right]^l}{m+q}l+\frac{1}{m+\delta}\left[\rho_2(q+1)\right]^l\\
&\qquad\left.-\frac{1}{m+\delta}\rho_2^l+\frac{4\left[\rho_2(q+1)\right]^l}{m+q}l(l-1)\right]\\
&\qquad+\rho_2^l\left[\frac{2\rho_1^{k-1}(p+1)^k}{n+p}k(k-1)+\frac{4r\left[\rho_1(p+1)\right]^k}{n+p}k\right.\\
&\qquad\left.\left.+\frac{1}{n}\left[\rho_1(p+1)\right]^k-\frac{1}{n}\rho_1^k+\frac{4\left[\rho_1(p+1)\right]^k}{n+p}k(k-1)\right]\right\},
\end{aligned}$$

this demonstrates (i).

(ii) The simultaneous approximation convergence rate is now available. Let $1 \leq \rho_1 < \rho_1^* \leq \rho_1(p+1)$, $1 \leq \rho_2 < \rho_2^* \leq \rho_2(q+1)$ and $\rho_1(q+1) < R_1$, $\rho_2(q+1) < R_2$. Using Cauchy's formula, we obtain

$$\left|\frac{\partial^{j_1+j_2} L_{n,m,p,q}^{r,s}(f)}{\partial z_1^{j_1} \partial z_2^{j_2}}(z_1, z_2) - \frac{\partial^{j_1+j_2} f}{\partial z_1^{j_1} \partial z_2^{j_2}}(z_1, z_2)\right|$$

$$\leq \frac{j_1! j_2!}{(2\pi i)^2} \int\limits_{|v_2-z_2|=\rho_2^*} \int\limits_{|v_1-z_1|=\rho_1^*} \frac{\left|L_{n,m,p,q}^{r,s}(f)(v_1, v_2) - f(v_1, v_2)\right|}{|v_1 - z_1|^{j_1+j_2} |v_2 - z_2|^{j_1+j_2}} dv_1 dv_2$$

the use of passing to absolute value $|z_1| \leq \rho_1$, $|z_2| \leq \rho_2$ and taking into consideration that $|v_1 - z_1| = \rho_1^* - \rho_1$, $|v_2 - z_2| = \rho_2^* - \rho_2$, by using the estimate from (i), we are able to derive

$$\left|\frac{\partial^{j_1+j_2} L_{n,m,p,q}^{r,s}(f)}{\partial z_1^{j_1} \partial z_2^{j_2}}(z_1, z_2) - \frac{\partial^{j_1+j_2} f}{\partial z_1^{j_1} \partial z_2^{j_2}}(z_1, z_2)\right|$$

$$\leq M_{\rho_1^*, \rho_2^*, n,m,p,q}^{r,s}(f) \frac{j_1!}{\left(\rho_1^* - \rho_1\right)^{j_1+1}} \frac{j_2!}{\left(\rho_2^* - \rho_2\right)^{j_2+1}}$$

of which the theorem is proved.

The second finding relates to the Voronovskaja-type theorem for operator (3). For this, first, let's give the theorem in the form of two variables in real spaces by considering the main result of the study [1], which will be used in the proof.

Theorem 2.2 *Let $L_{n,n,p,q}^{r,s}(f; x, y)$ be real form of the operators given with the definition (3), $f \in C^{2,2}([0, 1+p] \times [0, 1+q])$. Then for all $(x, y) \in C^{2,2}([0, 1+p] \times [0, 1+q])$ we have*

$$\lim_{n\to\infty} n\left(L_{n,n,p,q}^{r,s}(f; x, y) - f(x, y)\right)$$

$$= px\frac{\partial f}{\partial x}(x, y) + qy\frac{\partial f}{\partial y}(x, y)$$

$$+ \frac{1}{2}x(1 - x)\frac{\partial^2 f}{\partial x^2}(x, y) + y(1 - y)\frac{\partial^2 f}{\partial y^2}(x, y).$$

Now, we present the following qualitative asymptotic formula by applying the asymptotic Voronovskaja-type theorem for the bivariate complex Schurer–Stancu operator.

An explanation of qualitative Voronovskaja theorem is provided below.

Theorem 2.3 *For fixed* $p, q \in N \cup \{0\}$, *and* $R_1 > p+1$, $R_2 > q+1$ *suppose that in* D_R, $f : D_R \to C$ *is analytical with* $f(z_1, z_2) = \sum_{k=0}^{\infty}\sum_{l=0}^{\infty} c_{k,l} z_1^k z_2^l$. *Then for all* $n \in N$ *and* $|z_1| \le \rho_1$, $|z_2| \le \rho_2$ *where* $1 \le \rho_1 \le \rho_1(p+1)$, $1 \le \rho_2 \le \rho_2(q+1)$ *and* $\rho_1(p+1) < R_1$, $\rho_2(q+1) < R_2$ *we get*

$$\begin{aligned}
&\lim_{n\to\infty} n\Big(L^{r,s}_{n,n,p,q}(f; z_1, z_2) - f(z_1, z_2)\Big)\\
&= p z_1 \frac{\partial f}{\partial z_1}(z_1, z_2) + q z_2 \frac{\partial f}{\partial z_2}(z_1, z_2)\\
&+ \frac{1}{2}\left\{ z_1(1 - z_1)\frac{\partial^2 f}{\partial z_1^2}(z_1, z_2) + z_2(1 - z_2)\frac{\partial^2 f}{\partial z_2^2}(z_1, z_2)\right\},
\end{aligned}$$

uniformly.

Proof $\left\{n\Big(L^{r,s}_{n,n,p,q}(f; z_1, z_2) - f(z_1, z_2)\Big)\right\}_{n\in N}$ is the sequence of analytic functions that holds uniformly in any compact disk $\overline{D_\rho} = \overline{D_{\rho_1}} \times \overline{D_{\rho_2}}$ with $1 \le \rho_1(p+1) < R$, $1 \le \rho_2(q+1) < R_2$. In fact, the following derives from (i) of Theorem 2.1.

$$\left|n\Big(L^{r,s}_{n,n,p,q}(f; z_1, z_2) - f(z_1, z_2)\Big)\right| \le n M^{r,s}_{\rho_1,\rho_2,n,n,p,q}(f) := \mathcal{M}^{r,s}_{\rho_1,\rho_2,n,n,p,q}(f)$$

for all $n \in N$ and $(z_1, z_2) \in \overline{D_\rho}$ with $1 \le \rho_1(p+1) < R_1$, $1 \le \rho_2(q+1) < R_2$, here

$$\begin{aligned}
\mathcal{M}^{r,s}_{\rho_1,\rho_2,n,n,p,q}(f) =& \frac{2n\rho_1(p+1)^2}{n+p}\sum_{k=2}^{\infty}|c_k| k(k-1)\big[\rho_1(p+1)\big]^{k-2}\\
&+ \frac{4rn\rho_1(p+1)}{n+p}\sum_{k=1}^{\infty}|c_k| k\big[\rho_1(p+1)\big]^{k-1} + \frac{1}{n}\sum_{k=0}^{\infty}|c_k|\big[\rho_1(p+1)\big]^{k}\\
&- \sum_{k=0}^{\infty}|c_k|\rho_1^k + \frac{4n\big[\rho_1(p+1)\big]^2}{n+p}\sum_{k=2}^{\infty}|c_k| k(k-1)\big[\rho_1(p+1)\big]^{k-2}\\
&+ \frac{2n\rho_2(q+1)^2}{n+q}\sum_{l=2}^{\infty}|c_l| l(l-1)\big[\rho_2(q+1)\big]^{l-2}\\
&+ \frac{4sn\rho_2(q+1)}{n+q}\sum_{l=1}^{\infty}|c_l| l\big[\rho_2(q+1)\big]^{l-1} + \sum_{l=0}^{\infty}|c_l|\big[\rho_2(q+1)\big]^{l}\\
&- \sum_{l=0}^{\infty}|c_l|\rho_2^l + \frac{4n\big[\rho_2(q+1)\big]^2}{n+q}\sum_{l=2}^{\infty}|c_l| l(l-1)\big[\rho_2(q+1)\big]^{l-2} < \infty.
\end{aligned}$$

Then, in accordance with the Vitali Theorem (e.g., p. 1 of [4]) $\left\{n\Big(L^{r,s}_{n,n,p,q}(f; z_1, z_2) - f(z_1, z_2)\Big)\right\}_{n\in N}$ is uniformly convergent in any $\overline{D_\rho}$, this proves the theorem.

To determine the precise order of approximation by $L^{r,s}_{n,m,p,q}(f)$, the following theorem will be helpful.

Theorem 2.4 *For fixed $p, q \in N \cup \{0\}$, and $R_1 > p+1$, $R_2 > q+1$ suppose that $f : D_R \to C$ is analytic in D_R with $f(z_1, z_2) = \sum_{k=0}^{\infty}\sum_{l=0}^{\infty} c_{k,l} z_1^k z_2^l$ for all $(z_1, z_2) \in \overline{D_\rho}$. Denoting $\|f\|_{\rho_1,\rho_2} = sup\{|f(z_1, z_2)|; |z_1| \le \rho_1; |z_2| \le \rho_2\}$ and the following complex partial differential equation has no solution for f.*

$$pz_1 \frac{\partial f}{\partial z_1}(z_1, z_2) + \frac{z_1(1-z_1)}{2}\frac{\partial^2 f}{\partial z_1^2}(z_1, z_2) + qz_2 \frac{\partial f}{\partial z_2}(z_1, z_2) + \frac{z_2(1-z_2)}{2}\frac{\partial^2 f}{\partial z_2^2}(z_1, z_2) = 0, \quad |z_1| \le \rho_1, \quad |z_2| \le \rho_2, \tag{4}$$

then we have

$$\left\| L^{r,s}_{n,n,p,q}(f) - f \right\|_{\rho_1,\rho_2} \ge \frac{\mathcal{K}^{r,s}_{\rho_1,\rho_2 f}}{n}$$

where $\mathcal{K}^{r,s}_{\rho_1,\rho_2 f}$ depends only on f, r, s, ρ_1, ρ_2.

Proof For all $|z_1| \le \rho_1$, $|z_2| \le \rho_2$ and $n \in N$, we are able to write

$$\begin{aligned}
&L^{r,s}_{n,n,p,q}(f)(z_1, z_2) - f(z_1, z_2)\\
&= \frac{1}{n}\Bigg\{ \frac{pz_1}{2}\frac{\partial f}{\partial z_1}(z_1, z_2) + \frac{z_1(1-z_1)}{4}\frac{\partial^2 f}{\partial z_1^2}(z_1, z_2)\\
&+ \frac{pz_2}{2}\frac{\partial f}{\partial z_2}(z_1, z_2) + \frac{z_2(1-z_2)}{4}\frac{\partial^2 f}{\partial z_2^2}(z_1, z_2)\\
&+ n\Big[L^{r,s}_{n,n,p,q}(f; z_1, z_2) - f(z_1, z_2)\\
&- pz_1 \frac{\partial f}{\partial z_1}(z_1, z_2) - qz_2 \frac{\partial f}{\partial z_2}(z_1, z_2)\\
&- \frac{1}{2} z_1(1-z_1)\frac{\partial^2 f}{\partial z_1^2}(z_1, z_2) - z_2(1-z_2)\frac{\partial^2 f}{\partial z_2^2}(z_1, z_2)\Big]\Bigg\},
\end{aligned}$$

taking into the inequalities

$$\|S+R\|_{\rho_1,\rho_2} \ge \left| \|S\|_{\rho_1,\rho_2} - \|R\|_{\rho_1,\rho_2} \right| \ge \|S\|_{\rho_1,\rho_2} - \|R\|_{\rho_1,\rho_2}$$

it follows

$$\begin{aligned}
\left\| L^{r,s}_{n,n,p,q}(f) - f \right\|_{\rho_1,\rho_2} = \Bigg\| \frac{1}{n}\Bigg\{ &\frac{pz_1}{2}\frac{\partial f}{\&\partial z_1}(z_1, z_2) + \frac{z_1(1-z_1)}{4}\frac{\partial^2 f}{\partial z_1^2}(z_1, z_2)\\
&+ \frac{pz_2}{2}\frac{\partial f}{\partial z_2}(z_1, z_2) + \frac{z_2(1-z_2)}{4}\frac{\partial^2 f}{\partial z_2^2}(z_1, z_2)
\end{aligned}$$

$$+ n\Big[L^{r,s}_{n,n,p,q}(f; z_1, z_2) - f(z_1, z_2) - pz_1$$

$$\frac{\partial f}{\partial z_1}(z_1, z_2) - qz_2 \frac{\partial f}{\partial z_2}(z_1, z_2) - \frac{1}{2} z_1(1 - z_1)$$

$$\frac{\partial^2 f}{\partial z_1^2}(z_1, z_2) - z_2(1 - z_2)\frac{\partial^2 f}{\partial z_2^2}(z_1, z_2)\Big]\Big\}\Big\|_{\rho_1, \rho_2}$$

$$\geq \frac{1}{n}\Big\{\Big\|\frac{pz_1}{2}\frac{\partial f}{\partial z_1}(z_1, z_2) + \frac{z_1(1 - z_1)}{4}\frac{\partial^2 f}{\partial z_1^2}(z_1, z_2)$$

$$+ \frac{pz_2}{2}\frac{\partial f}{\partial z_2}(z_1, z_2) + \frac{z_2(1 - z_2)}{4}\frac{\partial^2 f}{\partial z_2^2}(z_1, z_2)\Big\|_{\rho_1, \rho_2}$$

$$- n\Big\|\Big[L^{r,s}_{n,n,p,q}(f; z_1, z_2) - f(z_1, z_2) - pz_1\frac{\partial f}{\partial z_1}(z_1, z_2)$$

$$- qz_2\frac{\partial f}{\partial z_2}(z_1, z_2) - \frac{1}{2} z_1(1 - z_1)$$

$$\frac{\partial^2 f}{\partial z_1^2}(z_1, z_2) - z_2(1 - z_2)\frac{\partial^2 f}{\partial z_2^2}(z_1, z_2)\Big]\Big\|_{\rho_1, \rho_2}\Big\}.$$

By denoting $S(z_1, z_2) := \frac{pz_1}{2}\frac{\partial f}{\partial z_1}(z_1, z_2) + \frac{z_1(1-z_1)}{4}\frac{\partial^2 f}{\partial z_1^2}(z_1, z_2) + \frac{pz_2}{2}\frac{\partial f}{\partial z_2}(z_1, z_2) + \frac{z_2(1-z_2)}{4}\frac{\partial^2 f}{\partial z_2^2}(z_1, z_2)$ and if we prove that $|S|_\rho > 0$, by Theorem 2.2, there exists an index n_0 depending on $f, \rho_1, \rho_2, r, s, p, q$ such that for all $n \geq n_0$ we have

$$\Big\|L^{r,s}_{n,n,p,q}(f) - f\Big\|_{\rho_1, \rho_2} \geq \frac{1}{n}\frac{|S|_{\rho_1, \rho_2}}{2}.$$

Also, for $n \in \{1, 2, \ldots, n_0 - 1\}$ we obviously write $\Big\|L^{r,s}_{n,n,p,q}(f) - f\Big\|_\rho \geq \frac{H^{r,s}_{n,\rho_1,\rho_2,p,q}}{n}$, with $H^{r,s}_{n,\rho_1,\rho_2,p,q}(f) := n\Big\|L^{r,s}_{n,n,p,q}(f) - f\Big\|_\rho > 0$. Finally, it follows that $|L^{r,s}_{n,n,p,q}(f) - f|_\rho \geq \frac{\mathcal{K}^{r,s}_{\rho_1,\rho_2,p,q,}(f)}{n}$, for all $n \in N$, where

$$\mathcal{K}^{r,s}_{\rho_1,\rho_2,p,q}(\mathrm{f}) := min\Big\{H^{r,s}_{1,\rho_1,\rho_2,p,q}, H^{r,s}_{2,\rho_1,\rho_2,p,q}, \ldots, H^{r,s}_{n_0-1,\rho_1,\rho_2,p,q}, \frac{|S|_{\rho_1,\rho_2}}{2}\Big\}$$

and the proof is now completed.

We get the following exact order by combining Theorem 2.1 with Theorem 2.4.

Corollary *The exact order is approximated by the bivariate complex Stancu–Schurer operator* $L^{r,s}_{n,n,p,q}(f)$ *in* $\frac{1}{n}$ *if f is not a solution to Eq.* (4).

References

1. Barbosu, D.: Schurer–Stancu type operator. Stud. Univ. Babes-Bolyai Math. **48** (3) (2003)
2. Bernstein, S.N.: Sur la convergence de certaines suites des polynomes. J. Math. Pures Appl. **15**(9), 345–358 (1935)
3. Çetin, N.: A new complex generalized Bernstein–Schurer operator. Carpathian J. Math. **37**(1), 81–89 (2021)
4. Gal, S.G.: Approximation by Complex Bernstein and Convolution Type Operators. World Scientific (2009)
5. Kantorovich, L.V.: Sur la convergence de la suite de polynomes de S. Bernstein en dehors de lfinterval fundamental. Bull. Acad. Sci. URSS **8**, 1103–1115 (1931)
6. Lorentz, G.G.: Bernstein Polynomials. University of Toronto Press, Toronto (1953)
7. Stancu, D.D.: Approximation of functions by means of a new generalized Bernstein operator. Calcolo **20**(2), 211–229 (1983)
8. Wright, E.M.: The Bernstein approximation polynomials in the complex plane. J. Lond. Math. Soc. **5**, 265–269 (1930)

Research in Best Approximation for Fixed Points in Generalized Metric Spaces

Nesrin Manav Tatar

Abstract Takahashi (Kodai Math Sem Rep 22:142–149, 1970) demonstrated convex structures in metric spaces that were not embedded in Banach spaces in 1970, sparking additional investigation. Extending to metric spaces, our work develops and extends Branciari's theorem (Branciari in Int J Math Math Sci 29:531–536, 2002) to show common fixed point theorems for mappings meeting a specific integral condition. We address the best approximation method for exhibiting practical significance, emphasizing its relevance in both theoretical and practical situations as indicated by the authors (Manav Tatar and Agarwal in Mathematics 11(21):4455, 2023; Manav Tatar in Best approximation of Fixed Point results in generalized metric spaces, 2023). This study presents a concept that is consistent across several metric spaces.

Keywords Fixed point theorem · Generalized metric space · Modular metric space · Integral type contraction · Best approximation

1 Background and Preliminary Concepts

In Banach's traditional theory of contraction mappings, it is postulated that if $F : W \to W$ is a contraction mapping in the entire metric space (W, u), then there exists a unique fixed point. This basic principle has sparked intellectual interest, resulting in a thorough investigation and improvement of fixed point theory in the metric setting.

Takahashi [33] was the first to introduce convex structures in metric spaces, showing that they exist outside of Banach spaces in 1970, sparking more investigation in this topic. He proved fixed point theorems for maps having invariant properties. This foundational study has fueled ongoing research in the subject.

N. M. Tatar (✉)
Department of Mathematics, Erzincan Binali Yildirim University, Erzincan, Turkey
e-mail: nmanav@erzincan.edu.tr

O. Castillo et al. (eds.), *Advanced Mathematics for the Modeling and Solution of Challenging Problems in Engineering*, Studies in Systems, Decision and Control 582,
https://doi.org/10.1007/978-3-031-82342-8_22

Branciari [8] published a fixed point theorem for a single mapping in 2002, proving a correlation for this inequality similar to the Banach Contraction Principle. In [23], an illustrated example of such investigation is offered in a graphical style, emphasizing the practical importance of integral-type rationals and providing a remarkable application instance.

To reach this goal, scholars embarked on a trip across many abstract spaces in order to widen the boundaries of metric fixed point theory. One notable research includes investigating modular metric spaces, which Chistyakov first described as an extended idea derived from modular spaces in 2003, 2005, and in a book [10–12].

Azadifar et al. [4] gave insight on an essential aspect of the provided fixed point for integral-type compatible mappings and their existence-uniqueness in modular metric spaces in 2013.

Gupta et al. [16] approach for a shared fixed point theorem relating the contraction of integral types was advanced by the researchers [20] in 2021. Their [24] academic investigation aims to expand the applicability of the Banach fixed point theorem in modular metric spaces by examining contractive conditions that involve integral types. They methodically offer a series of theorems explaining the existence and uniqueness of a common fixed point among self-mappings that satisfy requirements of integral variety contractions. To demonstrate the strength of their findings, they provide a wide range of corollaries and instances.

However, an essential classical theorem on best approximation, attributed to Fan [13], provides assurance about the presence of an element within a given set.

O'Regan and Shahzad [28] investigated Fan's result on approximation in a multivalued form in 2003. The author [17] then investigated cases with non-star-shaped input values in 2005. They investigated frequent fixed point discoveries for operators on domains with no star-shaped features, proposing Brosowski-Meinardus type approximation theorems as practical applications.

Malih [22] further on the idea of optimal approximation inside metric spaces in 2021, addressing contractions of integral types using various research techniques. The investigation went beyond integral-type contractions to include the examination of optimum proximity sites inside metric spaces. Scholarly references [14, 15] demonstrate this expansion of reach.

Reich's work [32] entailed combining conclusions from previous sections, resulting in the creation of ground-breaking fixed point theorems. In addition, he developed a unique iteration of Fan's fixed point theorem, elegantly incorporating ideas from approximation theory into the domain of fixed point theory.

A theorem outlining proximity pairs was put out by the authors [6], who explained the necessary circumstances for proving the existence of an element. In addition, they proved that the proximity pair theorem and the best approximation theorem are related.

Also, as described in [2], Asadi et al. have significantly advanced fixed point theory in the context of modular metric spaces. They presented the idea of w-proximal quasi-contraction mappings and the optimum w-proximity point within the context of modular metric spaces. Importantly, their groundbreaking research has not only advanced modular metric spaces but also cleared the path for optimum proximity point solutions to be applied to generalized modular metric spaces.

We explain Branciari [8] contractions to prove the validity of fixed point theorems in the domain of generalized modular metric spaces given in [24, 25]. This endeavor not only expands but also deepens Branciari's initial theorems. We analyze several integral contractions to discover and define fixed points within this environment, building on Branciari's basic work given in [24, 25].

To emphasize our research's practical ramifications, we prioritize the idea of the best proximity pair, which leads to the formulation of the best approximation theorem. We highlight this theoretical concept to a specific situation, led by Fan's prior research's [13] best approximation approach given in [24, 25]. As a result, we include and use their discoveries, bolstering the robustness and application of our findings within the larger landscape of generalized modular metric spaces given in [24, 25].

1.1 Expanded Context of Generalized Metric Spaces

The researchers [34] introduced a new interpretation of the modular metric notion, which can be applied to any abstract set. This interpretation allows for a more dynamic and diverse use of this mathematical framework. Furthermore, they provided fundamental definitions that improve and enrich understanding of the extended modular metric concept. This section thoroughly covers these issues and offers in-depth explanations.

Definition 1.1 [34] Let W be a set, and let $\theta : (0, \infty) \times W \times W \to [0, \infty]$ be a function with the following properties:

(θ_1) *If* $\theta_p(\zeta, \varrho) = 0$ for all $p > 0$, then ζ is equal to ϱ for all values of ζ and ϱ in the set W.

(θ_2) $\theta_p(\zeta, \varrho) = \theta_p(\varrho, \zeta)$ is true for every p greater than zero and for all values of ζ and ϱ in the set W.

(θ_3) Let denote the following statement: There is a positive integer L such that if (ζ, ϱ) belongs to $W \times W$ and $\{\zeta_n\}$ is a sequence in W such that the limit as n approahes infinity of $\theta_p(\zeta_n, \zeta)$ is 0 for some p greater than 0, then.

$$\theta_p(\zeta, \varrho) \leq L \operatorname{limsup}_{n\to\infty}\theta_p(\zeta_n, \varrho).$$

The combination of (W, θ) constitutes a generalized modular metric space.

If there are ζ and ϱ in W such that the sequence $\{\zeta_n\}$ is a subset of W and the limit as n approaches infinity of $\theta_p(\zeta_n, \zeta)$ is 0 for some $p > 0$, and $\theta_p(\zeta, \varrho)$ is finite, then it is necessary that K be greater than or equal 1.

Let us assume that ζ_0 belongs to the set W, whereas θ_p (abbreviated as θ) represents a generalized modular metric on W.

Browse the set.

$$W_\theta = W_\theta(\zeta_0) = \{\zeta \in W : \theta_p(\zeta, \zeta_0) \to 0 \ \text{ as } p \to \infty\}.$$

This collection is sometimes referred to as a generalized modular set. Below, we offer supplementary and valuable definitions.

Definition 1.2 [32] Consider a generalized modular metric space denoted as (W_θ, θ). The subsequent notions are explicitly delineated:

(1) A sequence $\{\zeta_n\}_{n\in N}$ in W_θ is considered θ-convergent to $\zeta \in W_\theta$ if and only if the value of $\theta_p(\zeta_n, \zeta)$ approaches 0 as n approaches infinity, where p is a positive number.

(2) A sequence $\{\zeta_n\}_{n\in N}$ in W_θ is considered θ-Cauchy if the limit of $\theta_p(\zeta_m, \zeta_n)$ approaches 0 as m and n approach infinity with p being greater than 0.

(3) A subset A of W_θ is considered θ-closed if, for each sequence $\{\zeta_n\}_{n\in N}$ from A that converges to ζ (where ζ is an element of A), and the set of all nonempty θ-closed subsets of W_θ is denoted as $V(W_\theta)$.

(4) A subset A of W_θ is considered θ-complete if there is a point $\zeta \in A$ such that as n approaches infinity, $\theta_p(\zeta_n, \zeta)$ approaches 0 for any $\{\zeta_n\}_{n\in N}$ that forms a θ-Cauchy sequence, and as both n and m approach infinity, $\theta_p(\zeta_m, \zeta_n)$ approaches 0 for some p.

A mapping that is consistent with the omega (ω-compatible) is defined as follows.: Provided two self-mappings, G and F on W_ω space, if the condition $\omega_p(GF(\zeta_n), FG(\zeta_n)) \to 0$ holds true, where $\{\zeta_n\}_{n=1}^{\infty}$ is a sequence and $G\zeta_n \to \varrho, F\zeta_n \to \varrho$ for any $\varrho \in W_\omega$ for p greater than 0.

Theorem 1.1 [4] *Consider a complete modular metric space denoted by W_ω. Let l and m be real numbers, where l is greater than m.*

Let's examine the self-compatible mappings F and G: $W_\omega \to W_\omega$ that meet the inclusion condition: $F(W_\omega) \subseteq G(W_\omega)$.

Furthermore, if there exists a value p in the interval (0,1) such that for any Z and P in the set W_ω and P is greater than 0, the following inequality is satisfied:

$$\int_0^{\omega_{p/l}(F\zeta, F\varrho)} \chi(\vartheta)d\vartheta \leq \int_0^{\omega_{p/m}(G\zeta, G\varrho)} \chi(\vartheta)d\vartheta, \tag{1}$$

Let χ: $R^+ \to R^+$ be a nonnegative Lebesgue integrable and summable function such that $\int_0^\epsilon \chi(\vartheta)d\vartheta > 0$ for all $\epsilon > 0$. Moreover, in the case where either G or F is continuous, there is a unique common fixed point that occurs for both mappings.

Kerim et al. [20] made significant progress in the topic by introducing common fixed point theorems in modular metric spaces in 2021. In addition, they included specific instances that demonstrate contractions defined by integral types.

Theorem 1.2 [20] *Consider W_ω as a fully equipped modular metric space with the modular metric ω. Let F and G be mappings from W_ω to W_ω that are self-compatible and meet the following conditions:*

$$F(W_\omega) \subseteq G(W_\omega)$$

$$\int_0^{\omega_p(F\zeta,F\varrho)} \chi(\vartheta)d\vartheta \le \beta\big(\omega_p(G\zeta,G\varrho)\big) \int_0^{\omega_p(G\zeta,G\varrho)} \chi(\vartheta)d\vartheta - \psi\left(\int_0^{\omega_p(G\zeta,G\varrho)} \chi(t)dt\right), \tag{2}$$

For every ζ and ϱ belonging to the set Wω, where (χ, ψ) belongs to the set (Φ_1, Φ_2) and β is a function from the real numbers to the interval [0, 1) such that the limit supremum of β (s) as s approaches ϑ is less than 1, for any positive ϑ. Therefore, there exists a unique common fixed point u in the set W_ω for both F and G.

Φ_1 is defined in the text as the set of all functions $\chi : [0, \infty) \to [0, \infty)$ that satisfy the conditions

(1) χ is continuous and does not decrease.
(2) The function $\chi(\vartheta)$ equals zero if and only if ϑ equals zero.

The set Φ_2 represents all functions $\chi : [0, \infty) \to [0, \infty)$ that satisfy the following conditions: χ is a Lebesgue integrable function with a positive value $\int_0^{\upsilon} \chi(\vartheta)d\vartheta > 0$ for all $\upsilon > 0$, and the sum of its nonnegative values is finite. The function ψ maps the interval $[0, \infty)$ to itself, with the condition that $\psi(0) = 0$. The set of all such functions is denoted as Φ_3.

Jungck [18] introduced a more comprehensive concept called compatibility, which extends beyond weak commutativity in 1976. Then, they propose a new definition for a generalized modular metric space [19].

Definition 1.3 [19] Consider (W,u) be regarded as a metric space. The self-mappings G and F are considered compatible if, for every sequence $\{\zeta_n\}$ in W such that the limit as n approaches infinity of $F(\zeta_n)$ and $G(\zeta_n)$ equals z for any z in W and p is greater than 0, the limit as n approaches infinity of $u(GF(\zeta_n), FG(\zeta_n))$ is equal to 0.

In his work, Jungck [18] introduced a fixed point theorem that extended the concept of Banach's Principle to commuting maps. Subsequently, other authors have further developed and built upon Jungck's discoveries. Additionally, Sessa [29] introduced the concept of weak commutativity.

Definition 1.4 [29] Consider a metric space denoted by (W, u). The self-mappings G and F are regarded to weakly commute if

$$u(GF(\zeta), FG(\zeta)) \leq u(F(\zeta), G(\zeta))$$

for all $\zeta \in W$ and for $p > 0$.

The authors [1] devised the (E.A.) property and observed that when weakly compatible self-mappings exhibit this feature, they also have common fixed points.

Definition 1.5 [1] Let's examine two self-mappings, G and F, of a metric space (W, u). If there is a sequence $\{\zeta_n\}$ that satisfies the following conditions:

$$\lim_{n\to\infty} G\zeta_n = \lim_{n\to\infty} F\zeta_n = \vartheta. \tag{3}$$

If there exists a ϑ in W such that both G and F fulfill criterion (E.A).

Theorem 1.3 [1] *Consider ζ and ϑ be members of W Let G and F be two self-mappings of a metric space (W, u). They are considered weakly compatible if,*

(i) *G and F fulfill the condition of property (E.A).*
(ii) $u(G\zeta, G\varrho) < max\{u(F\zeta, F\varrho), [u(G\zeta, Fx) + u(G\varrho, F\varrho)]/2, [u(G\varrho, F\zeta) + u(G\zeta, F\varrho)]/2\}$,
(iii) $GW \subseteq FW$

If the subspaces GW or FW are both complete within the larger space W, then the operators G and F possess a single, distinct fixed point that they both unique common fixed point.

Malih [22] demonstrated that weakly compatible self-mappings that fulfill the criterion of integral type contractions possess a unique common fixed point.

Theorem 1.4 [22] *Consider G and F as weakly compatible self-mappings of a subset W in the metric space (A, u). They must meet the following condition: The set clG(W) is a subset of F(W), and both cl(G(W)) and F(W) are complete,*

$$\int_0^{u(G(\zeta),G(\varrho))} \chi(\vartheta)d\vartheta \leq \varrho \int_0^{u(F(\zeta),F(\varrho))} \chi(\vartheta)d\vartheta, \tag{4}$$

For any ζ and ϱ belonging to the set W, with ϱ ranging from 0 to 1 (excluding 1), and χ being a Lebesgue-integrable function that is summable on every compact subset of $[0, \infty)$, nonnegative, and satisfying the condition that for all ϵ greater than 0. Therefore, G and F possess a unique common fixed point.

Khan and Khan [21] provided the subsequent definitions:

Definition 1.6 [21] Consider a metric space (W, u) where V is a nonempty subset of W and ζ is an element of W. The following ideas are defined:

(i) The distance from ζ to V is defined as the infimum of the set of distances between ζ and all points ϱ in V, denoted as $U(\zeta, V) = \inf\{U(\zeta, \varrho); \varrho \in V\}$.
(ii) A vector ϱ belongs to V and a vector ζ belongs to W are said to have a ϱ-approximation if $U(Z, \varrho) = u(\zeta, V) = \inf\{U(\zeta, z); z \in V\}$ According to V.
(iii) The optimal estimation of ζ is defined as the principal value (PV) of ζ, denoted as PV(ζ), which is the set of elements ϱ in V that satisfy the condition $U(\mathrm{Z}, \varrho) = u(\zeta, V)$.
(iv) V is considered proximinal when the set PV(ζ) is nonempty for every ζ in W.
(v) Proximinality is a property that occurs when the set PV (ζ) is limited and the set V is not open, which implies that the set PV (ζ) is also not open.
(vi) A mapping $G : W \to W$ is said to have a G-invariant subset V if $G(V)$ is a subset of W.

Furthermore, they provided an illustration of invariant approximation by presenting the following theorem in metric spaces [21].

Theorem 1.5 [21] *Consider a metric space (W,u) with a non-expansive mapping* $\mathrm{G} : \mathrm{W} \to \mathrm{W}$. *Let* $\Gamma \in W$ *be a fixed point of G. If V is a closed subset of W that is invariant under the action of G, and if* $\mathrm{G} : \mathrm{W} \to \mathrm{W}$ *has a restriction* $G|V : V \to W$ *that is a compact mapping, so the set* $PV(\Gamma)$ *of best approximations is not empty.*

Malih [22] combined Theorems 1.4 and 1.5 from metric spaces to form Theorem 2.2, which is currently recognized.

2 Central Outcomes

This section centers on presenting and providing evidence for theorems in generalized modular metric spaces, using Branciari's methods. Theorems, beginning with Theorem 2.1 and onwards, were presented in [26] without any associated demonstrations.

Theorem 2.1 [26] *L Consider a* θ*-complete generalized modular metric space* (W, θ_p), *where* ϱ *is an element in the open interval (0, 1). Consider* $G : W \to W$ *as a Branciari contraction (BC) where* ζ *and* ϱ *are elements of W,*

$$\int_0^{\theta_p(G(\zeta),G(\varrho))} \chi(\vartheta)d\vartheta \leq \varrho \int_0^{\theta_p(\zeta,\varrho)} \chi(\vartheta)d\vartheta, \tag{5}$$

Lebesgue-integrable mapping summable(LIMS, for short) (i.e., with finite integral) on all compact subset of $[0, \infty)$ *nonnegative, and for all* $\nu > 0$, $\int_0^{\nu} \chi(\vartheta)d\vartheta > 0$ *is* $\chi : R^+ \to R^+$; *let G be a function that has a single fixed point an in the set W, for any* ζ *in W, the limit as n approaches infinity of* $G^n(\zeta)$ *is equal to a. This holds true for all values of p greater than zero.*

Proof Let us acquire

$$\int_0^{\theta_p(G^n(\zeta),G^{n+1}(\zeta))} \chi(\vartheta)d\vartheta \le \varrho \int_0^{\theta_p(G^{n-1}(\zeta),G^n(\zeta))} \chi(\vartheta)d\vartheta \le \cdots \le \varrho^n \int_0^{\theta_p(\zeta,G(\zeta))} \chi(t)dt. \tag{6}$$

It involves doing the operation (5) repeatedly for a total of n times;

$$\int_0^{\theta_p(G^n(\zeta),G^{n+1}(\zeta))} \chi(\vartheta)d\vartheta \le \varrho^n \int_0^{\theta_p(\zeta,G(\zeta))} \chi(\vartheta)d\vartheta. \tag{7}$$

Given that ϱ is an element of the interval (0, 1), we may conclude that

$$\int_0^{\theta_p(G^n(\zeta),G^{n+1}(\zeta))} \chi(\vartheta)d\vartheta \to 0_+ \text{ as } n \to \infty. \tag{8}$$

As n approaches infinity, the value of $\theta_p\big(G^n(\zeta), G^{n+1}(\zeta)\big)$ tends to 0. We choose.

$$\lim_{n\to\infty}\sup\big(\theta_p\big(G^n(\zeta), G^{n+1}(\zeta)\big)\big) = \nu > 0, \tag{9}$$

There exists a natural number Q_ν and a sequence $\{G^{nQ}(\zeta)\}_{Q\ge Q_\nu}$ such that the limit of $(\theta_p\big(G^{nQ}\big((\zeta), G^{nQ+1}(\zeta)\big)\big)$ as Q approaches infinity is greater than zero, and $(\theta_p\big(G^{nQ}\big((\zeta), G^{nQ+1}(\zeta)\big)\big) \to \nu \ge \nu/2$ for all Q greater than or equal to Q_ν, then we possess:

$$0 < \int_0^{\nu/2} \chi(\vartheta)d\vartheta \le \lim_{Q\to\infty} \int_0^{\theta_p\left(G^{nQ}(\zeta),G^{nQ+1}(\zeta)\right)} \chi(\vartheta)d\vartheta = 0 \tag{10}$$

This statement is contradictory.

For each complex number ζ in the set W and the sequence $\{G^n(\zeta)\}_{n\in\mathrm{N}}$ considered as a θ-Cauchy series, it is true that for any natural numbers m and n, where m is greater than Q_v and less than n, the value of $\theta_p(G^n(\zeta), G^m(\zeta))$ is smaller than v for every positive v and at least one Q_v in the set of natural numbers. Let's consider the existence of a positive value v, such that for any Q in the set of natural numbers, there are two other numbers n_Q and m_Q in the set of natural numbers, where c is less than m_Q which is less than n_Q, and v is less than or equal to $\theta_p\big(G^r(\zeta), G^{mQ}(\zeta)\big)$.

Next, let's examine the smallest values of m_Q and m_Q in the set of natural numbers such that v is less than or equal to $\theta_p\big(G^r(\zeta), G^{mQ}(\zeta)\big)$, but $\theta_p\big(G^r(\zeta), G^{mQ}(\zeta)\big)$ is smaller than v for any r in the range from $m_Q + 1$ to $n_Q - 1$. This applies to the sequences $\{n_Q\}_{Q>0}$ and $\{m_Q\}_{Q>0}$ elements in N.

Currently, we examine the nature of $\theta_p\big(G^{nQ}(\zeta), G^{mQ}(\zeta)\big)$ and $\theta_p\big(G^{nQ+1}(\zeta), G^{mQ+1}(\zeta)\big)$. First and foremost, we possess $\theta_p\big(G^{nQ}(\zeta), G^{mQ}(\zeta)\big) \to v$ during the time that $Q \to \infty$, inevitably through (θ_3) and (6) and (7).

$$v \le \theta_p\big(G^{nQ}(\zeta), G^{mQ}(\zeta)\big) \le \theta_p\big(G^{nQ-1}(\zeta), G^{nQ}(\zeta)\big) \tag{11}$$

$$v \le \theta_p\big(G^{nQ}(\zeta), G^{mQ}(\zeta)\big) \le \theta_p\big(G^{nQ-1}(\zeta), G^{mQ}(\zeta)\big) \tag{12}$$

due to the calculations provided below

$$\theta_p\big(G^{nQ-1}(\zeta), G^{nQ}(\zeta)\big) + v \to v_+, \tag{13}$$

as Q approaches infinity and there is a natural number ϱ such that for any ϱ less than Q, at least one of them satisfies the condition$\theta_p(G^{nQ+1}(\zeta), G^{mQ+1}(\zeta))$ is less than v; if there is a subsequence $\big(Q_\varrho\big)_{\varrho\in N}$ that is a subset of N and satisfies$v \le \theta_p\big(G^{nQ_\varrho+1}(\zeta), G^{mQ_\varrho+1}(\zeta)\big)$, then it follows from Eqs. (1), (9), and (10)

$$\int_0^{\theta_p\left(G^{nQ_\varrho+1}(\zeta), G^{mQ_\varrho+1}(\zeta)\right)} \chi(\vartheta)d\vartheta \le \varrho \int_0^{\theta_p\left(G^{nQ_\varrho}(\zeta), G^{mQ_\varrho}(\zeta)\right)} \chi(\vartheta)d\vartheta \tag{14}$$

if we admit $\varrho \to \infty$ in both sides of (12), we obtain for $v > 0$, $\int_0^v \chi(\vartheta)d\vartheta \le \varrho \int_0^v \chi(\vartheta)d\vartheta$ is a contradiction for $\varrho \in (0, 1)$ and the integral is positive.

For a given positive integer ϱ, it is true that $\theta_p\big(G^{nQ_\varrho+1}(\zeta), G^{mQ_\varrho+1}(\zeta)\big)$ is less than v for all positive integers ϱ that are less than Q. Finally, we demonstrate the strong feature that there is a β_v in the range of $(0, v)$ and a Q_v in the set of natural numbers such that for any Q_v less than Q, where Q is also a natural number, we have the inequality $\theta_p\big(G^{nQ_\varrho+1}(\zeta), G^{mQ_\varrho+1}(\zeta)\big)$ is less than or equal to $v - \beta_v$. We estimate the existence of a subsequence $\big(Q_\varrho\big)_{\varrho\in N}$ that is a subset of the set of natural numbers, such that $\theta_p\big(G^{nQ_\varrho+1}(\zeta), G^{mQ_\varrho+1}(\zeta)\big)$ approaches $v -$ as ϱ approaches infinity. We begin with this

$$\int_0^{\theta_p\left(G^{nQ_\varrho+1}(\zeta), G^{mQ_\varrho+1}(\zeta)\right)} \chi(\vartheta)d\vartheta \le k \int_0^{\theta_p\left(G^{nQ_\varrho}(\zeta), G^{mQ_\varrho}(\zeta)\right)} \chi(\vartheta)d\vartheta \tag{15}$$

giving $\varrho \to \infty$ we have once more the contradiction that $\nu > 0$, $\int_0^{\nu} \chi(\vartheta)d\vartheta \leq \varrho \int_0^{\nu} \chi(\mathrm{t})dt$. Finally, at the conclusion of this stage, we can demonstrate the Cauchy framework$\{G^n(\zeta)\}$, $\zeta \in W$; truly, for every individual $Q_\nu \in \mathrm{N}$ that means for every individual $Q_\nu < Q$, $Q \in \mathrm{N}$

$$\nu \leq \theta_{\mathrm{p}}\big(\mathrm{G}^{\mathrm{nQ}}(\zeta), \mathrm{G}^{\mathrm{mQ}}(\zeta)\big) \leq \theta_{\mathrm{p}}\big(\mathrm{G}^{\mathrm{nQ+1}}(\zeta), \mathrm{G}^{\mathrm{nQ}}(\zeta)\big) \tag{16}$$

$$\nu \leq \theta_{\mathrm{p}}\big(\mathrm{G}^{\mathrm{nQ}}(\zeta), \mathrm{G}^{\mathrm{mQ}}(\zeta)\big) \leq \theta_{\mathrm{p}}\big(\mathrm{G}^{\mathrm{mQ+1}}(\zeta), \mathrm{G}^{\mathrm{mQ}}(\zeta)\big) \tag{17}$$

$$\nu \leq \theta_{\mathrm{p}}\big(\mathrm{G}^{\mathrm{nQ}}(\zeta), \mathrm{G}^{\mathrm{mQ}}(\zeta)\big) \leq \theta_{\mathrm{p}}\big(\mathrm{G}^{\mathrm{nQ+1}}(\zeta), \mathrm{G}^{\mathrm{mQ+1}}(\zeta)\big) \tag{18}$$

as a result of (14)–(16) we have

$$\theta_p\big(G^{mQ+1}(\zeta), G^{mQ}(\zeta)\big) + \theta_p\big(G^{nQ+1}(\zeta), G^{nQ}(\zeta)\big) + \nu - \beta_\nu < \nu - \beta_\nu \tag{19}$$

while $Q \to \infty$ hence $\nu \leq \nu - \beta_\nu$ which is a contradiction. This step has been empirically validated. Presently, we demonstrate the presence of a stationary point.

Given that W is a θ-complete generalized modular metric space, there exists a point an in W that can be identified as a fixed point. For example, let's examine the expression$a = \lim_{n\to\infty} G^n(\zeta)$. Indeed, if the value $\theta_p(a, G(a))$ is greater than zero,

$$\int_0^{\omega\mathrm{p}/\mathrm{l}\,(\mathrm{F}\zeta,\mathrm{F}\varrho)} \chi(\vartheta)\mathrm{d}\vartheta \leq \int_0^{\omega\mathrm{p}/\mathrm{m}(\mathrm{G}\zeta,\mathrm{G}\varrho)} \chi(\vartheta)\mathrm{d}\vartheta, (1) \tag{20}$$

$$0 < \theta_p(a, G(a)) \leq \varrho \lim_{n\to\infty} \theta_p\big(G^{n+1}(\zeta), \mathrm{G(a)}\big) \to 0 \tag{21}$$

as n approaches infinity, we have both $\theta_p\big(G^{n+1}(\zeta), G(a)\big)$ and $\theta_p\big(G^{n+1}(\zeta), a\big)$ approach zero, we have

$$\int_0^{\theta_p(G^{n+1}(\zeta),\mathrm{G}(a))} \chi(\vartheta)d\vartheta \leq k \int_0^{\theta_p(G^{n+1}(\zeta),\mathrm{a})} \chi(\vartheta)d\vartheta \to 0 \tag{22}$$

as n approaches infinity, we have established Eq. (19). Presently, if $\theta_p\big(G^{n+1}(\zeta), G(a)\big)$ does not converge as n approaches infinity, it means that there exists a subsequence $G^{nQ+1}(\zeta)_{Q\in\mathrm{N}} \subseteq G^{n+1}(\zeta)_{n\in\mathrm{N}}$ such that $\nu \leq \theta_p\big(G^{nQ+1}(\zeta), G(a)\big)$ for a given $v > 0$. This leads to a disagreement:

$$0 < \int_0^{\nu} \chi(\mathrm{t})d\mathrm{t} \leq k \int_0^{\theta_p(G^{nQ_\varrho+1}(\zeta),G(a))} \chi(\vartheta)d\vartheta \to 0 \tag{23}$$

as Q approaches infinity.

Now we demonstrate the singularity of the fixed point. Suppose there are two distinct points, a_1, a_2, belonging to the set W, such that $G(a_1) = a_1$ and $G(a_2) = a_2$. Applying Eq. (1), we arrive at a contradiction

$$0 < \int_0^{\theta_p(a_1,a_2)} \chi(\vartheta)d\vartheta = \int_0^{\theta_p(G(a_1),G(a_2))} \chi(\vartheta)d\vartheta \leq \varrho \int_0^{\theta_p(a_1,a_2)} \chi(\vartheta)d\vartheta < \int_0^{\theta_p(a_1,a_2)} \chi(\vartheta)d\vartheta \tag{24}$$

tends towards zero. The final stage further demonstrates that for any ζ in the set W, the value of an is equal to *G(a),* which is the limit as n approaches infinity of $G^n(\zeta)$ such as$a = G(a) = \lim_{n\to\infty} G^n(\zeta)$. The proof has been concluded. ■

Authors [24] contended that Theorem 2.1 loses its validity when we assume that the mapping χ has a practically negligible value near zero. We substantiate this claim using the following example, where we are unable to exhibit a negative value for χ.

Example 2.1 [24] Let G: N → N and χ: R⁺ → R⁺ are defined by.

$$G(\zeta) = \begin{cases} 1 & \text{if } \zeta \neq 1, \\ 2 & \text{if } \zeta = 1, \end{cases}, \quad \chi(\zeta) = \begin{cases} e^{\frac{1}{1-\vartheta}} & \text{if } \vartheta > 1, \\ 0 & \text{if } \vartheta \in [0, 1], \end{cases}$$

We select θ as a function that maps from the set $(0, \infty) \times N \times N$ to the interval $[0, \infty]$. The θ-complete generalized modular metric space is defined as $\lim_{n\to\infty} \theta_p(G(\zeta), G(\varrho)) \leq 1$ for each $\zeta, \varrho \in$ N, for all $p > 0$ and casual constant in (0, 1)

$$0 < \int_0^{\theta_p(G(\zeta),G(\varrho))} \chi(\vartheta)d\vartheta \leq \varrho \int_0^{1} \chi(\vartheta)d\vartheta = 0 \leq \varrho \int_0^{\theta_p(\zeta,\varrho)} \chi(\vartheta)d\vartheta \tag{25}$$

Theorem 1.1 is true for all values of ϱ between 0 and 1, however G does not have a fixed point. Similarly, if we select $\chi \equiv 1$, then for every positive p and any ϱ in the range (0, 1), the following holds:

$$0 < \int_0^{\theta_p(G(\zeta),G(\varrho))} \chi(\vartheta)d\vartheta \leq \theta_p(G(\zeta), G(\varrho)) \leq 1 \leq \varrho\theta_p(\zeta, \varrho) = \varrho \int_0^{\theta_p(\zeta,\varrho)} \chi(\vartheta)d\vartheta \tag{26}$$

However, ıt is ımportant to note that g does not have any fixed points.

Additional outcomes are also relevant for a generalized modular metric space [4, 20].

Theorem 2.2 [24] *Consider a complete generalized modular metric space* (W, θ_p) *and allow* $a, b, c \in R$*, where an is greater than c. allow F and G be self-compatible mappings from W to W such that F(W) is a subset of G(W).*

$$\int_0^{\theta_{p/a}(F\zeta, F\varrho)} \chi(\vartheta)d\vartheta \le \int_0^{\theta_{p/c}(G\zeta, G\varrho)} \chi(\vartheta)d\vartheta, \tag{27}$$

for some $\varrho \in (0, 1)$ *and for* $p > 0$ *, all* $\zeta, \varrho \in W$*, LIMS with* $\int_0^{\epsilon} \chi(\vartheta)d\vartheta > 0$ *for all* $\in > 0$ *is* $\chi : \mathrm{R}^+ \to \mathrm{R}^+$*. If either G or F is continuous, then there is a unique common fixed point between them.*

Theorem 2.3 [24] *Let* (W, θ_p) *be a complete generalized modular metric space and let* $F, G : W \to W$ *be self compatible mappings which satisfy* $F(W) \subseteq G(W)$

$$\int_0^{\theta_p(F\zeta, F\varrho)} \chi(\vartheta)d\vartheta \le \beta\big(\theta_p(G\zeta, G\varrho)\big) \int_0^{\theta_p(G\zeta, G\varrho)} \chi(\vartheta)d\vartheta - \psi\left(\int_0^{\theta_p(G\zeta, G\varrho)} \chi(t)dt\right), \tag{28}$$

For all $\zeta, \varrho \in W$*, where* $(\chi, \psi) \in (\Phi_1, \Phi_2)$ *and* $\beta : \mathrm{R} \to [0, 1)$ *is a function with* $\limsup_{s \to \vartheta} \beta(s) < 1$ *, for all* $\vartheta > 0$*. Then F and G have a unique common fixed point* $u \in W$*.*

Jungck [18] proposed a fixed point theorem for maps that commute, so expanding upon Banach's Principle. Several scholars have since extended and elaborated on Jungck's results in diverse ways. In addition, Sessa [29] introduced the concept of weak commutativity, while other writers presented a new variation of it for generalized modular metric spaces as described in [24].

Definition 2.1 [29] In a generalized modular metric space (W, θ_p), the self-mappings G and F are said to be θ-weakly commuting if the inequality $\theta_p(GF(\zeta), FG(\zeta)) < \theta_p(F(\zeta), G(\zeta))$ holds true for any $\zeta \in W$ and for $p > 0$.

Jungck [19] later introduced a more comprehensive concept called compatibility, which encompasses weak commutativity. Then authors offered a new definition for a generalized modular metric space in [24].

Definition 2.2 [24] In a generalized modular metric space (W, θ_p), the self-mappings G and F are considered θ-compatible if the limit as n approaches infinity of $\theta_p(GF(\zeta_n), FG(\zeta_n))$ is equal to 0. This holds true when ζ_n is a sequence in W such that the limits as n approaches infinity of $F(\zeta_n)$ and $G(\zeta_n)$ both equal z, where z is an element of W and p is greater than 0.

It is straightforward to demonstrate that mappings that weakly commute are compatible, but it is important to note that neither implication can be reversed.

Vijayaraju et al. [33] investigated the presence of a distinct shared fixed point theorem for a collection of maps that adhere to a general contractive condition of integral nature. Razani and Moradi [27] have established the common fixed point theorem in modular spaces. The authors [24] expanded and refined the objectives put forward by Jungck [18] and the findings obtained by Branciari [8] in the realm of generalized modular metric spaces, with a special emphasis on compatible mappings. This section investigates the presence of a common fixed point for mappings that are θ-compatible and fulfill a contractive condition in modular metric spaces.

Now, let us examine the demonstration of the theorem presented in their research [24].

Theorem 2.4 [24] . *Let W be a θ-complete generalized modular metric space,* $\varrho \in$ R$^+$ *and* $G, F : W \to W$ *be two θ-compatible mappings such that*$G(W) \subseteq F(W)$*,*

$$\int_0^{\theta_p(G(\zeta),G(\varrho))} \chi(\vartheta)d\vartheta \leq \varrho \int_0^{\theta_p(F(\zeta),F(\varrho))} \chi(\vartheta)d\vartheta, \tag{29}$$

LIMS (i.e., with finite integral) on all compact subset of $[0, \infty)$ *nonnegative, and such that for all* $\chi : \mathrm{R}^+ \to \mathrm{R}^+, > 0 \int_0^{\upsilon} \chi(\vartheta)d\vartheta > 0$*. If either G or F is continuous, there is a unique common fixed point to both G and F.*

Proof Let ζ be any point in W and generate the sequence such that $G(\zeta_n) = F(\zeta_{n+1})$ for every$n \in N$, with $\zeta_1 = \zeta$. This is achievable because $G(W)$ is a subset of $F(W)$. For every value of n greater than or equal to 1, and for any positive value of p, it follows from Eq. (25) that

$$\int_0^{\theta_p(G(\zeta_{n+1}),G(\zeta_n))} \chi(\vartheta)d\vartheta \leq \varrho \int_0^{\theta_p(F(\zeta_{n+1}),F(\zeta_n))} \chi(\vartheta)d\vartheta$$
$$\leq \varrho \int_0^{\theta_p(G(\zeta_n),G(\zeta_{n-1}))} \chi(\vartheta)d\vartheta$$
$$\leq \varrho^2 \int_0^{\theta_p(F(\zeta_n),F(\zeta_{n-1}))} \chi(\vartheta)d\vartheta$$

by implementing this formula, we have successfully installed it

$$\int_0^{\theta_p(G(\zeta_{n+1}),G(\zeta_n))} \chi(\vartheta)d\vartheta \leq \varrho \int_0^{\theta_p(G(\zeta),\zeta)} \chi(\vartheta)d\vartheta \tag{30}$$

and taking the limit while $n \to \infty$ we take

$$\lim_{n\to\infty} \int_0^{\theta_p(G(\zeta_{n+1}),G(\zeta_n))} \chi(\vartheta)d\vartheta \leq 0 \tag{31}$$

While for each, $\upsilon > 0$ $\int_0^{\upsilon} \chi(\vartheta)d\vartheta > 0$ remarks

$$\lim_{n\to\infty}\theta_p(G(\zeta_{n\to\infty}), G(\zeta_n)) = 0. \tag{32}$$

We demonstrate that $\{G(\zeta_n)\}_{n=1}^{\infty}$ is θ-Cauchy. For any positive v, there is a natural number n_v such that the difference between $\theta_p(G(\zeta_{n+1}), G(\zeta_n))$ is less than v for all natural numbers n greater than or equal to n_v and p is greater than 0. Assuming n and n_1 are natural numbers, with n being less than n_1, we may conclude that $\theta_p(G(\zeta_{n+1}), G(\zeta_n)) < \frac{\upsilon}{n_1-\mathrm{n}}$ for all n less than n_1. We now manufacture

$$\begin{aligned} &\theta_p\big(G\big(\zeta_{n_1}\big), G(\zeta_n)\big) \leq \theta_p(G(\zeta_n), G(\zeta_{n+1})) \\ &+ \theta_p(G(\zeta_{n+1}), G(\zeta_{n+2})) + \cdots + \theta_p\big(G\big(\zeta_{n_1-1}\big), G\big(\zeta_{n_1}\big)\big) \\ &< \frac{\nu}{n_1 - n} + \frac{\nu}{n_1 - n} + \cdots + \frac{\nu}{n_1 - n} = \nu \end{aligned}$$

for all $n < n_1$.

Then $\{G(\zeta_n)\}_{n=1}^{\infty}$ is θ-Cauchy. Since W is θ-complete, there exists $a \in W$ such that $\theta_p(G(\zeta_n), a) \to 0$ as $n \to \infty$ If G is continuous, then $G^2(\zeta_n) \to G(a)$ and$GF(\zeta_n) \to G(a)$. From the θ-compatibility of W, we have $\theta_p(FG(\zeta_n), GF(\zeta_n)) \to 0$ as $n \to \infty$ for $0 < p$. Besides, $FG(\zeta_n) \to G(a)$ considering that $\theta_p(FG(\zeta_n), G(a)) \leq \theta_p(FG(\zeta_n), GF(\zeta_n))$ and $\theta_p(FG(\zeta_n), G(a)) \leq \theta_p(GF(\zeta_n), G(a))$, then in the latter part we show a is a common fixed point of G and F. From (25),

$$\int_0^{\theta_p\left(G^2(\zeta_n),G(\zeta_n)\right)} \chi(\vartheta)d\vartheta \leq \varrho \int_0^{\theta_p(G(\zeta_n),F(\zeta_n))} \chi(\vartheta)d\vartheta \tag{33}$$

If we take the limit $n \to \infty$, we have

$$\int_0^{\theta_p(G(a),a)} \chi(\vartheta)d\vartheta \leq \varrho \int_0^{\theta_p(G(a),a)} \chi(\vartheta)d\vartheta \tag{34}$$

answers that $\theta_p((G(a), a)) = 0$ for $0 < p$. As a result $G(a) = a$, then from $G(W) \subseteq F(W)$ there exists a point a_1 such as $G(a) = a = a_1$ and from (25) we have

$$\int_0^{\theta_p\left(G^2(\zeta_n),G(a_1)\right)} \chi(\vartheta)d\vartheta \le \varrho \int_0^{\theta_p(G(\zeta_n),F(a_1))} \chi(\vartheta)d\vartheta \tag{35}$$

if we take the limit while $n \to \infty$ we have

$$\int_0^{\theta_p(G(a),G(a_1))} \chi(\vartheta)d\vartheta \le \varrho \int_0^{\theta_p(G(a),F(a_1))} \chi(\vartheta)d\vartheta \tag{36}$$

and

$$\int_0^{\theta_p(a,G(a_1))} \chi(\vartheta)d\vartheta \le \varrho \int_0^{\theta_p(a,F(a_1))} \chi(\vartheta)d\vartheta \le \varrho \int_0^{\theta_p(a,a)} \chi(\vartheta)d\vartheta \tag{37}$$

Therefore, we may conclude that an is equal to $G(a_1)$ which is also equal to $F(a_1)$. Additionally, $F(a)$ is equal to $FG(a_1)$, $GF(a_1)$, $G(a)$, and a. Similarly, if we consider F to be continuous (instead of G), we may demonstrate that $F(a) = G(a) = a$ in a comparable manner. Let's assume that a and b are two shared fixed points of G and F. Then

$$\begin{aligned}\int_0^{\theta_p(a,b)} \chi(\vartheta)d\vartheta &\le \varrho \int_0^{\theta_p(G(a),G(b))} \chi(\vartheta)d\vartheta \\ &\le \varrho \int_0^{\theta_p(F(a),F(b))} \chi(\vartheta)d\vartheta \\ &\le \varrho \int_0^{\theta_p(a,b)} \chi(\vartheta)d\vartheta\end{aligned}$$

which asserts that $\theta_p(a, b) = 0$ for $0 < p$ and as a result $a = b$. ■

The following theorem is a supplemental revision of Theorem 2.1, using a condition $G, F : W_1 \to W_1$, where W_1 is a subset of W that is both θ-closed and θ-bounded.

Theorem 2.5 [26] *Let W be a θ-complete generalized modular metric space and W_1 be a θ-closed and θ-bounded subset of W, $\varrho \in \mathrm{R}^+$ and $G, F : W \to W$ two θ-compatible mappings such that $G(W) \subseteq F(W)$,*

$$\int_0^{\theta_p(G(\zeta),G(\varrho))} \chi(\vartheta)d\vartheta \le \varrho \int_0^{\theta_p(F(\zeta),F(\varrho))} \chi(\vartheta)d\vartheta \tag{38}$$

for all $\zeta, \varrho \in W_1$ *and* $0 < p$ *LIMS (i.e., with finite integral) on all compact subset of* $[0,\infty)$ *nonnegative, and such that for all* $\chi : \mathrm{R}^+ \to \mathrm{R}^+, > 0 \int_0^{\upsilon} \chi(\vartheta)d\vartheta > 0$. *If one of G or F is continuous, then there exists a unique common fixed point of G and F.*

Rhoades [29] proved the existence of two fixed-point theorems for mappings that satisfy a general contractive inequality of integral type in the context of a generalized modular metric space version [26]. These observations align with Branciari's [8] known theorem on contraction mapping:

Theorem 2.6 [26] *Let G be a mapping from complete metric space* (W, θ_p) *into itself satisfying that for all* $\zeta, \varrho \in W$.

$$\int_0^{\theta_p(G(\zeta),G(\varrho))} \chi(\vartheta)d\vartheta \le \varrho \int_0^{m(\zeta,\varrho)} \chi(\vartheta)d\vartheta \tag{39}$$

where

$$m(\xi, \varrho) = \max\left\{\theta_p(\zeta, \varrho), \theta_p(\zeta, G(\zeta)), \theta_p(\varrho, G(\varrho)), \frac{\theta_p(\zeta, G(\varrho)) + \theta_p(\varrho, G(\zeta))}{2}\right\} \tag{40}$$

and for all $\zeta, \varrho \in W$

$$\int_0^{\theta_p(G(\zeta),G(\varrho))} \chi(\vartheta)d\vartheta \le \varrho \int_0^{W(\zeta,\varrho)} \chi(\vartheta)d\vartheta \tag{41}$$

where

$$W(\zeta, \varrho) = \max\{\theta_p(\zeta, \varrho), \theta_p(\zeta, G(\zeta)), \theta_p(\varrho, G(\varrho)), \theta_p(\zeta, G(\varrho)), \theta_p(\varrho, G(\zeta))\} \tag{42}$$

The values of ϱ*, which belong to the interval [0, 1), and the function* $\chi : R+ \to R+$ *are provided in Theorem 1.1. Let G have a single fixed point an in the set W. This fixed point satisfies the condition that for any* ζ *in W, the limit as n approaches infinity of* $G^n(\zeta)$ *is equal to a.*

Berinde [7] presented a definition that serves as a source of inspiration for our subsequent one.

Definition 2.3 In a generalized modular metric space W, a mapping G is determined a weak contraction or a (V, ψ)-weak contraction when there exist two constants $V \geq 0$ and $\psi \in [0, 1)$ such that the following inequality holds for all $\zeta, \varrho \in W$:

$$\theta_p(G(\zeta), G(\varrho)) \leq \psi\theta_p(\zeta, \varrho) + V\big(\theta_p(\varrho, G(\zeta))\big).$$

In 2010, Olatinwo [25] extended Branciari's findings by applying Definition 2.3 and adopting the same methodology as Berinde and Berinde [7] in their research. Through Olatinwo's method, we present a fixed point theorem for generalized modular metric.

Definition 2.4 [7] A function $V : \mathrm{R}^+ \to \mathrm{R}^+$ is considered a comparison function if it accurately represents subsequent occurrences.

(i) If the value of $V(\vartheta)$ is less than ϑ for some positive ϑ, then ψ is a monotonically growing function.
(ii) The initial value of V at time 0 is 0.
(iii) The limit of $V^n(\vartheta)$ as n approaches infinity is 0 for any values of ϑ greater than or equal to 0.

Definition 2.5 [7] An altering distance function is a mapping $\Psi : \mathrm{R}^+ \to \mathrm{R}^+$ that satisfies the following two properties:

(i) The function Ψ is both continuous and monotonically increasing.
(ii) The function $\Psi(\vartheta)$ equals zero if and only if ϑ equals zero.

Here is a comparable feature for a generalized modular metric.

Theorem 2.7 [24] *Let (W, θ) be a complete generalized modular metric space, and let*

G: $W \to W$ satisfy a (V, ψ)-weak contraction of integral type, defined as follows:

$$\int_0^{\theta_p(G(\zeta),G(\varrho))} \chi(\vartheta)d\vartheta \leq V\left(\int_0^{\theta_p(\zeta,G(\zeta))} \chi(\vartheta)d\vartheta\right)^y\left(\int_0^{\theta_p(\varrho,G(\zeta))} \chi(\vartheta)d\vartheta\right) + \psi\left(\int_0^{\theta_p(\zeta,\varrho)} \chi(\vartheta)d\vartheta\right)$$

For all values of ζ and ϱ belonging to the set W, Given that V is greater than or equal to 0 and y is greater than or equal to 0. Let's suppose that.

(i) *The function $v : R^+ \to R^+$ is monotonically growing, while the function $\psi : R^+ \to R^+$ is continuous and used for comparison.*
(ii) *The Lebesgue-integrable mapping summable (LIMS)-Stieltjes function $\chi : R^+ \to R^+$ is a nonnegative function defined on the positive real numbers.*

Furthermore, G contains a solitary fixed point u that belongs to the set W. This fixed point u has the property that for each ζ in W, the limit as n approaches infinity of $G^n(\zeta)$ is equal to u.

The authors presented a weak contraction theorem of integral type version [26] for a generalized modular metric.

Theorem 2.8 [26] *Consider a θ-complete generalized modular metric space denoted by W. Let $G : W \to W$ be a mapping that satisfies a $(\zeta i, \psi)$-weak contraction of integral type, as given below:*

$$\int_0^{\theta_p(G(\zeta),G(\varrho))} \chi(\vartheta)d\vartheta \leq \zeta_i\left(\int_0^{\theta_p(\zeta,G(\zeta))} \chi(\vartheta)d\vartheta\right)\left(\int_0^{\theta_p(\varrho,G(\zeta))} \chi(\vartheta)d\vartheta\right) + \psi\left(\int_0^{\theta_p(\zeta,\varrho)} \chi(\vartheta)d\vartheta\right)$$

For any ζ and ϱ that belong to the set W. The following criteria are assumed:

(i) *ζ_i is a continuous function that maps positive real numbers to positive real numbers.*
(ii) *The function ψ is continuous and has the property $\psi(0) = 0$. Additionally, both ψ and v, which are functions mapping from the positive real numbers to the real numbers, are monotone increasing.*
(iii) *The Lebesgue-integrable mapping summable (LIMS)-Stieltjes function χ: $R^+ \to R^+$ is a nonnegative function, and it holds for all x in the domain of χ.*

Given these circumstances, G has a distinct fixed point u in the set W, such that for each ζ in W, the limit as n approaches infinity of $G^n(\zeta)$ is equal to u.

Aydi [3] proposed a formulation and fixed point theory for the contractive condition of integral type with variable distances. They employed [26] the identical methodology for the generalized modular metric.

Theorem 2.9 [26] $V : [0, \infty) \to [0, \infty)$ *is subadditive on all* $[m, n] \subset [0, \infty)$ *if*

$$\int_0^{m+n} V(\vartheta)d\vartheta \leq \int_0^{m} V(\vartheta)d\vartheta + \int_0^{n} V(\vartheta)d\vartheta \tag{43}$$

Theorem 2.10 [26] *Let W be a θ-complete generalized modular metric space and $G : W \to W$ such that.*

$$\psi\left(\int_0^{\theta_p(G(\zeta),G(\varrho))} \chi(\vartheta)d\vartheta\right) \leq \psi(\theta(\zeta,\varrho)) - \Phi(\theta(\zeta,\varrho)), \tag{44}$$

for all $\zeta, \varrho \in W$ *with non-negative real numbers* θ, γ, ϱ *such that* $2\theta + \gamma + 2\varrho < 1$, *where* ψ, Φ *are altering distances, and*

$$\theta(\zeta, \varrho) = \theta \int_0^{\theta_p(\zeta, G(\zeta)) + \theta_p(\varrho, G(\zeta))} \chi(\vartheta)d\vartheta + \gamma \int_0^{\theta(\zeta, \varrho)} \chi(\vartheta)d\vartheta$$
$$+ \varrho \int_0^{\max\{\theta_p(\varrho, G(\zeta)), \theta_p(\varrho, G(\zeta))\}} \chi(\vartheta)d\vartheta$$

LIMS for $\chi(\vartheta) : [0, \infty) \to [0, \infty)$ *is subadditive on all subset of* R^+, *non-negative such that for all* $\upsilon > 0$, $\int_0^{\upsilon} \chi(\vartheta)d\vartheta > 0. \int_0^{\upsilon} \chi(\vartheta)d\vartheta > 0$.

Then G has a unique fixed point in W.

3 Best Approximation Results and Their Practical Implications

We provide fixed point outcomes for optimal approximation in an extended modular metric space using Malih's angle [22], relying on the framework established by Theorems 2.1 and 2.2.

Theorem 3.1 [24] *Let* (A, θ_p) *be a subset of generalized modular metric space* W *with* $G, F : W \to W$ *weakly compatible self mappings satisfying* $clG(W) \subseteq F(W)$, $cl(G(W))$*or* $F(W)$ *be complete,*

$$\int_0^{\theta_p(G(\zeta), G(\varrho))} \chi(\vartheta)d\vartheta \le \varrho \int_0^{\theta_p(F(\zeta), F(\varrho))} \chi(\vartheta)d\vartheta \tag{45}$$

LIMS for $\zeta, \varrho \in W, \varrho \in [0, 1)$ *and* $\chi : R^+ \to R^+$ *nonnegative function with* $\int_0^{\epsilon} \chi(\vartheta)d\vartheta > 0$ *for al l*$\in > 0$. *Then G and F have a unique common fixed point.*

Example for better understanding of the theorem is given below:

Example 3.1 [24] Let W be the set $\{1, 2, 3, 4, 5\}$. They gave the following values for the function $\theta_p(1, 2) = \theta_p(2, 1) = 5, \theta_p(2, 3) = \theta_p(3, 2) = \theta_p(1, 3) = \theta_p(3, 1) = 1$, $\theta_p(1, 4) = \theta_p(4, 1) = \theta_p(2, 4) = \theta_p(4, 2) = \theta_p(3, 4) = \theta_p(4, 3) = 4, \theta_p(1, 5) = \theta_p(5, 1) = \theta_p(2, 5) = \theta_p(5, 2) = \theta_p(3, 5) = \theta_p(5, 3) = \theta_p(4, 5) = \theta_p(5, 4) = 2$.

They also gave $G(5) = 1$ and $G(x) = 2$ for all other values of x. Additionally, they hgave the function $F(\zeta) = \zeta$. Given the function$\chi(\vartheta) = 2\vartheta$, it holds true for all $\epsilon > 0$. If G and F are weakly compatible, then 2 is a fixed point for all ϱ.

Theorem 3.2 [24] *Let W be generalized modular metric space and* $G : W \to W$ *satisfy a fixed point* $z \in W$ *satisfying*

$$\int_0^{\theta_p(G(\zeta),G(\varrho))} \chi(\vartheta)d\vartheta \le \varrho \int_0^{\theta_p(\zeta,\varrho)} \chi(\vartheta)d\vartheta \tag{46}$$

The LIMS is defined for ζ *and* ϱ *belonging to the set W, where* ϱ *belongs to the interval [0, 1) and* χ *is a nonnegative function mapping from the positive real numbers to the positive real numbers, satisfying the condition for all* ϵ *greater than zero. If V is a closed subset of W that is invariant under the group G, and the mapping G|V is compact, then the set PV(z) of best approximations is non-empty.*

Samet's work [30] serves as inspiration for the following example:

Example 3.2 [24] Let W be the set $\{1, 2, 3, 4, 5\}$. They defined $\theta_p(1, 2) = \theta_p(2, 1) = 5/p$, $\theta_p(2, 3) = \theta_p(3, 2) = \theta_p(1, 3) = \theta_p(3, 1) = 1/p$, $\theta_p(1, 4) = \theta_p(4, 1) = \theta_p(2, 4) = \theta_p(4, 2) = \theta_p(3, 4) = \theta_p(4, 3) = 4/p$, $\theta_p(1, 5) = \theta_p(5, 1) = \theta_p(2, 5) = \theta_p(5, 2) = \theta_p(3, 5) = \theta_p(5, 3) = \theta_p(4, 5) = \theta_p(5, 4) = 2/p$ $\theta_{2p}(1, 2) \le \theta_p(1, 3) + \theta_p(3, 2)$ demonstrates that θ_p is not a modular function. $G(5)$ is equal to 1, but for all other numbers it is equal to 2, and for any positive ϵ. Let G be a set containing the elements 1 and 2. V is a closed subset of W that is invariant by the action of G. If V is a compact mapping, then the set PV (2) equals zero.

The applicability of our results may be extended to include two to four finite self-mappings for shared fixed points, specifically in symmetric spaces as described in Ref. [9]. Set-valued mappings yield comparable findings that serve as valuable guidance for optimization and other domains, facilitating further exploration in these areas [5]. Consideration should also be given to the application of this method to a system of initial value problems, as discussed in Ref. [31], as well as its use in matrix equations using the (α, β)-generalized contraction approach, as discussed in Ref. [32].

References

1. Aamri, M., El Moutawakil, D.: Some new common fixed point theorems under strict contractive conditions. J. Math. Anal. Appl. **270**, 181–188 (2002). https://doi.org/10.1016/S0022-247X(02)00059-8
2. Asadi, R., Karimi, L., Feizi, E.: Best 'omega'-proximity point for 'omega'-proximal quasi contraction mappings in modular metric spaces. MACO **1**(2), 95–101 (2020). https://doi.org/10.29252/maco.1.2.918
3. Aydi, H.: A fixed point theorem for a contractive condition of integral type involving altering distances. Int. J. Nonlinear Anal. Appl. **3**(1) (2012). ISSN: 2008-6822
4. Azadifar, B., Sadeghi, U., Saadati, R., Park, C.: Integral type contractions in modular metric spaces. J. Inequal. Appl. **2013**, 483 (2013)

5. Balaj, M., Khamsi, M.A.: Common fixed point theorems for set-valued mappings in normed spaces. RACSAM **113**, 1893–1905 (2019)
6. Basha, S.S., Veeramani, P.: Best approximations and best proximity pairs. Acta Sci. Math. (Szeged) **63**, 289–300 (1997)
7. Berinde, V.: Approximating fixed points of weak contractions using Picard iteration. Nonlinear Anal. Forum **9**(1), 43–53 (2004)
8. Berinde, M., Berinde, V.: On a general class of multi-valued weakly Picard mappings. J. Math. Anal. Appl. **326**, 772–782 (2007)
9. Branciari, A.: A fixed point theorem for mappings satisfying a general contractive condition of integral type. Int. J. Math. Math. Sci. **29**, 531–536 (2002)
10. Chauhan, S., Karapinar, E.: Some integral type common fixed point theorems satisfying ϕ-contractive conditions. Bull. Belg. Math. Soc. Simon Stevin **21**, 593–612 (2014)
11. Chistyakov, V.V.: Modular metric spaces, I: Basic concepts. Nonlinear Anal. **72**(1), 1–14 (2010)
12. Chistyakov, V.V.: Modular metric spaces, II: Application to superposition operators. Nonlinear Anal. **72**(1), 15–30 (2010)
13. Chistyakov, V.V.: Metric Modular Spaces: Theory and Applications. Springer Briefs in Mathematics. ISSN: 2191-8198 (electronic) Library of Congress Control Number: 2015956774 73 (2015), https://doi.org/10.1007/978-3-319-25283-4
14. Fan, K.: Extensions of two fixed point theorems of F. E. Browder. Math. Z. **112**, 234–240 (1969)
15. Fallahi, K., Rad, G.S., Fulga, A.: Best proximity points for $(\phi - \psi)$-weak contractions and some applications. Filomat **37**(6), 1835–1842 (2023). https://doi.org/10.2298/FIL2306835F
16. Fallahi, K., Ghahramani, H., Rad, G.S.: Integral type contractions in partially ordered metric spaces and best proximity point. Iran. J. Sci. Technol. Trans. Sci. **44**, 177–183 (2020). https://doi.org/10.1007/s40995-019-00807-0
17. Gupta, V., Mani, N., Gulati, N.: A common fixed point theorem satisfying contractive condition of integral type. IJREAS **2**(2), 2249–3905 (2012)
18. Hussain, N., O'Regan, D., Agarwal, R.P.: Common fixed point and invariant approximation results on non-star-shaped domains. Georgian Math. J. **12**(4), 659–669 (2005)
19. Jungck, U.: Commuting mappings and fixed point. Am. Math. Mon. **83**, 261–263 (1976)
20. Jungck, U.: Compatible mappings and common fixed points. Int. J. Math. Math. Sci. **11**, 771–779 (1986)
21. Kerim, H., Shatanawi, M., Tallafha, A.: Common fixed point theorems via integral type contraction in modular metric space. G.P.B. Sci. Bull. Ser. A **83**(3) (2021). https://doi.org/10.1186/s13663-018-0650-3
22. Khan, L.A., Khan, A.R.: An extension of Brosowski-Meinardus theorem on invariant approximation approx. Theory Appl. **11**(4), 12 (1995)
23. Malih, S.H.: Best approximation in metric space for contractive mapping of integral type and applications of fixed point to approximation theory. J. Coll. Basic Educ. **20**(84) (2014)
24. Naveen Mani, N., Sharma, A., Shukla, R.: Fixed point results via real-valued function satisfying integral type rational contraction. Abstr. Appl. Anal., Article ID 2592507, 6 p. (2023). https://doi.org/10.1155/2023/2592507
25. Manav Tatar, N., Agarwal, R.P.: Best approximation of fixed-point results for Branciari contraction of integral type on generalized modular metric space. Mathematics **11**(21), 4455 (2023)
26. Manav Tatar, N.: Best approximation of fixed point results in generalized metric spaces. In: Çuvalcıoğlu, G., Sarper, B., Tuğru, F. (eds.) 19th International IFS and Contemporary Mathematics and Engineering Conference, 08–11 July 2023, Tarsus, Mersin, Turkey (2023). https://acikerisim.tarsus.edu.tr/xmlui/bitstream/handle/20.500.13099/172/IFSCOM-E-2023-Abstract-Book.pdf?sequence=1&isAllowed=y
27. Manav, N., Turkoglu, D., Abdeljawad, T.: Common fixed point results for general contractive inequality of integral type on generalized modular metric space. In: Fourth International Conference of Mathematical Sciences (ICMS 2020). AIP Conf. Proc. 2334, 050002-1–050002-4 (2021). https://doi.org/10.1063/5.0042239

28. Olatinwo, M.O.: Some fixed point theorems for weak contraction conditions of integral type. Acta Universitatis Apulensis **24**, 331–338 (2010)
29. Razani, A., Moradi, R.: Common fixed point theorems of integral type in modular spaces. Bull. Iran. Math. Soc. **35**(11–24), 20 (2009)
30. Rhoades, B.: Two fixed-point theorems for mappings satisfying a general contractive condition of integral type. IJMMS **63**, 4007–4013 (2003). https://doi.org/10.1155/S0161171203208024
31. Sessa, S.: On a weak commutativity conditions of mappings in fixed point consideration. Publ. Inst. Wath. (Belgr.) **32**, 146–153 (1982)
32. Shukla, S., Dubey, N., Shukla, R.: Fixed point theorems in graphical cone metric spaces and application to a system of initial value problems. J. Inequal. Appl. **2023**, 91 (2023). https://doi.org/10.1186/s13660-023-03002-3
33. Shukla, R., Sinkala, W.: Convex (α, β)-generalized contraction and its applications in matrix equations. Axioms **12**, 859 (2023). https://doi.org/10.3390/axioms12090859
34. Reich, S.: Approximate selections, best approximations, fixed points and invariant sets. Math. Anal. Appl. **62**, 104–113 (1978)

Generalized Kantorovich–Schurer-Type Operators

Nursel Çetin

Abstract In this paper, we consider Kantorovich extension of generalized Bernstein–Schurer operators depending on a non-negative integer parameter. We prove approximation theorems in the space of continuous functions and L_p-space. Moreover, we obtain some estimates for the rate of convergence by using modulus of continuity and L_p modulus of smoothness of the first order.

Keywords Kantorovich operators · Schurer operators · Stancu operators · L_p-convergence · Modulus of continuity

1 Introduction

Stancu constructed a linear positive polynomial operator of Bernstein type, depending on a non-negative integer parameter ρ,

$$L_{\eta,\rho}(f,x) = \sum_{i=0}^{\eta} \omega_{\eta,i,\rho}(x) f\left(\frac{i}{\eta}\right), \quad x \in [0,1] \tag{1}$$

for $f \in C[0,1]$, $\eta \in \mathbb{N}$ such that $\eta > 2\rho$, where

$$\omega_{\eta,i,\rho}(x) = \begin{cases} (1-x)P_{\eta-\rho,i}(x); & 0 \le i < \rho \\ (1-x)P_{\eta-\rho,i}(x) + xP_{\eta-\rho,i-\rho}(x); & \rho \le i \le \eta-\rho \\ xP_{\eta-\rho,i-\rho}(x); & \eta-\rho < i \le \eta \end{cases},$$

that is a generalization of Bernstein's fundamental functions defined by

N. Çetin (✉)
Department of Mathematics, Polatlı Faculty of Science and Letters, Ankara Hacı Bayram Veli University, Ankara, Türkiye
e-mail: nurselcetin07@gmail.com

O. Castillo et al. (eds.), *Advanced Mathematics for the Modeling and Solution of Challenging Problems in Engineering*, Studies in Systems, Decision and Control 582,
https://doi.org/10.1007/978-3-031-82342-8_23

$$P_{\eta,i}(x)=\begin{cases}\binom{\eta}{i}x^i(1-x)^{\eta-i}; & 0\le i\le \eta \\ 0; & i<0 \quad or \quad i>\eta\end{cases}, \quad x\in[0,1],$$

and studied some approximation properties in [15] and also in [16]. The operator $L_{\eta,\rho}$ is named as ηth Stancu operator. We note that the special cases $\rho=0$ and $\rho=1$ give the classical Bernstein operators. After that, some authors have researched into Stancu operators $L_{\eta,\rho}$ and their some generalizations in the real and complex cases intensively (see, e.g., [3–7, 10, 12, 17]).

Since Bernstein operators are not appropriate for approximation of discontinuous functions, Kantorovich [11] introduced a linear positive operator known in the literature as Kantorovich operators so as to obtain approximation of Lebesgue integrable function on [0, 1]. Considering a non-negative integer γ, Schurer defined and studied any other generalization of Bernstein operators in [13] which is called as Schurer operators. In [2], Bărbosu introduced the Kantorovich form of Schurer-type operators $\tilde{K}_\eta^\gamma : L_1[0,1]\to C[0,1]$ as

$$\tilde{K}_\eta^\gamma(f,x)=\sum_{i=0}^{\eta+\gamma}\tilde{P}_{\eta,i}(x)(\eta+\gamma+1)\int_{\frac{i}{\eta+\gamma+1}}^{\frac{i+1}{\eta+\gamma+1}} f(t)dt, \quad \eta\in\mathbb{N}, \tag{2}$$

where $f\in L_1[0,1]$, $x\in[0,1]$, a non-negative integer γ and $\tilde{P}_{\eta,i}(x)$ is given by

$$\tilde{P}_{\eta,i}(x)=\begin{cases}\binom{\eta+\gamma}{i}x^i(1-x)^{\eta+\gamma-i}; & 0\le i\le \eta+\gamma \\ 0; & i<0 \quad or \quad i>\eta+\gamma\end{cases}, \tag{3}$$

and investigated some approximation properties of these operators.

Inspired by the above works, we consider a new generalization of Kantorovich–Schurer-type operators depending on a non-negative integer parameter ρ, defined by

$$\tilde{K}_{\eta,\rho}^\gamma(f,x)=\sum_{i=0}^{\eta+\gamma}\tilde{\omega}_{\eta,i,\rho}(x)(\eta+\gamma+1)\int_{\frac{i}{\eta+\gamma+1}}^{\frac{i+1}{\eta+\gamma+1}} f(t)dt, \quad x\in[0,1], \tag{4}$$

where $f\in L_1[0,1]$, γ is a non-negative integer, $\eta\in\mathbb{N}$ such that $\eta+\gamma>2\rho$ and

$$\tilde{\omega}_{\eta,i,\rho}(x) = \begin{cases}(1-x)\tilde{P}_{\eta-\rho,i}(x); & 0\le i<\rho \\ (1-x)\tilde{P}_{\eta-\rho,i}(x)+x\tilde{P}_{\eta-\rho,i-\rho}(x); & \rho\le i\le \eta+\gamma-\rho \\ x\tilde{P}_{\eta-\rho,i-\rho}(x); & \eta+\gamma-\rho<i\le\eta+\gamma\end{cases} \tag{5}$$

with $\tilde{P}_{\eta,i}(x)$ given as in (3). We call $\tilde{K}^{\gamma}_{\eta,\rho}$ as generalized Kantorovich–Schurer operators. By using the definition of $\tilde{\omega}_{\eta,i,\rho}(x)$, the generalized Kantorovich–Schurer operators can be written as

$$\tilde{K}^{\gamma}_{\eta,\rho}(f,x) = \sum_{i=0}^{\eta+\gamma-\rho} \tilde{P}_{\eta-\rho,i}(x)(\eta+\gamma+1)\left[(1-x)\int_{\frac{i}{\eta+\gamma+1}}^{\frac{i+1}{\eta+\gamma+1}} f(t)dt + x\int_{\frac{i+\rho}{\eta+\gamma+1}}^{\frac{i+\rho+1}{\eta+\gamma+1}} f(t)dt\right]. \tag{6}$$

We note that for $\rho = 0$ and $\rho = 1$, the generalized Kantorovich–Schurer operators $\tilde{K}^{\gamma}_{\eta,\rho}$ reduce to the Kantorovich–Schurer-type operators studied in [2]. When $\rho = 0$ and $\rho = 1$ with $\gamma = 0$, the operators $\tilde{K}^{\gamma}_{\eta,\rho}$ become Kantorovich operators [11]. Here, firstly we obtain approximation theorems in the space of continuous functions and L_p-space. Then, we establish some estimates for the rate of convergence by using modulus of continuity and L_p modulus of smoothness of the first order. Finally, we give some graphics which show the relevance of the results.

2 Approximation by the Generalized Kantorovich–Schurer Operators

Let us denote by $e_k(x) = x^k$, $k \in \mathbb{N} \cup \{0\}$, the test functions and $\varphi^j_x(t) := (t-x)^j$, $j \in \mathbb{N}$.

In what follows, we give the first three moments of the generalized Kantorovich–Schurer operators.

Lemma 2.1 *For the operators $\tilde{K}^{\gamma}_{\eta,\rho}$ given by* (4), *one has*

$$\tilde{K}^{\gamma}_{\eta,\rho}(e_0,x) = 1, \tag{7}$$

$$\tilde{K}^{\gamma}_{\eta,\rho}(e_1,x) = x + \frac{1-2x}{2(\eta+\gamma+1)}, \tag{8}$$

$$\begin{aligned}\tilde{K}^{\gamma}_{\eta,\rho}(e_2,x) = x^2 + \frac{1}{3(\eta+\gamma+1)^2}\big\{&-3x^2\big(3\eta+3\gamma+1-\rho+\rho^2\big)\\ &+3x\big(2\eta+2\gamma-\rho+\rho^2\big)+1\big\}.\end{aligned} \tag{9}$$

Lemma 2.2 *For the central moments of the operators $\tilde{K}^{\gamma}_{\eta,\rho}$ given by* (4), *one has*

$$\tilde{K}^{\gamma}_{\eta,\rho}\big(\varphi^1_x, x\big) = \frac{1-2x}{2(\eta+\gamma+1)},$$

$$\tilde{K}^{\gamma}_{\eta,\rho}\left(\varphi_x^2, x\right) = \frac{\eta+\gamma-1+\rho(\rho-1)}{(\eta+\gamma+1)^2}x(1-x) + \frac{1}{3(\eta+\gamma+1)^2}.$$

Applying the classical Korovkin Theorem to the sequence of $\tilde{K}^{\gamma}_{\eta,\rho}$, from (7)–(9) we have the following convergence theorem in the space of real-valued and continuous functions.

Theorem 2.1 *If $f \in C\left[0, 1+\gamma\right]$ and ρ, γ are non-negative integers, then*

$$\lim_{\eta\to\infty} \tilde{K}^{\gamma}_{\eta,\rho}(f) = f$$

uniformly on $[0, 1]$.

L_p-approximation by the sequence of $\tilde{K}^{\gamma}_{\eta,\rho}$ is presented in the following.

Theorem 2.2 *If $f \in L_p[0, 1]$, $1 \le p < \infty$, and ρ, γ are non-negative integers, then we have*

$$\lim_{\eta\to\infty} \tilde{K}^{\gamma}_{\eta,\rho}(f) = f$$

in $L_p[0, 1]$.

Proof Let $\left\|\tilde{K}^{\gamma}_{\eta,\rho}\right\|$ denote the operator norm of $\tilde{K}^{\gamma}_{\eta,\rho}$ acting from $L_p[0, 1]$ into $C[0, 1]$, where ρ, γ are non-negative integers and $\eta \in \mathbb{N}$ such that $\eta+\gamma > 2\rho$. Then, it suffices to prove that there exists a $C > 0$ such that $\left\|\tilde{K}^{\gamma}_{\eta,\rho}\right\| \le C$. Firstly, we consider the cases $\rho = 0$ and $\rho = 1$, which give the Kantorovich–Schurer-type operators investigated in [2], that is

$$\tilde{K}^{\gamma}_{\eta}(f, x) = \tilde{K}^{\gamma}_{\eta,0}(f, x) = \tilde{K}^{\gamma}_{\eta,1}(f, x).$$

From (2), we can write

$$\left|\tilde{K}^{\gamma}_{\eta}(f, x)\right|^p \le \sum_{i=0}^{\eta+\gamma} \tilde{P}_{\eta,i}(x)\left|(\eta+\gamma+1)\int_{\frac{i}{\eta+\gamma+1}}^{\frac{i+1}{\eta+\gamma+1}} f(t)dt\right|^p.$$

Taking into consideration the facts that $\psi(s) = |s|^p$, $1 \le p < \infty$, $s \in [0, 1]$, is convex and $\sum_{i=0}^{\eta+\gamma} \tilde{P}_{\eta,i}(x) = 1$ for $\tilde{P}_{\eta,i}(x) \ge 0$, by using Jensen's inequality and integral Jensen's inequality (see, e.g., [1]) we obtain

$$\left|\tilde{K}^{\gamma}_{\eta}(f, x)\right|^p \le \sum_{i=0}^{\eta+\gamma} \tilde{P}_{\eta,i}(x)(\eta+\gamma+1)\int_{\frac{i}{\eta+\gamma+1}}^{\frac{i+1}{\eta+\gamma+1}} |f(t)|^p dt$$

$$= \sum_{i=0}^{\eta+\gamma} \binom{\eta+\gamma}{i} x^i (1-x)^{\eta+\gamma-i} (\eta+\gamma+1) \int_{\frac{i}{\eta+\gamma+1}}^{\frac{i+1}{\eta+\gamma+1}} |f(t)|^p dt,$$

which follows that

$$\int_0^1 \left|\tilde{K}_\eta^\gamma(f,x)\right|^p dx \leq \sum_{i=0}^{\eta+\gamma} \binom{\eta+\gamma}{i}$$

$$\times \int_0^1 x^i (1-x)^{\eta+\gamma-i} dx (\eta+\gamma+1) \int_{\frac{i}{\eta+\gamma+1}}^{\frac{i+1}{\eta+\gamma+1}} |f(t)|^p dt.$$

By the well-known Beta integral, we have

$$\int_0^1 \left|\tilde{K}_\eta^\gamma(f,x)\right|^p dx \leq \sum_{i=0}^{\eta+\gamma} \int_{\frac{i}{\eta+\gamma+1}}^{\frac{i+1}{\eta+\gamma+1}} |f(t)|^p dt = \int_0^1 |f(t)|^p dt.$$

Thus, passing to the L_p-norm, we have $\left\|\tilde{K}_\eta^\gamma\right\|_{L_p[0,1]} \leq \|f\|_{L_p[0,1]}$ for every $f \in L_p[0,1]$. That is, $\tilde{K}_\eta^\gamma$ is a bounded operator with $\left\|\tilde{K}_\eta^\gamma\right\| \leq 1$. Now, we consider the case $\rho > 1$. From (4), we get

$$\left|\tilde{K}_{\eta,\rho}^\gamma(f,x)\right|^p \leq \sum_{i=0}^{\eta+\gamma} \tilde{\omega}_{\eta,i,\rho}(x) \left|(\eta+\gamma+1) \int_{\frac{i}{\eta+\gamma+1}}^{\frac{i+1}{\eta+\gamma+1}} f(t)dt\right|^p.$$

Taking into account of the facts that $\psi(s) = |s|^p$, $1 \leq p < \infty$, $s \in [0,1]$, is convex and $\sum_{i=0}^{\eta+\gamma} \tilde{\omega}_{\eta,i,\rho}(x) = 1$ for $\tilde{\omega}_{\eta,i,\rho}(x) \geq 0$, by Jensen's inequality and integral Jensen's inequality (see, e.g., [1]) we can write

$$\left|\tilde{K}_{\eta,\rho}^\gamma(f,x)\right|^p \leq \sum_{i=0}^{\eta+\gamma} \tilde{\omega}_{\eta,i,\rho}(x)(\eta+\gamma+1) \int_{\frac{i}{\eta+\gamma+1}}^{\frac{i+1}{\eta+\gamma+1}} |f(t)|^p dt$$

$$= \left\{ \sum_{i=0}^{\eta+\gamma-\rho} (1-x)\tilde{P}_{\eta-\rho,i}(x) + \sum_{i=\rho}^{\eta+\gamma} x\tilde{P}_{\eta-\rho,i-\rho}(x) \right\}$$

$$\times\,(\eta+\gamma+1)\int\limits_{\frac{i}{\eta+\gamma+1}}^{\frac{i+1}{\eta+\gamma+1}}|f(t)|^p dt,$$

which follows that

$$\int\limits_0^1 \left|\tilde{K}_{\eta,\rho}^{\gamma}(f,x)\right|^p dx$$

$$\leq \sum_{i=0}^{\eta+\gamma-\rho}\binom{\eta+\gamma-\rho}{i}\int\limits_0^1 x^i(1-x)^{\eta+\gamma-\rho-i+1}dx(\eta+\gamma+1)\int\limits_{\frac{i}{\eta+\gamma+1}}^{\frac{i+1}{\eta+\gamma+1}}|f(t)|^p dt$$

$$+\sum_{i=\rho}^{\eta+\gamma}\binom{\eta+\gamma-\rho}{i-\rho}\int\limits_0^1 x^{i-\rho+1}(1-x)^{\eta+\gamma-\rho}dx(\eta+\gamma+1)\int\limits_{\frac{i}{\eta+\gamma+1}}^{\frac{i+1}{\eta+\gamma+1}}|f(t)|^p dt.$$

From the well-known Beta integral, we obtain

$$\int\limits_0^1 \left|\tilde{K}_{\eta,\rho}^{\gamma}(f,x)\right|^p dx$$

$$\leq \sum_{i=0}^{\eta+\gamma-\rho}\frac{(\eta+\gamma-\rho-i+1)(\eta+\gamma+1)}{(\eta+\gamma-\rho+2)(\eta+\gamma-\rho+1)}\int\limits_{\frac{i}{\eta+\gamma+1}}^{\frac{i+1}{\eta+\gamma+1}}|f(t)|^p dt$$

$$+\sum_{i=\rho}^{\eta+\gamma}\frac{(\eta-\rho+1)(\eta+\gamma+1)}{(\eta+\gamma-\rho+2)(\eta+\gamma-\rho+1)}\int\limits_{\frac{i}{\eta+\gamma+1}}^{\frac{i+1}{\eta+\gamma+1}}|f(t)|^p dt.$$

Because $\eta+\gamma>2\rho$ for $\rho\in\mathbb{N}\cup\{0\}$, we get $\eta+\gamma-\rho>\rho$. Therefore, we can write

$$\int\limits_0^1 \left|\tilde{K}_{\eta,\rho}^{\gamma}(f,x)\right|^p dx$$

$$\leq \frac{\eta+\gamma+1}{\eta+\gamma-\rho+2}\left\{\sum_{i=0}^{\rho-1}+\sum_{i=\rho}^{\eta+\gamma-\rho}\right\}\frac{\eta+\gamma-\rho-i+1}{\eta+\gamma-\rho+1}\int\limits_{\frac{i}{\eta+\gamma+1}}^{\frac{i+1}{\eta+\gamma+1}}|f(t)|^p dt$$

$$+\frac{\eta+\gamma+1}{\eta+\gamma-\rho+2}\left\{\sum_{i=\rho}^{\eta+\gamma-\rho}+\sum_{i=\eta+\gamma-\rho+1}^{\eta+\gamma}\right\}\frac{\eta-\rho+1}{\eta+\gamma-\rho+1}\int\limits_{\frac{i}{\eta+\gamma+1}}^{\frac{i+1}{\eta+\gamma+1}}|f(t)|^p dt$$

$$\leq\frac{\eta+\gamma+1}{\eta+\gamma-\rho+2}\left\{\sum_{i=0}^{\rho-1}\frac{\eta+\gamma-\rho-i+1}{\eta+\gamma-\rho+1}+\sum_{i=\rho}^{\eta+\gamma-\rho}\frac{\eta+\gamma-2\rho+2}{\eta+\gamma-\rho+1}\right.$$

$$\left.+\sum_{i=\eta+\gamma-\rho+1}^{\eta+\gamma}\frac{\eta-\rho+1}{\eta+\gamma-\rho+1}\right\}\int\limits_{\frac{i}{\eta+\gamma+1}}^{\frac{i+1}{\eta+\gamma+1}}|f(t)|^p dt. \tag{10}$$

Taking into account of the facts that we get $\eta+\gamma-\rho-i+1\leq\eta+\gamma-\rho+1$ for $0\leq i\leq\rho-1$, we obtain $\eta+\gamma-2\rho+2\leq\eta+\gamma-\rho+1$ when $\eta+\gamma-2\rho+1<\eta+\gamma-\rho+1$, and we have $i-\rho+1\leq\eta+\gamma-\rho+1$ for $\eta+\gamma-\rho+1\leq i\leq\eta+\gamma$, (10) reduces to

$$\int\limits_0^1\left|\tilde{K}_{\eta,\rho}^{\gamma}(f,x)\right|^p dx$$

$$\leq\frac{\eta+\gamma+1}{\eta+\gamma-\rho+2}\left\{\sum_{i=0}^{\rho-1}+\sum_{i=\rho}^{\eta+\gamma-\rho}+\sum_{i=\eta+\gamma-\rho+1}^{\eta+\gamma}\right\}\int\limits_{\frac{i}{\eta+\gamma+1}}^{\frac{i+1}{\eta+\gamma+1}}|f(t)|^p dt$$

$$=\frac{\eta+\gamma+1}{\eta+\gamma-\rho+2}\sum_{i=0}^{\eta+\gamma}\int\limits_{\frac{i}{\eta+\gamma+1}}^{\frac{i+1}{\eta+\gamma+1}}|f(t)|^p dt.$$

$$=\frac{\eta+\gamma+1}{\eta+\gamma-\rho+2}\int\limits_0^1|f(t)|^p dt. \tag{11}$$

Taking $\sup\limits_{\eta+\gamma>2\rho}\frac{\eta+\gamma+1}{\eta+\gamma-\rho+2}$ for $\rho>1$, we get

$$\sup_{\eta+\gamma>2\rho}\frac{\eta+\gamma+1}{\eta+\gamma-\rho+2}=\frac{2\rho+2}{\rho+3}:=C_1.$$

Hence, (11) gives

$$\int\limits_0^1\left|\tilde{K}_{\eta,\rho}^{\gamma}(f,x)\right|^p dx\leq C_1\int\limits_0^1|f(t)|^p dt.$$

Passing to the L_p-norm, we have $\left\|\tilde{K}_{\eta,\rho}^{\gamma}\right\|_{L_p[0,1]} \leq C_1^{1/p}\|f\|_{L_p[0,1]}$ for every $f \in L_p[0,1]$. It implies that $\tilde{K}_{\eta,\rho}^{\gamma}$ is a bounded operator with $\left\|\tilde{K}_{\eta,\rho}^{\gamma}\right\| \leq C_1^{1/p}$ for every $\eta \in \mathbb{N}$ such that $\eta + \gamma > 2\rho$. Let $\varepsilon > 0$ be given. Since the density of $C[0,1]$ in $L_p[0,1]$, there exists a $\varphi \in C[0,1]$ such that $\|f - \varphi\|_{L_p[0,1]} < \varepsilon$. Then, from Theorem 2.1, there exists an $\eta_0 \in \mathbb{N}$ with $\eta > \eta_0$ such that $\left\|\tilde{K}_{\eta,\rho}^{\gamma}(\varphi) - \varphi\right\|_{C[0,1]} < \varepsilon$. By the following inequality

$$\begin{aligned}\left\|\tilde{K}_{\eta,\rho}^{\gamma}(f) - f\right\|_{L_p[0,1]} &\leq \left\|\tilde{K}_{\eta,\rho}^{\gamma}(f) - \tilde{K}_{\eta,\rho}^{\gamma}(\varphi)\right\|_{L_p[0,1]} \\ &\quad + \left\|\tilde{K}_{\eta,\rho}^{\gamma}(\varphi) - \varphi\right\|_{C[0,1]} + \|\varphi - f\|_{L_p[0,1]},\end{aligned}$$

we arrive at

$$\begin{aligned}&\left\|\tilde{K}_{\eta,\rho}^{\gamma}(f) - f\right\|_{L_p[0,1]} \\ &\leq \left\|\tilde{K}_{\eta,\rho}^{\gamma}\right\|\|f - \varphi\|_{L_p[0,1]} + \left\|\tilde{K}_{\eta,\rho}^{\gamma}(\varphi) - \varphi\right\|_{C[0,1]} + \|\varphi - f\|_{L_p[0,1]} \\ &< \left(2 + C_1^{1/p}\right)\varepsilon + \varepsilon + \varepsilon = \left(4 + C_1^{1/p}\right)\varepsilon,\end{aligned}$$

which completes the proof.

Next, we give the rate of convergence for the sequence $\left\{\tilde{K}_{\eta,\rho}^{\gamma}(f)\right\}_{\eta\in\mathbb{N}}$.

Now, we recall that if $I \subset \mathbb{R}$ is a given interval and f is a real valued function defined on I and bounded on this interval, modulus of continuity of f, is the function ω_1 defined by for any $\delta > 0$

$$\omega_1(f, \delta) = \sup\{|f(t) - f(x)| : t, x \in I, |t - x| \leq \delta\}.$$

Theorem 2.3 *If $f \in C\left[0, 1 + \gamma\right]$ and ρ, γ are non-negative integers, then for every $\eta \in \mathbb{N}$ such that $\eta + \gamma > 2\rho$ and $x \in [0,1]$, we have*

$$\left|\tilde{K}_{\eta,\rho}^{\gamma}(f, x) - f(x)\right| \leq 2\omega_1\left(f, \sqrt{\delta_{\eta,\rho}^{\gamma}(x)}\right),$$

where

$$\delta_{\eta,\rho}^{\gamma}(x) = \tilde{K}_{\eta,\rho}^{\gamma}\left(\varphi_x^2, x\right)$$

and $\tilde{K}_{\eta,\rho}^{\gamma}\left(\varphi_x^2, x\right)$ is given as in Lemma 2.2.

Proof Using the following well-known property of modulus of continuity

$$|f(t) - f(x)| \leq \left(1 + \frac{(t - x)^2}{\delta^2}\right)\omega_1(f, \delta)$$

for $t, x \in [0, 1]$ and $\delta > 0$, we get

$$\left|\tilde{K}^{\gamma}_{\eta,\rho}(f, x) - f(x)\right| \leq \tilde{K}^{\gamma}_{\eta,\rho}(|f(t) - f(x)|, x)$$
$$\leq \left(1 + \frac{\tilde{K}^{\gamma}_{\eta,\rho}\left(\varphi_x^2, x\right)}{\delta^2}\right)\omega_1(f, \delta).$$

Choosing $\delta = \sqrt{\delta^{\gamma}_{\eta,\rho}(x)} = \sqrt{\tilde{K}^{\gamma}_{\eta,\rho}\left(\varphi_x^2, x\right)}$, we obtain the desired result.

In the following, we present an estimate regarding global approximation via modulus of continuity.

Corollary 2.1 *If $f \in C\big[0, 1 + \gamma\big]$ and ρ, γ are non-negative integers, then for every $\eta \in \mathbb{N}$ such that $\eta + \gamma > 2\rho$ and $x \in [0, 1]$, we have*

$$\left|\tilde{K}^{\gamma}_{\eta,\rho}(f, x) - f(x)\right| \leq 2\omega_1\left(f, \sqrt{\delta^{\gamma}_{\eta,\rho}}\right),$$

where

$$\delta^{\gamma}_{\eta,\rho} = \max_{x \in [0,1]} \delta^{\gamma}_{\eta,\rho}(x).$$

Below, we prove the rate of approximation via modulus of continuity in the space of all real-valued continuous functions on $[0, 1]$ such that $f' \in C\big[0, 1 + \gamma\big]$.

Theorem 2.4 *If $f \in C^1\big[0, 1 + \gamma\big]$, and ρ, γ are non-negative integers, then for every $\eta \in \mathbb{N}$ such that $\eta + \gamma > 2\rho$ and $x \in [0, 1]$, we have*

$$\left|\tilde{K}^{\gamma}_{\eta,\rho}(f, x) - f(x)\right| \leq \left|f'(x)\right|\left|\frac{1 - 2x}{2(\eta + \gamma + 1)}\right|$$
$$+ 2\sqrt{\delta^{\gamma}_{\eta,\rho}(x)}\,\omega_1\left(f', \sqrt{\delta^{\gamma}_{\eta,\rho}(x)}\right),$$

where

$$\delta^{\gamma}_{\eta,\rho}(x) = \tilde{K}^{\gamma}_{\eta,\rho}\left(\varphi_x^2, x\right).$$

Proof From the result in [14], we have

$$\left|\tilde{K}^{\gamma}_{\eta,\rho}(f, x) - f(x)\right|$$
$$\leq |f(x)|\left|\tilde{K}^{\gamma}_{\eta,\rho}(e_0, x) - 1\right| + \left|f'(x)\right|\left|\tilde{K}^{\gamma}_{\eta,\rho}\left(\varphi_x^1, x\right)\right|$$
$$+ \sqrt{\tilde{K}^{\gamma}_{\eta,\rho}\left(\varphi_x^2, x\right)}\left\{\sqrt{\tilde{K}^{\gamma}_{\eta,\rho}(e_0, x)} + \frac{1}{\delta}\sqrt{\tilde{K}^{\gamma}_{\eta,\rho}\left(\varphi_x^2, x\right)}\right\}\omega_1\left(f', \delta\right).$$

By Lemmas 2.1 and 2.2, taking $\delta = \sqrt{\delta^{\gamma}_{\eta,\rho}(x)} = \sqrt{\tilde{K}^{\gamma}_{\eta,\rho}(\varphi^2_x, x)}$ we arrive at the result.

The following theorem estimates the order of global approximation.

Theorem 2.5 *If $f \in C^1[0, 1+\gamma]$, and ρ, γ are non-negative integers, then for every $\eta \in \mathbb{N}$ such that $\eta + \gamma > 2\rho$ and $x \in [0, 1]$, we have*

$$\left|\tilde{K}^{\gamma}_{\eta,\rho}(f, x) - f(x)\right| \le \frac{1}{2(\eta+\gamma+1)} M + 2\sqrt{\delta^{\gamma}_{\eta,\rho}}\,\omega_1\left(f', \sqrt{\tau^{\gamma}_{\eta,\rho}}\right),$$

where

$$M = \max_{x\in[0,1]} \left|f'(x)\right| \text{ and } \tau^{\gamma}_{\eta,\rho} = \max_{x\in[0,1]} \sqrt{\delta^{\gamma}_{\eta,\rho}(x)}.$$

In what follows, we prove a theorem which provides an upper estimate for the L_p-norm of the approximation error by the new operator $\tilde{K}^{\gamma}_{\eta,\rho}$.

Let us consider L_p modulus of smoothness of the first order is given by for all $f \in L_p[0, 1-h]$

$$\omega_1(f, \delta)_p = \sup_{0<h\le\delta} \|f(\cdot + h) - f(\cdot)\|_{L_p[0,1-h]}, \quad (1 \le p < \infty),$$

where $\|\cdot\|_{L_p[0,1-h]}$ is the L_p-norm defined over $[0, 1-h]$ (see, e.g., [8]).

Peetre's K-functional is very useful tool for the error of the convergence in L_p-norm. Now, we recall Peetre's K-functional defined for functions in $L_p[0, 1]$. Let, $L^{(1)}_p[0, 1]$, $1 \le p < \infty$, denote the space

$$L^{(1)}_p[0, 1] = \{f \in L_p[0, 1] : f \text{ is absolutely continuous function on } [0, 1] \text{ and} f' \in L_p[0, 1]\}.$$

For any $f \in L_p[0, 1]$, $1 \le p < \infty$, and $\delta > 0$, Peetre's K-functional is defined by

$$K_{1,p}(f, \delta) := \inf\left\{\|f - g\|_{L_p[0,1]} + \delta\left\|g'\right\|_{L_p[0,1]} : g \in L^{(1)}_p[0, 1]\right\}.$$

The connection between Peetre's K-functional and L_p modulus of smoothness of the first order is given by

$$M_1\omega_1(f, \delta)_p \le K_{1,p}(f, \delta) \le M_2\omega_1(f, \delta)_p,$$

where M_1, M_2 are positive constants not depending on f and p (see, e.g., [8, 9]).

To prove the rate of convergence in L_p-space, we need the following result.

Theorem 2.6 *If $\psi \in L^{(1)}_p[0, 1]$, $p > 1$, and ρ, γ are non-negative integers, then for every $\eta \in \mathbb{N}$ such that $\eta + \gamma > 2\rho$ and $x \in [0, 1]$, the operators $\tilde{K}^{\gamma}_{\eta,\rho}$ satisfy the* *following*

$$\left\|\tilde{K}^{\gamma}_{\eta,\rho}(\psi)-\psi\right\|_{L_p[0,1]}\leq 2^{1/p}\frac{p}{p-1}\max_{x\in[0,1]}\sqrt{\tilde{K}^{\gamma}_{\eta,\rho}\left(\varphi_x^2,x\right)}\left\|\psi'\right\|_{L_p[0,1]},$$

where $\tilde{K}^{\gamma}_{\eta,\rho}\left(\varphi_x^2,x\right)$ *is given as in Lemma* 2.2.

Proof Using (4), we get

$$\begin{aligned}\left|\tilde{K}^{\gamma}_{\eta,\rho}(\psi,x)-\psi(x)\right|&=(\eta+\gamma+1)\left|\sum_{i=0}^{\eta+\gamma}\tilde{\omega}_{\eta,i,\rho}(x)\int_{\frac{i}{\eta+\gamma+1}}^{\frac{i+1}{\eta+\gamma+1}}[\psi(t)-\psi(x)]dt\right|\\&\leq(\eta+\gamma+1)\sum_{i=0}^{\eta+\gamma}\tilde{\omega}_{\eta,i,\rho}(x)\int_{\frac{i}{\eta+\gamma+1}}^{\frac{i+1}{\eta+\gamma+1}}\int_{x}^{t}\left|\psi'(s)\right|dsdt\\&\leq\Omega_{\psi'}(x)(\eta+\gamma+1)\sum_{i=0}^{\eta+\gamma}\tilde{\omega}_{\eta,i,\rho}(x)\int_{\frac{i}{\eta+\gamma+1}}^{\frac{i+1}{\eta+\gamma+1}}|t-x|dt,\end{aligned}$$

where $\Omega_{\psi'}(x)=\sup_{t\in[0,1]}\frac{1}{t-x}\int_x^t\left|\psi'(s)\right|ds\ (t\neq x)$ is the Hardy–Littlewood majorant of ψ'. By the Cauchy–Schwarz's inequality, we obtain that

$$\begin{aligned}\left|\tilde{K}^{\gamma}_{\eta,\rho}(\psi,x)-\psi(x)\right|&\leq\Omega_{\psi'}(x)(\eta+\gamma+1)^{1/2}\left(\sum_{i=0}^{\eta+\gamma}\tilde{\omega}_{\eta,i,\rho}(x)\right)^{1/2}\\&\quad\times\left(\sum_{i=0}^{\eta+\gamma}\tilde{\omega}_{\eta,i,\rho}(x)\int_{\frac{i}{\eta+\gamma+1}}^{\frac{i+1}{\eta+\gamma+1}}(t-x)^2dt\right)^{1/2}\\&\leq\Omega_{\psi'}(x)\max_{x\in[0,1]}\sqrt{\tilde{K}^{\gamma}_{\eta,\rho}\left(\varphi_x^2,x\right)}.\end{aligned}\tag{12}$$

From the Hardy–Littlewood theorem (see, [18]), for $p>1$ the following inequality holds

$$\int_0^1\Omega^p_{\psi'}(x)dx\leq 2\left(\frac{p}{p-1}\right)^p\int_0^1\left|\psi'(x)\right|^pdx.$$

Hence, (12) immediately follows that

$$\left\| \tilde{K}_{\eta,\rho}^{\gamma}(\psi) - \psi \right\|_{L_p[0,1]} \leq 2^{1/p} \frac{p}{p-1} \max_{x\in[0,1]} \sqrt{\tilde{K}_{\eta,\rho}^{\gamma}\left(\varphi_x^2, x\right)} \left\| \psi' \right\|_{L_p[0,1]}.$$

Theorem 2.7 *If $f \in L_p[0,1]$, $p > 1$, and ρ, γ are non-negative integers, then for every $\eta \in \mathbb{N}$ such that $\eta + \gamma > 2\rho$ and $x \in [0,1]$, the operators $\tilde{K}_{\eta,\rho}^{\gamma}$ hold*

$$\left\| \tilde{K}_{\eta,\rho}^{\gamma}(f) - f \right\|_{L_p[0,1]} \leq \left(3 + C_1^{1/p}\right) C_2 \left(1 + \frac{2^{1/p}}{3 + C_1^{1/p}} \frac{p}{p-1}\right) \omega_1\left(f, \tau_{\eta,\rho}^{\gamma}\right)_p,$$

where $C_1 = \frac{2\rho+2}{\rho+3}$, $C_2 > 0$ and $\tau_{\eta,\rho}^{\gamma} = \max_{x\in[0,1]} \sqrt{\tilde{K}_{\eta,\rho}^{\gamma}\left(\varphi_x^2, x\right)}$.

Proof We have

$$\left\| \tilde{K}_{\eta,\rho}^{\gamma}(f) - f \right\|_{L_p[0,1]} = \begin{cases} \left(3 + C_1^{1/p}\right) \|f\|_{L_p[0,1]}, & f \in L_p[0,1] \\ 2^{1/p} \frac{p}{p-1} \tau_{\eta,\rho}^{\gamma} \left\| f' \right\|_{L_p[0,1]}, & f \in L_p^{(1)}[0,1] \end{cases},$$

where $C_1 = \frac{2\rho+2}{\rho+3}$ and $\tau_{\eta,\rho}^{\gamma} = \max_{x\in[0,1]} \sqrt{\tilde{K}_{\eta,\rho}^{\gamma}\left(\varphi_x^2, x\right)}$ is as in Theorem 2.6. For arbitrary $\psi \in L_p^{(1)}[0,1]$, one has

$$\begin{aligned}
&\left\| \tilde{K}_{\eta,\rho}^{\gamma}(f) - f \right\|_{L_p[0,1]} \\
&\leq \left(3 + C_1^{1/p}\right) \|f - \psi\|_{L_p[0,1]} + 2^{1/p} \frac{p}{p-1} \tau_{\eta,\rho}^{\gamma} \left\| \psi' \right\|_{L_p[0,1]} \\
&\leq \left(3 + C_1^{1/p}\right) \left\{ \|f - \psi\|_{L_p[0,1]} + 2^{1/p} \frac{p}{(p-1)} \frac{1}{\left(3 + C_1^{1/p}\right)} \tau_{\eta,\rho}^{\gamma} \left\| \psi' \right\|_{L_p[0,1]} \right\} \\
&\leq \left(3 + C_1^{1/p}\right) K_{1,p}\left(f, \frac{2^{1/p}}{3 + C_1^{1/p}} \frac{p}{p-1} \tau_{\eta,\rho}^{\gamma}\right) \\
&\leq \left(3 + C_1^{1/p}\right) C_2 \left(1 + \frac{2^{1/p}}{3 + C_1^{1/p}} \frac{p}{p-1}\right) \omega_1\left(f, \tau_{\eta,\rho}^{\gamma}\right)_p, \quad C_2 > 0.
\end{aligned}$$

Now we give some numerical examples to demonstrate the approximation process by the generalized Kantorovich–Schurer operators $\tilde{K}_{\eta,\rho}^{\gamma}(f)$ with the help of Maple.

Example 2.1 The convergence of the generalized Kantorovich–Schurer operators $\tilde{K}_{\eta,\rho}^{\gamma}(f)$ and the Kantorovich operators $K(f)$ defined in [11] to $f(x) = \frac{1}{15^{\pi x+1}}$ is shown in Fig. 1 for $\rho = 1$, $\eta = 3$ and $\gamma = 1$.

Example 2.2 The convergence of the generalized Kantorovich–Schurer operators $\tilde{K}_{\eta,\rho}^{\gamma}(f)$ and the Kantorovich operators $K(f)$ defined in [11] to $f(x) = \cos(3\pi x) - \sin(9x)$ is demonstrated in Fig. 2 for $\rho = 1$, $\eta = 2$ and $\gamma = 1$.

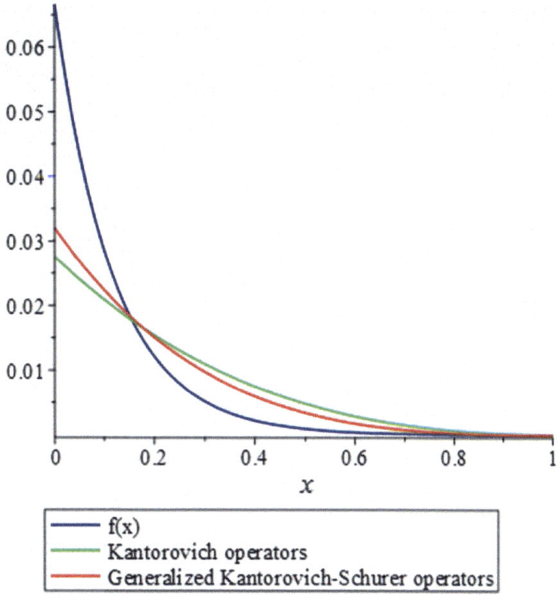

Fig. 1 Approximation process of the operators $\tilde{K}^{\gamma}_{\eta,\rho}(f)$ and $K(f)$ to $f(x) = \frac{1}{15^{\pi x+1}}$

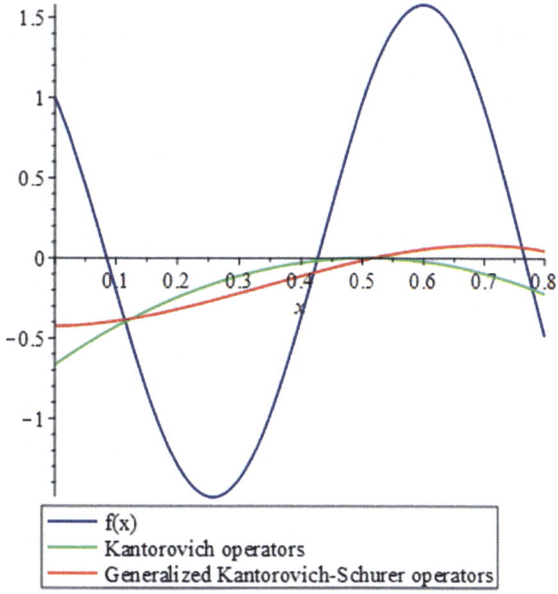

Fig. 2 Approximation process of the operators $\tilde{K}^{\gamma}_{\eta,\rho}(f)$ and $K(f)$ to $f(x) = \cos(3\pi x) - \sin(9x)$

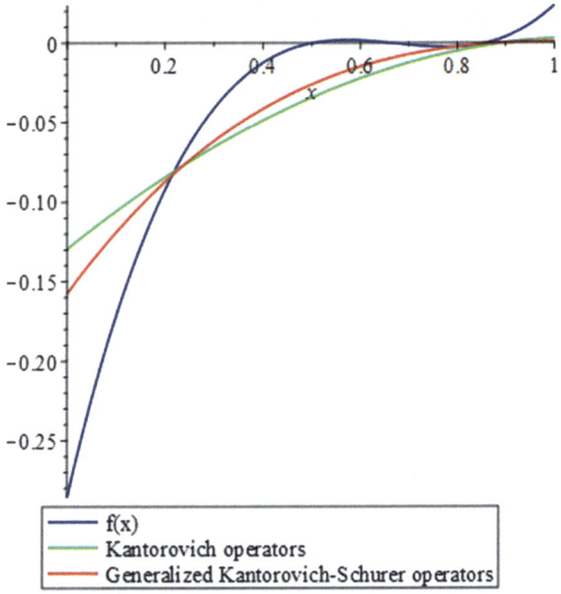

Fig. 3 Approximation process of the operators $\tilde{K}_{\eta,\rho}^{\gamma}(f)$ and $K(f)$ to $f(x) = \left(x-\frac{1}{2}\right)\left(x-\frac{2}{3}\right)\left(x-\frac{6}{7}\right)$

Example 2.3 The convergence of the generalized Kantorovich–Schurer operators $\tilde{K}_{\eta,\rho}^{\gamma}(f)$ and the Kantorovich operators $K(f)$ defined in [11] to $f(x) = \left(x-\frac{1}{2}\right)\left(x-\frac{2}{3}\right)\left(x-\frac{6}{7}\right)$ is shown in Fig. 3 for $\rho = 1$, $\eta = 2$ and $\gamma = 1$.

Example 2.4 The convergence of the generalized Kantorovich–Schurer operators $\tilde{K}_{\eta,\rho}^{\gamma}(f)$ to $f(x) = \cos(5\pi x) + \sin(9x)$ is illustrated in Fig. 4 for different values of η.

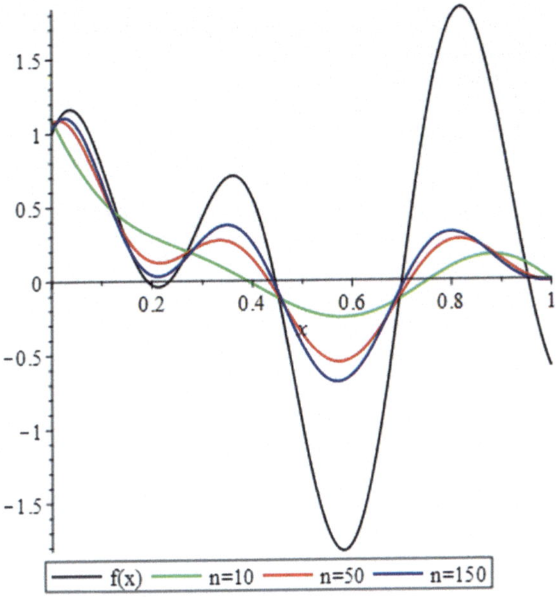

Fig. 4 Approximation process of the operators $\tilde{K}^{\gamma}_{\eta,\rho}(f)$ to $f(x) = \cos(5\pi x) + \sin(9x)$

References

1. Altomare, F.: Korovkin-type theorems and approximation by positive linear operators. Surv. Approx. Theory **5**, 92–164 (2010)
2. Bărbosu, D.: The Kantorovich form of Schurer–Stancu operators. Demonstr. Math. **XXXVII**(2), 383–391 (2004)
3. Bostancı, T., Başcanbaz-Tunca, G.: On Stancu operators depending on a non-negative integer. Filomat **36**(18), 6129–6138 (2022)
4. Bustamante, J., Quesada, J.M.: A property of Ditzian–Totik second order moduli. Appl. Math. Lett. **23**(5), 576–580 (2010)
5. Çetin, N.: A new generalization of complex Stancu operators. Math. Methods Appl. Sci. **42**, 5582–5594 (2019)
6. Çetin, N., Başcanbaz-Tunca, G.: Approximation by a new complex generalized Bernstein operator. An. Univ. Oradea Fasc. Mat. **26**(2), 129–141 (2019)
7. Çetin, N.: A new complex generalized Bernstein–Schurer operator. Carpathian J. Math. **37**(1), 81–89 (2021)
8. Devore, R.A., Lorentz, G.G.: Constructive Approximation. Springer, Berlin (1993)
9. Johnen, H.: Inequalities connected with the moduli of smoothness. Mat. Vesnik **9**(24), 289–303 (1972)
10. Kajla, A.: The Kantorovich variant of an operator defined by D. D. Stancu. Appl. Math. Comput. **316**, 400–408 (2018)
11. Kantorovich, L.V.: Sur certains développements suivant les polynômes de la forme de S. Bernstein. I, II. C. R. Acad. Sci. URSS 563–568, 595–600 (1930)
12. Kumar, A.: A new kind of variant of the Kantorovich type modification operators introduced by D. D. Stancu. Results Appl. Math. **11**, 100158 (2021)
13. Schurer, F.: Linear positive operators in approximation theory. Math. Inst. Techn. Univ. Delft Rep. (1962)
14. Shisha, O., Bond, B.: The degree of convergence of sequences of linear positive operators. Proc. Natl. Acad. Sci. U.S.A. **60**, 1196–1200 (1968)

15. Stancu, D.D.: Quadrature formulas constructed by using certain linear positive operators. In: Numerical Integration (Proceedings Conference, Oberwolfach, 1981), ISNM 57, pp. 241–251. Birkhäuser Verlag, Basel (1982)
16. Stancu, D.D.: Approximation of functions by means of a new generalized Bernstein operator. Calcolo **20**(2), 211–229 (1983)
17. Yang, R., Xiong, J., Cao, F.: Multivariate Stancu operators defined on a simplex. Appl. Math. Comput. **138**, 189–198 (2003)
18. Zygmund, A.: Trigonometric Series I, II. Cambridge University Press, Cambridge, UK (1959)

Approximation by Generalization of Bernstein–Schurer Operators

Nursel Çetin and Nesibe Manav Mutlu

Abstract In this study, we investigate some approximation properties of a new generalization of Bernstein–Schurer operators. Firstly, we give a uniform approximation result and the rate of convergence by means of modulus of continuity. Then, we obtain Voronovskaja and Grüss–Voronovskaja type results. Finally, we demonstrate some numerical examples relevant to our results.

Keywords Bernstein–Schurer operators · Stancu operators · Modulus of continuity · Voronovskaja type theorem · Grüss–Voronovskaja type theorem

1 Introduction

Stancu introduced a new family of linear positive operators named as Stancu operators in the literature, as follows

$$L_{\eta,\tau}(g,u)=\sum_{i=0}^{\eta} w_{\eta,i,\tau}(u)g\left(\frac{i}{\eta}\right), \quad u\in[0,1] \tag{1}$$

for $g \in C[0,1]$, a non-negative integer parameter $\tau,\ \eta \in \mathbb{N}$ such that $\eta > 2\tau$, where

$$w_{\eta,i,\tau}(u)=\begin{cases}(1-u)p_{\eta-\tau,i}(u); & 0\le i<\tau\\ (1-u)p_{\eta-\tau,i}(u)+up_{\eta-\tau,i-\tau}(u); & \tau\le i\le\eta-\tau\\ up_{\eta-\tau,i-\tau}(u); & \eta-\tau<i\le\eta\end{cases}$$

and

N. Çetin
Department of Mathematics, Ankara Hacı Bayram Veli University, Ankara, Türkiye

N. Manav Mutlu (✉)
Department of Management Information Systems, İstanbul Nişantaşı University, İstanbul, Türkiye
e-mail: nesibe.manavm@gmail.com

O. Castillo et al. (eds.), *Advanced Mathematics for the Modeling and Solution of Challenging Problems in Engineering*, Studies in Systems, Decision and Control 582,
https://doi.org/10.1007/978-3-031-82342-8_24

$$p_{\eta,i}(u) = \begin{cases} \binom{\eta}{i} u^i (1-u)^{\eta-i}; & 0 \le i \le \eta \\ 0; & i < 0 \text{ or } i > \eta \end{cases}, \quad u \in [0,1]$$

is the well-known Bernstein basis polynomials (see, [12, 13]). For the special cases $\tau = 0$ and $\tau = 1$, Stancu operators defined by (1) give the classical Bernstein operators. Bustamante and Quesada [2] presented an asymptotic property related to Voronovskaja-type formula for the Stancu operators $L_{\eta,\tau}$. Yang et al. [14] considered multivariate setting of the Stancu operators on a simplex and gave that Lipschitz property of the original function is preserved by these operators. Very recently, Kantorovich type generalizations of Stancu operators were studied in the papers [1, 9, 10]. On the other hand, approximation properties of complex forms of the Stancu operators and their some generalizations were investigated in [3–5].

Considering a non-negative integer γ, Schurer [11] defined a generalization of Bernstein operators which is called as Bernstein–Schurer operators, given by

$$B_{\eta}^{\gamma}(g, u) = \sum_{i=0}^{\eta+\gamma} \tilde{p}_{\eta,i}(u) g\left(\frac{i}{\eta}\right), \quad u \in [0,1] \tag{2}$$

for any $\eta \in \mathbb{N}$ and $g \in C\big[0, 1+\gamma\big]$, the space of all real valued continuous functions on $\big[0, 1+\gamma\big]$, and $\tilde{p}_{\eta,i}(u)$ is given by

$$\tilde{p}_{\eta,i}(u) = \begin{cases} \binom{\eta+\gamma}{i} u^i (1-u)^{\eta+\gamma-i}; & 0 \le i \le \eta+\gamma \\ 0; & i < 0 \text{ or } i > \eta+\gamma \end{cases}. \tag{3}$$

Inspired by the above works, for $g \in C\big[0, 1+\gamma\big]$, τ, γ non-negative integers, $\eta \in \mathbb{N}$ such that $\eta + \gamma > 2\tau$ and

$$\tilde{w}_{\eta,i,\tau}(u) = \begin{cases} (1-u)\tilde{p}_{\eta-\tau,i}(u); & 0 \le i < \tau \\ (1-u)\tilde{p}_{\eta-\tau,i}(u) + u\tilde{p}_{\eta-\tau,i-\tau}(u); & \tau \le i \le \eta+\gamma-\tau \\ u\tilde{p}_{\eta-\tau,i-\tau}(u); & \eta+\gamma-\tau < i \le \eta+\gamma \end{cases} \tag{4}$$

with $\tilde{p}_{\eta,i}$ given as in (3), we introduce a new generalization of Bernstein–Schurer operators, given by

$$L_{\eta,\tau}^{\gamma}(g, u) = \sum_{i=0}^{\eta+\gamma} \tilde{w}_{\eta,i,\tau}(u) g\left(\frac{i}{\eta}\right), \quad u \in [0,1]. \tag{5}$$

We call $L_{\eta,\tau}^{\gamma}$ as generalized Bernstein–Schurer operators. By using the definition of $\tilde{w}_{\eta,i,\tau}(u)$ given by (4), the generalized Bernstein–Schurer operators defined by (5) can be written as

$$L_{\eta,\tau}^{\gamma}(g,u)=\sum_{i=0}^{\eta+\gamma-\tau}\tilde{p}_{\eta-\tau,i}(u)\left[(1-u)g\left(\frac{i}{\eta}\right)+ug\left(\frac{i+\tau}{\eta}\right)\right]. \quad (6)$$

For the cases $\tau=0$ and $\tau=1$ with $\gamma=0$, the generalized Bernstein–Schurer operators $L_{\eta,\tau}^{\gamma}$ reduce to the well-known Bernstein operators. When $\tau=0$ and $\tau=1$, these operators become Bernstein–Schurer operators studied in [11]. The case $\gamma=0$ gives Stancu operators $L_{\eta,\tau}$. Here, we investigate some approximation properties of the generalized Bernstein–Schurer operators $L_{\eta,\tau}^{\gamma}$. Firstly, we give a uniform approximation result and the rate of convergence by means of modulus of continuity. Then, we obtain Voronovskaja and Grüss–Voronovskaja type results. Finally, we demonstrate some numerical examples relevant to our results.

2 Auxiliary Results

In the following theorem, we give the real form of recurrence relation in Lemma 2.2 from [5] to calculate higher order moments of the new operator.

Lemma 2.1 *Suppose that τ is a non-negative integer, $\eta\in\mathbb{N}$ such that $\eta+\gamma>2\tau$ and $\gamma\in\mathbb{N}\cup\{0\}$. Then, we get the recurrence formula*

$$\begin{aligned}L_{\eta,\tau}^{\gamma}(e_{k+1},u)&=\frac{u(1-u)}{\eta}\left(L_{\eta,\tau}^{\gamma}(e_k,u)\right)'\\&+\frac{(\eta+\gamma-\tau+1)u+\tau-1}{\eta}L_{\eta,\tau}^{\gamma}(e_k,u)\\&+\frac{(1-\tau)(1-u)}{\eta}\left(\frac{\eta-\tau}{\eta}\right)^{k}B_{\eta-\tau}^{\gamma}(e_k,u),\end{aligned}$$

where $e_k(u)=u^k,\ k\in\mathbb{N}\cup\{0\}$ and B_{η}^{γ} is the Bernstein–Schurer operator given by (2).

In the following, we give moments and central moments of the generalized Bernstein–Schurer operators $L_{\eta,\tau}^{\gamma}$.

Lemma 2.2 *Considering the operators (6), we have the following moments*

$$L_{\eta,\tau}^{\gamma}(e_0,u)=1,$$

$$L_{\eta,\tau}^{\gamma}(e_1,u)=\left(1+\frac{\gamma}{\eta}\right)u,$$

$$L_{\eta,\tau}^{\gamma}(e_2,u)=\frac{1}{\eta^2}\left\{(\eta+\gamma-\tau)(\eta+\gamma+\tau-1)u^2+\left(\eta+\gamma-\tau+\tau^2\right)u\right\},$$

$$L_{\eta,\tau}^{\gamma}(e_3,u) = \frac{1}{\eta^3}\{(\eta+\gamma-\tau)(\eta+\gamma-\tau-1)(\eta+\gamma+2\tau-2)u^3$$
$$+3(\eta+\gamma-\tau)(\eta+\gamma+\tau^2-1)u^2+(\eta+\gamma-\tau+\tau^3)u\},$$

$$L_{\eta,\tau}^{\gamma}(e_4,u) = \frac{1}{\eta^4}\{(\eta+\gamma-\tau)(\eta+\gamma-\tau-1)(\eta+\gamma-\tau-2)$$
$$\times(\eta+\gamma+3\tau-3)u^4+6(\eta+\gamma-\tau)(\eta+\gamma-\tau-1)$$
$$\times(\eta+\gamma+\tau+\tau^2-2)u^3+(\eta+\gamma-\tau)[7(\eta+\gamma-\tau-1)$$
$$+4\tau^3+6\tau^2+4\tau]u^2+(\eta+\gamma-\tau+\tau^4)u\}.$$

Lemma 2.3 *For the operators (6), we have central moments as below*

$$L_{\eta,\tau}^{\gamma}(t-u,u) = \frac{\gamma}{\eta}u,$$

$$L_{\eta,\tau}^{\gamma}\left((t-u)^2,u\right) = \frac{\gamma^2}{\eta^2}u^2+\frac{[\eta+\gamma+\tau(\tau-1)]}{\eta^2}u(1-u),$$

$$L_{\eta,\tau}^{\gamma}\left((t-u)^4,u\right) = \frac{1}{\eta^4}\{[3\eta^2+\eta(6\tau^2-6\tau-6)+\gamma^4$$
$$-6\gamma^3+\gamma^2(-6\tau^2+6\tau+11)+\gamma(8\tau^3+6\tau^2-14\tau-6)$$
$$-3\tau^4-6\tau^3+3\tau^2+6\tau]u^4+[-6\eta^2$$
$$+\eta(6\gamma^2-24\gamma-12\tau^2+12\tau+12)$$
$$+6\gamma^3+\gamma^2(6\tau^2-6\tau-18)$$
$$+\gamma(-12\tau^3-12\tau^2+24\tau+12)+6\tau^4$$
$$+12\tau^3-6\tau^2-12\tau]u^3+[3\eta^2$$
$$+\eta(10\gamma+6\tau^2-6\tau-7)+7\gamma^2$$
$$+\gamma(4\tau^3+6\tau^2-10\tau-7)-4\tau^4$$
$$-6\tau^3+3\tau^2+7\tau]u^2+[\eta-\tau+\gamma+\tau^4]u\}.$$

Lemma 2.4 *From the previous Lemma, we have*

$$\lim_{\eta\to\infty}\eta L_{\eta,\tau}^{\gamma}(t-u,u) = \gamma u, \tag{7}$$

$$\lim_{\eta\to\infty}\eta L_{\eta,\tau}^{\gamma}\left((t-u)^2,u\right) = u(1-u), \tag{8}$$

$$\lim_{\eta\to\infty}\eta L_{\eta,\tau}^{\gamma}\left((t-u)^4,u\right) = 0. \tag{9}$$

3 Main Results

In this section, we show that the new operators satisfy the Korovkin type theorem and we give some direct results for all real-valued continuous functions on $\left[0, 1+\gamma\right]$.

Theorem 3.1 *For any* $g \subset C\left[0, 1+\gamma\right]$ *and* γ, τ *non-negative integers, then the sequence* $\left\{L_{\eta,\tau}^{\gamma}(g, u)\right\}$ *converges to* g *uniformly on* $[0, 1]$.

Proof For sufficiently large η, from Lemma 2.2, clearly $L_{\eta,\tau}^{\gamma}(e_0, u)$, $L_{\eta,\tau}^{\gamma}(e_1, u)$, $L_{\eta,\tau}^{\gamma}(e_2, u)$, respectively converges uniformly to 1, u and u^2 on $[0, 1]$. Then, by the well-known Korovkin theorem we get $L_{\eta,\tau}^{\gamma}(g, u)$ converges to $g(u)$ uniformly on $[0, 1]$ as $\eta \to \infty$. □

In what follows, we will give the rate of convergence for the sequence $\left\{L_{\eta,\tau}^{\gamma}(g, u)\right\}$. Firstly we recall some definitions.

For $g \in C\left[0, 1+\gamma\right]$ and $\delta > 0$, the first modulus of smoothness, modulus of continuity, of g is defined by

$$\omega(g, \delta) = \sup_{|k|\leq\delta} \sup_{u, u+k \in \left[0,1+\gamma\right]} |g(u+k) - g(u)|$$

and second modulus of smoothness of g is defined by

$$\omega_2(g, \delta) = \sup_{|k|\leq\delta} \sup_{u, u+2k \in \left[0,1+\gamma\right]} |g(u+2k) - 2g(u+k) + g(u)|.$$

For convenience, we need the following Peetre's K-functional defined as

$$K(g, \delta) = \inf_{h \in C^2\left[0,1+\gamma\right]} \left\{\|g-h\| + \delta\left\|h''\right\|\right\},$$

where $\delta > 0$ and $C^2\left[0, 1+\gamma\right] = \left\{h \in C\left[0, 1+\gamma\right] : h', h'' \in C\left[0, 1+\gamma\right]\right\}$ and $\|\cdot\|$ is the uniform norm on $C\left[0, 1+\gamma\right]$. We note that there is a relation between the modulus of smoothness and Peetre's K-functional as in the following: there exist positive constants M_1 and M_2 such that

$$M_1\omega_2(g, \delta) \leq K\left(g, \delta^2\right) \leq M_2\omega_2(g, \delta) \tag{10}$$

(see, e.g., Theorem 2.4 in p. 177 of [6], also [8]).

Now, we present some quantitative estimates concerning the rate of approximation by the sequence $\left\{L_{\eta,\tau}^{\gamma}(g, u)\right\}$.

Theorem 3.2 *Suppose that* $g \in C\left[0, 1+\gamma\right]$ *and* γ, τ *are non-negative integers. Then, for every* $\eta \in \mathbb{N}$ *such that* $\eta+\gamma > 2\tau$ *and* $u \in [0, 1]$, *the following inequality holds*

$$\left|L_{\eta,\tau}^{\gamma}(g, u) - g(u)\right| \leq 2\omega\left(g, \delta_{\eta,\tau}^{\gamma}(u)\right),$$

where $\delta^{\gamma}_{\eta,\tau}(u) = \sqrt{L^{\gamma}_{\eta,\tau}\big((t-u)^2, u\big)}$ *and* $L^{\gamma}_{\eta,\tau}\big((t-u)^2, u\big)$ *is given as in Lemma* 2.3.

Proof By using the well-known following property of modulus of continuity

$$|g(t) - g(u)| \leq \left(1 + \frac{(t-u)^2}{\delta^2}\right)\omega(g, \delta)$$

for $u, t \in [0, 1]$ and $\delta > 0$, we get that

$$\begin{aligned}\left|L^{\gamma}_{\eta,\tau}(g, u) - g(u)\right| &\leq L^{\gamma}_{\eta,\tau}(|g(t) - g(u)|, u) \\ &\leq \left(1 + \frac{1}{\delta^2}L^{\gamma}_{\eta,\tau}\big((t-u)^2, u\big)\right)\omega(g, \delta).\end{aligned}$$

Here, choosing $\delta = \delta^{\gamma}_{\eta,\tau}(u) = \sqrt{L^{\gamma}_{\eta,\tau}\big((t-u)^2, u\big)}$, we obtain the desired result.

Theorem 3.3 *Assume that* $g \in C\big[0, 1+\gamma\big]$ *and* γ, τ *are non-negative integers. Then, for every* $\eta \in \mathbb{N}$ *such that* $\eta + \gamma > 2\tau$ *and* $u \in [0, 1]$, *we have*

$$\left|L^{\gamma}_{\eta,\tau}(g, u) - g(u)\right| \leq M\omega_2\big(g, \psi^{\gamma}_{\eta,\tau}(u)\big) + \omega\left(g, \frac{\gamma}{\eta}u\right),$$

where $M > 0$ *is an absolute constant and*

$$\psi^{\gamma}_{\eta,\tau}(u) = \frac{1}{2}\sqrt{\frac{\gamma^2}{\eta^2}u^2 + \frac{\eta + \gamma + \tau(\tau - 1)}{2\eta^2}u(1-u)}. \tag{11}$$

Proof Let us define the following auxiliary operator

$$\overline{L}^{\gamma}_{\eta,\tau}(g, u) := L^{\gamma}_{\eta,\tau}(g, u) + g(u) - g\left(\left(1 + \frac{\gamma}{\eta}\right)u\right) \tag{12}$$

for $g \in C\big[0, 1+\gamma\big]$ and $u \in [0, 1]$. Thus, the operator $\overline{L}^{\gamma}_{\eta,\tau}$ is linear and positive and each operator preserves the linear functions:

$$\overline{L}^{\gamma}_{\eta,\tau}((t-u), u) = 0. \tag{13}$$

Let $h \in C^2\big[0, 1+\gamma\big]$. For every $u \in [0, 1]$, by Taylor's formula about an arbitrary fixed point u, we get

$$h(t) = h(u) + (t-u)h'(u) + \int_u^t (t-v)h''(v)dv.$$

Applying the operator $\overline{L}_{\eta,\tau}^{\gamma}$ to both sides of the last equality, we can write

$$\overline{L}_{\eta,\tau}^{\gamma}(h,u)-h(u)=h'(u)\overline{L}_{\eta,\tau}^{\gamma}(t-u,u)+\overline{L}_{\eta,\tau}^{\gamma}\left(\int_{u}^{t}(t-v)h''(v)dv,u\right). \tag{14}$$

From (12) and (13) for $g(t)=\int_u^t(t-v)h''(v)dv$, (14) reduces to

$$\begin{aligned}\overline{L}_{\eta,\tau}^{\gamma}(h,u)-h(u)&=L_{\eta,\tau}^{\gamma}\left(\int_{u}^{t}(t-v)h''(v)dv,u\right)\\&\quad-\int_{u}^{\left(1+\frac{\gamma}{\eta}\right)u}\left[\left(1+\frac{\gamma}{\eta}\right)u-v\right]h''(v)dv.\end{aligned} \tag{15}$$

Using

$$\left|\int_{u}^{\left(1+\frac{\gamma}{\eta}\right)u}\left[\left(1+\frac{\gamma}{\eta}\right)u-v\right]h''(v)du\right|\le\frac{1}{2}\left(L_{\eta,\tau}^{\gamma}(t-u,u)\right)^{2}\|h''\|$$

and Lemma 2.3, (15) follows that

$$\begin{aligned}\left|\overline{L}_{\eta,\tau}^{\gamma}(h,u)-h(u)\right|&\le L_{\eta,\tau}^{\gamma}\left(\left|\int_{u}^{t}(t-v)h''(v)dv\right|,u\right)\\&\quad+\left|\int_{u}^{\left(1+\frac{\gamma}{\eta}\right)u}\left[\left(1+\frac{\gamma}{\eta}\right)u-v\right]h''(v)dv\right|\\&\le\frac{\|h''\|}{2}\left[L_{\eta,\tau}^{\gamma}\left((t-u)^{2},u\right)+\left(L_{\eta,\tau}^{\gamma}(t-u,u)\right)^{2}\right]\\&\le\frac{\|h''\|}{2}\left[\frac{2\gamma^{2}}{\eta^{2}}u^{2}+\frac{\eta+\gamma+\tau(\tau-1)}{\eta^{2}}u(1-u)\right].\end{aligned} \tag{16}$$

Furthermore, considering (12), (13) and Lemma 2.2, for $g\in C[0,1+\gamma]$ we can write

$$\left|\overline{L}_{\eta,\tau}^{\gamma}(g,u)\right|\le\|g\|L_{\eta,\tau}^{\gamma}(e_{0},u)+\|g\|=3\|g\|. \tag{17}$$

By (12), (16) and (17), for $g, h \in C\big[0, 1+\gamma\big]$ we have

$$\begin{aligned}\left|L_{\eta,\tau}^{\gamma}(g,u)-g(u)\right| &\leq \left|\overline{L}_{\eta,\tau}^{\gamma}((g-h),u)-(g-h)(u)\right| \\ &\quad+\left|\overline{L}_{\eta,\tau}^{\gamma}(h,u)-h(u)\right|+\left|g\left(1+\frac{\gamma}{\eta}u\right)-g(u)\right| \\ &\leq 4\Bigg\{\|g-h\|+\frac{1}{8}\left\|h''\right\| \\ &\quad\times\left[\frac{2\gamma^2}{\eta^2}u^2+\frac{\eta+\gamma+\tau(\tau-1)}{\eta^2}u(1-u)\right]\Bigg\} \\ &\quad+\omega\left(g,\frac{\gamma}{\eta}u\right).\end{aligned}$$

Taking infimum over all $h \in C^2\big[0, 1+\gamma\big]$ and applying of (10), we find

$$\begin{aligned}\left|L_{\eta,\tau}^{\gamma}(g,u)-g(u)\right| &\leq 4K\left(g,\left(\psi_{\eta,\tau}^{\gamma}(u)\right)^2\right)+\omega\left(g,\frac{\gamma}{\eta}u\right) \\ &\leq M\omega_2\big(g,\psi_{\eta,\tau}^{\gamma}(u)\big)+\omega\left(g,\frac{\gamma}{\eta}u\right),\end{aligned}$$

where $\psi_{\eta,\tau}^{\gamma}(u)$ is given by (11).

Now, we give Voronovskaja and Grüss–Voronovskaja type theorems, respectively.

Theorem 3.4 *Assume that $g \in C\big[0, 1+\gamma\big]$ and g has the second order derivative at $u \in [0, 1]$. Then, we have*

$$\lim_{\eta\to\infty}\eta\big[L_{\eta,\tau}^{\gamma}(g,u)-g(u)\big]=\gamma u g'(u)+\frac{u(1-u)}{2}g''(u).$$

Proof By using Taylor's formula, we have

$$g(t)=g(u)+(t-u)g'(u)+\frac{1}{2}(t-u)^2g''(u)+r(t,u)(t-u)^2, \tag{18}$$

at the fixed point $u \in [0, 1]$, where $r(t, u)$ is a continuous function on $\big[0, 1+\gamma\big]$ and $r(t, u) \to 0$ as $t \to u$. Applying the operator $L_{\eta,\tau}^{\gamma}$ to (18), we get

$$\begin{aligned}L_{\eta,\tau}^{\gamma}(g,u)-g(u) &= L_{\eta,\tau}^{\gamma}(t-u,u)g'(u) \\ &\quad+\frac{1}{2}L_{\eta,\tau}^{\gamma}\big((t-u)^2,u\big)g''(u) \\ &\quad+L_{\eta,\tau}^{\gamma}\big(r(t,u)(t-u)^2,u\big).\end{aligned}$$

Then, taking the limit of both sides and using (7)–(8), we obtain

$$\lim_{\eta\to\infty}\eta\big[L^{\gamma}_{\eta,\tau}(g,u)-g(u)\big]=\gamma u g'(u)+\frac{u(1-u)}{2}g''(u)$$
$$+\lim_{\eta\to\infty}L^{\gamma}_{\eta,\tau}\big(r(t,u)(t-u)^2,u\big).$$

Since $r(t,u)\to 0$ as $t\to u$, for each $\varepsilon>0$ there exists a $\zeta>0$ such that $|r(t,u)|<\varepsilon$ whenever $|t-u|<\zeta$ for all t. On the other hand, since $r(t,u)$ is bounded on $[0,1+\gamma]$, there is a $M>0$ such that $|r(t,u)|\le M$ for all t. Therefore, we may write $|r(t,u)|\le M\frac{(t-u)^2}{\zeta^2}$ when $|t-u|\ge\zeta$. So, these arguments enable us to write $|r(t,u)|\le\varepsilon+M\frac{(t-u)^2}{\zeta^2}$ for all t. Thus, from the monotonicity and linearity of the operator $L^{\gamma}_{\eta,\tau}$, we get

$$\eta L^{\gamma}_{\eta,\tau}\big(r(t,u)(t-u)^2,u\big)\le\varepsilon\eta L^{\gamma}_{\eta,\tau}\big((t-u)^2,u\big)+\frac{M}{\zeta^2}\eta L^{\gamma}_{\eta,\tau}\big((t-u)^4,u\big).$$

Then, using (8) and (9), for all $u\in[0,1]$ one has

$$\lim_{\eta\to\infty}\eta L^{\gamma}_{\eta,\tau}\big(r(t,u)(t-u)^2,u\big)=0.$$

This proves our theorem.

In the sense of Gal and Gonska [7], we establish Grüss–Voronovskaja type theorem for the operator $L^{\gamma}_{\eta,\tau}$.

Theorem 3.5 *Let $g,h\in C^2\big[0,1+\gamma\big]$ and $\gamma\in\mathbb{N}\cup\{0\}$. Then, for each $u\in[0,1]$ the following holds*

$$\lim_{\eta\to\infty}\eta\big[L^{\gamma}_{\eta,\tau}(gh,u)-L^{\gamma}_{\eta,\tau}(g,u)L^{\gamma}_{\eta,\tau}(h,u)\big]=u(1-u)g'(u)h'(u).$$

Proof Since $(gh)(u)=g(u)h(u)$, $(gh)'(u)=g'(u)h(u)+g(u)h'(u)$ and $(gh)''(u)=g''(u)h(u)+2g'(u)h'(u)+g(u)h''(u)$, by simple calculations we have

$$\begin{aligned}
&L^{\gamma}_{\eta,\tau}(gh,u)-L^{\gamma}_{\eta,\tau}(g,u)L^{\gamma}_{\eta,\tau}(h,u)\\
&\quad=L^{\gamma}_{\eta,\tau}(gh,u)-g(u)h(u)-(gh)'(u)L^{\gamma}_{\eta,\tau}(t-u,u)\\
&\quad-\frac{(gh)''(u)}{2}L^{\gamma}_{\eta,\tau}\big((t-u)^2,u\big)-h(u)\big[L^{\gamma}_{\eta,\tau}(g,u)-g(u)\\
&\quad-g'(u)L^{\gamma}_{\eta,\tau}(t-u,u)-\frac{g''(u)}{2}L^{\gamma}_{\eta,\tau}\big((t-u)^2,u\big)\Big]\\
&\quad-L^{\gamma}_{\eta,\tau}(g,u)\big[L^{\gamma}_{\eta,\tau}(h,u)-h(u)-h'(u)L^{\gamma}_{\eta,\tau}(t-u,u)\\
&\quad-\frac{h''(u)}{2}L^{\gamma}_{\eta,\tau}\big((t-u)^2,u\big)\Big]+\frac{1}{2}L^{\gamma}_{\eta,\tau}\big((t-u)^2,u\big)\\
&\qquad\times\big[g(u)h''(u)+2g'(u)h'(u)-h''(u)L^{\gamma}_{\eta,\tau}(g,u)\big]
\end{aligned}$$

$$+ h'(u)L^{\gamma}_{\eta,\tau}(t-u,u)\big[g(u) - L^{\gamma}_{\eta,\tau}(g,u)\big].$$

By using (7) and (8), we get

$$\begin{aligned}
&\lim_{\eta\to\infty} \eta\big[L^{\gamma}_{\eta,\tau}(gh,u) - L^{\gamma}_{\eta,\tau}(g,u)L^{\gamma}_{\eta,\tau}(h,u)\big] \\
&= \lim_{\eta\to\infty} \eta\big[L^{\gamma}_{\eta,\tau}(gh,u) - g(u)h(u)\big] - (gh)'(u)\gamma u - \frac{(gh)''(u)}{2}u(1-u) \\
&- h(u)\left\{\lim_{\eta\to\infty} \eta\big[L^{\gamma}_{\eta,\tau}(g,u) - g(u)\big] - g'(u)\gamma u - \frac{g''(u)}{2}u(1-u)\right\} \\
&- \lim_{\eta\to\infty} L^{\gamma}_{\eta,\tau}(g,u)\left\{\lim_{\eta\to\infty} \eta\big[L^{\gamma}_{\eta,\tau}(h,u) - h(u)\big]\right. \\
&\left.- h'(u)\gamma u - \frac{h''(u)}{2}u(1-u)\right\} \\
&+ \frac{u(1-u)}{2}\left\{h''(u)\lim_{\eta\to\infty}\big[g(u) - L^{\gamma}_{\eta,\tau}(g,u)\big] + 2g'(u)h'(u)\right\} \\
&+ h'(u)\gamma u \lim_{\eta\to\infty}\big[g(u) - L^{\gamma}_{\eta,\tau}(g,u)\big].
\end{aligned}$$

Considering Theorems 3.1 and 3.4, the desired result can be achieved.

4 Graphical Examples

Now, using visuals, we first show the approximation of the operator $L^{\gamma}_{\eta,\tau}(g,u)$ to certain functions for varying η values. Then, we compare the approximation of the operator $L^{\gamma}_{\eta,\tau}(g,u)$ with Bernstein polynomials $B_{\eta}(g,u)$ for certain functions under some conditions.

Example 4.1 In Fig. 1, for $\eta = 20,\ 50,\ 100$ (respectively red, green and blue) values the convergence of $L^{\gamma}_{\eta,\tau}(g,u)$ to $g(u) = \left(u - \frac{1}{3}\right)\sin(\pi u)$ (black) is illustrated.

Example 4.2 In Fig. 2, for $\eta = 20,\ 50,\ 100$ (respectively red, green and blue) values the convergence of $L^{\gamma}_{\eta,\tau}(g,u)$ to $g(u) = \sin(2\pi u)$ (black) is illustrated.

Example 4.3 In Fig. 3, for $\eta = 20,\ 50,\ 100$ (respectively red, green and blue) values the convergence of $L^{\gamma}_{\eta,\tau}(g,u)$ to $g(u) = \cos(2\pi u) + 2\sin(\pi u)$ (black) is illustrated.

Example 4.4 In Fig. 4, for $\eta = 20$ the convergence of $L^{\gamma}_{\eta,\tau}(g,u)$ (green) and $B_{\eta}(g,u)$ (Bernstein polynomials) (blue) to $g(u) = \left(u - \frac{1}{2}\right)\left(u - \frac{2}{3}\right)\left(u - \frac{3}{4}\right)$ (black) is illustrated.

Example 4.5 In Fig. 5, for $\eta = 10$ the convergence of $L^{\gamma}_{\eta,\tau}(g,u)$ (green) and $B_{\eta}(g,u)$ (Bernstein polynomials) (blue) to $g(u) = \tanh(\pi u)$ (black) is illustrated.

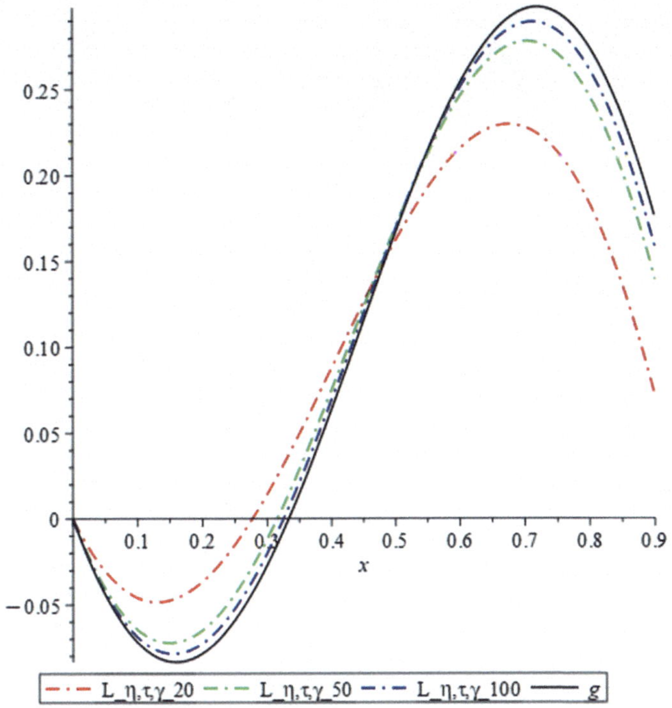

Fig. 1 The convergence $L^{\gamma}_{\eta,\tau}(g,u)$ operators to $g(u) = \left(u - \frac{1}{3}\right)\sin(\pi u)$ for $\eta = 20,\ 50,\ 100$

Example 4.6 In Fig. 6, for $\eta = 10$ the convergence of $L^{\gamma}_{\eta,\tau}(g,u)$ (green) and $B_{\eta}(g,u)$ (Bernstein polynomials) (blue) to $g(u) = -\sin(\pi u) - \cos(\pi u)$ (black) is illustrated.

As seen from Figs. 4, 5 and 6, the operator $L^{\gamma}_{\eta,\tau}(g,u)$ showed a better approximation in the given conditions and range from the polynomial $B_{\eta}(g,u)$.

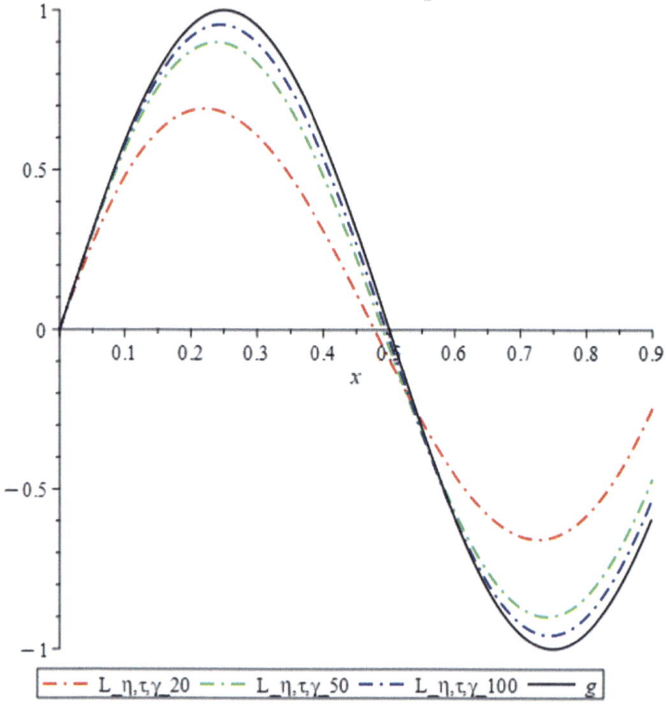

Fig. 2 The convergence $L_{\eta,\tau}^{\gamma}(g,u)$ operators to $g(u) = \sin(2\pi u)$ for $\eta = 20,\ 50,\ 100$

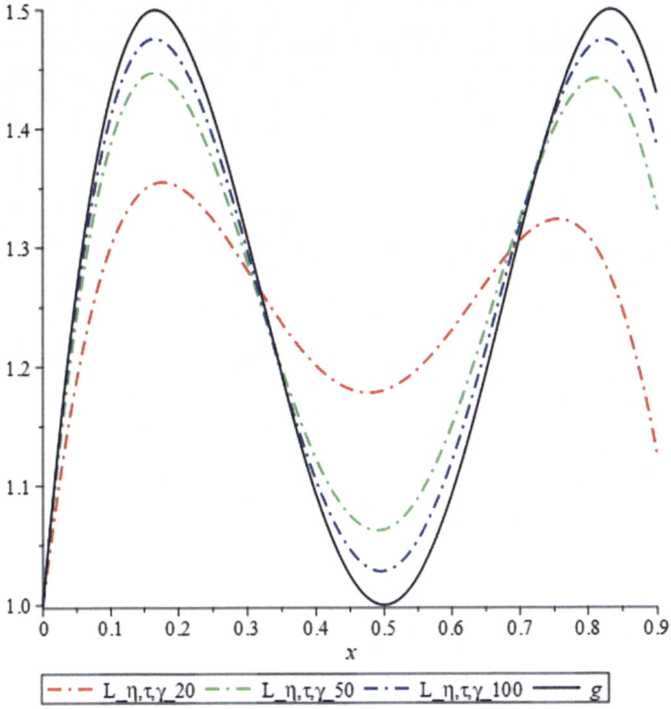

Fig. 3 The convergence $L^{\gamma}_{\eta,\tau}(g, u)$ operators to $g(u) = \cos(2\pi u) + 2\sin(\pi u)$ for $\eta = 20,\ 50,\ 100$

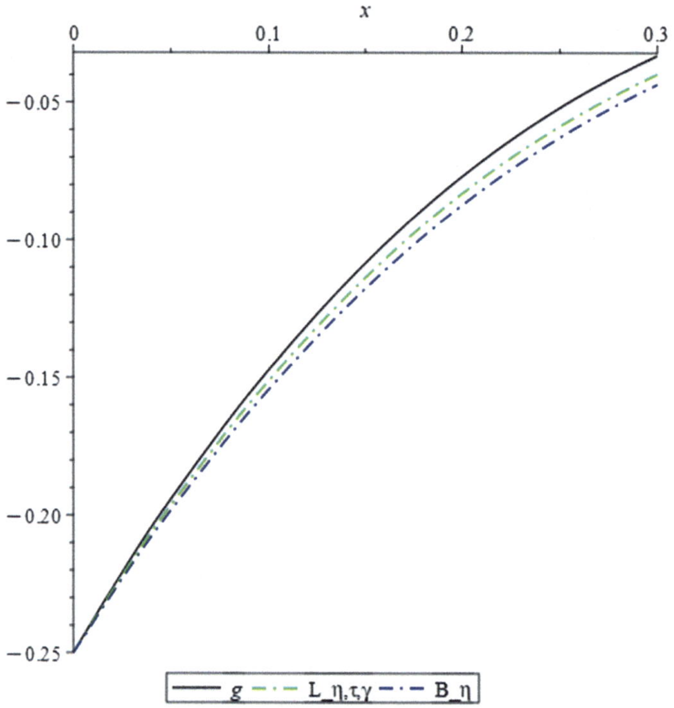

Fig. 4 The convergence $L^{\gamma}_{\eta,\tau}(g,u)$ and $B_\eta(g,u)$ operators to $g(u) = \left(u-\frac{1}{2}\right)\left(u-\frac{2}{3}\right)\left(u-\frac{3}{4}\right)$ for $\eta = 20$

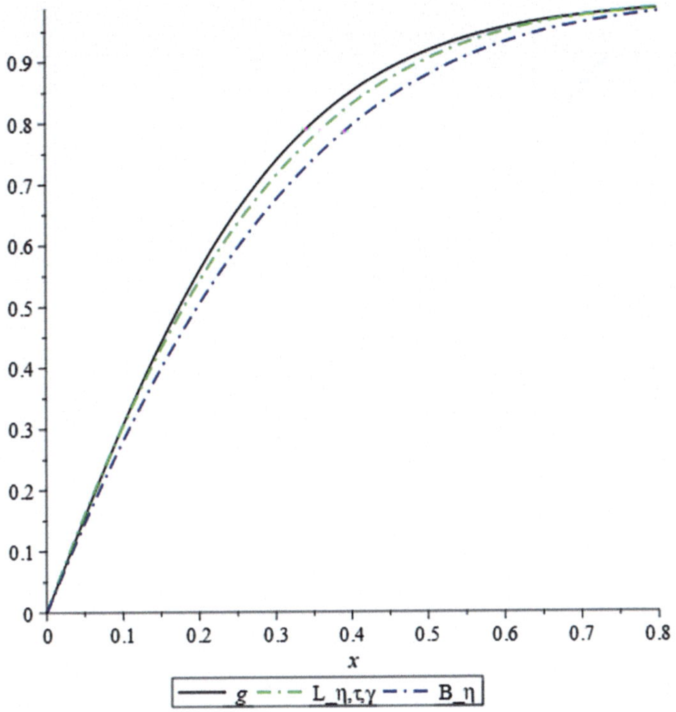

Fig. 5 The convergence $L^{\gamma}_{\eta,\tau}(g, u)$ and $B_{\eta}(g, u)$ operators to $g(u) = \tanh(\pi u)$ for $\eta = 10$

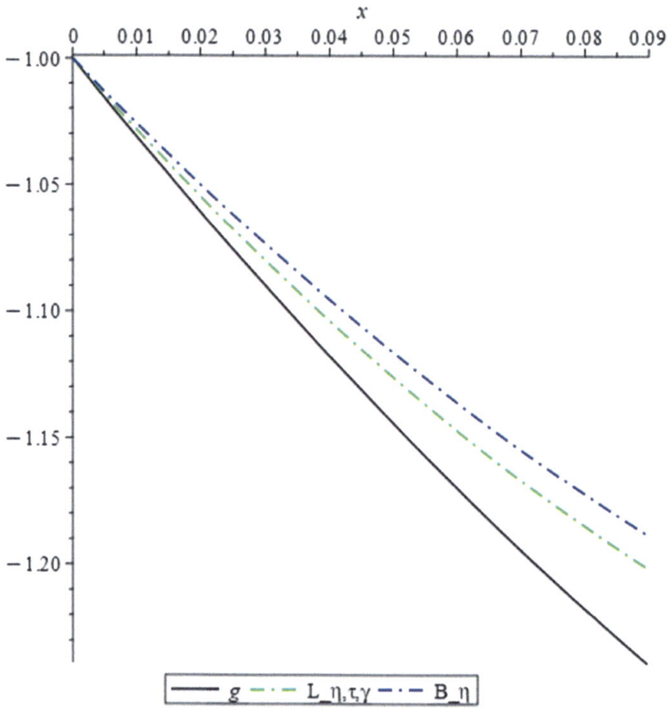

Fig. 6 The convergence $L^{\gamma}_{\eta,\tau}(g,u)$ and $B_\eta(g,u)$ operators to $g(u) = -\sin(\pi u) - \cos(\pi u)$ for $\eta = 10$

References

1. Bostancı, T., Başcanbaz-Tunca, G.: On Stancu operators depending on a non-negative integer. Filomat **36**(18), 6129–6138 (2022)
2. Bustamante, J., Quesada, J.M.: A property of Ditzian–Totik second order moduli. Appl. Math. Lett. **23**(5), 576–580 (2010)
3. Çetin, N.: A new generalization of complex Stancu operators. Math. Methods Appl. Sci. **42**, 5582–5594 (2019)
4. Çetin, N., Başcanbaz-Tunca, G.: Approximation by a new complex generalized Bernstein operator. An. Univ. Oradea Fasc. Mat. **26**(2), 129–141 (2019)
5. Çetin, N.: A new complex generalized Bernstein–Schurer operator. Carpathian J. Math. **37**(1), 81–89 (2021)
6. DeVore, R.A., Lorentz, G.G.: Constructive Approximation. Springer, Berlin (1993)
7. Gal, S.G., Gonska, H.: Grüss and Grüss–Voronovskaya-type estimates for some Bernstein-type polynomials of real and complex variables. Jaen J. Approx. **7**(1), 97–122 (2015)
8. Gupta, V., Agarwal, R.P.: Convergence Estimates in Approximation Theory. Springer, New York (2014)
9. Kajla, A.: The Kantorovich variant of an operator defined by D. D. Stancu. Appl. Math. Comput. **316**, 400–408 (2018)
10. Kumar, A.: A new kind of variant of the Kantorovich type modification operators introduced by D. D. Stancu. Results Appl. Math. **11**, 100158 (2021)

11. Schurer, F.: Linear positive operators in approximation theory. Math. Inst. Techn. Univ. Delft Rep. (1962)
12. Stancu, D.D.: Quadrature formulas constructed by using certain linear positive operators. In: Numerical Integration (Proceedings Conference, Oberwolfach, 1981), ISNM 57, pp. 241–251. Birkhauser Verlag, Basel (1982)
13. Stancu, D.D.: Approximation of functions by means of a new generalized Bernstein operator. Calcolo **20**(2), 211–229 (1983)
14. Yang, R., Xiong, J., Cao, F.: Multivariate Stancu operators defined on a simplex. Appl. Math. Comput. **138**, 189–198 (2003)

Once Again About the Pythagorean Triples

A. B. Urdaletova, S. K. Kydyraliev, and A. A. Doolotova

Abstract The enormous popularity of finding the natural solutions of the Fermat–Pythagorean equation $x^2 + y^2 = z^2$ has given rise to a large number of works devoted to finding formulas that allow one to directly write out solutions of the equation. Nevertheless, the magic contained in this problem makes us refer to it again and again. Another such case is described in this paper. The starting point is the formula for the difference of squares. It turns out to be sufficient to obtain a rather interesting result: if we take any natural odd number, square it, and consider the resulting number as the sum of two consecutive natural numbers, these numbers together with the original number give the Fermat–Pythagorean triplet. The proposed work is devoted to the proof of this and other more complex and interesting statements.

Keywords The Fermat–Pythagorean equation · Fermat's Last Theorem · Pythagorean triple

1 Introduction

The enormous popularity of finding the natural solutions of the Fermat–Pythagorean equation $x^2 + y^2 = z^2$ has given rise to a large number of works devoted to finding formulas that allow one to directly write out solutions of the equation [1–5]. This

A. B. Urdaletova
Department of Mathematics, Kyrgyz-Turkish Manas University, Bishkek, Kyrgyzstan
e-mail: anarkul.urdaletova@manas.edu.kg

S. K. Kydyraliev
Department of Applied Mathematics and Informatics, American University of Central Asia, Bishkek, Kyrgyzstan
e-mail: kydyraliev_s@auca.kg

A. A. Doolotova (✉)
Department of Business Administration, American University of Central Asia, Bishkek, Kyrgyzstan
e-mail: doolotova_a@auca.kg

O. Castillo et al. (eds.), *Advanced Mathematics for the Modeling and Solution of Challenging Problems in Engineering*, Studies in Systems, Decision and Control 582,
https://doi.org/10.1007/978-3-031-82342-8_25

paper shows how using the formula of square difference $a^2 - b^2 = (a - b)(a + b)$, all natural solutions of the equation $x^2 + y^2 = z^2$ can be obtained.

2 Main Results

1. The problem of finding solutions of the equation $x^2 + y^2 = z^2$ in the case when the length of the hypotenuse is one unit more than the length of the leg is solved quite simply. So, let $x^2 + y^2 = (y + 1)^2$. We rewrite the equation in the form $x^2 = (y + 1)^2 - y^2$, and use the formula for the difference of the squares: $x^2 = \left[(y + 1) - y\right] \cdot \left[(y + 1) + y\right] \Leftrightarrow x^2 = 1 \cdot \left[2y + 1\right]$. The last equality shows that the length of the smaller leg should be an odd number, and that every odd number x, where $x = 3, 5, 7, \ldots$ defines a solution to the equation—triple $\left(x; \left(x^2 - 1\right)/2; \left(x^2 + 1\right)/2\right)$.

Therefore, in this case, in order to get the Fermat–Pythagorean triple, it is enough to take any natural odd number, square it and consider the resulting number as the sum of two consecutive natural numbers.

In this way,

when $x = 3 \Rightarrow x^2 = 9 \Rightarrow$ we get a triple (3; 4; 5);
when $x = 5 \Rightarrow x^2 = 25 \Rightarrow$ we get a triple (5; 12; 13);
…;
when $x = 17 \Rightarrow x^2 = 289 \Rightarrow$ we get a triple (17, 144; 145);
….

2. Let us try to use this approach for the case when the length of the hypotenuse more than the length of the leg for two units.

So, $x^2 + y^2 = (y + 2)^2$. We rewrite the equation in the form $x^2 = (y + 2)^2 - y^2$, and use the formula for the difference of the squares: $x^2 = \left[(y + 2) - y\right] \cdot \left[(y + 2) + y\right] \Leftrightarrow x^2 = 2 \cdot \left[2y + 2\right] \Leftrightarrow x^2 = 2^2 \cdot \left[y + 1\right]$. Therefore, the number x can be represented in the form $x = 2x_1$, where x_1 is a natural number. Then, $(x_1)^2 = y + 1 \Leftrightarrow y = (x_1)^2 - 1$.

Taking x_1 odd, we get an even value of y. As, by the equality $x = 2x_1$, the number x is even, in this case all the members of the Pythagorean triple are even. For example, if $x_1 = 3$, then $x = 6$, $y = 8$ and $z = y + 2 = 10$. Dividing all these numbers by 2, we get the most famous Pythagorean triple: (3, 4, 5).

It is clear that multiplying all the terms of any Pythagorean triple by an arbitrary natural number, we get another Pythagorean triple. Consequently, the interesting problem is to obtain primitive Pythagorean triples—triples whose members do not have common factors. In particular, from the previous arguments it is clear that for any odd value of x_1 the Pythagorean triple will not be primitive.

At the same time, choosing natural even numbers as x_1, we obtain primitive Pythagorean triples $x = 2x_1,\ y = (x_1)^2 - 1,\ z = y + 2$.

In this way,

when $x_1 = 2$ we get a triple $x = 2x_1 = 4,\ y = (x_1)^2 - 1 = 3,\ z = y + 2 = 5$;
when $x_1 = 4$ we get a triple $x = 2x_1 = 8,\ y = (x_1)^2 - 1 = 15,\ z = y + 2 = 17$;
...;
when $x_1 = 10$ we get a triple $x = 2x_1 = 20,\ y = (x_1)^2 - 1 = 99,\ z = y + 2 = 101$;
....

3. We turn to the case when the length of the hypotenuse is 3 units longer than the length of the leg.

So, $x^2 + y^2 = (y + 3)^2$, and consequently: $x^2 = (y + 3)^2 - y^2 \Leftrightarrow x^2 = \big[(y + 3) - y\big] \cdot \big[(y + 3) + y\big] \Leftrightarrow x^2 = 3 \cdot \big[2y + 3\big]$. Therefore, x^2 is divisible by 3, and this only happens when x is divisible by 3. Thus, the number x can be represented as $x = 3x_1$, where x_1 is a positive integer. Then, $(x_1)^2 = 2y/3 + 1$. It follows from the last equality that the integer value x_1 and, accordingly, x is possible only in the case when y has a factor of 3. Since x also has a factor of 3, it turns out that a primitive Pythagorean triple not exist.

So, there are no primitive Pythagorean triples in which the length of the hypotenuse is 3 units longer than the length of the leg.

4. Consider the case when the length of the hypotenuse is four units longer than the length of the leg.

So $x^2 + y^2 = (y + 4)^2$. We rewrite the equation in the form $x^2 = (y + 4)^2 - y^2$, and use the formula for the difference of squares: $x^2 = \big[(y + 4) - y\big] \cdot \big[(y + 4) + y\big] \Leftrightarrow x^2 = 4 \cdot \big[2y + 4\big] \Leftrightarrow x^2 = 2^2\big[2y + 4\big]$. The last equality shows that the number x can be represented as $x = 2x_1$, where x_1 is a positive integer. Then, $(x_1)^2 = 2y + 2^2$. Hence, $y = \big[(x_1)^2 - 2^2\big]/2$. The number y will be an integer only for even x_1. Then, u will also be even. So, it turned out that when the length of the hypotenuse is four times the length of the leg, the Pythagorean triple will turn out only when x and y are even—that is, such triples will not be primitive.

Of course, you can go on and sort through all the natural numbers, one by one. But we all know that this road has no end. Therefore, we suggest looking at two more special cases, and then moving on to generalizations.

5. Let the length of the hypotenuse is nine units longer than the length of the leg.

We solve the equation $x^2 + y^2 = (y + 9)^2$. We rewrite it in the form $x^2 = (y + 9)^2 - y^2$, and use the formula for the difference of the squares: $x^2 = \big[(y + 9) - y\big] \cdot \big[(y + 9) + y\big] \Leftrightarrow x^2 = 9\big[2y + 9\big] \Leftrightarrow x^2 = 3^2\big[2y + 9\big]$. The number x can be represented in the form $x = 3x_1$, where x_1 is a natural number. Then, $(x_1)^2 = 2y + 3^2$. Hence, $y = \big[(x_1)^2 - 3^2\big]/2$. The number y will be an integer only for odd x_1. Then x will be odd, and y will be even. So, it turned out that with the length of the hypotenuse nine more than the length of the leg, you can get primitive Pythagorean triples.

To do this, we take natural odd numbers as x_1 and obtain primitive Pythagorean triples $x = 3x_1, y = [(x_1)^2 - 3^2]/2, z = y + 9$.

For example,

when $x_1 = 5$ we get: $x = 3x_1 = 15,\ y = \big[(x_1)^2 - 3^2\big]/2 = 8,\ z = y + 9 = 17$;

when $x_1 = 7$ we get: $x = 3x_1 = 21$, $y = \left[(x_1)^2 - 3^2\right]/2 = 20$, $z = y + 9 = 29$;
when $x_1 = 15$ we get: $x = 3x_1 = 45$, $y = \left[(x_1)^2 - 3^2\right]/2 = 108$, $z = y + 9 = 117$. Oh, it's not a primitive three. But this is easy to understand—we took $x_1 = 15$, and the number 15 has a factor of 3. We conclude: choosing x_1, we need to make sure that this odd number does not have the number 3 in the expansion.

...;

when $x_1 = 17$ we get: $x = 3x_1 = 51$, $y = \left[(x_1)^2 - 3^2\right]/2 = 140$, $z = y + 2 = 149$;

....

6. Let the length of the hypotenuse is eighteen units longer than the length of the leg.

We solve the equation $x^2 + y^2 = (y + 18)^2$. We rewrite it in the form $x^2 = (y + 18)^2 - y^2$, and use the formula for the difference of the squares: $x^2 = \left[(y + 18) - y\right] \cdot \left[(y + 18) + y\right] \Leftrightarrow x^2 = 18\left[2y + 18\right] \Leftrightarrow x^2 = 36\left[y + 9\right] \Leftrightarrow x^2 = 6^2\left[y + 9\right]$.

So, the number x can be represented in the form $x = 6x_1$, where x_1 is a natural number. Then, $(x_1)^2 = y + 3^2$. Hence, $y = [(x_1)^2 - 3^2]$. The number y will be even with odd x_1. Since $x = 6x_1$ is even, the corresponding Pythagorean triple will not be primitive. Therefore, we consider only even x_1. In this case, y will be odd. So, it turned out that with the length of the hypotenuse eighteen longer than the length of the leg, you can get primitive Pythagorean triples. For this, we take natural even numbers as x_1 and obtain primitive Pythagorean triples $x = 6x_1$, $y = \left[(x_1)^2 - 3^2\right]$, $z = y + 18$.

For example,

when $x_1 = 4$ we get: $x = 6x_1 = 24$, $y = \left[(x_1)^2 - 3^2\right] = 7$, $z = y + 18 = 25$;
when $x_1 = 6$ we get: $x = 6x_1 = 36$, $y = \left[(x_1)^2 - 3^2\right] = 27$, $z = y + 18 = 45$. Again, we did not get a primitive triple. But we already went through this—the fact is that $x_1 = 6$, and the number 6 has a factor of 3, that is, as x_1 you need to consider only natural even numbers that do not have the number 3 in the expansion.

...;

when $x_1 = 10$ we get: $x = 6x_1 = 60$, $y = \left[(x_1)^2 - 3^2\right] = 91$, $z = y + 18 = 109$;

....

7. It's time to move on to generalizations. Any natural number can be written in the form kp^2, that is, if the decomposition of the number into prime factors contains squares of numbers, then their product is written in the form p^2, the product of all other factors will be k. So, we solve the equation $x^2 + y^2 = \left(y + kp^2\right)^2$ in natural numbers.

We rewrite it in the form $x^2 = \left(y + kp^2\right)^2 - y^2$, and use the formula for the difference of the squares: $x^2 = \left[\left(y + kp^2\right) - y\right]\left[\left(y + kp^2\right) + y\right] \Leftrightarrow x^2 = kp^2\left[2y + kp^2\right]$.

Therefore, the number x can be represented in the form $x = px_1$, where x_1 is a natural number. In this case, $(x_1)^2 = k[2y + kp^2]$. Next, you need to consider three different cases.

I. $k = 1$. Then, $(x_1)^2 = \left[2y + p^2\right] \Rightarrow y = \left[(x_1)^2 - p^2\right]/2$.

For even p, the number x_1 must be even, and as a result, x and y will be even, which excludes the primitiveness of the Pythagorean triple. Example: paragraph 4. For odd p, the number x_1 must be odd, and as a result, x will be odd, and y will be even.

Therefore, choosing odd numbers as x_1 that do not have common factors with p, we obtain primitive Pythagorean triples: $x = px$; $y = \left[(x_1)^2 - p^2\right]/2$; $z = y + p^2$, satisfying the equation $x^2 + y^2 = \left(y + p^2\right)^2$.

Example: paragraph 5.

II. $k = 2$. Then, $(x_1)^2 = 2\left[2y + 2p^2\right] \Leftrightarrow (x_1)^2 = 2^2\left[y + p^2\right]$. Therefore, $x_1 = 2x_2 \Rightarrow x = px_1 = 2px_2 \Rightarrow y = \left[(x_2)^2 - p^2\right]$.

The number x will always be even. Also, the number $2p^2$ is even. Therefore, if the number y is even, that is, x_2 and p have the same parity, then the primitive Pythagorean triple will fail. The same will happen if x_2 has a common factor with p.

Therefore, in order to obtain primitive Pythagorean triples, as x_2 and p you need to choose numbers of different parity that do not have common factors. In this case, primitive triples of numbers satisfying the equation $x^2 + y^2 = (y + 2p^2)^2$ will be obtained at the form: $x = 2px_2$; $y = \left[(x_2)^2 - p^2\right]$; $z = y + 2p^2$.

Example: paragraphs 2, 5.

III. $k > 2$. Then, $(x_1)^2 = k\left[2y + kp^2\right] \Leftrightarrow (x_1)^2 = k^2\left[2y/k + p^2\right]$.

Therefore, x_1 will be integer only if y is divisible by k. But if x_1 is divided by k, then $x = px_1$ is divided by k. Thus we can conclude: if $k > 2$ the solution of the equation $x^2 + y^2 = (y + kp^2)^2$, cannot be a primitive Pythagorean triple.

Example: paragraph 3.

3 Conclusion

The problem of finding the solution to the equation $\mathrm{x^n + y^n = z^n}$ in natural numbers is probably the most popular mathematical problem. For cases where the exponent is greater than two, it is known as Fermat's Last Theorem. It is important to note, however, that our great fellows have contributed to the solution of this problem. For example, the works of Abu Mahmud Hamid ibn al-Khidr Al-Khujandi, who lived in the tenth century AD, proved that the sum of the cubes of natural numbers cannot be a natural number [6, 7]. Of course, in the case of squares, the problem is incomparably easier, but this does not detract from its appeal, especially for those who make their first steps in mathematics. We hope that the material presented in our paper will be of interest to all lovers of mathematics, especially to those who like short formulations and are not burdened by complicated techniques of proofs.

References

1. Eves, H.: An Introduction to the History of Mathematics. Saunders College Pub., Philadelphia (1990)
2. Hall, A.: Genealogy of Pythagorean triads. Math. Gaz. **54**, 377–379 (1970)
3. Neugebauer, O.: The Exact Sciences in Antiquity. Dover Publications, New York (1969)
4. Weil, A.: Number Theory: An Approach Through History from Hammurapi to Legendre, p. 21. Birkhauser, Boston (1984)
5. Beauregard, R.A., Suryanarayan, E.R.: Pythagorean triples: the hyperbolic view. Coll. Math. J. **27**, 170–181 (1996)
6. Marcorini, E.: The History of Science and Technology: A Narrative Chronology, vol. 91, 889 pp. (1988)
7. Turing, D.: The Story of Computing. Arcturus Publishing (2018)

An Example of an Innovative Approach for Enhancing Traffic Flow with Using Decentralized Traffic Signal Split Control Method

Serap Ergün

Abstract Urban regions across the world face a serious problem with traffic congestion, which results in a number of problems like financial losses, air pollution, and an increase in traffic accidents. Numerous studies are being actively done in response to this issue to address one of its major causes, particularly the operation of traffic lights at junctions, with the goal of reducing congestion. The traffic control system, a centralized control system, or established control parameters are used in France's flow method to traffic signal control. The latter strategy, however, has certain drawbacks, including poor reactivity in dynamic contexts and limited flexibility to extend or decrease the control region. In order to improve traffic flow, this study suggests a split control approach for traffic signals that makes use of a split model and an autonomous decentralized control method based on said model. By using the split balance equation, the split model describes the connection between traffic flow and splits. The outcomes show that the suggested approach is successful at precisely regulating crossings with a variety of geometries, and that it can smooth out traffic more effectively than the flow control approach.

Keywords Traffic flow optimization · Flow control method · Split control method · Traffic signals

1 Introduction

Facilitating traffic flow refers to the process of improving the movement and efficiency of vehicles on roads, reducing congestion, and ensuring smoother traffic operations. It involves implementing strategies and measures that optimize the flow of vehicles, minimize delays, and enhance overall transportation efficiency. Facilitating traffic flow can include various approaches such as optimizing traffic signal timings, implementing intelligent transportation systems, improving road infrastructure, and

S. Ergün (✉)
Department of Computer Engineering, Isparta University of Applied Sciences, Isparta, Turkey
e-mail: serapbakioglu@isparta.edu.tr

O. Castillo et al. (eds.), *Advanced Mathematics for the Modeling and Solution of Challenging Problems in Engineering*, Studies in Systems, Decision and Control 582,
https://doi.org/10.1007/978-3-031-82342-8_26

employing effective traffic management techniques. The ultimate goal is to create a safe, efficient, and seamless traffic environment for all road users [11, 17].

France has a sophisticated signal control system to manage traffic flow and guarantee road safety. The main components of the system are traffic lights and other technologies to control signal timings based on traffic circumstances [13, 18].

Traffic lights are frequently employed in metropolitan areas to regulate the flow of automobiles and pedestrians at junctions. In order to monitor the presence and movement of cars and alter the signal timings accordingly, these signals are frequently fitted with sensors or cameras [1]. By dynamically adjusting the signal phases based on flow conditions, it is intended to maximize traffic flow and decrease congestion [9, 20].

In addition, France has been making significant investments in Intelligent Transport Systems (ITS) to improve its signal control capabilities [8]. This entails the incorporation of technology including centralized control systems, adaptive signal control, and vehicle-to-traffic signal communication systems [18].

Advanced algorithms and real-time data are used by adaptive signal control systems to dynamically change signal timings according to traffic demand. This strategy prioritizes high-traffic movements and reduces needless delays during times of low traffic in order to increase the efficiency of traffic flow [13].

At present, in France, fixed-time traffic signal control, as a simple and centralized control method, is being implemented to regulate traffic lights at major intersections. However, this method is characterized by delayed information reflection and limited responsiveness to dynamically changing traffic conditions at individual intersections. Consequently, there exists a problem that needs to be addressed [7, 22].

To tackle these challenges, a split control method has been developed in order to align with the concept of an autonomous decentralized control approach. This method has been generalized to accommodate intersections comprising two or three presentations, thereby enabling its application without the need for manual design of control laws for each specific intersection.

Simulation experiments have demonstrated the effectiveness of the proposed method in optimizing traffic flow within existing road networks, as well as its capability to dynamically respond to and control situations characterized by rapid changes in traffic flow. Currently, demonstration experiments are underway to validate the efficacy of the proposed method by implementing traffic signal control on actual roads.

The organization of this paper is as follows: The proposed method is presented in Sect. 2 with the representation of traffic flow and split methods. In Sect. 3, the experiments are given. The results of the study are presented in Sect. 4, and finally the conclusion and outlook part is given in Sect. 5.

2 The Proposed Method

This research puts forward an autonomous decentralized signal control model that is easily implementable, considering real-world social implementation scenarios. In this method, control parameters are computed for each cycle using a split model, which considers the relationship between the control parameters and the load amount akin to the relationship between a physical split and the force it exerts. This study represents an extension of the split control model originally proposed by Soon et al. [15], which solely focused on a model specific to a one-lane crossroad with two-stage control, as represented in the Cellular Automaton model. However, in this research, the split model has been generalized to accommodate various intersection shapes and configurations, including complex intersections such as multi-way intersections and intersections featuring dedicated traffic lights for right turns [4].

This section begins by providing an explanation of the split, which serves as the object of control in this paper, followed by a detailed exposition of the proposed method.

2.1 *Traffic Flow Representation*

The concept of "traffic flow" in the context of road traffic can be explained in relation to traffic signal divisions [10]. In this context, "traffic flow" primarily refers to the visible flow paths indicated by traffic signals. For example, in Fig. 1, the flow paths represented by 1ϕ and 2ϕ respectively illustrate the flow state. When the traffic lights corresponding to 1ϕ and 2ϕ show a green signal, vehicles can follow the depicted flow paths.

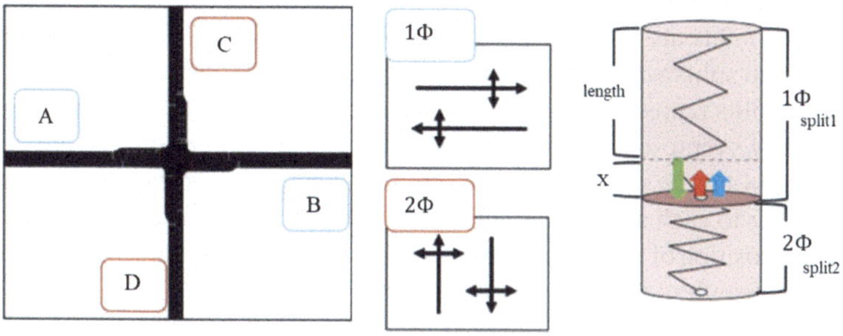

Fig. 1 The crossroad overview and split model

2.2 *Split Method*

At present, the operational traffic signal control system in France commonly employs a cyclical pattern, where the flow state of the traffic signal is conveyed through the progression of red, green, and yellow signals [2, 14]. This control approach involves two signal control parameters for each intersection [3, 16, 21]:

- Cycle length: The duration of one complete cycle.
- Split: The temporal ratio of each manifestation within the cycle.

In this paper, for a control method for traffic flow this split is proposed.

2.3 *Preferences*

The environment assumed in the proposed method is elucidated. In this method, as an autonomous decentralized control approach, a control unit is positioned at each signalized intersection to be controlled. Subsequently, this control unit is referred to as an agent. Traffic information in the vicinity of its respective intersection is acquired by each agent through the utilization of sensors, and subsequently, it is controlled. The value is derived from the nominal recognition distance of 175 m provided by smartmicro's UMRR-11 Type 132, which is intended to be utilized for demonstration experiments.

2.4 *Two-Point View Control*

The proposed method and its application in two-point control are described in this section. The points are denoted as 1ϕ and 2ϕ, similar to "Split", demonstrating the balance of two splits within a single cylinder.

The two splits represent the splits of 1ϕ and 2ϕ respectively, while the green–red and blue arrows represent the external force, the elastic force of the 1ϕ split, and the elastic force of the 1ϕ split, respectively. In this method, the traffic flow of vehicles corresponding to each situation is considered as the external force, determining the equilibrium position of the splits. Consequently, the split is determined and controlled based on the length of the splits in the balanced state [21].

In this research, the index that signifies the traffic conditions of the vehicles utilized in the proposed method is referred to as the load amount, distinct from the general term "traffic volume" associated with overall traffic. Assuming that the length of each split is 4ϕ and considering that the length of the tube in the split model does not influence the outcomes, the following equation holds:

$$\sum_{i=1,2} \Phi_{\mathrm{i}} = 1 \tag{1}$$

Under this assumption, the natural lengths of the two splits are both 1/2. Assuming that the split constant of the split model is k and the external force corresponding to the load difference between the flows is F, the displacement x from the natural length of the equilibrium position can be calculated as:

$$x = \frac{F}{2k} \tag{2}$$

Therefore, from Eqs. (1) and (2),

$$\begin{aligned} \Phi_1 &= \frac{1}{2} + \frac{F}{2k} \\ \Phi_2 &= \frac{1}{2} - \frac{F}{2k} \end{aligned} \tag{3}$$

The split value is determined by setting the split length as per the proposed method. Consequently, the upper and lower limits for the split value are established by the system.

The calculation of the load difference F between presentations is then explained. It involves taking the average number of vehicles entering the intersection for each lane of road l, which is obtained by dividing the number of vehicles entering the intersection from road l by the number of lanes [6]. Based on these observations, the load on road l is assessed.

$$q_1 = q_{l,in} + a^{q_{l,res}} \tag{4}$$

The second term on the right side of Eq. (4) is an exponential function with the coefficient a as the base, and this term takes on a larger value as the number of remaining scraps increases. Let R_i be the set of incoming roads to the intersection corresponding to flow i. Using this, the load quantity Q_i can be calculated as:

$$Q_i = \max(q_l | l \in R_i) \tag{5}$$

Using the above definitions, the load difference between the two-phase control can be expressed by Eq. (6).

$$F = \frac{Q_1 - Q_2}{Q_1 + Q_2} \tag{6}$$

The fractional term on the right side of Eq. (6) is normalized using the sum to make it unitless. Note that k and a are parameters set in advance by the system administrator.

Therefore, in the system, the split of each appearance is determined by the agent using Eqs. (3) and (6) with the observed values obtained from the sensors. It is considered preferable to perform the calculation in units of cycles to ensure equal aggregation for each traffic flow [21].

Next, the application of the method will be explained using an actual example. An example is employed, as depicted in Fig. 1. In Fig. 1, the roads extending in four directions are denoted as A, B, C, and D, respectively.

This is because of the correspondence between traffic flows and roads (Eq. 7).

$$\begin{aligned} R_1 &= \{A, B\} \\ R_2 &= \{C, D\} \end{aligned} \tag{7}$$

Regarding the aggregation of the load, for example, on road A, vehicles entering the intersection from road A are observed, resulting in Eq. (4).

$$q_A = q_{A,in} + a^{q_{A,res}} \tag{8}$$

In the same way, loads on roads B, C, and D are calculated from observations and aggregation, and splits are determined from Eqs. (3) and (6).

2.5 n-Presentation (Display) Control

In this section, the case the case of n-presentation control is explained. The split for each indication is derived by employing the same calculation method as in "Two-Point View Control", as indicated by the following equation:

$$\Phi_i = \frac{1}{n} + \frac{\sum_{j=i}^{n-1} nF_j - \sum_{j=i}^{n-1} jF_j}{nk} \quad (1 \le i \le n) \tag{9}$$

In Formula (9), $\sum_{i=a}^{b} i = 0\,(a > b)$ should not be ignored. Here, the flow load amount difference is calculated by the following formula.

$$F_j = \frac{Q_j - Q_{j+1}}{\sum_i Q_i} \quad (1 \le i \le n-1) \tag{10}$$

The definition of the set of inflow channels R_i and load quantity Q_i is the same as in "Two-Point View Control".

2.6 Compatible with Right-Turn Only Lanes

In this section, in addition to the control explained so far, the case where there is a dedicated right-turn lane and a corresponding indication are explained.

Figure 2 shows an example of a road with an assumed right turn only lane. The road in Fig. 2 shows that lane A•B can be used when a right-turn right-of-way is granted, and lane C can be used when a right-turn right-of-way is granted. There is a corresponding signal traffic flow.

This section assumes this road and explains the control method.

In this method, a nested structure is introduced into the split model to perform control that takes into account right-turn lanes.

In determining the right-turn lane split, it is assumed that the road including the right-turn-only lane is assigned the split 1Φ in advance by the method described in "Two-Point View Control" and "n-Presentation (Display) control".

At this time, the following equation holds for the straight left turn split $\Phi_{A.B}$ and the right turn split Φ_C.

$$\Phi_1 = \Phi_{A.B} + \Phi_C \tag{11}$$

Since Eq. (2) holds in the state of split equilibrium, each split can be calculated from Eqs. (2) and (13) as follows:

$$\begin{aligned} \Phi_{A.B} &= \Phi_1 \times \left(\frac{1}{2} + \frac{F}{2k}\right) \\ \Phi_C &= \Phi_1 \times \left(\frac{1}{2} - \frac{F}{2k}\right) \end{aligned} \tag{12}$$

When performing the calculation of Eq. (12), the load amount measured for each lane is used. The load amount difference between flows can be calculated in Eq. (12) using the following equation.

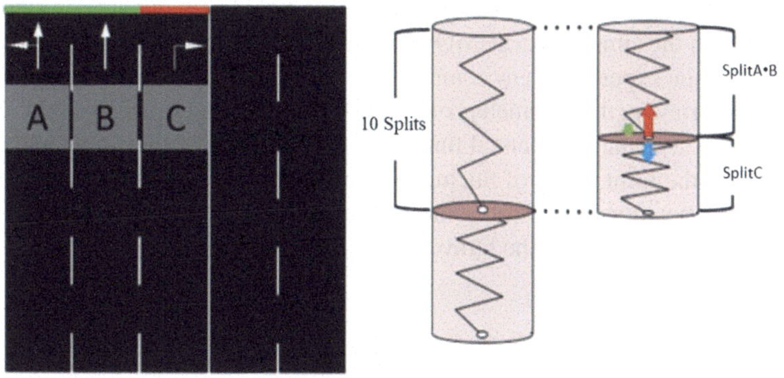

Fig. 2 The right turn lane

$$F = \frac{\max(Q_A, Q_B) - Q_C}{\max(Q_A, Q_B) + Q_{C_s}} \tag{13}$$

Therefore, if there is a dedicated right-turn lane, the split for the dedicated right-turn lane can be calculated and controlled using Eqs. (12) and (13). For example, if there are dedicated right-turn lanes in both the "AB" and "CD" directions in Fig. 1, the processing described above for the right-turn-only lanes can be performed in each of 1Φ and 2Φ, that means control is possible.

2.7 Fluctuation Suppression Control

In this section, fluctuation suppression control for stable operation of the split model is mainly explained when the traffic flow is close to saturation.

When the load difference between the periods becomes large in the short term, the split model preferentially distributes the split to the period where the load is dominant.

However, when there is a continuously high demand for any presentation, the demand for the prioritized presentation is temporarily eased, while the demand for the non-prioritized presentation is temporarily increased.

As a result, in the next cycle, splits are preferentially allocated to events that are not given priority in the previous cycle. There are cases.

Appropriate control becomes impossible when such vibrations occur. Therefore, empirically, by setting limits on the range of split fluctuations in each cycle, stable control can be achieved even when traffic demand is high. In this study, this limit width is set by the ratio to the cycle length.

3 Experiments

In this section, the contents of the experiment on the road network are described and constructed on the simulation. SUMO (Simulation of urban mobility) is used as the simulator for the experiment environment.

In the experiment, the parameters of the split model are the split constant $k = 0.6$, the base $a = 1.5$ of the exponential function for the number of remaining unhandled vehicles, and the limit width of fluctuation suppression is 10% of the cycle length. All the inflow positions and travel routes are the same, and the simulation continued until all the vehicles had finished traveling.

3.1 Experiment in Real Area

In the experiment in the actual area, Scenario 1 is used, in which the experiment is performed on the road network shown in Fig. 3, which reproduces the planned demonstration site on the simulation.

In the real area, the 16 signalized intersections shown in Fig. 4 are controlled. In this experiment, vehicles flowed into the road network and evaluated based on the following scenario. This corresponds to time. In the following, letters A to G are used to indicate the position in Fig. 3 for the purpose of explaining the scenario. Areas A and C are urban areas.

From 0 to 10,800 s is defined as commuter rush to urban areas. $B \rightarrow F \rightarrow A$, $D \rightarrow G \rightarrow C$, $E \rightarrow G \rightarrow F \rightarrow A$ are the main traffic flows. From 10,801 to 21,600 s there is no major traffic flow. From 21,601 to 32,400 s, the rush hour for returning home from the urban area is assumed, and $A \rightarrow F \rightarrow B$, $C \rightarrow G \rightarrow D$, $A \rightarrow F \rightarrow G \rightarrow E$ are the main traffic flows.

A total of 1500 vehicles/h are run for major traffic flows at each time. In addition, 500 vehicles/h are run at random inflow points and travel routes in the network at all times.

For the experimental comparison method in this section, control using the centralized control system, which is currently used in the field, is used.

In addition, in the experiments in this section, the actual centralized control system parameters are used for the cycle length for all methods.

Fig. 3 The simulation road network

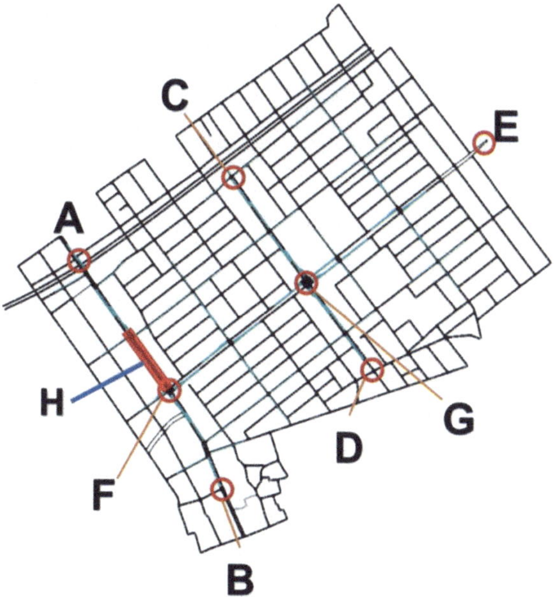

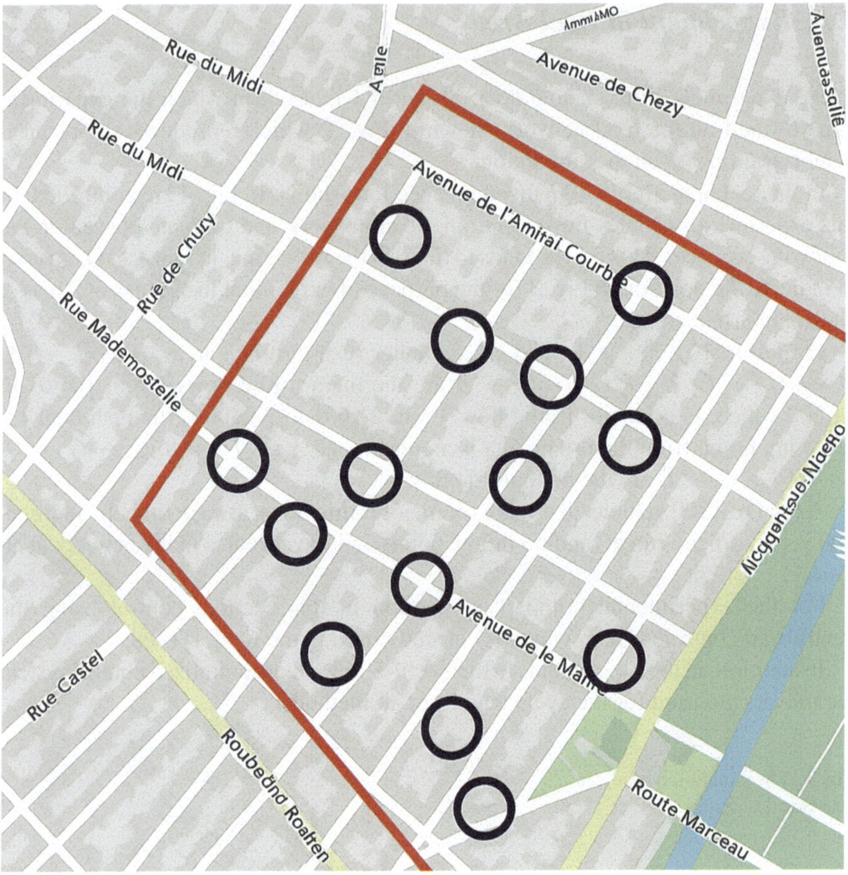

Fig. 4 The controlled intersection taken from OpenStreetMap

The direction of the green light at the start is the same between the comparison method and the proposed method at all intersections.

3.2 Experiments in Environments with Accidents and Construction

An experiment is conducted assuming a situation in which the road environment deteriorated due to construction work or an accident, and large traffic jams occurred around the site of the construction accident.

In this section Scenario 2 is presented. The environment is the same as Scenario 1, with the road network shown in Fig. 3. Work is carried out in area H, which is coloured red in Fig. 3.

In this accident area, the effects of construction work and accidents are reproduced by lowering the legal speed limit from 50 to 15 km/h in area H.

For the vehicle inflow scenario, an inflow scenario is used for the commuter rush hour portion of Scenario 1, which is three hours of real time. Other conditions are the same as Scenario 1.

4 Results

Figures 5 and 6 show graphs of the experimental results for each scenario. These are obtained by dividing vehicles into bins at regular intervals according to departure time and calculating the average lost time.

For Scenario 1, Fig. 5 shows that the split model has the effect of reducing the average loss time when there is a major traffic flow compared to the flow control.

On the other hand, when the main traffic flow is small, the average loss time is the same as the flow control. In the entire scenario, the split model with limit reduced the average loss time by 9.4% compared to the current control.

In Scenario 2, experiments are conducted considering the effects of construction work and accidents. Such a situation is a situation in which a sudden traffic flow that is different from the normal traffic flow occurs.

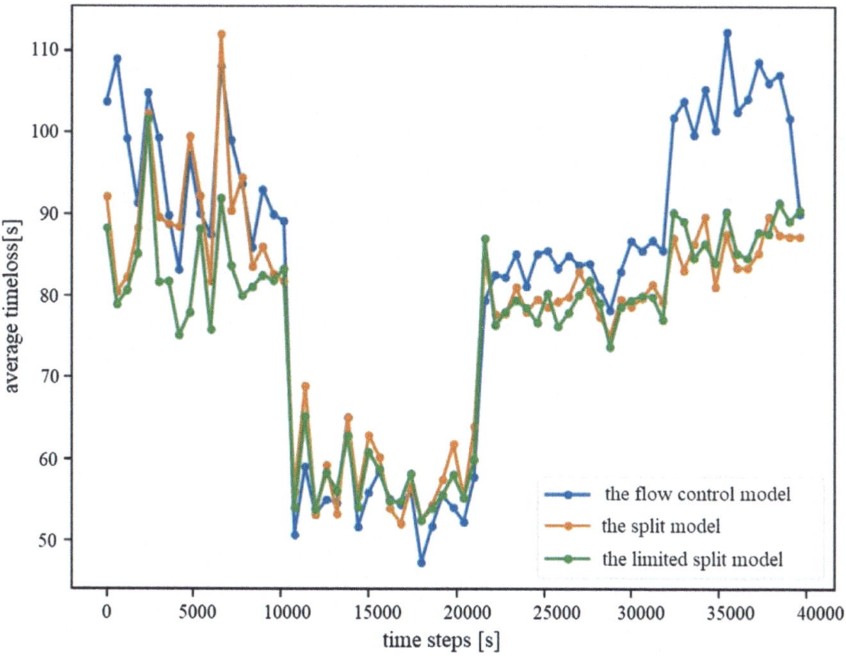

Fig. 5 Time transition of average loss time of Scenario 1

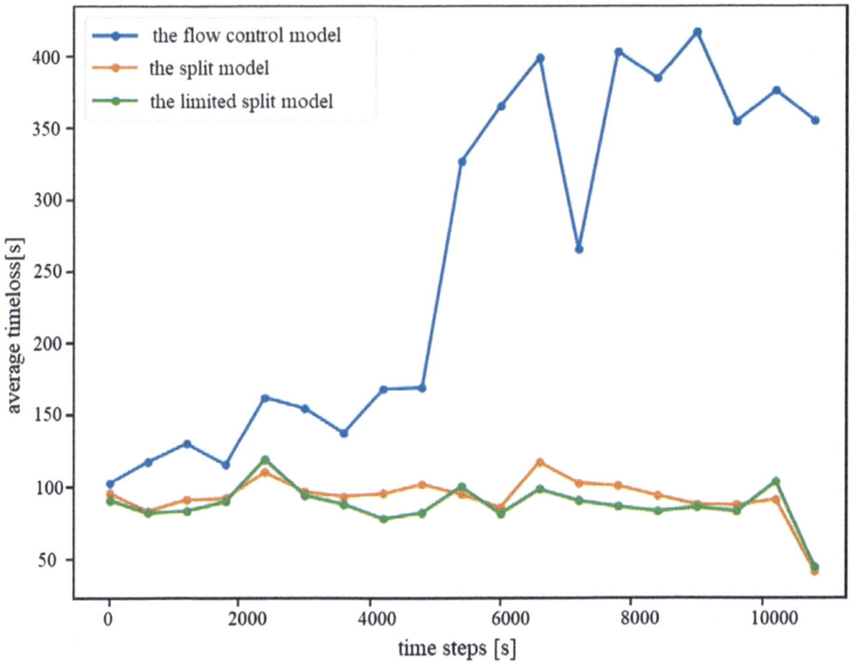

Fig. 6 Time transition of average loss time of Scenario 2

From Fig. 6, it can be seen that with the flow control, the loss time increases as time passes, and the traffic congestion situation worsens. It can be seen that the system is able to cope with sudden traffic flows.

In the entire scenario, the limited split model reduced the average loss time by 64.0% compared to the flow control.

From the experimental results of wide-area control in Scenario 1 and Scenario 2, autonomous distributed control using a split model can be controlled even in a wide control area with many signalized intersections and in a real environment with complex intersection shapes. In addition, it has been demonstrated that the system can cope with sudden traffic flows caused by construction work and accidents, and has the effect of facilitating traffic flow.

5 Conclusion and Outlook

In this study, an autonomous decentralized control method using a split model that models the relationship between traffic flow and split based on a split balance formula are proposed, constructed an existing road network on a simulator, and verified the usefulness of the proposed method. As a result of verification, it is shown that the proposed method can autonomously control a wide area with many intersections and has the effect of facilitating traffic flow.

The proposed decentralized traffic signal split control method holds significant promise for enhancing traffic flow in urban areas. However, there are several avenues for further exploration and development that can shape its future implementation and impact.

One key aspect to consider is the integration of emerging technologies such as connected and autonomous vehicles (CAVs) [5, 16]. By leveraging the connectivity and advanced capabilities of CAVs, the decentralized control method can be further optimized for seamless coordination between traffic signals and vehicles. This integration can lead to more efficient traffic flow, reduced congestion, and enhanced safety on the roadways [12, 19].

Additionally, the advancement of data analytics techniques presents an exciting opportunity. By harnessing the power of machine learning, artificial intelligence, and real-time data analysis, the decentralized split control method can adaptively respond to dynamic traffic conditions and make accurate predictions for optimizing signal timing and splits. This will enable the system to continually learn and improve its performance over time.

Furthermore, the scalability and generalization of the decentralized control method should be explored. As urban areas continue to grow and evolve, it is crucial to develop approaches that can handle complex networks with a large number of intersections. By addressing challenges related to scalability and generalization, the method can be applied effectively in diverse urban environments, from bustling city centres to suburban neighbourhoods.

An important aspect of future work lies in evaluating the societal impacts of implementing the decentralized split control method. Conducting comprehensive assessments of its economic, environmental, and social benefits will provide valuable insights for policymakers and urban planners. Such evaluations can inform decision-making processes and support the adoption of innovative traffic management strategies that improve the quality of life for residents and contribute to sustainable urban development.

Lastly, cost-effectiveness analysis should be undertaken to evaluate the economic viability of the decentralized control method. Understanding the initial implementation costs, long-term operational expenses, and potential savings derived from improved traffic flow will assist transportation authorities and stakeholders in making informed investment decisions.

In summary, the future of the decentralized traffic signal split control method lies in the integration of cutting-edge technologies, scalable implementation, comprehensive evaluation of societal impacts, and cost-effectiveness analysis. By addressing these aspects, the method can pave the way for smarter, more efficient, and sustainable transportation systems, ultimately enhancing the quality of life for urban residents and shaping the cities of the future. Future tasks include verifying the actual effect based on the demonstration experiment scheduled for next fiscal year, and identifying problems with autonomous distributed control operation in a real environment.

References

1. Ahn, S., Choi, J.: Internet of vehicles and cost-effective traffic signal control. Sensors (Switzerland) **19**(6) (2019). https://doi.org/10.3390/s19061275
2. Eom, M., Kim, B.I.: The traffic signal control problem for intersections: a review. Eur. Transp. Res. Rev. **12**(1) (2020). https://doi.org/10.1186/s12544-020-00440-8
3. Ergün, S.: Dynamic traffic signal split control method at pedestrian crossings. Eur. J. Sci. Technol. (2022). https://doi.org/10.31590/ejosat.1216804
4. Feng, C., Zhu, Z.: Separate dual-left turn signal optimization timing method. In: ICTIS 2019—5th International Conference on Transportation Information and Safety (2019). https://doi.org/10.1109/ICTIS.2019.8883587
5. Goulet, N., Ayalew, B.: Distributed maneuver planning with connected and automated vehicles for boosting traffic efficiency. IEEE Trans. Intell. Transp. Syst. **23**(8) (2022). https://doi.org/10.1109/TITS.2021.3096878
6. Jafari, S., Savla, K.: A decentralized feedback approach for flow control in highway traffic networks. Automatica **146** (2022). https://doi.org/10.1016/j.automatica.2022.110661
7. Ji, Z., Shen, G., Wang, J., Collotta, M., Liu, Z., Kong, X.: Multi-vehicle trajectory tracking towards digital twin intersections for internet of vehicles. Electronics **12**(2), 275 (2023)
8. Kim, D., Jeong, O.: Cooperative traffic signal control with traffic flow prediction in multi-intersection. Sensors (Switzerland) **20**(1) (2020). https://doi.org/10.3390/s20010137
9. Kolat, M., Kővári, B., Bécsi, T., Aradi, S.: Multi-agent reinforcement learning for traffic signal control: a cooperative approach. Sustainability **15**(4), 3479 (2023)
10. Mohajerpoor, R., Cai, C., Ramezani, M.: Optimal traffic signal control of isolated oversaturated intersections using predicted demand. IEEE Trans. Intell. Transp. Syst. **24**(1) (2023). https://doi.org/10.1109/TITS.2022.3209606
11. Mushtaq, A., Haq, I.U., Sarwar, M.A., Khan, A., Khalil, W., Mughal, M.A.: Multi-agent reinforcement learning for traffic flow management of autonomous vehicles. Sensors **23**(5), 2373 (2023)
12. Nguyen, C.H.P., Hoang, N.H., Vu, H.L.: A joint trajectory planning and signal control framework for a network of connected and autonomous vehicles. IEEE Trans. Intell. Transp. Syst. (2023). https://doi.org/10.1109/TITS.2023.3241281
13. Poliak, M., Jurecki, R., Buckner, K.: Autonomous vehicle routing and navigation, mobility simulation and traffic flow prediction tools, and deep learning object detection technology in smart sustainable urban transport system. Contemp. Read. Law Soc. Justice **14**, 25 (2022)
14. Rafter, C.B., Anvari, B., Box, S., Cherrett, T.: Augmenting traffic signal control systems for urban road networks with connected vehicles. IEEE Trans. Intell. Transp. Syst. **21**(4) (2020). https://doi.org/10.1109/TITS.2020.2971540
15. Soon, K.L., Lim, J.M.Y., Parthiban, R.: Coordinated traffic light control in cooperative green vehicle routing for pheromone-based multi-agent systems. Appl. Soft Comput. J. **81** (2019). https://doi.org/10.1016/j.asoc.2019.105486
16. Sotoudeh, S.M., HomChaudhuri, B.: Velocity optimization and robust energy management of connected power-split hybrid electric vehicles. J. Dyn. Syst. Meas. Control Trans. ASME **144**(1) (2022). https://doi.org/10.1115/1.4052946
17. Storani, F., Di Pace, R., de Luca, S.: A hybrid traffic flow model for traffic management with human-driven and connected vehicles. Transportmetrica B **10**(1) (2022). https://doi.org/10.1080/21680566.2022.2026836
18. Wu, X., Wang, P.: Intelligent transportation information interaction technology. In: Intelligent Road Transport Systems: An Introduction to Key Technologies, pp. 151–192. Springer (2022)
19. Xue, Y., Zhang, X., Cui, Z., Yu, B., Gao, K.: A platoon-based cooperative optimal control for connected autonomous vehicles at highway on-ramps under heavy traffic. Transp. Res. Part C Emerg. Technol. **150** (2023). https://doi.org/10.1016/j.trc.2023.104083
20. Yao, Q., Li, T., Yan, C., Deng, Z.: Accident responsibility identification model for internet of vehicles based on lightweight blockchain. Comput. Intell. **39**(1), 58–81 (2023)

21. Ye, B.L., Wu, W., Mao, W.: Distributed model predictive control method for optimal coordination of signal splits in urban traffic networks. Asian J. Control **17**(3) (2015). https://doi.org/10.1002/asjc.1011
22. Zhang, H., Lai, Y., Chen, Y.: Authentication methods for internet of vehicles based on trusted connection architecture. Simul. Model. Pract. Theory **122**, 102681 (2023)

Analysis of a Multi-response Experimental Data Set: Modeling with SUR and Multi-objective Optimization by Using NSGA-II and MODE

Serhan Tunçel and Özlem Türkşen

Abstract In this study, analysis of a multi-response experimental data set was considered. The multi-response experimental data set, which have linearly correlated responses, was analyzed in three stage. In the first stage, called modeling, the multi-response experimental data set was modeled by using Seemingly Unrelated Regression (*SUR*) due to linear relationship between the responses. The second stage, called multi-objective optimization, was achieved by using multi-objective metaheuristic methods, Non-dominated Sorting Genetic Algorithm-II (NSGA-II) and Multi Objective Differential Evolution (MODE). In the last stage, called decision making, the obtained results were discussed and compromise experimental conditions were determined by using Multi-Criteria Decision Making (MCDM) methods, called TOPSIS, MABAC, CODAS. It is seen from the results that parameter estimates of the *SUR* method has more reliable than the ordinary least square (*OLS*) estimates. Pareto solution set of the MODE is more prefable according to the performance metrics and the TOPSIS and the CODAS give same compromise solution among the Pareto solutions. The obtained compromise solution can be used as an experiment condition by the researcher confidently.

Keywords Seemingly Unrelated Regression (*SUR*) · Non-dominated Sorting Genetic Algorithm II (NSGA II) · Multi Objective Differential Evolution (MODE) · Multi-criteria Decision Making (MCDM)

S. Tunçel
Hacettepe University, Ankara, Turkey
e-mail: serhan.tuncel@hacettepe.edu.tr

Ö. Türkşen (✉)
Ankara University, Ankara, Turkey
e-mail: turksen@ankara.edu.tr

O. Castillo et al. (eds.), *Advanced Mathematics for the Modeling and Solution of Challenging Problems in Engineering*, Studies in Systems, Decision and Control 582,
https://doi.org/10.1007/978-3-031-82342-8_27

1 Introduction

Analysis of experimental data has importance in the field of science and engineering. The experimental data composes response and input variables. If the experimental data contains more than one response variable, generally, multivariate multiple regression analysis is applied for modeling of the data. Model parameter estimations are obtained by using Ordinary Least Square (*OLS*) method to get predicted response models in polynomial form. It is also necessary to examine that if the responses are correlated to each other or not during the modeling of multi-response experimental data with the regression analysis. The *OLS* method cannot take into consideration the linear relationship between the responses. It is clear that ignoring the linear relationship between the responses will cause loss of information for the data. In this case it would be appropriate to use the Seemingly Unrelated Regression (*SUR*) method for modeling of the linearly correlated responses.

The *SUR* method is obtained by applying the regression coefficients of Aitken's generalized *OLS* method to the entire system of equations simultaneously. The first theoretical studies about the *SUR* is done by [43]. It is suggested in [43] that estimating the regression coefficients with a single equation with Aitken Generalized Least Squares rather than estimating them separately. In [45], additional properties of Aitken estimators for the *SUR* are determined. It is showed that Aitken's estimate has minimal mean square estimation error and unbiased disturbance covariances of *OLS* residuals with single equations. The finite sample density function and exact moments of the coefficients of the predictions in the *SUR* equation for the two equations model are derived in [44]. Finite sample properties of the two equations *SUR* model were examined by [32] and it has been confirmed that when the relationship between two dependent variables is greater than %50, even in small samples, the *SUR* estimation is more efficient than the *OLS*. In [23], it is showed that the *SUR* estimators have smaller variance than the *OLS* if the sample size is large enough. The *SUR* method has been applied in different fields, such as economy-finance [25, 29], tourism [1], agriculture and animal husbandry [8, 12, 19], transportation [33, 41], engineering [27, 39], health [18], sociology [22].

The predicted response models should be optimized under the constraints to determine the appropriate experiment conditions. Each predicted response model can be considered as an objective function in the optimization stage of the experimental data analysis. It is not possible to obtain a single optimal solution if the objective functions are more than one. In this case, the problem in the optimization stage becomes Multi-Objective Optimization (MOO) problem. The solution of the MOO problem will be a solution set which is composed with many alternative solutions, called Pareto solution set. Although, many classical methods are applied to the MOO problem, multi-objective metaheuristic (MOM) methods are often preferred in order to obtain Pareto solution set for multi-objective functions in a short time.

There are wide variety of the MOM methods defined in the literature. In this study, Non-dominated Sorting Genetic Algorithm-II (NSGA-II) and Multi-Objective

Differential Evolution (MODE) are preferred for the MOO purpose. The NSGA-II, introduced by [10] and the MODE, introduced by [35] are population based MOM methods. These MOM methods, which have stochastic search mechanism with genetic operators, are also called Artificial Intelligence Optimization Algorithms [6, 20]. There have been several studies about applications of the NSGA-II to the response surface problems. In [38], Pareto solution set is obtained by applying the restricted NSGA-II method in the optimization stage of the multi-response surface problem. Compromise experiment conditions are chosen through FCM and TOPSIS decision making methods. The other applications can be seen in the studies of [11, 15, 21, 31, 36].

The MODE has been developed in recently by [35]. In [4], an adaptive version of the MODE is proposed. Performance of the MODE was compared with the NSGA-II for four test problems. It has been observed that the Pareto solutions obtained through the MODE are very diverse on the Pareto front. Other studies about the MODE and applications to the response surface problems can be seen in the studies of [5, 14, 26, 34, 46].

It is necessary to decide for a compromise solution from the Pareto solution set obtained by the MOM methods for decision making. Each response is considered as a criteria which wanted to be minimized or maximized in the experimental data set. According to these criteria, Multi-Criteria Decision Making (MCDM) methods are applied to get a the compromise solution. There have been many MCDM methods in the literature. Detailed information about the application of the MCDM methods with MATLAB® implementation can be seen in [28]. In this study, TOPSIS (Technique for Order Preference by Similarity), MABAC (Multi-Attributive Border Approximation area Comparison) and CODAS (Combinative Distance-based Assessment) are preferred to use as MCDM methods to determine a compromise solution among the Pareto solutions. These MCDM methods differ according to the normalization method of the decision matrix. Sum based, max/min value based and range based normalization methods are used for TOPSIS, MABAC and CODAS, respectively. Compromise solutions obtained by these methods will be considered as suitable experiment conditions in the decision making stage for the analysis of experimental data.

One of the main aim of this study is considering the linear relationship of the multi-responses of experimental data sets during the modeling stage. For this purpose, the *SUR* method is used. The predicted model performance of the *SUR* and the *OLS* were compared by using performance metrics, e.g. R^2 (Coefficient of Determination), *MAPE* (Mean Absolute Percentage Error), *RMSE* (Root Mean Square Error) and *MAE* (Mean Absolute Error). It is seen that the parameter estimates of the *SUR* model have smaller standard error than the *OLS* model. The optimization results become more reliable when the modeling stage is achieved with minimum error by the *SUR* method. Another aim of this study is performing MOO by using the NSGA-II and the MODE to decide a compromise solution with MCDM methods.

The paper is organized as follows. In Sect. 2, the *SUR* method is introduced for the modeling of linearly correlated responses and theoretical explanations about the *SUR* method are given. In Sect. 3, the structure of the MOO problem is defined.

The NSGA-II and the MODE methods are introduced and performance metrics are presented. In Sect. 4, the algorithmic steps of the MCDM methods, TOPSIS, MABAC and CODAS, are given. In Sect. 5, application is done on a chemical data set from the literature and obtained results are discussed in detail. Finally, conclusion is given in Sect. 6.

2 Seemingly Unrelated Regression

The relationship, actually exist between the responses, is needed to be taken into account to perform statistical analysis of multi-response experimental data with minimum error. If there is a linear or non-linear relationship between the responses, estimated response models should be obtained by taking this relationship into account. In this study, analysis of linearly correlated responses will be discussed.

An experimental dataset with n observations, consists of p input variables and r response variables, can be defined as in Table 1. Let the functional relationship between the responses and the input variables can be expressed by a linear regression model. Accordingly, the response model can be written as

$$\boldsymbol{Y}_i = \boldsymbol{X\beta} + \boldsymbol{\varepsilon}_i, \quad i = 1, 2, \ldots, r. \tag{1}$$

Here, $\boldsymbol{Y_i} = \left[y_{1i}\, y_{2i} \ldots y_{ni}\right]^T$ is a $n \times 1$ vector of responses; $\boldsymbol{X} : n \times p$ is of rank p; $\boldsymbol{\beta} = \left[\beta_1\, \beta_2 \ldots \beta_p\right]'$ is a regression coefficient vector of dimension $p \times 1$ and $\boldsymbol{\varepsilon}_i : n \times 1$, $\boldsymbol{\varepsilon}_i = [\varepsilon_{1i} : \varepsilon_{2i} : \ldots : \varepsilon_{ni}]'$, $i = 1, 2, \ldots, r$. When there is no linear relationship between the responses, parameter estimates can be obtained for each individual response with the *OLS* by using following formula

$$\hat{\boldsymbol{\beta}}^{OLS} = \left(\boldsymbol{X}^T\boldsymbol{X}\right)^{-1}\boldsymbol{X}^T\boldsymbol{Y}_i, \quad i = 1, 2, \ldots, r. \tag{2}$$

The predicted response model, using estimator $\hat{\boldsymbol{\beta}}^{OLS}$ can be written as

Table 1 Multi-response experimental data

No	Input variables				Responses			
	X_1	X_2	…	X_p	Y_1	Y_2	…	Y_r
1	x_{11}	x_{12}	…	x_{1p}	y_{11}	y_{12}	…	y_{1r}
2	x_{21}	x_{22}	…	x_{2p}	y_{21}	y_{22}	…	y_{2r}
⋮	⋮	⋮	⋱	⋮	⋮	⋮	⋱	⋮
n	x_{n1}	x_{n2}	…	x_{np}	y_{n1}	y_{n2}	…	y_{nr}

$$\hat{Y}_i = \boldsymbol{X}\hat{\boldsymbol{\beta}}^{OLS}, \quad i = 1, 2, \ldots, r. \tag{3}$$

The linear regression model, given in Eq. (1) can be expressed in matrix from as

$$\boldsymbol{Y} - \boldsymbol{X}\boldsymbol{\beta} + \boldsymbol{\varepsilon}. \tag{4}$$

Here, $\boldsymbol{Y}$, response vector of dimension $rn \times 1$, $\boldsymbol{X}$, design matrix of dimension $rn \times p^*$ $(p^* = \sum_{i=1}^{r} p_i)$, $\boldsymbol{\beta}$, regression coefficient vector of dimension, $p^* \times 1$, $\boldsymbol{\varepsilon}$, error vector of dimension $rn \times 1$. More clearly, Eq. (4) can be written as

$$\begin{bmatrix} y_{11} \\ y_{21} \\ \vdots \\ y_{n1} \\ y_{12} \\ y_{22} \\ \vdots \\ y_{n2} \\ \vdots \\ y_{1r} \\ y_{2r} \\ \vdots \\ y_{nr} \end{bmatrix} = \begin{bmatrix} \underset{n \times p_1}{X_1} & 0 & \cdots & 0 \\ 0 & \underset{n \times p_2}{X_2} & \cdots & 0 \\ \vdots & \vdots & \ddots & \vdots \\ 0 & 0 & \cdots & \underset{n \times p_r}{X_r} \end{bmatrix} \begin{bmatrix} \beta_{01} \\ \beta_{11} \\ \vdots \\ \beta_{p_1 1} \\ \beta_{02} \\ \beta_{12} \\ \vdots \\ \beta_{p_2 2} \\ \vdots \\ \beta_{0r} \\ \beta_{1r} \\ \vdots \\ \beta_{p_r r} \end{bmatrix} + \begin{bmatrix} \varepsilon_{11} \\ \varepsilon_{21} \\ \vdots \\ \varepsilon_{1n} \\ \varepsilon_{12} \\ \varepsilon_{22} \\ \vdots \\ \varepsilon_{n2} \\ \vdots \\ \varepsilon_{1r} \\ \varepsilon_{2r} \\ \vdots \\ \varepsilon_{nr} \end{bmatrix} \tag{5}$$

The assumptions for Eq. (4) are given below

$$E(\boldsymbol{\varepsilon}) = \boldsymbol{0}_s$$

and

$$Var(\boldsymbol{\varepsilon}) = \begin{bmatrix} \sigma_{11}I & . & . & \sigma_{1r}I \\ . & . & . & . \\ . & . & . & . \\ \sigma_{r1}I & . & . & \sigma_{rr}I \end{bmatrix} = \psi = \sum \otimes I_n.$$

Parameter estimation with the *SUR* method consists of three main steps as follows [37]:

Step 1: The predicted response model is obtained separately by using the *OLS* method, $\hat{\boldsymbol{Y}}_i$ and residuals are calculated, $\boldsymbol{e}_i$, $i = 1, 2, \ldots, r$.

Step 2: The residual covariance matrix for the responses is obtained, $\hat{\sum} = [\hat{\sigma}_{il}]$. Calculate the covariance matrix components, $\hat{\sigma}_{il} = \frac{e_i^T e_l}{n}$, $i, l = 1, 2, \ldots, r$.

Step 3: The parameter estimator vector of the SUR is obtained as

$$\hat{\beta}^{SUR} = \left(X^T\left(\hat{\sum}^{-1} \otimes I_n\right)X\right)^{-1} X^T\left(\hat{\sum}^{-1} \otimes I_n\right)Y. \tag{6}$$

3 Multi-objective Optimization

In general, mathematical formulation of a multi-objective optimization (MOO) problem can be written as

$$\begin{aligned} &min/max\, f_1 = \hat{Y}_1(\boldsymbol{x}) \\ &min/max\, f_2 = \hat{Y}_2(\boldsymbol{x}) \\ &\quad\vdots \\ &min/max\, f_r = \hat{Y}_r(\boldsymbol{x}) \\ &\boldsymbol{x} \in S. \end{aligned} \tag{7}$$

Here, $\boldsymbol{x} = \left[x_1\ x_2\ \ldots\ x_p\right]^T$ is the vector of input variable values, $f(\boldsymbol{x})$ represents the objective function and S is the constraint set. In Eq. (7), each predicted response model $\hat{Y}_i(\boldsymbol{x})$, $i = 1, 2, \ldots, r$, is considered as an objective function, f_i, $i = 1, 2, \ldots, r$. A single solution does not provide optimality for MOO. Therefore, alternative solutions are needed to obtain which are called Pareto solutions.

There are various optimization methods defined in the literature for the solution of MOO problem, given in Eq. (8). Multi-Objective Metaheuristic (MOM) methods, which produce Pareto solutions in a single run, are used for the MOO when the classical methods are not sufficient. The MOM methods are quite successful to get global optimum solution without getting stuck with local solutions in short time. Fast non-dominated sorting and crowding distance approaches are two main principles for obtaining Pareto solution set in the MOM methods. The fast non-dominated sorting approach ensures that solutions are sorted by separating them into surfaces according to their dominance. Further, the crowding distance provides the measurement of the distances between the solutions on each surface and each other.

Figure 1 shows the Pareto surfaces obtained as a result of the fast non-dominated sorting approach. In Fig. 2, the crowding distance is shown. In this study, the NSGA-II and the MODE methods are used to obtain Pareto solution set.

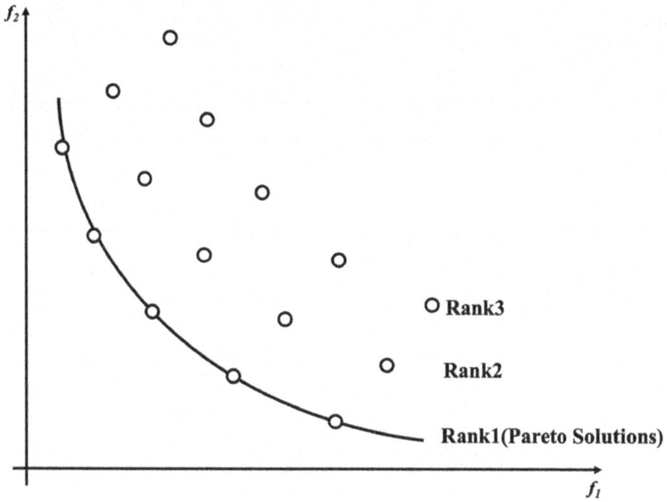

Fig. 1 Surfaces determined by fast non-dominated sorting

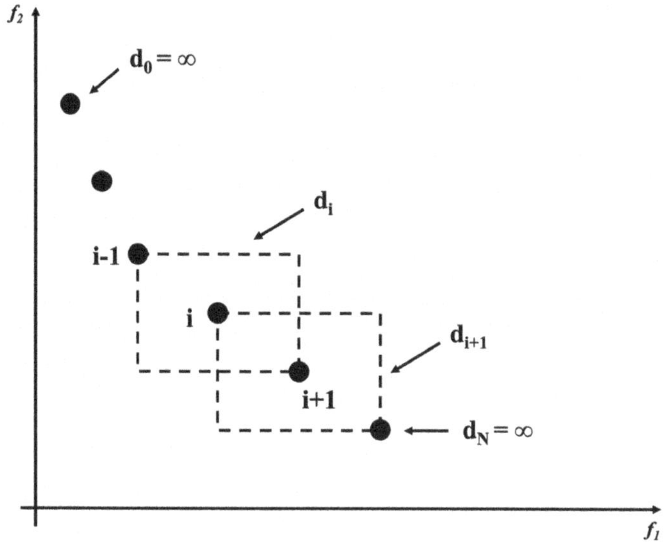

Fig. 2 Crowding distances on the surface [2]

3.1 NSGA-II (Non-dominated Sorting Genetic Algorithm-II)

The Non-dominated Sorting Genetic Algorithm-II (NSGA-II), introduced in [10], can achieve faster search on the research domain of the MOO problem. Simulations with difficult problems showed that the solutions spread well and convergence is strong on the NSGA-II Pareto optimal surface. The algorithmic steps of the NSGA-II method are given below:

Step 1: Determine the tuning parameters, population size (n_{pop}), maximum generation number ($maxn_{gen}$), genetic operators which are selection, crossover and mutation. Generate initial population (P) with size n_{pop} randomly and $n_{gen} = 0$.
Step 2: Create a new generation population (Q) with size n_{pop} by applying genetic operators to P.
Step 3: Combine parent and new generation population, $R = P \cup Q$.
Step 4: Obtain and sort non-dominated surface.
Step 5: Calculate crowding distances of the solutions on each surface. Sort the surfaces in ascending order and crowding distances within the surface in descending order.
Step 6: Carry out crossover and mutation operators. Create new generation solutions (Q) according to crossover and mutation rates. Update number of generation, $n_{gen} = n_{gen} + 1$.
Step 7: If number of generations is not reached, Return to Step 2, otherwise algorithm is terminated.

3.2 MODE (Multi Objective Differential Evolution)

Differential Evolution Algorithm was developed by [35]. Simple implementation and good results against various problems make the algorithm stand out. The algorithmic steps of the Multi Objective Differential Evolution (MODE) method are given as follows [4, 20]:

Step 1: Determine the tuning parameters, number of variables (D), population size ($n_{pop} \geq 4$), maximum generation number ($maxn_{gen}$), scaling vector (F) and crossover ratio (CR).
Step 2: Determine initial population.

$$\forall_i \leq n_{pop} \wedge \forall_j \leq D : x_{j,i,G=0} = lower(x_j) + rand_j[0, 1] \cdot \left(upper(x_j) - lower(x_j)\right)$$

$lower(x_j)$, $upper(x_j)$: The lower and upper bounds of the variables
r_1, r_2, r_3: Randomly selected chromosomes from the population ($r_1 \neq r_2 \neq r_3 \neq i$)
$x_{j,i,G}$: jth gene of ith chromosome in G generation.

Step 3: Compute mutation operator.

$$\forall_j \leq D : n_{j,i,G+1} = x_{j,r_3,G} + F \times \left(x_{j,r_1,G} - x_{j,r_2,G}\right)$$

$n_{j,i,G+1}$: Mutated intermediate chromosome.

Step 4: Carry out crossover operator.

$$\forall_j \leq D,\ x_{j,u,G+1} = \begin{cases} n_{j,i,G+1} & \textit{if } \left(rand_j[0,1]\right) < CR \vee j = j_{rand} \\ x_{j,i,G} & \textit{other} \end{cases}$$

$x_{j,u,G+1}$: Chromosome formed after crossover.

Step 5: Perform selection.

$$\forall_i \leq n_{pop} : x_{i,G+1} = \begin{cases} x_{u,G+1} & \textit{if } f\left(x_{u,G+1}\right) \leq f\left(x_{i,G}\right) \\ x_{i,G} & \textit{other} \end{cases}$$

Step 6: Repeat Steps 3–5 until the $maxn_{gen}$ is reached.

Step 7: Apply non-dominated sorting.

Step 8: Calculate crowding distance.

Step 9: Sort surfaces in ascending order, crowding distances within the surface in descending order.

Fast-acting results can be obtained in the MODE due to the simplicity of the crossover and mutation processes. It is also stated that the MODE algorithm performs well in converging to the real Pareto front and the spread is good [5, 16, 26]. As a result of the MOO, the Pareto solution set can be presented as a matrix (P) given below:

$$P = \begin{array}{c} \\ 1. \\ 2. \\ \vdots \\ n_{pop} \end{array} \begin{array}{c} \begin{array}{cccc} f_1 & f_2 & \cdots & f_r \end{array} \\ \begin{bmatrix} d_{11} & d_{12} & \ldots & d_{1r} \\ d_{21} & d_{22} & \ldots & d_{2r} \\ \vdots & \vdots & \ddots & \vdots \\ d_{n_{pop}1} & d_{n_{pop}2} & \ldots & d_{n_{pop}r} \end{bmatrix} \end{array} \tag{8}$$

In Eq. (8), the $n_{pop} \times r$ dimensional matrix consisting of n_{pop} alternatives and r criteria will be considered as the decision matrix in the decision process. In this study, it is considered that each criteria has equal importance.

3.3 Performance Metrics

Different algorithms can be applied to find Pareto solutions for the MOO problems. After obtaining Pareto solutions with the MOO methods, it is desired to see the performances of the algorithms. For this purpose, various performance metrics can be calculated based on numerical data between the obtained Pareto solutions and the real Pareto front. The following situations should be satisfied to consider that the MOO algorithms perform well [47].

- It is desired that the obtained Pareto solution has less distance from the real Pareto front (Convergency).
- The obtained Pareto solutions should be close to each other (Well-distribution).
- On a Pareto surface, solutions should be spread over a wide range (Diversity).

Convergence metrics evaluate how far the obtained solutions are from the true Pareto surface. For example, Generational Distance. The diversity metric deals with the distribution and spread of the elements of the Pareto solution set obtained by the MOO algorithm. For example, Spacing and Spread. The sample metrics given for these metrics can be summarized as follows.

3.3.1 Generational Distance

The generational distance (GD), defined by [40], refers to the distance between the created Pareto surface and the real Pareto front. The closest Pareto solution to the real Pareto solution is determined with the GD metric. Algorithms with the lowest GD have the best convergence with the optimal Pareto surface [9]. The formula is given as

$$GD = \frac{1}{n}\left(\sum_{i=1}^{n} d_i\right) \tag{9}$$

where n is the number of Pareto solutions and

$$d_i = \min_{j=1}^{m} \sqrt{\sum_{k=1}^{r} \left(f_{ik} - f_{jk}^{*}\right)^2} \tag{10}$$

in which f_{jk}^{*} is the jth solution of the kth objective function of the observed Pareto optimal front.

3.3.2 Spacing

The spacing metric shows how well the resulting solutions are distributed among each other. The method with the smallest metric value provides the most successful distribution.

$$S = \sqrt{\frac{1}{n}\sum_{i=1}^{n} \left(d_i - \overline{d}\right)^2} \tag{11}$$

where $d_i = \min_{l \in nl \neq i} \sum_{k=1}^{r} |f_k^i - f_k^l|$ and $\overline{d} = \sum_{i=1}^{n} \frac{di}{n}$.

3.3.3 Spread

The spread metric indicates the degree of spread achieved by non-dominant solutions obtained from a system. Spread metric [3] is given by

$$\Delta = \frac{d_u + d_l + \sum_{i=1}^{n} \left| d_i - \overline{d} \right|}{d_u + d_l + (n-1)\overline{d}} \tag{12}$$

where d_l is the Euclidean distance between left extreme solution in real Pareto solution set and generated Pareto left extreme solution. d_u is the Euclidean distance between right extreme solution in real Pareto solution set and generated Pareto right extreme solution, $d_i = \sqrt{\sum_{k=1}^{r} \sum_{i=2}^{n} \left(f_{ik} - f_{(i-1)k}\right)^2}$. A lower value of the metric indicates that solutions spread better on the surface.

4 Multi-criteria Decision Making

It is necessary to decide on a compromise solution from the Pareto solutions obtained by the MOM methods. For this purpose, the MCDM methods are applied to the Pareto set. Among the MCDM methods, TOPSIS, MABAC and CODAS are selected for decision making stage in this study. The algorithmic structure of these MCDM methods can be seen below.

4.1 TOPSIS

TOPSIS (Technique for Order Preference by Similarity) is a multi-criteria decision-making method discovered by Hwang and Yoon in 1981 [17]. While this method creates the benefit matrix defined as a positive ideal, the negative ideal can be considered as a cost matrix. The method is based on determining the alternative solution that is shortest from the positive ideal and furthest from the negative ideal. There must be at least two alternatives. The method can be examined in six steps [42].

Step 1: Construct decision matrix as given in Eq. (8).
Step 2: Construct normalized decision matrix.

$$Z = \begin{bmatrix} z_{11} & z_{12} & \cdots & z_{1r} \\ z_{21} & z_{22} & \cdots & z_{2r} \\ \vdots & \vdots & \ddots & \vdots \\ z_{n1} & z_{n2} & \cdots & z_{nr} \end{bmatrix} \tag{13}$$

To normalize the decision matrix, it is obtained by dividing each element in the matrix by the square root of the sum of squares of that column. The normalized matrix is shown in Eq. (13). It is calculated with $z_{ij} = \frac{d_{ij}}{\sqrt{\sum_{i=1}^{n} d_{ij}^2}}$, $i = 1, 2, \ldots, n,\ j = 1, 2, \ldots r$.

Step 3: Obtain weight vector

$$V = \begin{bmatrix} w_1 z_{11} & w_2 z_{12} & \ldots & w_r z_{1r} \\ w_1 z_{21} & w_2 z_{22} & \ldots & w_r z_{2r} \\ \vdots & \vdots & \ddots & \vdots \\ w_1 z_{n1} & w_2 z_{n2} & \cdots & w_r z_{nr} \end{bmatrix} = \begin{bmatrix} v_{11} & v_{12} & \ldots & v_{1r} \\ v_{21} & v_{22} & \ldots & v_{2r} \\ \vdots & \vdots & \ddots & \vdots \\ v_{n1} & v_{n2} & \cdots & v_{nr} \end{bmatrix} \tag{14}$$

The importance of evaluation criteria for the decision maker may differ from each other. In this case, a weighting process is applied for the criteria. The weighting process is shown in Eq. (14). Based on the more important criterion, TOPSIS can choose the most suitable alternative. Weight values are defined as $w_j, j = 1, 2 \ldots, r$. Here $\sum_{j=1}^{r} w_j = 1$.

Step 4: Determine positive ideal solution and negative ideal solution.

For each element of the V matrix, a positive ideal and a negative ideal are determined according to the purpose of the criterion in the column of that element. In this case, if the determined criterion is defined as benefit, the greatest values of the solutions in the column will be included in the positive ideal solutions, and the smallest values will be included in the negative ideal solution. If the criteria are defined as costs, they will consist of the smallest values of the solutions in the columns in the positive ideal solution and the greatest values of the solutions in the negative ideal solution. Positive ideal solution and negative ideal solution are defined as respectively $V^* = \{v_1^*, v_2^*, \ldots, v_r^*\}$, $V^- = \{v_1^-, v_2^-, \ldots, v_r^-\}$.

Step 5: Compute distance value.

The distance of each alternative in the V matrix to the positive ideal and negative ideal solution determined according to the evaluation criterion to which the solutions belong should be calculated. The Euclidean approach is used for this distance. Distance values are calculated as follows

$$S_i^* = \sqrt{\sum_{j=1}^{r} \left(v_{ij} - v_j^*\right)^2} \quad i = 1, 2, \ldots, n \tag{15}$$

$$S_i^- = \sqrt{\sum_{j=1}^{r} \left(v_{ij} - v_j^-\right)^2} \quad i = 1, 2, \ldots, n \tag{16}$$

Step 6: Calculate closeness coefficients of each alternatives as follows:

$$C_i^* = \frac{S_i^-}{S_i^* + S_i^-} \quad i = 1, 2, \ldots, n \tag{17}$$

Since the result is the ratio of the distance to the negative ideal solution to the total distance, the C_i^* values of the alternatives that are far from the negative ideal solution will be close to 1. In this case, as a result of the calculation, alternatives close to 1 are preferred.

4.2 MABAC

MABAC (Multi-Attributive Border Approximation area Comparison) is an MCDM method developed by [30]. With this method, the best alternative is determined according to their distance from the defined border proximity area. The MABAC steps are defined as follows [7].

Step 1: Construct decision matrix as given in Eq. (8).

Step 2: Construct normalized decision matrix.

The elements of the matrix containing the normalized values in Eq. (13) are calculated as follows

$$z_{ij} = \frac{d_{ij} - d_i^-}{d_i^+ - d_i^-} \tag{18}$$

$$z_{ij} = \frac{d_{ij} - d_i^+}{d_i^- - d_i^+} \tag{19}$$

Equation (18) includes the normalization process in terms of benefit, and Eq. (19) includes the normalization calculation in terms of cost. Here, $d_i^+ = \max(d_1, d_2, \ldots, d_n)$ is the maximum value among the alternatives in the relevant criterion, and $d_i^- = \min(d_1, d_2, \ldots, d_n)$ is the minimum value among the alternatives in the relevant criterion.

Step 3: Apply weighting process.

$$v_{ij} = w_j\left(z_{ij} + 1\right) \tag{20}$$

$$V = \begin{bmatrix} w_1(z_{11}+1) & w_2(z_{12}+1) & \ldots & w_r(z_{1r}+1) \\ w_1(z_{21}+1) & w_2(z_{22}+1) & \ldots & w_r(z_{2r}+1) \\ \vdots & \vdots & \ddots & \vdots \\ w_1(z_{n1}+1) & w_2(z_{n2}+1) & \ldots & w_r(z_{nr}+1) \end{bmatrix} = \begin{bmatrix} v_{11} & v_{12} & \ldots & v_{1r} \\ v_{21} & v_{22} & \ldots & v_{2r} \\ \vdots & \vdots & \ddots & \vdots \\ v_{n1} & v_{n2} & \ldots & v_{nr} \end{bmatrix} \tag{21}$$

Weighting is obtained with Eq. (20). The weighted decision matrix is shown with Eq. (21).

Step 4: Calculate the border proximity area.

$$g_j = \left(\prod_{i=1}^{n} v_{ij}\right)^{(1/n)} \tag{22}$$

$$G = \left[g_1 \ g_2 \ \dots \ g_r \right] \tag{23}$$

The border proximity area for each criterion is calculated with Eq. (22). The vector in Eq. (23) was created by combining the criterion boundary proximity areas.

Distances are calculated by subtracting the vector from each alternative in the weighted decision matrix.

$$Q = V - G = \begin{bmatrix} q_1 \ q_2 \cdots q_r \\ q_1 \ q_2 \cdots q_r \\ \vdots \ \vdots \ \ddots \ \vdots \\ q_1 \ q_2 \cdots q_r \end{bmatrix} \tag{24}$$

Being an alternative in the decision matrix A_i, the border may be in the positive region (G^+) or negative region (G^-) in the proximity field.

$$A_i = \begin{cases} G^+, \ \mathrm{q}_{ij} > 0 \\ G, \ \ \mathrm{q}_{ij} = 0 \\ G^-, \ \mathrm{q}_{ij} < 0 \end{cases} \tag{25}$$

In Eq. (25), it is shown in which region the alternative solution is located through the distance matrix. In order for alternative A_i to be the best alternative, it should be in the positive region as much as possible in its criteria [30].

Step 5: Sort the alternatives.

$$S_i = \sum_{j=1}^{r} q_{ij}, \quad i = 1, 2, \dots, n, \ j = 1, 2, \dots, r \tag{26}$$

Equation (27) is calculated by the sum of the values in each row of the Q matrix. Here, score values are obtained for each alternative. The alternative with the highest score among the scores is selected as the best alternative.

4.3 CODAS

CODAS (Combinative Distance-Based Assessment) is a MCDM method introduced to the literature by [13]. This method evaluates the negative ideal solution distances of the alternatives in the decision matrix. The alternative furthest from the negative ideal solutions is chosen as the best alternative. Euclidean distance is first used for distance measurements. In cases where alternatives cannot be compared with Euclidean distance, Taxicab distance is preferred. The CODAS steps can be defined as follows [13].

Step 1: Construct decision matrix as given in Eq. (8).

Step 2: Construct normalized decision matrix.

The elements of the matrix containing the normalized values in Eq. (13) are calculated as follows

$$z_{ij} = \begin{cases} \frac{d_{ij}}{\max_i d_{ij}}, & j \in N_b \\ \frac{\min_i d_{ij}}{d_{ij}}, & j \in N_c \end{cases} \tag{27}$$

Here, N_b represents the criteria in terms of benefit and N_c represents the criteria in terms of cost.

Step 3: Apply weighting process.

The weighting process in Eq. (14) is applied similarly for the CODAS.

Step 4: Calculate the negative ideal solution.

$$ns_j = \min_j v_{ij} \tag{28}$$

The negative ideal solution in the normalized and weighted decision matrix is shown in Eq. (28). Here, a vector is created from the alternatives that give the lowest value for the relevant criterion.

Step 5: Calculate Euclidean and Taxicab distances.

$$E_i = \sqrt{\sum_{j=1}^{r} \left(v_{ij} - ns_j\right)^2} \tag{29}$$

$$T_i = \sum_{j=1}^{r} \left|v_{ij} - ns_j\right| \tag{30}$$

Step 6: A relative evaluation matrix is created.

$$h_{ik} = (E_i - E_k) + (\psi(E_i - E_k) \times (T_i - T_k)) \tag{31}$$

In Eq. (31), h_{ik} is a matrix of size $n \times n$. ψ is a boundary function. This function is defined as follows.

$$\psi(d) = \begin{cases} 1, & |\mathrm{d}| \geq \tau \\ 0, & |\mathrm{d}| < \tau \end{cases} \quad (32)$$

The boundary parameter τ must be determined by the decision maker. It is recommended to be in the range [0.01, 0.05].

Step 7: Calculate score values of alternatives.

$$H_i = \sum_{k=1}^{n} h_{ik} \quad (33)$$

With Eq. (33), each alternative is summed on a row basis in the h_{ik} matrix to create overall scores. The alternative with the highest value is chosen as the best alternative.

5 Application

The data for the modeling and the optimization application was selected from the field of chemistry. It has been seen that there are three input variables in the data set as temperature (K), pressure (MPa) and co-solvent (%). The dependent variables were defined as the extraction efficiency of carotenoids and the extraction efficiency of chlorophylls. There are 15 observations [24]. In the modeling and optimization parts of this data set, calculations were made through MATLAB2021b and R programming languages. Table 2 contains the Chemistry dataset. In this data set, the input variables are coded with Box–Behnken design.

The coded levels of the input variables and corresponding real values were presented in Table 3.

If there is a linear relationship between the responses in multi-response experimental data, the choice of the parameter method becomes important. For this purpose, the linear relationship between the responses was checked. The Pearson correlation between the responses was found %56.6. This value is statistically significant at the 95% confidence level.

Since there was a correlation between the responses, it was thought that model estimation with the *SUR* method would be more effective than model estimation with *OLS*. The predicted response models were obtained with the *SUR* method by the following algorithm steps of the *SUR*. Firstly, the *OLS* models are obtained. The results were given in Table 4 for the Response-1 (Y_1) and Response-2 (Y_2), respectively.

When Table 4 is examined, it is seen that the model estimates obtained by *OLS* are significant at the 95% confidence level. In order to make model estimation with the

Table 2 Chemistry dataset [24]

No	X_1	X_2	X_3	Y_1	Y_2
1	+ 1	+ 1	0	0.206	2.122
2	0	− 1	+ 1	0.107	1.888
3	− 1	0	− 1	0.138	1.519
4	0	− 1	− 1	0.159	1.750
5	− 1	+ 1	0	0.123	1.468
6	0	+ 1	+ 1	0.196	2.473
7	+ 1	− 1	0	0.185	1.612
8	+ 1	0	− 1	0.167	1.956
9	0	0	0	0.244	2.113
10	0	0	0	0.238	1.987
11	− 1	− 1	0	0.058	1.348
12	− 1	0	+ 1	0.014	1.485
13	+ 1	0	+ 1	0.098	2.286
14	0	+ 1	− 1	0.251	1.992
15	0	0	0	0.226	2.031

Table 3 Chemistry data input variables and coded levels

Input variables	Coded levels		
	− 1	0	1
X_1: Temperature (K)	303	318	333
X_2: Pressure	5	11	17
X_3: Co-solvent (%)	2	4	6

Table 4 ANOVA tables of the *OLS* models obtained for Y_1 and Y_2

OLS	df	Y_1				Y_2			
		S.S.	M.S.	F	*p*-value	S.S.	M.S.	F	*p*-value
Regression	9	0.067	0.007	10.91	0.009	1.469	0.163	21.01	0.002
Residuals	5	0.003	0.001			0.039	0.008		
Total	14	0.070				1.508			

SUR, the variance–covariance matrix estimation was calculated from the residuals of the *OLS* response estimations as $\hat{\sum} = \begin{bmatrix} 0.0002 & 0.0002 \\ 0.0002 & 0.0026 \end{bmatrix}$.

By using $\hat{\sum}$, residual covariance matrix of the responses, the *SUR* method is applied the modeling results Y_1 and Y_2 given in Table 5.

When Table 5 is examined, the models created with the *SUR* are statistically significant at the 95% confidence level. The *OLS* and the *SUR* estimation results can

Table 5 ANOVA tables of the *SUR* models obtained with Y_1 and Y_2

SUR	df	Y_1				Y_2			
		S.S.	M.S.	F	*p*-value	S.S.	M.S.	F	*p*-value
Regression	9	0.067	0.007	10.91	0.009	1.469	0.163	21.01	0.002
Residuals	5	0.003	0.001			0.039	0.008		
Total	14	0.070				1.508			

be seen in Table 6, for the $\hat{Y}_1$.

$$\hat{Y}_1^{OLS} = 0.236 + 0.040X_1 + 0.033X_2 - 0.038X_3 - 0.084X_1^2 - 0.048X_3^2 \tag{34}$$

$$\hat{Y}_1^{SUR} = 0.236 + 0.040X_1 + 0.033X_2 - 0.038X_3 - 0.084X_1^2 - 0.048X_3^2 \tag{35}$$

It is seen that the model estimates obtained by the two methods are formed with the same variables for the *OLS* and the *SUR*, respectively, as given in Eqs. (34) and (35). Similar calculations are done for the $\hat{Y}_2$. The obtained results is given in Table 7.

$$\hat{Y}_2^{OLS} = 2.044 + 0.270X_1 + 0.182X_2 + 0.114X_3 - 0.310X_1^2 \tag{36}$$

$$\begin{aligned}\hat{Y}_2^{SUR} &= 2.044 + 0.270X_1 + 0.182X_2 + 0.114X_3 - 0.310X_1^2 \\ &- 0.096X_2^2 + 0.078X_3^2 + 0.098X_1X_2 + 0.091X_1X_3 + 0.086X_2X_3\end{aligned} \tag{37}$$

Table 6 The *OLS* and *SUR* model parameter estimates, standard errors and significance (*p*-value) for the $\hat{Y}_1$

Parameters	*OLS*			*SUR*		
	Parameter estimates	*p*-value	Std. error	Parameter estimates	*p*-value	Std. error
β_0	**0.236**	**0.0000**	0.015	**0.236**	**0.0000**	0.009
β_1	**0.040**	**0.0072**	0.009	**0.040**	**0.0006**	0.005
β_2	**0.033**	**0.0152**	0.009	**0.033**	**0.0015**	0.005
β_3	**– 0.038**	**0.0100**	0.009	**– 0.038**	**0.0009**	0.005
β_4	**– 0.084**	**0.0017**	0.014	**– 0.084**	**0.0001**	0.008
β_5	– 0.010	0.5149	0.014	– 0.010	0.2793	0.008
β_6	**– 0.048**	**0.0163**	0.014	**– 0.048**	**0.0016**	0.008
β_7	– 0.011	0.4371	0.013	– 0.011	0.2039	0.008
β_8	0.014	0.3397	0.013	0.014	0.1272	0.008
β_9	– 0.001	0.9564	0.013	– 0.001	0.9250	0.008

The values shown in bold are the parameters that are statistically significant at the 95% confidence level

Table 7 The *OLS* and *SUR* parameter estimates, standard errors and significance (*p*-value) for the $\hat{Y}_2$

Parameters	*OLS*			*SUR*		
	Parameter estimates	*p*-value	Std. error	Parameter estimates	*p*-value	Std. error
β_0	**2.044**	**(0.0000)**	0.051	**2.044**	**(0.0000)**	0.029
β_1	**0.270**	**(0.0003)**	0.031	**0.270**	**(0.0000)**	0.018
β_2	**0.182**	**(0.0020)**	0.031	**0.182**	**(0.0002)**	0.018
β_3	**0.114**	**(0.0145)**	0.031	**0.114**	**(0.0014)**	0.018
β_4	**– 0.310**	**(0.0011)**	0.046	**– 0.310**	**(0.0001)**	0.026
β_5	– 0.096	(0.0908)	0.046	**– 0.096**	**(0.0152)**	0.026
β_6	0.078	(0.1497)	0.046	**0.078**	**(0.0320)**	0.026
β_7	0.098	(0.0779)	0.044	**0.098**	**(0.0122)**	0.025
β_8	0.091	(0.0939)	0.044	**0.091**	**(0.0160)**	0.025
β_9	0.086	(0.1094)	0.044	**0.086**	**(0.0199)**	0.025

The values shown in bold are the parameters that are statistically significant at the 95% confidence level

It can be seen from Table 7 that the standard errors of the *OLS* and *SUR* parameter estimations for the second response, respectively, are lower in the *SUR* estimations. For the second response, an estimated model was created with more input variables than the *OLS* method with the *SUR* method, which takes into account the linear relationship of the responses. The predicted responses were given in Eqs. (36) and (37) for the *OLS* and then *SUR*, respectively.

It can be seen from Table 8 that the first response has similar performance according to the *OLS* and the *SUR* methods. However, the *SUR* method has better performance than the *OLS* method for the second response. It is possible to say that the *SUR* models are suitable for the optimization stage.

In the study, it is desired to maximize the responses. Thus, the MOO problem can be given as

Table 8 Performance metrics comparison of the *OLS* and the *SUR* methods

Performance metrics	$\hat{Y}_1$				$\hat{Y}_2$			
	R^2	*MAPE*	*RMSE*	*MAE*	R^2	*MAPE*	*RMSE*	*MAE*
OLS	0.9237	17.027	0.0189	0.0163	0.8660	5.6085	0.1161	0.0997
SUR	0.9237	17.027	0.0189	0.2391	**0.9742**	**2.2823**	**0.0509**	**0.0416**

The better values in terms of performance metrics are shown in bold

$$\begin{aligned} max\, f_1 &= \hat{Y}_1^{SUR}(\boldsymbol{x}) \\ max\, f_2 &= \hat{Y}_2^{SUR}(\boldsymbol{x}) \\ -\mathbf{1} \le & \boldsymbol{x} \le \mathbf{1} \end{aligned} \tag{38}$$

The MOO process was carried out with the NSGA-II and the MODE algorithms. The tuning parameters of the NSGA-II and the MODE are given in Table 9.

In Appendix, Pareto solution sets which are obtained by using the NSGA-II and the MODE are given. The obtained Pareto solutions can be seen clearly in Fig. 3.

The observed data, given in Table 2, were chosen as a reference Pareto data set to compare the optimization algorithms. Among the observed data, with Rank1 solutions were preferred and considered as reference Pareto data in Fig. 4.

The Pareto solution sets, obtained through the NSGA-II and the MODE, with Rank1 solutions are given in Fig. 5.

In order to get statistical results, 100 independent runs are performed with the NSGA-II and the MODE algorithms. The calculated statistics are given in Table 10. The MODE can be preferred according to the GD metric which is smaller than the GD value of NSGA-II. Besides, for the spacing metric, the MODE with a lower value provides an advantage in terms of distribution. As a spread, the NSGA-II algorithm

Table 9 Tuning parameters of the NSGA-II and the MODE

	Tuning parameters
NSGA-II	$n_{pop} = 53,\ maxn_{gen} = 600$
MODE	$n_{pop} = 53,\ maxn_{gen} = 500,\ CR = 0.8,\ F = 0.6$

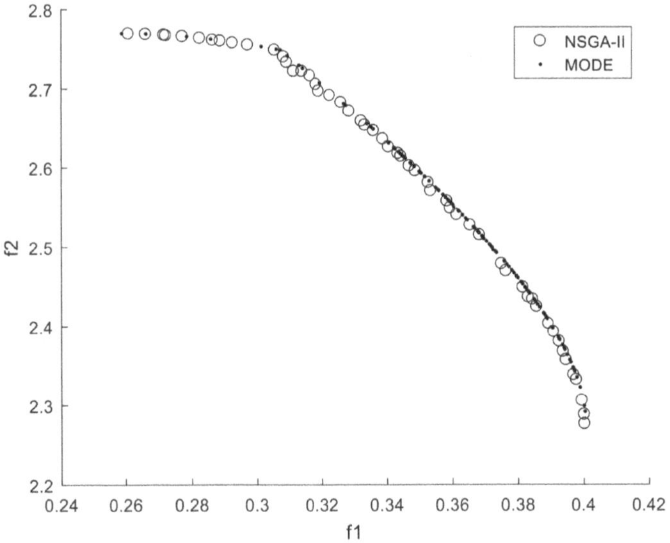

Fig. 3 Pareto solutions obtained with the NSGA-II and the MODE

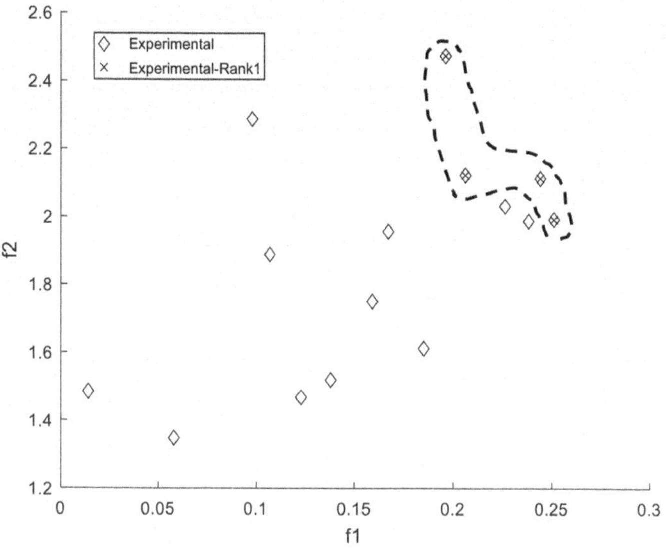

Fig. 4 Observations from the data set

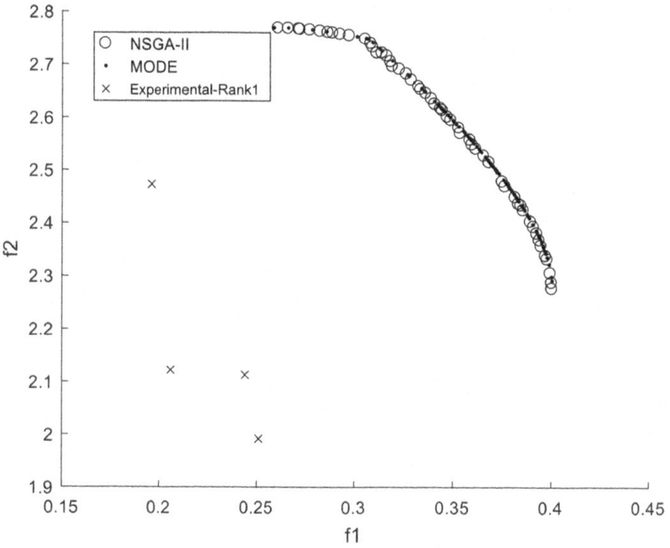

Fig. 5 Pareto solutions obtained with the NSGA-II and the MODE with Rank1 reference Pareto set

Table 10 The performance metrics of the NSGA-II and the MODE with statistics

Metrics	NSGA-II					MODE				
	Mean	Std.	Min.	Max.	Med.	Mean	Std.	Min.	Max.	Med.
GD	0.2333	0.0033	0.2251	0.2426	0.2333	**0.2255**	**0.0057**	**0.2142**	**0.2445**	**0.2249**
Spacing	0.0044	0.0006	0.0031	0.0063	0.0044	**0.0005**	**0.0007**	**0.0001**	**0.0043**	**0.0003**
Spread	**0.6276**	**0.0588**	**0.4638**	**0.7719**	**0.6288**	0.8174	0.0943	0.6233	1.0000	0.8000

The better values in terms of performance metrics are shown in bold

Table 11 The compromise solutions obtained by the MODE

Method	Input variables			Response-1	Response-2
	x_1	x_2	x_3	$\hat{Y}_1$	$\hat{Y}_2$
TOPSIS	1	1	− 0.333	0.400	2.292
MABAC	1	1	0.956	0.313	2.730
CODAS	1	1	− 0.333	0.400	2.292

stands out in terms of the width of the solutions. It is possible to say that the MODE has better results according the three metrics. For this reason, compromise solution can be chosen from the Pareto solution set which is obtained by using the MODE.

The TOPSIS, the MABAC and the CODAS methods are applied to the Pareto solution set to determine a compromise solution. The obtained results are given in Table 11. It can be said from Table 11 that the TOPSIS and the CODAS choose the same compromise solution as $\boldsymbol{x} = \begin{bmatrix} 1 & 1 & -0.333 \end{bmatrix}$. The experimental studies can be done by using this compromise solution.

6 Conclusions

In this study, modeling, optimization and decision-making stages of a multi-response experimental data, in which the responses were linearly correlated, was investigated. In the modeling stage, a statistically significant linear relationship was considered between the responses and the *SUR* was applied. The parameter estimates of the *SUR* were obtained with lower standard errors than the *OLS* estimates. It is possible to say that the parameter estimates of the *SUR* method were more reliable than the *OLS* estimates. When the performance metrics were examined, it was seen that the *SUR* method had better performance. In the MOO stage, predicted response models, obtained by the *SUR*, were considered as objective functions. Pareto solutions were obtained with the NSGA-II and the MODE. It was seen from the results that the MODE showed better performance than the NSGA-II according to the convergence and spacing metrics. Therefore, the Pareto set of the MODE was considered as a decision matrix. In the decision-making stage, compromise solution was chosen from the Pareto set of the MODE through the TOPSIS, the MABAC and the CODAS. It was observed that the TOPSIS and the CODAS, which have different normalization methods, chose the same compromise solution. Consequently, it can be said that decision making with MCDM methods helps the researchers to make decision about the experimental conditions objectively.

Acknowledgements This study is supported by Ankara University Scientific Research Projects (Project Number: 21L0430015).

Appendix: Pareto Solution Sets

No	MODE		NSGA-II	
	f_1	f_2	f_1	f_2
1	0.2870	2.7620	0.400	2.276
2	0.4010	2.2730	0.261	2.769
3	0.3210	2.7010	0.266	2.769
4	0.3580	2.5600	0.272	2.768
5	0.3460	2.6100	0.400	2.288
6	0.3100	2.7380	0.308	2.740
7	0.3680	2.5190	0.399	2.306
8	0.3020	2.7530	0.349	2.596
9	0.3770	2.4750	0.326	2.682
10	0.3800	2.4630	0.393	2.380
11	0.2880	2.7610	0.277	2.766
12	0.3280	2.6770	0.394	2.367
13	0.4000	2.3060	0.328	2.671
14	0.3350	2.6520	0.297	2.755
15	0.3850	2.4300	0.336	2.646
16	0.3300	2.6700	0.318	2.705
17	0.3420	2.6270	0.339	2.636
18	0.3950	2.3620	0.316	2.716
19	0.3980	2.3350	0.353	2.581
20	0.3900	2.3990	0.293	2.758
21	0.3870	2.4210	0.272	2.767
22	0.4000	2.2820	0.397	2.338
23	0.3380	2.6390	0.389	2.403
24	0.3930	2.3820	0.398	2.332

(continued)

(continued)

No	MODE		NSGA-II	
	f_1	f_2	f_1	f_2
25	0.3930	2.3790	0.322	2.692
26	0.3370	2.6430	0.347	2.602
27	0.3450	2.6150	0.314	2.722
28	0.3970	2.3470	0.282	2.764
29	0.3890	2.4080	0.375	2.478
30	0.3770	2.4750	0.289	2.760
31	0.3060	2.7500	0.361	2.540
32	0.3660	2.5260	0.368	2.515
33	0.3640	2.5360	0.332	2.658
34	0.3640	2.5340	0.309	2.733
35	0.3600	2.5530	0.376	2.469
36	0.3850	2.4300	0.395	2.357
37	0.3620	2.5450	0.286	2.761
38	0.3620	2.5450	0.344	2.614
39	0.3600	2.5520	0.305	2.749
40	0.3890	2.4060	0.358	2.557
41	0.3070	2.7490	0.386	2.424
42	0.3070	2.7490	0.343	2.618
43	0.3670	2.5230	0.383	2.436
44	0.3970	2.3430	0.353	2.570
45	0.3450	2.6130	0.365	2.527
46	0.3060	2.7500	0.340	2.626
47	0.3970	2.3430	0.319	2.697
48	0.3450	2.6140	0.391	2.392
49	0.3070	2.7490	0.384	2.434
50	0.3070	2.7490	0.359	2.549
51	0.3070	2.7490	0.381	2.449
52	0.3070	2.7490	0.333	2.653
53	0.3070	2.7490	0.311	2.722

References

1. Aksakal, M., Çilan, C.A.: Türkiye'ye yönelik turizm talebinin görünürde ilişkisiz regresyon modelleri ile incelenmesi. Uluslararası İktisadi İdari İncelemeler Derg. **7**(14), 235–256 (2015)
2. Araujo, F., Trindade, J., Borges, J., Klautau, A., Couto, I.: Multi-objective genetic algorithm to automatically estimating the input parameters of formant-based speech synthesizers. In: Real-World Application of Genetic Algorithm, pp. 283–302 (2012)
3. Audet, C., Bigeon, J., Cartier, D., Digabel, S.L., Salomon, L.: Performance indivators in multiobjective optimization. Eur. J. Oper. Res. **292**, 397–422 (2018)
4. Babu, B.V., Anbarasu, B.: Multi-objective differential evolution (MODE): an evolutionary algorithm for multi-objective optimization problems (MOOPs). In: Proceedings of International Symposium and 58th Annual Session of IIChE (2005)
5. Basu, M.: Economic environmental dispatch using multi-objective differential evolution. Appl. Soft Comput. **11**, 2845–2853 (2011)
6. Bektaş, G., Niğdeli, S.M., Yücel, M., Kayabekir, A.E.: Yapay Zeka Optimizasyon Algoritmaları ve Mühendislik Uygulamaları. Seçkin Yayıncılık, Ankara (2021)
7. Božanić, D., Pamucar, D., Karovic, S.: Application the MABAC method in support of decision-making on the use of force in a defensive operation. Tehnika **71**(1), 129–136 (2016)
8. Cadavez, V.A.P., Henningsen, A.: The use of seemingly unrelated regression to predict the carcass composition of lambs. Meat Sci. **92**, 548–553 (2012)
9. Deb, K.: Multi-Objective Optimization Using Evolutionary Algorithms. Wiley, New York (2001)
10. Deb, K., Pratab, A., Agarwal, S., Meyarivan, T.: Fast and elitist multiobjective genetic algorithm: NSGA-II. IEEE Trans. Evol. Comput. **6**(2), 182–197 (2002)
11. Esfe, M.H., Hajmohammad, H., Moradi, R., Arani, A.A.A.: Multi-objective optimization of cost and thermal performance of double walled carbon nanotubes/water nanofluids by NSGA-II using response surface method. Appl. Therm. Eng. **112**, 1648–1657 (2017)
12. Fu, L., Sharma, R.P., Wang, G., Tang, S.: Modelling a system of nonlinear additive crown width models applying seemingly unrelated regression for Prince Rupprecht larch in northern China. For. Ecol. Manage. **386**, 71–80 (2017)
13. Ghorabaee, M.K., Zavadskas, E.K., Turskis, Z., Antucheviciene, J.: A new combinative distance-based assessment (CODAS) method for multi-criteria decision-making. Econom. Comput. Econom. Cybernet. Stud. Res. **50**(3), 25–44 (2016)
14. Gaitonde, V.N., Manjaiah, M., Maradi, S., Karnik, S.R., Petkar, P.M., Davim, J.P.: Multiresponse optimization in wire electric discharge machining (WEDM) of HCHCr steel by integrating response surface methodology (RSM) with differential evolution (DE). In: Computational Methods and Production Engineering, pp. 199–221 (2017)
15. Hu, S., Zou, Y., Gai, Y., Huang, Z., Song, G.: Risk-based multiobjective optimal seismic design for RC piers using the response surface method and NSGA-II. Adv. Civ. Eng. **3**, 1–11 (2021)
16. Huang, V.L., Suganthan, P.N., Qin, K., Baskar, S.: Differential evolution with external archive and harmonic distance-based diversity measure. School of Electrical and Electronic Enginering Nanyang, Technological University Technical Report (2008)
17. Hwang, C.L., Yoon, K.: Multiple Attribute Decision Making: Methods and Applications. Springer-Verlag, New York (1981)
18. Jahanshad, N., Nir, T.M., Toga, A.W., Jack, C.R., Jr., Bernstein, M.A., Weiner, M.W., Thompson, P.M.: Seemingly unrelated regression empowers detection of network failure in dementia. Neurobiol. Aging **36**, 103–112 (2015)
19. Jaleta, M., Kassie, M., Erenstein, O.: Determinants of maize stover utilization as feed, fuel and soil amendment in mixed crop-livestock systems, Ethiopia. Agric. Syst. **134**, 17–23 (2015)
20. Karaboğa, D.: Yapay Zeka Optimizasyon Algoritmaları. Nobel Yayınları, Ankara (2020)
21. Li, C., Huang, J., Wang, K., Chen, Z., Liu, Q.: Optimization of processing parameters of laser skin welding in vitro combining the response surface methodology with NSGA-II. Infrared Phys. Technol. **103** (2019)

22. Liew, H.: Health and well-being of middle age Indonesians: an application of seemingly unrelated regression (SUR) models. Health Policy Technol. **6**, 323–327 (2017)
23. Liu, A.: Efficient estimation of two seemingly unrelated regression equations. J. Multivar. Anal. **82**(2), 445–456 (2002)
24. Lu, J., Feng, X., Han, Y., Xue, C.: Optimization of subcritical fluid extraction of carotenoids and chlorophyll a from *Laminaria japonica* Aresh by response surface methodology. J. Sci. Food Agric. **94**, 139–145 (2014)
25. Martin, S., Rice, N., Jacobs, R., Smith, P.: The market for elective surgery: joint estimation of supply and demand. J. Health Econ. **26**, 263–285 (2007)
26. Monsef, H., Naghashzadegen, M., Jamali, A., Farmani, R.: Comparison of evolutionary multi objective optimization algorithms in optimum design of water distribution network. Ain Shams Eng. J. **10**, 103–111 (2019)
27. Montoya, M.A., Haddock, J.E.: Estimating asphalt mixture volumetric properties using seemingly unrelated regression equations approaches. Constr. Build. Mater. **225**, 829–837 (2019)
28. Özçalıcı, M.: Matlab ile Çok Kriterli Karar Verme Teknikleri. Nobel Yayıncılık, Ankara (2021)
29. Parajuli, R., Zhang, D., Kosman, K.: Province specific impacts of the 2006 United States-Canada softwood lumber agreement: a seemingly unrelated regression approach. For. Policy Econ. **90**, 1–6 (2018)
30. Pamučar, D., Ćirović, G.: The selection of transport and handling resources in logistics centers using multi-attributive border approximation area comparison (MABAC). Expert Syst. Appl. **42**(6), 3016–3028 (2015)
31. Ravichandran, S., Kumudinidevi, R.P., Bharathidasan, S.G., Jeba, E.: Coordinated controller design of grid connected DFIG based wind turbine using response surface methodology and NSGA II. Sustain. Energy Technol. Assess. **8**, 120–130 (2014)
32. Revankar, N.S.: Some finite sample results in the context of two seemingly unrelated regression equations. J. Am. Stat. Assoc. **69**(345), 187–190 (1974)
33. Sheng, M., Sharp, B.: Aggregate road passenger travel demand in New Zealand: a seemingly unrelated regression approach. Transp. Res. Part A **124**, 55–68 (2019)
34. Singh, D., Kumar, V., Vaishali, Kaur, M.: Classification of COVID-19 patients from chest CT images using multi-objective differential evolution-based convolutional neural networks. Eur. J. Clin. Microbiol. Infect. Dis. **39**, 1379–1389 (2020)
35. Storn, R., Price, K.: Differential evolution—a simple and efficient adaptive scheme for global optimization over continuous spaces. J. Glob. Optim. **23**(1) (1995)
36. Torabi, S.H.R., Alibabaei, S., Bonab, B.B., Sadeghi, M.H., Faraji, G.: Design and optimization of turbine blade preform forging using RSM and NSGA II. J. Intell. Manuf. **28**, 1409–1419 (2017)
37. Tunçel, S.: Çok yanıtlı deneysel verilerin görünüşte ilişkisiz regresyon analizi ile modellenmesi ve optimal değişken değerlerinin belirlenmesi. Ankara Üniversitesi Yüksek Lisans Tezi, Ankara
38. Türkşen, Ö.: Optimization and decision making stages for multiple responses: an application of NSGA-II and FCM clustering algorithm. Gazi Univ. J. Sci. **28**(2), 321–330 (2015)
39. Türkşen, Ö., Tunçel, S., Vural, N.: A seemingly unrelated regression modeling for extraction process in green chemistry. In: y-BIS 2019 Conference Book: Recent Advances in Data Science and Business Analytics, pp. 129–133 (2019)
40. Van Veldhuizen, D.A., Lamont, G.B.: Multiobjective evolutionary algorithm research: a history and analysis. Technical Report TR-98-03. Air Force Institute of Technology, Wright-Patterson AFB, OH (1998)
41. Welki, A.M., Zlatoper, T.J.: How highway safety regulations and enforcement activities affect subcategories of motor vehicle fatalities. Transp. Res. Part E **45**, 1030–1038 (2009)
42. Yıldırım, B.F., Demirci, E.: Banka performansının TOPSIS-M uygulaması ile değerlendirilmesi. Priene Uluslararası Sosyal Bilimler Derg. **1**(1), 35–48 (2017)
43. Zellner, A.: An efficient method of estimating seemingly unrelated regressions and tests of aggregation bias. J. Am. Stat. Assoc. **57**, 348–368 (1962)

44. Zellner, A.: Estimators for seemingly unrelated regression equations: some exact finite sample results. J. Am. Stat. Assoc. **58**(304), 977–992 (1963)
45. Zellner, A., Huang, D.S.: Further properties of efficient estimators for seemingly unrelated regression equations. Int. Econ. Rev. **3**(3), 300–313 (1962)
46. Zhao, W., Ma, A., Ji, J., Chen, X., Yao, T.: Multiobjective optimization of a double-side linear Vernier PM motor using response surface method and differential evolution. IEEE Trans. Ind. Electron. **67**(1), 80–90 (2020)
47. Zitzler, E., Deb, K., Thiele, L.: Comparison of multiobjective evolutionary algorithms: empirical results. Evol. Comput. **8**(2), 173–195 (2000)

Advances in Symmetric Polynomials

Şehmus Fındık

Abstract Let $\mathbb{C}$ be the field of complex numbers and let $\mathbb{C}[W_n]$ be the polynomial algebra over $\mathbb{C}$ in the set $W_n = \{w_1, \ldots, w_n\}$ of n commuting variables. This is the free associative algebra of rank n in the variety of commutative associative algebras over the field $\mathbb{C}$. A polynomial $p \in \mathbb{C}[W_n]$ is called symmetric if $p(w_1, \ldots, w_n) = p\big(w_{\pi(1)}, \ldots, w_{\pi(n)}\big)$ for each permutation $\in S_n$, where S_n denotes the symmetric group n elements. It is folklorely known that such polynomials form a subalgebra of $\mathbb{C}[W_n]$, which is equal to the algebra $\mathbb{C}[W_n]^{S_n}$ of invariants of the symmetric group S_n in $\mathbb{C}[W_n]$. The algebra $\mathbb{C}[W_n]^{S_n}$ is finitely generated by algebraically independent elementary symmetric polynomials. The concept of symmetric polynomials can be generalized to other finitely generated free or relatively free objects in non-commutative and non-associative varieties such as Lie, Poisson, or Leibniz algebras. In this study, the symmetric polynomials of free metabelian Lie algebras, Poisson algebras, and Leibniz algebras were surveyed.

Keywords Lie algebra · Poisson algebra · Symmetric polynomial

1 Introduction

Famous mathematician David Hilbert posed twenty three unsolved problems in 1900. The problem number 14 was the birth of the invariant theory, which is searching for whether the algebra $\mathbb{C}[W_n]^G$ of invariants of any subgroup G of $GL_n(\mathbb{C})$ is finitely generated. Here $\mathbb{C}[W_n]$ stands for the algebra of polynomials in commuting variables $W_n = \{w_1, \ldots, w_n\}$ over $\mathbb{C}$.

The problem has been handled for partial cases in an affirmative direction. Let us fix a field K with characteristic 0. We may mention the work of Noether [1] in which she showed that $K[W_n]^G$ is finitely generated as an algebra for finite groups

Ş. Fındık (✉)
Çukurova University, Adana, Turkey
e-mail: sehfin@gmail.com

O. Castillo et al. (eds.), *Advanced Mathematics for the Modeling and Solution of Challenging Problems in Engineering*, Studies in Systems, Decision and Control 582,
https://doi.org/10.1007/978-3-031-82342-8_28

G. Later Weitzenböck [2] showed that the algebra $K[W_n]^{\exp\delta}$ of invariants is finitely generated, where δ is an arbitrary locally nilpotent linear derivation.

One of the most interesting subgroup G can be considered as the symmetric group S_n acting on the generators $w_1, \ldots, w_n$ linearly, and extended to polynomials canonically. Polynomials belonging to $K[W_n]^{S_n}$ are symmetric, that is

$$p(w_1, \ldots, w_n) = p\big(w_{\pi(1)}, \ldots, w_{\pi(n)}\big)$$

for each permutation π. $K[W_n]^{S_n}$ is generated by following polynomials that are the algebraically independent.

$$\begin{aligned} e_1^{(n)} &= w_1 + \cdots + w_n \\ e_2^{(n)} &= w_1 w_2 + \cdots + w_{n-1} w_n \\ &\vdots \\ e_n^{(n)} &= w_1 w_2 \cdots w_n \end{aligned}$$

2 Lie Algebras

A vector space L is a Lie algebra with multiplication $[.,.]$ called commutator with following properties.

$$[a, a] = 0$$

$$[[a, b], c] + [[b, c], a] + \big[[c, b], y\big] = 0$$

The first identity is equivalent to skew symmetry

$$\big[x, y\big] + \big[y, x\big] = 0$$

provided the characteristic of the field is not equal to 2. The second identity is called the Jacobi identity. The ideal $L' = [L, L]$ of the L generated by all commutators $\big[x, y\big]$, $x, y \in L$, is called the commutator ideal.

The identity

$$[[a, b], [c, d]] = 0$$

is known as the metabelian identity. A Lie algebra satisfying the metabelian identity is called metabelian. If a Lie algebra L is metabelian, then $L'' = \big[L', L'\big] = 0$.

We shall use the left normed commutators, that is

$$[z_1, z_2, z_3] = [[z_1, z_2], z_3]$$

Let F_n be the n-generated free algebra in the variety of metabelian Lie algebras generated by $W_n = \{w_1, \ldots, w_n\}$. We assume that all Lie algebras in this variety are over a field of characteristic 0. It is well known by Bahturin [3] that the commutator ideal F_n' of the algebra F_n possesses a basis with elements

$$\left[w_{l_1}, w_{l_2}, w_{l_3}, \ldots, w_{l_k}\right]$$

such that $l_1 > l_2 \leq l_3 \leq \cdots \leq l_k$, $k \geq 2$. We also have in the commutator ideal F_n' that

$$\left[w_i, w_l, w_{l_{\pi(1)}}, \ldots, w_{l_{\pi(k)}}\right] = \left[w_i, w_l, w_{l_1}, \ldots, w_{l_k}\right]$$

for each permutation $\pi \in S_k$ as a result of the metabelian identity. This gives rise to the fact that F_n' is a right $K[X_n]$-module:

$$uw_i = u(\mathrm{ad}w_i) = [u, w_i]$$

where $\mathrm{ad}w_i : F_n \to F_n$ is the adjoint operator.

It is known by Shmel'kin [4] that F_n can be embedded into a wreath product $W(U_n)$. Let us denote the embedding by

$$\xi : F_n \to W(U_n)$$

so that

$$\xi\left(\left[w_i, w_j, w_{j_1}, \ldots, w_{j_k}\right]\right) = \left(u_i w_j - u_i w_j\right) w_{j_1} \cdots w_{j_k}$$

and linearly extended on all linear combinations of Lie monomials.

The algebra $\left(F_n'\right)^{S_n}$ in F_n' is a $K[X_n]^{S_n}$-module. The generators of this module is given in the next theorem [5]. See also Fındık and Öğüşlü [6] for algebraic structure in the case $n = 2$.

Theorem 1 $\left(F_n'\right)^{S_n}$ *is generated by the set*

$$\mathcal{G}_n = \left\{\xi^{-1}\left(h_{ij}^{(n)}\right) : h_{ij}^{(n)} = j\varepsilon_i^{(n)} e_j^{(n)} - i\varepsilon_j^{(n)} e_i^{(n)}, 1 \leq i < j \leq n\right\}$$

as a $K[X_n]^{S_n}$*-module, such that*

$$\varepsilon_{q+1}^{(n)} = \sum_{i=1}^{n} u_i \sum w_{j1_1} \cdots w_{j_q}$$

where $1 \leq j_1 < \cdots < j_q \leq n$, $i \neq j_l$.

Example 1 We illustrate some results for small rank.

i. Let $n = 2$. The $K[W_2]^{S_2} = K\left[e_1^{(2)}, e_2^{(2)}\right]$-module, $\left(F_2'\right)^{S_2}$ is generated by

$$[w_2, w_1, w_2 - w_1]$$

ii. Let $n = 3$. The $K[W_3]^{S_3} = K\left[e_1^{(3)}, e_2^{(3)}, e_3^{(3)}\right]$-module, $\left(F_3'\right)^{S_3}$ is generated by

$$\begin{aligned}
&[w_2, w_1]ad(w_2 - w_1) + [w_3, w_1]ad(w_3 - w_1)\\
&\quad + [w_3, w_2]ad(w_3 - w_2)\\
&[w_2, w_1]ad((w_2 - w_1)w_3) + [w_3, w_1]ad((w_3 - w_1)w_2)\\
&\quad + [w_3, w_2]ad((w_3 - w_2)w_1)\\
&[w_2, w_1]ad((w_2 - w_1)w_3w_3) + [w_3, w_1]ad((w_3 - w_1)w_2w_2)\\
&\quad + [w_3, w_2]ad((w_3 - w_2)w_1w_1)
\end{aligned}$$

3 Poisson Algebra

A vector space P is a Poisson algebra with two operations; "associator" that is P is an associative and commutative algebra, and "commutator" that is P is a Lie algebra. The relation between associator and commutator is given as below.

$[a, bc] = [a, b]c + b[a, c]$.

If $xy = \left[x, y\right] = 0$ holds, then the given Poisson algebra is commutative, and it is metabelian if it can be expressed as an extension of an abelian Poisson algebra by itself.

Let P_n be the free metabelian Poisson algebra of rank n, and assume it is generated by W_n. Then the followings hold.

$$\begin{aligned}
&P_nP_nP_nP_n = 0\\
&[P_n, P_n][P_n, P_n] = 0\\
&[[P_n, P_n], [P_n, P_n]] = 0
\end{aligned}$$

In the work by Zhang et al. [7], a basis for P_n was proposed. Let

$$\begin{aligned}
B_{1n} &= X_n \cup \left\{\left[w_{j_1}, w_{j_2}, w_{j_3}, \ldots, w_{j_k}\right] : j_1 > j_2 \leq j_3 \leq \cdots \leq j_k\right\}\\
B_{2n} &= \left\{w_iw_j, w_iw_jw_k : i \leq j \leq k\right\}\\
B_{3n} &= \left\{\left[w_i, w_j\right]w_k : i > j\right\}\\
B_{4n} &= \left\{\left[w_i, w_j, w_k\right]w_l : j < i \leq l, j \leq k < l\right\}
\end{aligned}$$

Then $B_{1n} \cup B_{2n} \cup B_{3n} \cup B_{4n}$ is a linear basis for P_n. That is

$$P_n = (KB_{1n} = F_n) \oplus KB_{2n} \oplus KB_{3n} \oplus KB_{4n}$$

where F_n is the free metabelian Lie algebra of rank n generated by W_n. Observe that $(KB_{in})^{S_n} \subset KB_{2i}$, $i = 1, 2, 3, 4$, and thus

$$(P_n)^{S_n} = (F_n)^{S_n} \oplus (KB_{2n})^{S_n} \oplus (KB_{3n})^{S_n} \oplus (KB_{4n})^{S_n}$$

We have the explicit elements of $(F_n)^{S_n}$ from the previous section. On the other hand clearly $(KB_{2n})^{S_n}$ is spanned by elements

$$\begin{aligned} p_1^{(n)} &= w_1^2 + \cdots + w_n^2 \\ p_2^{(n)} &= w_1 w_2 + \cdots + w_{n-1} w_n \end{aligned}$$

of length 2, and of length 3 as follows;

$$\begin{aligned} p_3^{(n)} &= w_1^3 + \cdots + w_n^3 \\ p_4^{(n)} &= w_1^2 w_2 + \cdots + w_{n-1} w_n^2 \\ p_5^{(n)} &= w_1 w_2 w_3 + \cdots + w_{n-2} w_{n-1} w_n \end{aligned}$$

as a consequence of $K[W_n]^{S_n}$.

We know by Dushimirimana et al. [8] that the space $(KB_{3n})^{S_n}$ is spanned by

$$p_6^{(n)} = [w_2, w_1](w_2 - w_1) + \cdots + \left[w_n, w_{n-1}\right](w_n - w_{n-1})$$

and the vector space $(KB_{4n})^{S_n}$ is spanned by

$$\begin{aligned} p_7^{(n)} &= [w_2, w_1, w_1]w_2 + \cdots + \left[w_n, w_{n-1}, w_{n-1}\right]w_n \\ p_8^{(n)} &= \sum_{1 \le j < i < k \le n} \left(-\left[w_i, w_j, w_i\right]w_k + \left[w_i, w_j, w_j\right]w_k + \left[w_k, w_j, w_i\right]w_k\right) \end{aligned}$$

Example 2 We illustrate some results for small rank.

Let $n = 2$. $P_2^{S_2}$ is spanned by $F_2^{S_2}$ together with

$$\begin{aligned} &e_1^{(2)} = w_1 + w_2, p_1^{(2)} = w_1^2 + w_2^2, p_2^{(2)} = w_1 w_2, \\ &p_3^{(2)} = w_1^3 + w_2^3, p_4^{(2)} = w_1^2 w_2 + w_2^2 w_1, p_7^{(2)} = [w_2, w_1, w_1]w_2 \end{aligned}$$

as a vector space.

4 Leibniz Algebras

A Leibniz algebra L is furnished with a bilinear commutator $[.,.]$ with the identity

$$[x, y]\mathrm{ad}z = [[x, y], z] = [[x, z], y] + [x, [y, z]] = [x\mathrm{ad}z, y] + [x, y\mathrm{ad}z]$$

or simply

$$[x, y]r_z = [xr_z, y] + [x, yr_z]$$

denoting $\mathrm{ad}z = r_z$.

Consider the free metabelian Leibniz algebra L_n of rank n generated by X_n over a field K of characteristic zero. The metabelian property yields that $[L_n', L_n'] = 0$, where L_n' is the commutator ideal of L_n. Hence,

$$[x_i, x_j]r_{j_1} \cdots r_{j_k} = [x_i, x_j]r_{j_{\pi(1)}} \cdots r_{j_{\pi(k)}}$$

for every permutation $\pi \in S_k$. Similar to the Lie case, L_n' is a right $K[R_n]$-module, where $R_n = \{r_1 = \mathrm{ad}x_1, \ldots, r_n = \mathrm{ad}x_n\}$. Drensky and Piacentini Cattanneo [9] showed that L_n is of following basis elements.

$$x_{i_1}, [x_{i_1}, x_{i_2}], [x_{i_1}, x_{i_2}]r_{i_3} \cdots r_{i_k}$$

such that $1 \leq i_1, i_2 \leq n$, $1 \leq i_3 \leq \cdots \leq i_k$. This gives that L_n' is a free right $K[R_n]$-module generated by
$[x_i, x_j]$, $1 \leq i, j \leq n$.

Let

$$A_n = \left\{ \sum_{i=1}^{n} a_i p_i : p_i \in K[R_n],\ a_i = [x_i, x_i] \right\}$$

$$B_n = \left\{ \sum_{1 \leq j \neq k \leq n} b_{jk} q_{jk} : q_{jk} \in K[R_n],\ b_{jk} = [x_j, x_k] \right\}$$

Then, $L_n' = A_n \oplus B_n$. One may also check that $(A_n)^{S_n} \subset A_n$ and $(B_n)^{S_n} \subset B_n$. Thus,
$(L_n')^{S_n} = (A_n)^{S_n} \oplus (B_n)^{S_n}$.

The following two theorems [10] give forms of symmetric polynomials in the right $K[R_n]^{S_n}$-module $(L_n')^{S_n}$.

Theorem 2 *Let $p = \sum_{i=1}^{n} a_i p_i$ be a polynomial in A_n for some $p_i \in K[R_n]$. Then, p is symmetric if and only if*

$$p_1(r_1, r_2, \ldots, r_n) = \pi p_1(r_1, r_2, \ldots, r_n) = p_1\big(r_1, r_{\pi(2)}, \ldots, r_{\pi(n)}\big)$$

for every permutation $\pi \in S_n$ *fixing 1, and* $\sigma p_i = p_i$ *for every permutation* $\sigma \in S_n$ *fixing i, and*

$$p_i(r_1, r_2, \ldots, r_n) = (1i)p_1(r_1, r_2, \ldots, r_n)$$

for transpositions $(1i)$, $2 \leq i \leq n$.

Theorem 3 *Let* $q = \sum_{1\leq i\neq j\leq n} b_{ij}q_{ij}$ *be a polynomial in* B_n *for some* $q_{ij} \in K[R_n]$. *Then, q is symmetric if and only if* $q_{ij} = \sigma q_{kl}$ *for every permutation* σ *such that* $\sigma(i) = k$, $\sigma(j) = l$, *and in particular*

$$\begin{aligned} q_{1i} &= (2i)q_{12}, \\ q_{i2} &= (1i)q_{12}, \\ q_{2i} &= (1i)q_{21}, \\ q_{i1} &= (2i)q_{21} \\ q_{21} &= (12)q_{12}, \\ q_{ij} &= (1i)(2j)q_{12} \end{aligned}$$

$3 \leq i \neq j \leq n$ *and* $q_{ij} = \pi q_{ij}$ *for every permutation* π *fixing* i, j.

Example 3 In particular assume that $n = 2$. Symmetric polynomials in L_n are of the form

$$s = s_1 + \cdots + s_6$$

where

$$\begin{aligned} s_1 &= \sum_{i=1}^{n} \alpha x_i \\ s_2 &= \sum_{i=1}^{n} [x_i, x_i](1i)f \\ s_3 &= \sum_{3\leq i\neq j\leq n} [x_i, x_j](1i)(2j)g \\ s_4 &= \sum_{3\leq i\leq n} [x_1, x_i](2i)g + [x_i, x_2](1i)g \\ s_5 &= [x_2, x_1]h \\ s_6 &= [x_1, x_2]g \end{aligned}$$

$$s_7 = \sum_{3 \le i \le n} [x_i, x_1](2i)h + [x_2, x_i](1i)h$$

for some $\alpha \in K, f, g, h \in K[R_n]$ such that $\pi f = f$ for every permutation π fixing 1, $\sigma g = g$ for every permutation σ fixing 1 and 2, and $h = (12)g$.

5 Discussion

As seen in commutative and non-commutative analogues, the results feed the generalizations, and the findings in both cases are compatible. Starting by commutative algebras, we follow the path of non-commutative, then non-associative, non-commutative and non-associative (Lie), non-Lie algebras, etc.

For instance, free associative algebra is a non-commutative generalization of the commutative and associative algebra in the same number of generators. Evaluating the results obtained in the associative algebra modulo the ideal generated by all commutators defined in the usual way, i.e., the ideal generated by all elements of the form

$$[a, b] = ab - ba,$$

we get the elementary symmetric polynomials of the corresponding commutative algebra.

Similarly, Leibniz algebras and Poisson algebras are generalizations of Lie algebras. Hence, theorems for Leibniz algebra hold in the Lie case provided the skew symmetry holds; i.e., the commutator of each element by itself is zero. Also, the findings in the Poisson case, will immediately give rise to the results in Lie setting once the associator is trivial.

The next objective in this direction might be the description of the algebra of symmetric polynomials of well known classical free algebras. It is worth to mention the paper by Boumova et al. [11] where the authors give a finite generating set for the algebra of symmetric polynomials of free associative algebras consisting of an additional action. Also, the work by Fındık and Kelekci [12] discuss about the symmetric polynomials of the algebra of generic matrices.

Another approach will be algebraic structures of those (invariant) algebras, such as the minimal generating sets, free generating sets (when they are free), automorphism groups of them (some special automorphisms; e.g., inner-outer-pointwise inner, tame, wild), derivations and algebras of constants, and structures of newly obtained related groups and algebras. Finally, we may suggest the readers to have a try to algebras of invariants of subgroups of the symmetric group, which already consist of symmetric polynomials.

References

1. Noether, E.: Der Endlichkeitssatz der Invarianten endlicher Gruppen. Math. Ann. **77**, 89–92 (1916)
2. Weitzenböck, R.: Uber die Invarianten von linearen Gruppen. Acta Math. **58**, 231–293 (1932)
3. Bahturin, Yu. A.: Identical Relations in Lie Algebras (Russian), Nauka, Moscow (1985). Translation: VNU Science Press, Utrecht (1987)
4. Shmel'kin, A.L.: Wreath products of Lie algebras and their application in the theory of groups (Russian), Trudy Moskov. Mat. Obshch. **29**, 247–260 (1973). Translation: Trans. Moscow Math. Soc. **29**, 239–252 (1973)
5. Drensky, V., Fındık, Ş, Öğüşlü, N.Ş: Symmetric polynomials in the free metabelian Lie algebras. Mediterr. J. Math. **17**(5), 1–11 (2020)
6. Fındık, Ş, Öğüşlü, N.Ş: Palindromes in the free metabelian Lie algebras. Internat. J. Algebra Comput. **29**(05), 885–891 (2019)
7. Zhang, Z.Z., Chen, Y., Bokut, L.A.: Word Problem for Finitely Presented Metabelian Poisson Algebras. arXiv:1907.05953 (2019)
8. Dushimirimana, A., Fındık, Ş., Öğüşlü, N. Ş.: Symmetric polynomials in the free metabelian Poisson algebras. J. Algebr. Its Appl. 2350049 (2021)
9. Drensky, V., Piacentini Cattanneo, G.M.: Varieties of metabelian Leibniz algebras. J. Alg. Appl. **1**(1), 31–50 (2002)
10. Fındık, Ş, Özkurt, Z.: Symmetric polynomials in Leibniz algebras and their inner automorphisms. Turk. J. Math. **44**(6), 2306–2311 (2020)
11. Boumova, S., Drensky, V., Dzhundrekov, D., Kassabov, M.: Symmetric polynomials in free associative algebras. Turk. J. Math. **46**(5), 1674–1690 (2022)
12. Fındık, Ş, Kelekci, O.: Symmetric polynomials of algebras related with 2×2 generic traceless matrices. Internat. J. Algebra Comput. **31**(07), 1433–1442 (2021)

Advances in the CBH Formula

Şehmus Fındık

Abstract The solution to $z = \log(e^x e^y)$ is $x + y$ in some commutative algebra A. However, when A is not commutative, $z \neq x + y$. In general, z can be expressed as $x + y$ plus a sum of formal power series of x and y in commutators, that is called Campbell-Baker-Hausdorff (CBH) formula, or the Hausdorff series. Some of its terms are given as follows.

$$z = z_1 + z_2 + \cdots$$

where $z_1 = x + y$, $z_2 = \frac{1}{2}[x, y]$, $z_3 = \frac{1}{12}[x, [x, y]] + \frac{1}{12}[y, [y, x]], \ldots$

In this study, we survey the Hausdorff series evaluated in different monoid rings.

Keywords CBH formula · Hausdorff series · Monoid ring

1 Introduction

The Campbell-Baker-Hausdorff (CBH) formula gives an answer to the equation $z = \log(e^x e^y)$ in the associative algebra $A = K\langle x, y\rangle$ generated by non-commuting variables x and y over a field K of characteristic zero.

The CBH formula was initially given by Campbell in the end of the nineteenth century in his two papers [1, 2]. In these papers, he aimed to construct a Lie group making use of a given Lie algebra. In those works, he faced with convergence problems. The CBH formula were given by Baker [3] and independently by Hausdorff [4]. In these studies they worked abstractly power series removing convergence problems. However, they lost the touch with the Lie group issue.

Campbell, Baker and Hausdorff, roughly speaking, asked whether z is an infinite combination of x and y in commutators, with no desire of finding a closed formula. In 1947 Dynkin gave an explicit but not closed formula:

Ş. Fındık (✉)
Çukurova University, Adana, Turkey
e-mail: sehfin@gmail.com

O. Castillo et al. (eds.), *Advanced Mathematics for the Modeling and Solution of Challenging Problems in Engineering*, Studies in Systems, Decision and Control 582,
https://doi.org/10.1007/978-3-031-82342-8_29

$$\sum_{k\geq 1}\sum_{p_i,q_i}\frac{(-1)^k}{k}\frac{\left[x^{p_1}y^{q_1}\cdots x^{p_k}y^{q_k}\right]}{\left(\sum_{i=1}^{k}(p_i+q_i)\right)p_1!q_1!\cdots p_k!q_k!}$$

Let L be the Lie subalgebra of the free associative algebra A generated by x and y. Note that, the operation in L is $[x, y] = xy - yx$ that is isomorphic to the free Lie algebra with 2 generators.

Let $\hat{A}$ be the completion of A that consists of power series in the variables x, y.

The equation $e^z = e^x e^y$ is solvalble for some

$$z = H(x, y) = \sum z_i \in \hat{A}$$

where z_n stands for the homogeneous component of degree n. The classical Campbell-Baker-Hausdorff formula gives that z is in the completion $\hat{L}$. In other words, z_n is a Lie polynomial in x and y for every positive integer n; hence, $z \in \hat{L}$.

The series $H(x, y)$ (Hausdorff series) is in $\hat{L}$. Let us indicate the first several summands:

$$H(x, y) = z_1 + z_2 + z_3 + z_4 + z_5 + \cdots$$

where the first summand is $z_1 = x + y$ while the second is $z_2 = \frac{1}{2}[x, y]$, the third and the forth are

$$z_3 = \frac{1}{12}[x, [x, y]] + \frac{1}{12}[y, [y, x]] \text{ and } z_4 = -\frac{1}{24}[x, [y, [x, y]]]$$

and the fifth summand is

$$\begin{aligned} z_5 = &\frac{1}{720}[y, [y, [y, [y, x]]]] - \frac{1}{720}[x, [x, [x, [x, y]]]] \\ &+ \frac{1}{360}[x, [y, [y, [y, x]]]] + \frac{1}{360}[y, [x, [x, [x, y]]]] \\ &+ \frac{1}{120}[y, [x, [y, [x, y]]]] + \frac{1}{120}[x, [y, [x, [y, x]]]] \end{aligned}$$

which is not expressed as a closed formula. It is a natural trying to simplify the expressions for components z_n of $H(x, y)$ and get a formula for Hausdorff series.

2 Baker's Approach

The motivation of this section is a part of the paper by Baker [5], before exploring CBH formula, is. Consider the Lie algebra G with fixed basis $\{p_1, \ldots, p_r\}$ over K. The multiplication of G

$$p_i p_j = \sum_{k=1}^{r} c_{ijk} p_k$$

determines the structure constants c_{ijk} of G. Let

$$Ad_{p_j} : u \to [p_j, u]$$

be the adjoint operator. The negative of Ad_{p_j} as its matrix form with for the fixed basis of G is

$$P_{p_j} = \sum_{i,k=1}^{r} c_{ijk} E_{ki}$$

where E_{ki} are the usual matrix units in the $r \times r$ matrix algebra $M_r(K)$. Every element $v = v(X) \in G$ has the form

$$v = x_1 p_1 + \cdots + x_r p_r$$

for some $x_j \in K$, and the negative of the matrix of Ad_v is

$$P = P(X) = x_1 P_1 + \cdots + x_r P_r$$

By the Cayley-Hamilton theorem, P^r can be expressed as a combination of $I, P, \ldots, P^{r-1}$ with coefficients in $K[x_1, \ldots, x_r]$. Hence e^P is a combination of them, as well, with coefficients that are formal power series in $x_1, \ldots, x_r$. Evaluating $P(Z)$ from the equation

$$e^{P(Z)} = e^{P(X)} e^{P(Y)}$$

Baker shows how to find $v(Z) \in G$. He also gives such evaluations for concrete Lie algebras. He following example is stated in page 114–115 in the paper by Baker [5].

Example 1 Let $\{p_1, p_2, p_3\}$ be the basis of (three-dimensional) Lie algebra G_1. We assume $[p_1, p_2] = p_2$, $[p_1, p_3] = 0 = [p_2, p_3]$. Then the matrix $P(X)$ is of the form

$$P(X) = \begin{pmatrix} 0 & 0 & 0 \\ x_2 & -x_1 & 0 \\ 0 & 0 & 0 \end{pmatrix}$$

It satisfies the equality $P^2(X) = -x_1 P(X)$. Therefore,

$$e^{P(X)} = \begin{pmatrix} 1 & 0 & 0 \\ \frac{x_2}{x_1}\left(1 - e^{-x_1}\right) & e^{-x_1} & 0 \\ 0 & 0 & 1 \end{pmatrix}$$

Now $e^{P(X)}e^{P(Y)} = e^{P(Z)}$ gives

$$z_2 = \frac{x_1 + y_1}{1 - e^{-x_1 - y_1}}\left(\frac{x_2}{x_1}\left(1 - e^{-x_1}\right) + \frac{y_2}{y_1}e^{-x_1}\left(1 - e^{-y_1}\right)\right)$$

and that $z_1 = x_1 + y_1, z_3 = x_3 + y_3$.

The second example is stated on page 103 in the paper by Baker [5].

Example 2 Consider the four-dimensional Lie algebra G_2 with basis $\{p_1, p_2, p_3, p_4\}$ with the only nonzero multiplications $[p_1, p_2] = p_2$ and $[p_1, p_3] = p_4$.

Then

$$P(X) = \begin{pmatrix} 0 & 0 & 0 & 0 \\ x_2 & -x_1 & 0 & 0 \\ 0 & 0 & -x_1 & 0 \\ x_3 & 0 & -x_1 & 0 \end{pmatrix}$$

and

$$e^{P(X)} = \begin{pmatrix} 1 & 0 & 0 & 0 \\ a & e^{-x_1} & 0 & 0 \\ 0 & 0 & 1 & 0 \\ x_3 & 0 & -x_1 & 1 \end{pmatrix}$$

where

$$a = \left(1 - e^{-x_1}\right)\frac{x_2}{x_1}$$

Similar computations give

$$z_2 = \frac{x_1 + y_1}{1 - e^{-x_1 - y_1}}\left(\frac{x_2}{x_1}\left(1 - e^{-x_1}\right) + \frac{y_2}{y_1}e^{-x_1}\left(1 - e^{-y_1}\right)\right)$$

and that $z_1 = x_1 + y_1, z_3 = x_3 + y_3, z_4 = x_4 + y_4 + \frac{1}{2}(x_3y_1 - x_1y_3)$.
The third example is stated on page 116–119 in the paper by Baker [5].

Example 3 Let the Lie algebra G_3 be spanned by $\{p_1, p_2, p_3\}$ and

$$[p_1, p_2] = p_1, [p_1, p_3] = 2p_2, [p_2, p_3] = p_3$$

Then,

$$P(X) = \begin{pmatrix} x_2 & -x_1 & 0 \\ 2x_3 & 0 & -2x_1 \\ 0 & x_3 & -x_2 \end{pmatrix}$$

$$e^{P(X)} = 1 + P(X)\frac{\sinh\sqrt{x_2^2 - 4x_1x_3}}{\sqrt{x_2^2 - 4x_1x_3}} + P^2(X)\frac{\cosh\sqrt{x_2^2 - 4x_1x_3} - 1}{\sqrt{x_2^2 - 4x_1x_3}}$$

which provides relatively complicated expressions for z_1, z_2, z_3.

3 The CBH Formula in Monoid Rings

Let $\mathbb{R}$ stand for the set of real numbers. We denote $R_n = \{a_1, \ldots, a_n\}$ for a right zero semigroup, and A for the monoid ring that is spanned by 1 and R_n over $\mathbb{R}$.

The $\mathbb{R}$-vector space with basis $1, a_1, \ldots, a_n$ is isomorphic to A, with rules: $a_i a_j = a_j, i, j = 1, \ldots, n$.

Now let $u, v \in A$ be arbitrary elements such that

$$u = x_0 1 + \ldots + x_n a_n$$
$$v = y_0 1 + \ldots + y_n a_n$$

for $x_i, y_j \in \mathbb{R}$. Hence $uv \in A$ is as below.

$$uv - x_0 v - y_0 u = \sum_{i,j=1}^{n} x_i y_j a_j$$

The next lemmas and theorems provide the evoluation on CBH formula in A [6].

Lemma 1 *Let $e^w = e^u e^v$ such that $u, v, w \in A$ are in the form*

$$u = x_0 + \ldots + x_n a_n$$
$$v = y_0 + \ldots + y_n a_n$$
$$w = z_0 + \ldots + z_n a_n$$

Then $e^{w-z_0} = e^{u-x_0} e^{v-y_0}$ and $z_0 = x_0 + y_0$.

Lemma 2 *Let $u = x_1 a_1 + \ldots + x_n a_n \in A$. Then*

$$e^u = 1 + u\frac{e^x - 1}{x}$$

where $x = x_1 + \ldots + x_n$.

Theorem 1 *Let*

$$u = x_1 a_1 + \ldots + x_n a_n \in A$$
$$v = y_1 a_1 + \ldots + y_n a_n \in A$$
$$w = z_1 a_1 + \ldots + z_n a_n \in A$$

such that $e^w = e^u e^v$. *Then* $z = x + y$, *and*

$$z_j = \frac{x+y}{e^{x+y}-1}\left(\frac{e^x-1}{x}x_j + e^x\frac{e^y-1}{y}y_j\right)$$

for $j = 1, \ldots, n$, *where* $x = x_1 + \ldots + x_n$, $y = y_1 + \ldots + y_n$, *and* $z = z_1 + \ldots + z_n$.

Corollary 1 *Let*

$$u = x_0 + \ldots + x_n a_n \in A$$
$$v = y_0 + \ldots + y_n a_n \in A$$
$$w = z_0 + \ldots + z_n a_n \in A$$

such that $e^w = e^u e^v$. *Then*

$$x + y = z$$
$$x_0 + y_0 = z_0$$
$$\frac{x+y}{e^{x+y}-1}\left(x_j\frac{e^x-1}{x} + y_j e^x\frac{e^y-1}{y}\right) = z_j$$

for $j = 1, \ldots, n$. *Here* $x = \sum x_i$, $y = \sum y_i$, *and* $z = \sum z_i$.

Example 4 We handle T_n as the full transformation semigroup. It includes all possible self maps on the set $X_n = \{1, \ldots, n\}$. One may find details in the book by Howie [7]. Every $\alpha \in T_n$ can be expressed as $(\alpha_1, \ldots, \alpha_n)$ uniquely, i.e., $\alpha(i) = \alpha_i$. Let

$$R_n = \{r_1 = (1, \ldots, 1), \ldots, r_n = (n, \ldots, n)\}$$

be the subsemigroup of T_n. Clearly $r_i r_i = r_j$ in R_n that is it coincides with Green's D_1 class. Therefore, the semigroup R_n is right-zero.

Now consider the monoid algebra C with basis $\{1, r_1, \ldots, r_n\}$ over $\mathbb{R}$. Then we may apply Corollary 6 for elements of C.

We provide an example in order to illustrate it. Let

$$u = 1 + r_1 + 2r_2, v = 2 + r_1 + r_2$$

where we assume that $n = 2$. Then
the solution to the equation $e^w =$ for $w \in C$ is

$$w - \log(e^u e^v) = 3 + r_1\left(\frac{e^3-1}{3} + \frac{e^5-e^3}{2}\right)\frac{5}{e^5-1} + r_2\left(\frac{e^3-1}{3} + \frac{e^5-e^3}{4}\right)\frac{10}{e^5\ \ 1}$$

In the following lines, we consider the cartesian product of right zero and left zero semigroups; i.e., rectangular bands [8].

Now we consider a more general concept. Let $R_n = \{r_1, \ldots, r_n\}$ and $L_m = \{l_1, \ldots, l_m\}$ be a right and left zero semigroups, respectively. Then

$$A_{mn} = L_m \times R_n$$

is a rectangular band consisting of ordered pairs $a_{ij} = (l_i, r_j)$. The multiplication rule in A_{mn} is as follows:$a_{ij}a_{kl} = a_{il}$. In particular if $m = 1$ or $n = 1$, then A_{mn} becomes a left zero or a right zero semigroup, respectively. Since such trivial cases were already handled, we assume that $m, n \geq 2$.

Now let $\mathbb{R}(A_{mn})$ be the monoid ring spanned on the basis

$$\{1, a_{ij} : 1 \leq i \leq m, 1 \leq j \leq n\}$$

The multiplication in $\mathbb{R}(A_{mn})$ is as below. Let

$$u = x_{00} + \sum_{i,j} x_{ij}a_{ij}, \quad x = \sum_{i,j} x_{ij}, \quad r_{x_i} = \sum_{1 \leq j \leq n} x_{ij}, \quad c_{x_i} = \sum_{1 \leq i \leq m} x_{ij}$$
$$v = y_{00} + \sum_{i,j} y_{ij}a_{ij}, \quad y = \sum_{i,j} y_{ij}, \quad r_{y_i} = \sum_{1 \leq j \leq n} y_{ij}, \quad c_{y_i} = \sum_{1 \leq i \leq m} y_{ij}$$

Then

$$uv = x_{00}v + y_{00}u + \sum_{i,j} r_{x_i}c_{y_j}a_{ij}$$

The following theorems give the results of CBH formula in $\mathbb{R}(A_{mn})$.

Lemma 3 *Let u and v be defined as above without constant terms. Then, for $k \geq 3$,*

$$u^k = x^{k-2}u^2$$

and

$$u^2 = \sum_{i,j} r_{x_i}c_{x_j}a_{ij}, \quad v^2 = \sum_{i,j} r_{y_i}c_{x_j}a_{ij}, \quad uv = \sum_{i,j} r_{x_i}c_{y_j}a_{ij}$$
$$u^2v = \sum_{i,j} xr_{x_i}c_{y_j}a_{ij}, \quad uv^2 = \sum_{i,j} yr_{x_i}c_{y_j}a_{ij}, \quad u^2v^2 = \sum_{i,j} xyr_{x_i}c_{y_j}a_{ij}$$

Theorem 2 *Let $u, v, w \in A$ such that $e^w = e^u e^v$, where*

$$u = \sum_{i,j} x_{ij} a_{ij},\ v = \sum_{i,j} y_{ij} a_{ij},\ w = \sum_{i,j} z_{ij} a_{ij}$$

Then,

$$z_{ij} = t_{ij} - r_{z_i} c_{z_j} \frac{e^{x+y} - 1 - x - y}{(x+y)^2}$$

where

$$t_{ij} = x_{ij} + y_{ij} r_{x_i} c_{y_j} \frac{(e^x - 1)(e^y - 1)}{xy} + r_{x_i} c_{x_j} \frac{e^x - 1 - x}{x^2} + r_{y_i} c_{y_j} \frac{e^y - 1 - y}{y^2}$$

$$r_{z_i} = \frac{x+y}{e^{x+y} - 1} \left(r_{x_i} \left(1 + \frac{(e^x - 1)(e^y - 1)}{x} + \frac{e^x - 1 - x}{x} \right) + r_{y_i} \left(1 + \frac{e^y - 1 - y}{y} \right) \right)$$

$$c_{z_j} = \frac{x+y}{e^{x+y} - 1} \left(c_{x_j} \left(1 + \frac{e^x - 1 - x}{x} \right) + c_{y_j} \left(1 + \frac{(e^x - 1)(e^y - 1)}{y} + \frac{e^y - 1 - y}{y} \right) \right)$$

Example 5 Let $m = n = 2$, and

$$a_{11} = \begin{pmatrix} 1 & 1 & 0 \\ 0 & 0 & 0 \\ 0 & 0 & 0 \end{pmatrix}, a_{12} = \begin{pmatrix} 1 & 1 & 1 \\ 0 & 0 & 0 \\ 0 & 0 & 0 \end{pmatrix}$$

$$a_{21} = \begin{pmatrix} 0 & 0 & 0 \\ 1 & 1 & 0 \\ 0 & 0 & 0 \end{pmatrix}, a_{22} = \begin{pmatrix} 0 & 0 & 0 \\ 1 & 1 & 1 \\ 0 & 0 & 0 \end{pmatrix}$$

Then the set A_{22} of $a_{11}, a_{12}, a_{21}, a_{22}$ obeys the rules of rectangular band. Let

$$u = 2a_{11} - a_{12} + a_{21} + 3a_{22}$$
$$v = -a_{11} + a_{12} - 2a_{21} - a_{22}$$
$$w = z_{11}a_{11} + z_{12}a_{12} + z_{21}a_{21} + z_{22}a_{22}$$

such that $w = \log(e^u e^v)$. Then, we get that

$$x = 2 - 1 + 1 + 3 = 5$$
$$r_{x_1} = 2 - 1 = 1$$
$$r_{x_2} = 1 + 3 = 4$$
$$c_{x_1} = 2 + 1 = 3$$
$$c_{x_2} = -1 + 3 = 2$$
$$y = -1 + 1 - 2 - 1 = -3$$

$$\begin{aligned} r_{y_1} &= -1 + 1 = 0 \\ r_{y_2} &= -2 - 1 = -3 \\ c_{y_1} &= -1 - 2 = -3 \\ c_{y_2} &= 1 - 1 = 0 \end{aligned}$$

$$\begin{aligned}
t_{11} &= 1 + \frac{(e^5 - 1)(e^{-3} - 1)}{5} + \frac{3(e^5 - 6)}{25} \\
t_{12} &= \frac{2(e^5 - 6)}{25} \\
t_{21} &= -1 + \frac{4(e^5 - 1)(e^{-3} - 1)}{5} + \frac{12(e^5 - 6)}{25} + (e^{-3} + 2) \\
t_{22} &= 2 + \frac{8(e^5 - 6)}{25} \\
r_{z_1} &= \frac{2}{e^2 - 1}\left(1 + \frac{(e^5 - 1)(e^{-3} - 1)}{5} + \frac{(e^5 - 6)}{25}\right) \\
r_{z_2} &= \frac{2}{e^2 - 1}\left(4\left(\frac{(e^5 - 1)(e^{-3} - 1)}{5} + \frac{(e^5 - 6)}{5}\right) - 3\left(1 - \frac{e^{-3} + 2}{3}\right)\right) \\
c_{z_1} &= \frac{2}{e^2 - 1}\left(3\left(1 + \frac{e^5 - 6}{5}\right) - 3\left(1 - \frac{(e^5 - 1)(e^{-3} - 1)}{3} - \frac{e^{-3} + 2}{3}\right)\right) \\
c_{z_2} &= \frac{4}{e^2 - 1}\left(1 + \frac{e^5 - 6}{5}\right)
\end{aligned}$$

and finally

$$\begin{aligned}
z_{11} &= t_{11} - r_{z_1} c_{z_1} \frac{e^2 - 3}{4} \\
z_{11} &= t_{11} - r_{z_1} c_{z_1} \frac{e^2 - 3}{4} \\
z_{21} &= t_{21} - r_{z_2} c_{z_1} \frac{e^2 - 3}{4} \\
z_{22} &= t_{22} - r_{z_2} c_{z_2} \frac{e^2 - 3}{4}
\end{aligned}$$

with the values provided above.

4 Discussion

The formula mentioned in the whole paper has been investigated to have nice closed formulas in some particular cases. One may work in different algebras to find better results. This multiplication rule gives rise to construct new types of groups or rings, amongst there might exist even nonassociative cases. It might also be beneficial in the studies of inner automorphisms. For instance, inner automorphisms of nilpotent Lie algebras can be described by CBH formula for adjoint operators induced by their elements.

References

1. Campbell, J.E.: On a law of combination of operators bearing on the theory of continuous transformation groups. Proc. Lond. Math. Soc. **28**, 381–390 (1897a)
2. Campbell, J.E.: On a law of combination of operators (second paper). Proc. Lond. Math. Soc. **29**, 14–32 (1897b)
3. Baker, H.F.: Alternants and continuous groups. Proc. Lond. Math. Soc. **II**(3), 24–47 (1905)
4. Hausdorff, F.: Die symbolische Exponential Formel in der Gruppentheorie. Berichte Uber die Verhandlungen Sachsischen Akademie der Wisssenchaften zu Leipzig **58**, 19–48 (1906)
5. Baker, H.F.: On the exponential theorem for a simply transitive continuous group, and the calculation of the finite equations from the constants of structure. Proc. London Math. Soc. **34**, 91–127 (1901)
6. Fındık, Ş, Kelekci, O.: Hausdorff series in a semigroup ring. Internat. J. Algebra Comput. **30**(04), 853–859 (2020)
7. Howie, J.M.: Fundamentals of Semigroup Theory. Clarendon Press, Oxford (1995)
8. Kelekci, O.: Hausdorff series in semigroup rings of rectangular. Creat. Math. Inform. Accepted (2020)

Investigation of Fractional Order Genesio-Tesi System Using Laplace Transformation

Zafer Bekiryazici, Suleyman Sengul, and Mehmet Merdan

Abstract In this paper the fractional-order Genesio-Tesi chaotic system is investigated with Laplace transformation method. Genesio-Tesi system is a three-dimensional equation system with quadratic nonlinearity and is frequently used for exemplifying chaotic systems. In this study, Caputo fractional derivative is used to obtain the fractional system and Laplace transformation is applied in order to analyze the explicit solutions of the system.

Keywords Genesio-Tesi system · Caputo derivative · Laplace transform

1 Introduction

Mathematical modeling of real-life phenomena is used in many fields in a wide range that spans from engineering sciences such as electrical and civil engineering to social sciences such as economics and psychology. Many number systems and equation types are used in these models and a suitable selection can be made according to the problem of interest. For instance, financial models frequently contain stochastic equations whereas ordinary differential equation systems are the main tools for modeling the spread of infectious diseases. The use of fractional calculus should be pointed out specifically, since the current trend in mathematical modeling is to use various types of fractional derivatives with numerous solution methods to obtain the exact, approximate, or numerical solutions of the systems being analyzed.

Fractional calculus can be thought of as a generalization of ordinary calculus in which the orders of derivation and integration need not be integers. Fractional operators such as Caputo-Fabrizio fractional derivative or Riemann-Liouville fractional integral enable a more efficient modeling of certain real-life problems. Hence, the

Z. Bekiryazici (✉) · S. Sengul
Department of Mathematics, Recep Tayyip Erdogan University, Rize, Turkey
e-mail: zafer.bekiryazici@erdogan.edu.tr

M. Merdan
Department of Mathematical Engineering, Gumushane University, Gumushane, Turkey

O. Castillo et al. (eds.), *Advanced Mathematics for the Modeling and Solution of Challenging Problems in Engineering*, Studies in Systems, Decision and Control 582,
https://doi.org/10.1007/978-3-031-82342-8_30

topic has been drawing increasing attention in the last couple of decades with many studies on viscoelasticity, astrophysics, and biological systems. The emergence of new definitions for fractional operators also increases the intensity of research in this area.

In this study, the Genesio-Tesi system [1], which is a well-known chaotic model, is analyzed in the context of fractional calculus. Genesio-Tesi system has been extensively studied in the literature in its ordinary and fractional forms since it is a well known and basic example of a chaotic system. The system consists of three differential equations with a single quadratic nonlinear term and a certain set of parameter values causes chaotic behavior in the solutions. Recent studies on Genesio-Tesi system can listed as follows. Bataineh et al. have used Bernstein collocation method to solve the system [2]. Rohilla and Mittal have used B-spline functions to obtain the solutions [3]. Gandhi et al. have used semi-analytical homotopy analysis method to analyze the fractional order Genesio-Tesi system [4]. Razzaghian et al. have introduced an adaptive fuzzy fractional-order fast terminal sliding mode control procedure for the system [5].

Our study uses Caputo fractional derivation to obtain a system of fractional differential equations based on the deterministic system. The fractional Genesio-Tesi system is analyzed by fractional Laplace transformation method using the definition in [6] to obtain the explicit solutions. Fractional Laplace transformation has been used for analyzing various problems in the literature. For instance, Saadeh et al. have used this method to obtain the solutions to fractional Lane-Emden equations [7]. Sene has studied second-grade fluids with Newtonian heating using fractional Laplace transformation [8]. Sheikh et al. have used the method to analyze the fractal-fractional model for the MHD flow of Casson fluid in a channel [9]. Abdulhameed et al. have studied the flow and transport of Burgers fluid using this method [10]. Tanriverdi et al. have analyzed a fractional order atmosphere-soil-land plant carbon cycle system with the method [11]. In this study, the explicit solutions of fractional order Genesio-Tesi system are analyzed by using fractional Laplace transformation method for Caputo fractional derivatives. The proposed method can also be applied to various other equation systems in the literature to obtain the explicit solutions.

2 Preliminaries

There are numerous types of fractional derivatives in the literature, such as the Riemann-Liouville, Grünwald-Letnikov, Caputo-Fabrizio and Atangana-Baleanu derivatives. Some of the basic definitions necessary for this study are given below.

Definition The Riemann-Liouville fractional integral of order $\alpha > 0$ of an integrable function is defined as [6]

$$ {}_{0}^{RL}I_{x}^{\alpha} f(x) = \frac{1}{\Gamma(\alpha)} \int_{0}^{x} (x-t)^{\alpha-1} f(t) dt. $$

Definition The Riemann-Liouville fractional derivative of order $0 < \alpha < 1$ of a function is defined as [6]

$$ {}_0^{RL}D_x^\alpha f(x) = \frac{1}{\Gamma(1-\alpha)}\frac{d}{dx}\int_0^x (x-t)^{-\alpha} f(t)dt. $$

Definition The Caputo fractional derivative of order $0 < \alpha < 1$ of a function is defined as [12]

$$ {}_0^{C}D_x^\alpha f(x) = \frac{1}{\Gamma(1-\alpha)}\int_0^x (x-t)^{-\alpha} f'(t)dt. $$

Caputo fractional derivation is sometimes preferred in modeling studies since the initial conditions in fractional models with Caputo derivative need not be of fractional order. Hence, the same integer-ordered initial conditions can be used.

The fractional Laplace transformation is based on ordinary Laplace transform which is defined by Pierre-Simon Laplace.

Definition The Laplace transform of a function is defined as

$$ \mathcal{L}\{f(t)\} = F(s) = \int_0^\infty e^{-st} f(t)dt $$

for all real numbers $t \geq 0$ if the integral exists.

Definition The fractional Laplace transform of Caputo derivative defined as

$$ \mathcal{L}\{D^\alpha f(x)\} = s^\alpha F(s) - \sum_{k=0}^{n-1} s^{\alpha-k-1} f^{(k)}(0) \tag{1} $$

where the order of derivation is $n-1 < \alpha < n$ for $n \in \mathbb{Z}^+$ [6].

The inverse fractional Laplace transforms have been given for various functions in the literature [11, 13]. For instance, if $\left|\frac{b}{s^\alpha + a}\right| < 1$ and $\Re(s) > |a|^{\frac{1}{a}}$, then

$$ \begin{aligned} \mathcal{L}^{-1}\left[\frac{1}{s}\frac{s^\alpha + a}{(s^\alpha + a)^2 + b^2}\right] &= \sum_{k=0}^{\infty}(-1)^k b^{2k}\frac{t^{(2k+1)\alpha} E_{\alpha,\alpha+1}^{2k}(-at^\alpha)}{(2k)!} \\ \mathcal{L}^{-1}\left[\frac{1}{s}\frac{1}{(s^\alpha + a)^2 + b^2}\right] &= \sum_{k=0}^{\infty}(-1)^k b^{2k}\frac{t^{(2k+2)\alpha} E_{\alpha,\alpha+1}^{2k+1}(-at^\alpha)}{(2k+1)!} \end{aligned} \tag{2} $$

where the special Mittag-Leffler function

$$ E_{\alpha,\beta}^m(at^\alpha) = \sum_{k=0}^{\infty}\frac{(k+m)!}{k!}\frac{a^k t^{\alpha k}}{\Gamma(\alpha k + \alpha m + \beta)} $$

is used for the inverse transformation [13].

3 Genesio-Tesi System and Explicit Solutions

The deterministic Genesio-Tesi system is given as

$$\begin{aligned}\frac{dx}{dt} &= y,\\ \frac{dy}{dt} &= z,\\ \frac{dz}{dt} &= -az - by - cx + x^2,\end{aligned} \tag{3}$$

where x, y and z are the state variables [1]. The parameters a, b and c satisfy the condition $ab < c$ and many studies in the literature use the set of values $a = 1.2, b = 2.92$ and $c = 6$ for numerical analysis of the system since these values are known to cause chaotic behavior [14]. The initial values for the system are given as $x(0) = 0.2, y(0) = -0.3$ and $z(0) = 0.1$ [14]. The behavior of the system has been shown below for the set of parameter values $a = 1.2, b = 2.9, c = 6.1$ (see Fig. 1).

Using Caputo fractional derivative of order $0 < \alpha < 1$, the fractional model can be obtained as follows.

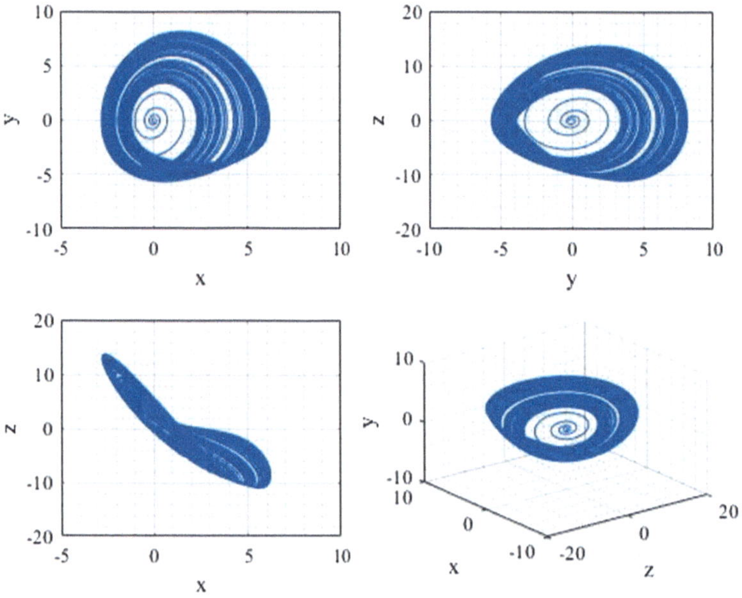

Fig. 1 Behavior of Genesio-Tesi system for parameter values $a = 1.2, b = 2.9, c = 6.1$

$$\frac{d^\alpha x}{dt^\alpha} = y,$$
$$\frac{d^\alpha y}{dt^\alpha} = z,$$
$$\frac{d^\alpha z}{dt^\alpha} = -az - by - cx + x^2. \tag{4}$$

The same parameter and initial values will be used for the system. The fractional system (4) will be linearized around its equilibrium point. The steady state is obtained for (4) by defining the system

$$\frac{d^\alpha x}{dt^\alpha} = f(x, y, z),$$
$$\frac{d^\alpha y}{dt^\alpha} = g(x, y, z),$$
$$\frac{d^\alpha z}{dt^\alpha} = h(x, y, z), \tag{5}$$

and finding the points that satisfy $f(x, y, z) = g(x, y, z) = h(x, y, z) = 0$. The equilibrium points are then found as $X_0 = (0, 0, 0)$ and $E_0 = (c, 0, 0)$. The nonlinear system (4) is linearized around E_0 and rewritten for $x = u - x_0 = u - c$, $y = v - y_0 = v - 0$ and $z = w - z_0 = w - 0$ using the Jacobian matrix

$$J(c, 0, 0) = \begin{bmatrix} \frac{\partial f}{\partial x}(c, 0, 0) & \frac{\partial f}{\partial y}(c, 0, 0) & \frac{\partial f}{\partial z}(c, 0, 0) \\ \frac{\partial g}{\partial x}(c, 0, 0) & \frac{\partial g}{\partial y}(c, 0, 0) & \frac{\partial g}{\partial z}(c, 0, 0) \\ \frac{\partial h}{\partial x}(c, 0, 0) & \frac{\partial h}{\partial y}(c, 0, 0) & \frac{\partial h}{\partial z}(c, 0, 0) \end{bmatrix} = \begin{bmatrix} 0 & 1 & 0 \\ 0 & 0 & 1 \\ c & -b & -a \end{bmatrix}$$

which gives

$$\begin{bmatrix} \frac{d^\alpha u}{dt^\alpha} \\ \frac{d^\alpha v}{dt^\alpha} \\ \frac{d^\alpha (w-c)}{dt^\alpha} \end{bmatrix} = \begin{bmatrix} 0 & 1 & 0 \\ 0 & 0 & 1 \\ c & -b & -a \end{bmatrix} \begin{bmatrix} u - c \\ v \\ w \end{bmatrix} \Rightarrow \begin{cases} \frac{d^\alpha u}{dt^\alpha} = v, \\ \frac{d^\alpha v}{dt^\alpha} = w, \\ \frac{d^\alpha w}{dt^\alpha} = c(u - c) - bv - aw. \end{cases}$$

Returning to the variables x, y and z, the linearized system is given as

$$\frac{d^\alpha x}{dt^\alpha} = y,$$
$$\frac{d^\alpha y}{dt^\alpha} = z, \tag{6}$$
$$\frac{d^\alpha z}{dt^\alpha} = cx - by - az.$$

The Laplace transform for the system above is obtained by using (1) such that $\mathcal{L}\{\mathrm{d}^\alpha \mathrm{x}/\mathrm{dt}^\alpha\} = \mathcal{L}\{\mathrm{y}\}$, $\mathcal{L}\{\mathrm{d}^\alpha \mathrm{y}/\mathrm{dt}^\alpha\} = \mathcal{L}\{\mathrm{z}\}$ and $\mathcal{L}\{d^\alpha z/dt^\alpha\} = \mathcal{L}\{cx - by - az\}$.

Hence, the linearity of the Laplace transform operator gives

$$\begin{aligned}
s^{\alpha}X(s) - s^{\alpha-1}x(0) &= Y(s),\\
s^{\alpha}Y(s) - s^{\alpha-1}y(0) &= Z(s),\\
s^{\alpha}Z(s) - s^{\alpha-1}z(0) &= cX(s) - bY(s) - aZ(s),
\end{aligned}$$

where $X(s) = \mathcal{L}[x(t)]$, $Y(s) = \mathcal{L}\big[y(t)\big]$ and $Z(s) = \mathcal{L}[z(t)]$. This system can be written in matrix form as

$$\begin{bmatrix} s^{\alpha} & -1 & 0 \\ 0 & s^{\alpha} & -1 \\ -c & b & s^{\alpha}+a \end{bmatrix}\begin{bmatrix} X(s) \\ Y(s) \\ Z(s) \end{bmatrix} = \begin{bmatrix} s^{\alpha-1}x(0) \\ s^{\alpha-1}y(0) \\ s^{\alpha-1}z(0) \end{bmatrix}$$

where the solutions $X(s)$, $Y(s)$ and $Z(s)$ are obtained for $x(0) = 0.2$, $y(0) = -0.3$, $z(0) = 0.1$, $a = 1.2$, $b = 2.92$ and $c = 6$ as

$$\begin{aligned}
X(s) &= \frac{0.2s^{3\alpha} - 0.06s^{2\alpha} + 0.324s^{\alpha}}{s\big(s^{3\alpha} + 1.2s^{2\alpha} + 2.92s^{\alpha} - 6\big)},\\
Y(s) &= \frac{-0.3s^{3\alpha} - 0.26s^{2\alpha} + 1.2s^{\alpha}}{s\big(s^{3\alpha} + 1.2s^{2\alpha} + 2.92s^{\alpha} - 6\big)},\\
Z(s) &= \frac{0.1s^{3\alpha} + 2.076s^{2\alpha} - 1.8s^{\alpha}}{s\big(s^{3\alpha} + 1.2s^{2\alpha} + 2.92s^{\alpha} - 6\big)}.
\end{aligned} \tag{7}$$

Using (2) for inverse transformation, we get

$$\begin{aligned}
x(t) &= 0.2 + 0.0598701157550353\,3t^{\alpha}E_{\alpha,\alpha+1}(1.100543924754674t^{\alpha})\\
&\quad - 0.3598701157550353A - 0.3798381964190669\,6B,\\
y(t) &= -0.3 + 0.06588969216856322t^{\alpha}E_{\alpha,\alpha+1}(1.100543924754674t^{\alpha})\\
&\quad + 0.03411030783143\,67A + 1.9227216303064905B,\\
z(t) &= 0.1 + 0.07251450042006778t^{\alpha}E_{\alpha,\alpha+1}(1.100543924754674t^{\alpha})\\
&\quad + 1.8834854995799313A - 2.3524848311828577B,
\end{aligned} \tag{8}$$

where A and B are given as:

$$A := \sum_{k=0}^{\infty}(-4.128724052588621)^{k}\frac{t^{(2k+1)\alpha}E^{2k}_{\alpha,\alpha+1}(-1.150271962377337t^{\alpha})}{(2k)!},$$

$$B := \sum_{k=0}^{\infty}(-4.128724052588621)^{k}\frac{t^{(2k+2)\alpha}E^{2k+1}_{\alpha,\alpha+1}(-1.150271962377337t^{\alpha})}{(2k+1)!}.$$

The special Mittag-Leffler functions $E^{2k}_{\alpha,\alpha+1}$ and $E^{2k+1}_{\alpha,\alpha+1}$ above are defined as

$$E^{2k}_{\alpha,\alpha+1}(-1.150271962377337t^{\alpha}) = \sum_{l=0}^{\infty} \frac{(l+2k)!}{l!} \frac{(-1.150271962377337)^l t^{\alpha l}}{\Gamma(\alpha l + 2\alpha k + \alpha + 1)}$$

$$E^{2k+1}_{\alpha,\alpha+1}(-1.150271962377337t^{\alpha}) = \sum_{l=0}^{\infty} \frac{(l+2k+1)!}{l!} \frac{(-1.150271962377337)^l t^{\alpha l}}{\Gamma(\alpha l + 2\alpha k + 2\alpha + 1)}$$

The results for the linearized system (6) given with Eq. (8) can be compared with the numerical results of the fractional system (4) obtained with an improved Adams-Bashforth-Moulton algorithm presented in [15]. This numerical method can be written for (5) as

$$x_{n+1} = x_0 + \frac{h^{\alpha}}{\Gamma(\alpha+2)}\left[f\left(x^p_{n+1}, y^p_{n+1}, z^p_{n+1}\right) + \sum_{j=0}^{n} \gamma_{1,j,n+1} f\left(x_j, y_j, z_j\right)\right]$$

$$y_{n+1} = y_0 + \frac{h^{\alpha}}{\Gamma(\alpha+2)}\left[g\left(x^p_{n+1}, y^p_{n+1}, z^p_{n+1}\right) + \sum_{j=0}^{n} \gamma_{2,j,n+1} g\left(x_j, y_j, z_j\right)\right]$$

$$z_{n+1} = z_0 + \frac{h^{\alpha}}{\Gamma(\alpha+2)}\left[h\left(x^p_{n+1}, y^p_{n+1}, z^p_{n+1}\right) + \sum_{j=0}^{n} \gamma_{3,j,n+1} h\left(x_j, y_j, z_j\right)\right] \quad (9)$$

where

$$x^p_{n+1} = x_0 + \frac{1}{\Gamma(\alpha)} \sum_{j=0}^{n} \beta_{1,j,n+1} f\left(x_j, y_j, z_j\right)$$

$$y^p_{n+1} = y_0 + \frac{1}{\Gamma(\alpha)} \sum_{j=0}^{n} \beta_{2,j,n+1} g\left(x_j, y_j, z_j\right)$$

$$z^p_{n+1} = z_0 + \frac{1}{\Gamma(\alpha)} \sum_{j=0}^{n} \beta_{3,j,n+1} h\left(x_j, y_j, z_j\right)$$

and $\beta_{i,j,n+1} = \frac{h^{\alpha}}{\alpha}((n+1-j)^{\alpha} - (n-j)^{\alpha})$ and $\gamma_{i,j,n+1} = n^{\alpha+1} - (n-\alpha)(n+1)^{\alpha}$ for $j = 0$, $\gamma_{1,j,n+1} = (n-j+2)^{\alpha+1} + (n-j)^{\alpha+1} - 2(n-j+1)^{\alpha+1}$ for $1 \le j \le n$ and $\gamma_{1,j,n+1} = 1$ for $j = n+1$ using the initial conditions (x_0, y_0, z_0)[15].

xy, *xz* and *yz*-phase planes for the fractional Genesio-Tesi system have been given below (Figs. 2, 3 and 4).

The effects of the change in the order of derivation can be seen in the figure above. The upper-left diagram has been obtained for $\alpha = 0.9$, the upper-right has been obtained for $\alpha = 0.92$, the lower-left for $\alpha = 0.95$ and the lower-right for $\alpha = 0.98$. The same placement is also valid for Figs. 3, 4 and 5.

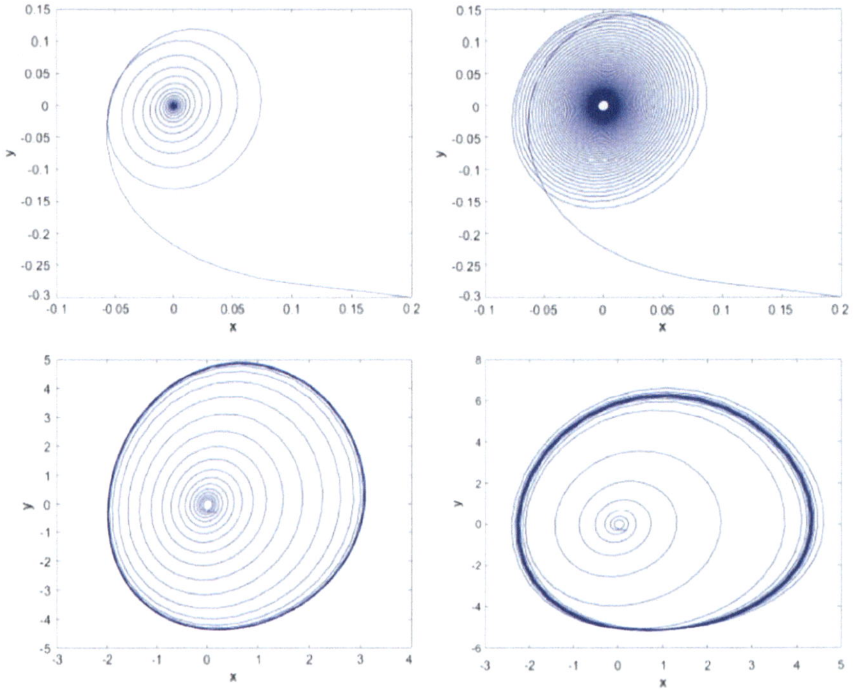

Fig. 2 xy-phase plane for $\alpha = 0.9, 0.92, 0.95, 0.98$

The phase portraits of Genesio-Tesi system with ordinary derivation for $a = 1.2$, $b = 2.92$, $c = 6$ and $x(0) = 0.2$, $y(0) = -0.3$, $z(0) = 0.1$ have been given in the literature [14, 16]. It is seen that as the order or derivation α approaches 1, the phase portraits in (Figs. 2, 3 and 4) become similar to the case with ordinary derivation.

The three-dimensional phase portrait has been given in Fig. 5. Similar arguments can be made for the three-dimensional phase portrait, as they become similar to the phase portraits of the system with integer-ordered derivatives.

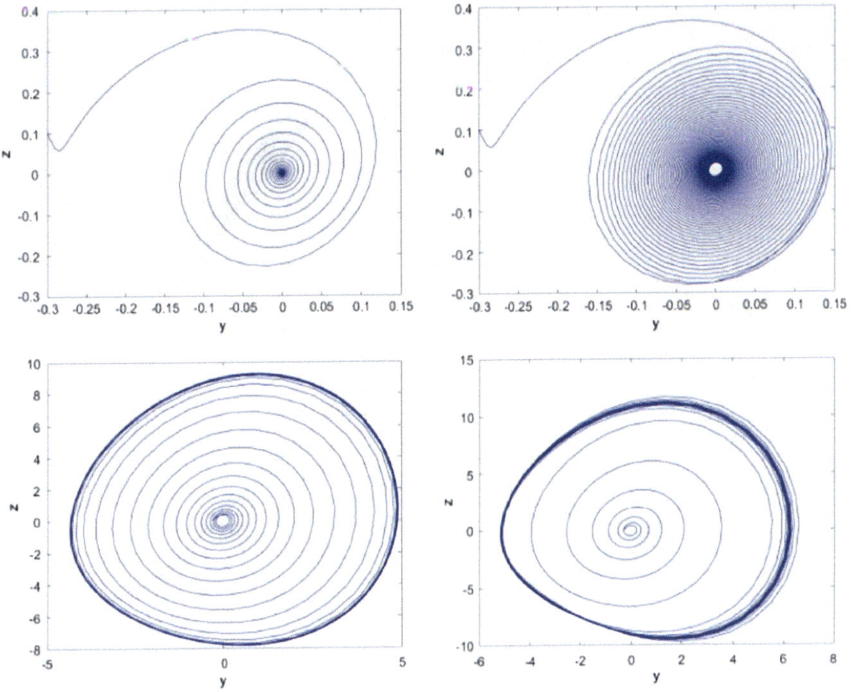

Fig. 3 yz-phase plane $\alpha = 0.9, 0.92, 0.95, 0.98$

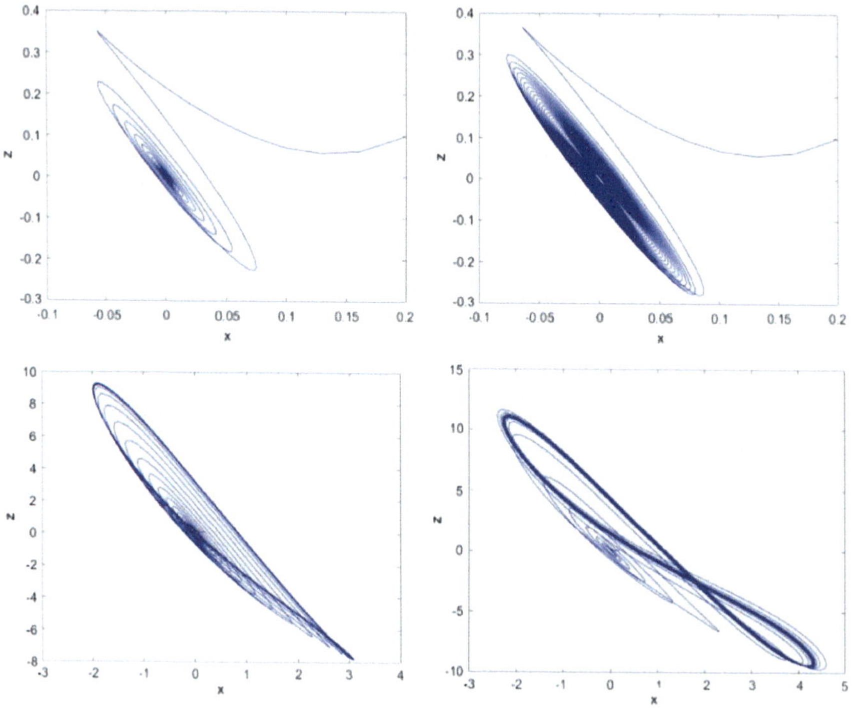

Fig. 4 xz-phase plane $\alpha = 0.9, 0.92, 0.95, 0.98$

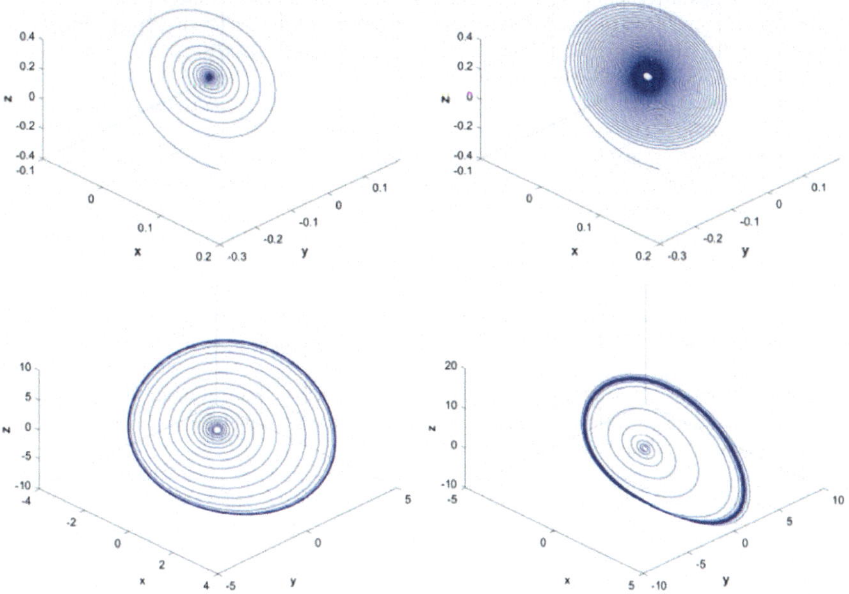

Fig. 5 xyz-phase plane for $\alpha = 0.9, 0.92, 0.95, 0.98$

4 Conclusion

In this study, the Genesio-Tesi system, which is one of the most well-known chaotic models, have been investigated with Caputo fractional derivative and fractional Laplace transformation method. The system of nonlinear fractional differential equations has been linearized around the equilibrium point $E_0 = (c, 0, 0)$ and explicit solutions of the linearized system have been obtained. The solutions, given in equation system (8), have been obtained in terms of Mittag-Leffler functions. An improved Adams-Bashforth-Moulton algorithm has also been employed to visualize the solutions of the fractional-ordered system. It is seen that the chaotic behavior of the fractional system becomes similar to the case with ordinary derivatives as the order or derivation approaches 1.

References

1. Genesio, R., Tesi, A.: Harmonic balance methods for the analysis of chaotic dynamics in nonlinear systems. Automatica **28**(3), 531–548 (1992)
2. Bataineh, A.S., Isik, O.R., Oqielat, M.A., Hashim, I.: An enhanced adaptive Bernstein collocation method for solving systems of ODEs. Mathematics **9**(4), 425 (2021)

3. Rohila, R., Mittal, R.C.: Analysis of chaotic behavior of three-dimensional dynamical systems by a B-spline differential quadrature algorithm. Asian.Eur. J. Math. **15**(04), 2250077 (2022)
4. Gandhi, H., Singh, D., Tomar, A.: The solution of system of time fractional ordinary differential equations by semi-analytical technique. AIP Conf. Proc. **2435**, 020010 (2022)
5. Razzaghian, A., Kardehi Moghaddam, R., Pariz, N.: Adaptive fuzzy fractional-order fast terminal sliding mode control for a class of uncertain nonlinear systems. Int. J. Ind. Electron. Control. Optim. **5**(1), 77–87 (2022)
6. Podlubny, I.: Fractional Differential Equations, Academic Press (1998)
7. Saadeh, R., Burqan, A., El-Ajou, A.: Reliable solutions to fractional Lane-Emden equations via Laplace transform and residual error function. Alex. Eng. J. **61**(12), 10551–10562 (2022)
8. Sene, N.: Second-grade fluid with Newtonian heating under Caputo fractional derivative: analytical investigations via Laplace transforms. Mathematical Modelling and Numerical Simulation with Applications **2**(1), 13–25 (2022)
9. Sheikh, N.A., Ching, D.L.C., Abdeljawad, T., Khan, I., Jamil, M., Nisar, K.S.: A fractal-fractional model for the MHD flow of Casson fluid in a channel. Comput, Mater Contin **67**(2), 1385–1398 (2021)
10. Abdulhameed, M., Tahiru Adamu, G., Dauda, G.Y.: Modeling electro-osmotic flow and thermal transport of Caputo fractional Burgers fluid through a micro-channel. Proc Inst Mech Eng, Part E: J Process Mech Eng **235**(6), 2254–2270 (2021)
11. Tanriverdi, T., Baskonus, H.M., Mahmud, A.A., Muhamad, K.A.: Explicit solution of fractional order atmosphere-soil-land plant carbon cycle system. Ecol. Complex. **48**, 100966 (2021)
12. Caputo, M., Fabrizio, M.: A new definition of fractional derivative without singular kernel. Prog. Fract. Differ. & Appl. **1**(2), 73–85 (2015)
13. Gorenflo, R., Kilbas, A.A., Mainardi, F., Rogosin, S.V.: Mittag-Leffler Functions, Related Topics and Applications. Springer (2020)
14. Gökdoğan, A., Merdan, M., Yildirim, A.: The modified algorithm for the differential transform method to solution of Genesio systems. Commun. Nonlinear Sci. Numer. Simul. **17**(1), 45–51 (2012)
15. Faieghi, M.R., Delavari, H.: Chaos in fractional-order Genesio-Tesi system and its synchronization. Commun. Nonlinear Sci. Numer. Simul. **17**(2), 731–741 (2012)
16. Dural, D., Uyaroglu, Y.: Genesio-Tesi kaotik sisteminin parametrik analizi. In: 2nd International Symposium on Innovative Technologies in Engineering and Science, pp. 18–20. Karabuk, Turkey (2014)

The manufacturer's authorised representative in the EU is Springer Nature Customer Service Centre GmbH, Europaplatz 3, 69115 Heidelberg, Germany. If you have any concerns regarding our products, please contact ProductSafety@springernature.com

Printed and bound by CPI Group (UK) Ltd, Croydon, CR0 4YY
01/10/2026
02225850-0020